Students and External Readers
DATE DUE FOR RETURN
Staff & Research Students
DATE OF ISSUE
UNIVERSITY LIBRARY
29 JUN 2001
ANN GGL 06
30 JUN 87
UNIVERSITY OF NOTTINGHAM
WITHDRAWN
FROM THE LIBRARY
N.B. All books must be returned for the Annual Inspection in June
Any book which you borrow remains your responsibility until the loan slip is cancelled

Methods in Enzymology

Volume LIV

BIOMEMBRANES

Part E

METHODS IN ENZYMOLOGY

Methods in Enzymology

Volume LIV

Biomembranes

Part E: Biological Oxidations Specialized Techniques

EDITED BY

Sidney Fleischer
DEPARTMENT OF MOLECULAR BIOLOGY
VANDERBILT UNIVERSITY, NASHVILLE, TENNESSEE

Lester Packer
MEMBRANE BIOENERGETICS GROUP
DEPARTMENT OF PHYSIOLOGY-ANATOMY
UNIVERSITY OF CALIFORNIA, BERKELEY, CALIFORNIA

1978

ACADEMIC PRESS New York San Francisco London

A Subsidiary of Harcourt Brace Jovanovich, Publishers

ACADEMIC PRESS, INC.
111 Fifth Avenue, New York, New York 10003

United Kingdom Edition published by
ACADEMIC PRESS, INC. (LONDON) LTD.
24/28 Oval Road, London NW1 7DX

Library of Congress Cataloging in Publication Data

Main entry under title:

Biomembranes.

(Methods in enzymology, v. 31–)
Pt. E: edited by Sidney Colowick & Nathan O. Kaplan; Sidney Fleischer/Lester Packer, volume editors.
Includes bibliographical references and indexes.
1. Cell membranes. 2. Cell fractionation. 3. Cell organelles. I. Fleischer, Sidney, ed. II. Packer, Lester, ed. III. Series.
[DNLM: 1. Cell membrane. W1 ME9615K v. 31 / QH601 B6193]
QP601.M49 vol. 31–32 [QH601] 574.1'925'08s [574.8'75]
ISBN 0–12–181954–X (v. 54) 54–9110

PRINTED IN THE UNITED STATES OF AMERICA

Table of Contents

Contributors to Volume LIV

Article numbers are in parentheses following the names of contributors.
Affiliations listed are current.

LYLE J. ARNOLD, JR. (14), *Department of Chemistry, School of Medicine, University of California, San Diego, La Jolla, California 92093*

DAVID P. BALLOU (7), *Department of Biological Chemistry, The University of Michigan, Ann Arbor, Michigan 48109*

OCTAVIAN BÂRZU (26), *Department of Biochemistry, Medical and Pharmaceutical Institute, 3400 Cluj-Napoca, Romania*

HELMUT BEINERT (10, 11, 24), *Institute for Enzyme Research, The University of Wisconsin, Madison, Wisconsin 53706*

MAURIZIO BRUNORI (6), *C. N. R. Centre for Molecular Biology, Institutes of Chemistry and Biochemistry, Faculty of Medicine, University of Rome, Rome, Italy*

WINSLOW S. CAUGHEY (18), *Department of Biochemistry, Colorado State University, Fort Collins, Colorado 80523*

SUNNEY I. CHAN (19), *A. A. Noyes Laboratory of Chemical Physics, California Institute of Technology, Pasadena, California 91125*

BRITTON CHANCE (9, 27), *Johnson Research Foundation, University of Pennsylvania School of Medicine, Philadelphia, Pennsylvania 19104*

RICHARD J. CHERRY (4), *Eidgenossische Technische Hochschule, Laboratorium für Biochemie, ETH-Zentrum, CH-8092 Zürich, Switzerland*

DON DEVAULT (3), *Department of Physiology and Biophysics, University of Illinois at Urbana-Champaign, Urbana, Illinois 61801*

P. LESLIE DUTTON (23), *Johnson Research Foundation and Department of Biochemistry and Biophysics, University of Pennsylvania, Philadelphia, Pennsylvania 19174*

HANS FRAUENFELDER (28), *Department of Physics, University of Illinois at Urbana–Champaign, Urbana, Illinois 61801*

LESLIE W. -M. FUNG (13), *Department of Life Sciences, University of Pittsburgh, Pittsburgh, Pennsylvania 15260*

RONALD C. GAMBLE (19), *A. A. Noyes Laboratory of Chemical Physics, California Institute of Technology, Pasadena, California 91125*

QUENTIN GIBSON (8), *Department of Biochemistry, Wing Hall, Cornell University, Ithaca, New York 14853*

RAJ K. GUPTA (12), *The Institute for Cancer Research, The Fox Chase Cancer Center, Philadelphia, Pennsylvania 19111*

CHIEN HO (13), *Department of Life Sciences, University of Pittsburgh, Pittsburgh, Pennsylvania 15260*

JOHN C. MAXWELL (18), *Department of Biochemistry, Colorado State University, Fort Collins, Colorado 80523*

ALBERT S. MILDVAN (12), *The Institute for Cancer Research, The Fox Chase Cancer Center, Philadelphia, Pennsylvania 19111*

THOMAS H. MOSS (21), *IBM Watson Research Center, Yorktown Heights, New York 10598 and Physics Department, Columbia University, New York, New York 10025*

ECKARD MÜNCK (20), *Freshwater Biological Institute, University of Minnesota, Navarre, Minnesota 55392*

YASH P. MYER (16), *Department of Chemistry, State University of New York at Albany, Albany, New York 12222*

WILLIAM H. ORME-JOHNSON (10), *Department of Biochemistry, University of Wisconsin, Madison, Wisconsin 53706*

NOZOMU OSHINO (27), *Nihon Schering K. K., Development and Research Department, Yodogawa-ku, Japan*

REIKO OSHINO (27), *Biochemistry Lab, Kobe Yamate Women's College, Ikuta-ku, Kobe, Japan*

GRAHAM PALMER (10), *Department of Biochemistry, Rice University, Houston, Texas 77001*

P. M. RENTZEPIS (2), *Bell Laboratories, Murray Hill, New Jersey 07974*

THOMAS G. SPIRO (15), *Department of Chemistry, Princeton University, Princeton, New Jersey 08540*

BERT L. VALLEE (25), *Biophysics Research Laboratory, Department of Biological Chemistry, Harvard Medical School, and The Peter Bent Brigham Hospital, Boston, Massachusetts 02115*

CLAUDE VEILLON (25), *Vitamin and Mineral Nutrition Laboratory, Nutrition Institute, Agricultural Research Center, U. S. Department of Agriculture, Beltsville, Maryland 20705*

LARRY E. VICKERY (17), *Department of Physiology, College of Medicine, University of California, Irvine, California 92717*

KAREN J. WIECHELMAN (13), *Department of Life Sciences, University of Pittsburgh, Pittsburgh, Pennsylvania 15260*

DAVID F. WILSON (1), *Department of Biochemistry and Biophysics, University of Pennsylvania Medical School, Philadelphia, Pennsylvania 19104*

GEORGE S. WILSON (22), *Department of Chemistry, University of Arizona, Tuscon, Arizona 85721*

H. T. WITT (5), *Max-Volmer-Institut für Physikalische Chemie und Molekularbiologie, Technische Universität Berlin, 1000 Berlin 12, Germany*

KWAN-SA YOU (14), *Department of Pediatrics, Duke University Medical School, Durham, North Carolina 27710*

Preface

A great deal of progress has taken place in biological oxidations and bioenergetics since "Oxidation and Phosphorylation" edited by Ronald W. Estabrook and Maynard E. Pullman (Volume X of "Methods of Enzymology") became available in 1967. To update this field five volumes on biomembranes (Volumes LII–LVI, Parts C–G, respectively) have been prepared, three dealing with biological oxidations and two with bioenergetics.

In this volume, Part E of "Biomembranes," subtitled "Specialized Techniques" we aim to bring together biochemical and biophysical techniques important to the study of biological oxidations.

We are pleased to acknowledge the good counsel of the members of our Advisory Board for these five volumes. Special thanks are also due Drs. E. Carafoli, G. Palmer, H. Penefsky, and A. Scarpa for their helpful comments on our outlines for these volumes. Valuable counsel for this volume was also provided by Drs. A. Azzi, H. Beinert, B. Chance, A. Mildvan, and G. Palmer. We were very gratified by the enthusiasm and cooperation of the participants in the field of biological oxidations and bioenergetics whose advice, comments, and contributions have enriched and made possible these volumes. The friendly cooperation of the staff of Academic Press is gratefully acknowledged.

SIDNEY FLEISCHER
LESTER PACKER

METHODS IN ENZYMOLOGY

EDITED BY

Sidney P. Colowick and Nathan O. Kaplan

VANDERBILT UNIVERSITY
SCHOOL OF MEDICINE
NASHVILLE, TENNESSEE

DEPARTMENT OF CHEMISTRY
UNIVERSITY OF CALIFORNIA
AT SAN DIEGO
LA JOLLA, CALIFORNIA

METHODS IN ENZYMOLOGY

EDITORS-IN-CHIEF

Sidney P. Colowick   Nathan O. Kaplan

VOLUME VIII. Complex Carbohydrates
Edited by ELIZABETH F. NEUFELD AND VICTOR GINSBURG

VOLUME IX. Carbohydrate Metabolism
Edited by WILLIS A. WOOD

VOLUME X. Oxidation and Phosphorylation
Edited by RONALD W. ESTABROOK AND MAYNARD E. PULLMAN

VOLUME XI. Enzyme Structure
Edited by C. H. W. HIRS

VOLUME XII. Nucleic Acids (Parts A and B)
Edited by LAWRENCE GROSSMAN AND KIVIE MOLDAVE

VOLUME XIII. Citric Acid Cycle
Edited by J. M. LOWENSTEIN

VOLUME XIV. Lipids
Edited by J. M. LOWENSTEIN

VOLUME XV. Steroids and Terpenoids
Edited by RAYMOND B. CLAYTON

VOLUME XVI. Fast Reactions
Edited by KENNETH KUSTIN

VOLUME XVII. Metabolism of Amino Acids and Amines (Parts A and B)
Edited by HERBERT TABOR AND CELIA WHITE TABOR

VOLUME XVIII. Vitamins and Coenzymes (Parts A, B, and C)
Edited by DONALD B. MCCORMICK AND LEMUEL D. WRIGHT

VOLUME XIX. Proteolytic Enzymes
Edited by GERTRUDE E. PERLMANN AND LASZLO LORAND

VOLUME XX. Nucleic Acids and Protein Synthesis (Part C)
Edited by KIVIE MOLDAVE AND LAWRENCE GROSSMAN

VOLUME XXI. Nucleic Acids (Part D)
Edited by LAWRENCE GROSSMAN AND KIVIE MOLDAVE

VOLUME XXII. Enzyme Purification and Related Techniques
Edited by WILLIAM B. JAKOBY

VOLUME XXIII. Photosynthesis (Part A)
Edited by ANTHONY SAN PIETRO

VOLUME XXIV. Photosynthesis and Nitrogen Fixation (Part B)
Edited by ANTHONY SAN PIETRO

VOLUME XXV. Enzyme Structure (Part B)
Edited by C. H. W. HIRS AND SERGE N. TIMASHEFF

VOLUME XXVI. Enzyme Structure (Part C)
Edited by C. H. W. HIRS AND SERGE N. TIMASHEFF

VOLUME XXVII. Enzyme Structure (Part D)
Edited by C. H. W. HIRS AND SERGE N. TIMASHEFF

VOLUME XXVIII. Complex Carbohydrates (Part B)
Edited by VICTOR GINSBURG

VOLUME XXIX. Nucleic Acids and Protein Synthesis (Part E)
Edited by LAWRENCE GROSSMAN AND KIVIE MOLDAVE

VOLUME XXX. Nucleic Acids and Protein Synthesis (Part F)
Edited by KIVIE MOLDAVE AND LAWRENCE GROSSMAN

VOLUME XXXI. Biomembranes (Part A)
Edited by SIDNEY FLEISCHER AND LESTER PACKER

VOLUME XXXII. Biomembranes (Part B)
Edited by SIDNEY FLEISCHER AND LESTER PACKER

VOLUME XXXIII. Cumulative Subject Index Volumes I–XXX
Edited by MARTHA G. DENNIS AND EDWARD A. DENNIS

Volume XXXIV. Affinity Techniques (Enzyme Purification: Part B)
Edited by William B. Jakoby and Meir Wilchek

Volume XXXV. Lipids (Part B)
Edited by John M. Lowenstein

Volume XXXVI. Hormone Action (Part A: Steroid Hormones)
Edited by Bert W. O'Malley and Joel G. Hardman

Volume XXXVII. Hormone Action (Part B: Peptide Hormones)
Edited by Bert W. O'Malley and Joel G. Hardman

Volume XXXVIII. Hormone Action (Part C: Cyclic Nucleotides)
Edited by Joel G. Hardman and Bert W. O'Malley

Volume XXXIX. Hormone Action (Part D: Isolated Cells, Tissues, and Organ Systems)
Edited by Joel G. Hardman and Bert W. O'Malley

Volume XL. Hormone Action (Part E: Nuclear Structure and Function)
Edited by Bert W. O'Malley and Joel G. Hardman

Volume XLI. Carbohydrate Metabolism (Part B)
Edited by W. A. Wood

Volume XLII. Carbohydrate Metabolism (Part C)
Edited by W. A. Wood

Volume XLIII. Antibiotics
Edited by John H. Hash

Volume XLIV. Immobilized Enzymes
Edited by Klaus Mosbach

Volume XLV. Proteolytic Enzymes (Part B)
Edited by Laszlo Lorand

Volume XLVI. Affinity Labeling
Edited by William B. Jakoby and Meir Wilchek

Volume XLVII. Enzyme Structure (Part E)
Edited by C. H. W. Hirs and Serge N. Timasheff

[1] Kinetic Measurements: An Overview

By David F. Wilson

The shortest time interval of interest to biochemists is the approximately 10^{-15} sec required for a chromophore to absorb a photon. This is the initial event in all photoreactions, and the resulting activated singlet states may have lifetimes of up to approximately 10^{-8} sec before they decay. Molecular motion begins at approximately 10^{-14} sec with vibration along the interatomic bonds and molecular rotations being near 10^{-12} sec for small molecules (see Fig. 1). Chemistry involving bimolecular reactions in solution becomes significant only at times greater than approximately 10^{-8} sec. For example, a diffusion-limited reaction with a second-order rate constant of 10^9 M^{-1} sec^{-1} will have a half-time of approximately 10^{-9} sec if both reactants are at 1 M concentrations. At more physiological concentrations of near 10^{-3} M or less such a reaction would have a half-time near 10^{-3} sec, and even simple protonation reactions ($K = 10^{10}$ M^{-1} sec^{-1} to 10^{11} M^{-1} sec^{-1}, pH 7.0) have half-times of greater than 10^{-4} sec.

An experimentalist is concerned with measuring both the chemical species which participate in a reaction and the time course of the reaction. Thus methods must be selected which are appropriate for measuring the chemical species and which can make the required measurements with an adequate signal-to-noise ratio. When the reactions involved are of short duration (less than a few seconds), it is also necessary to use special methods for initiating the reaction in order to measure its time course. Given a completely stable or reproducible sample, the signal-to-noise ratio of the measurement increases as the square root of the number of measurements made or the amount of time over which the signal is integrated until limited by the intrinsic sensitivity of the method and/or instabilities in the measuring system

New measuring methods are continuously being developed and known methods are continuously being improved, extending their sensitivity and the time range over which measurements can be made. In general a method with existing technology becomes applicable in a certain time range and is useful over a time range dependent on the characteristics of the method. In theory any method can be extended to any longer time period, but in practice the inherent instabilities limit measurements at longer times.

Kinetic methods used for measurements of reactions occurring with half-times of less than 3 or 4 sec are faced with the special problem of initiation of the reaction to be measured. Reagent mixing rates can be improved to give measurements in 10^{-3} to 10^{-4} sec but are unsuited for

ISBN 0-12-181954-X

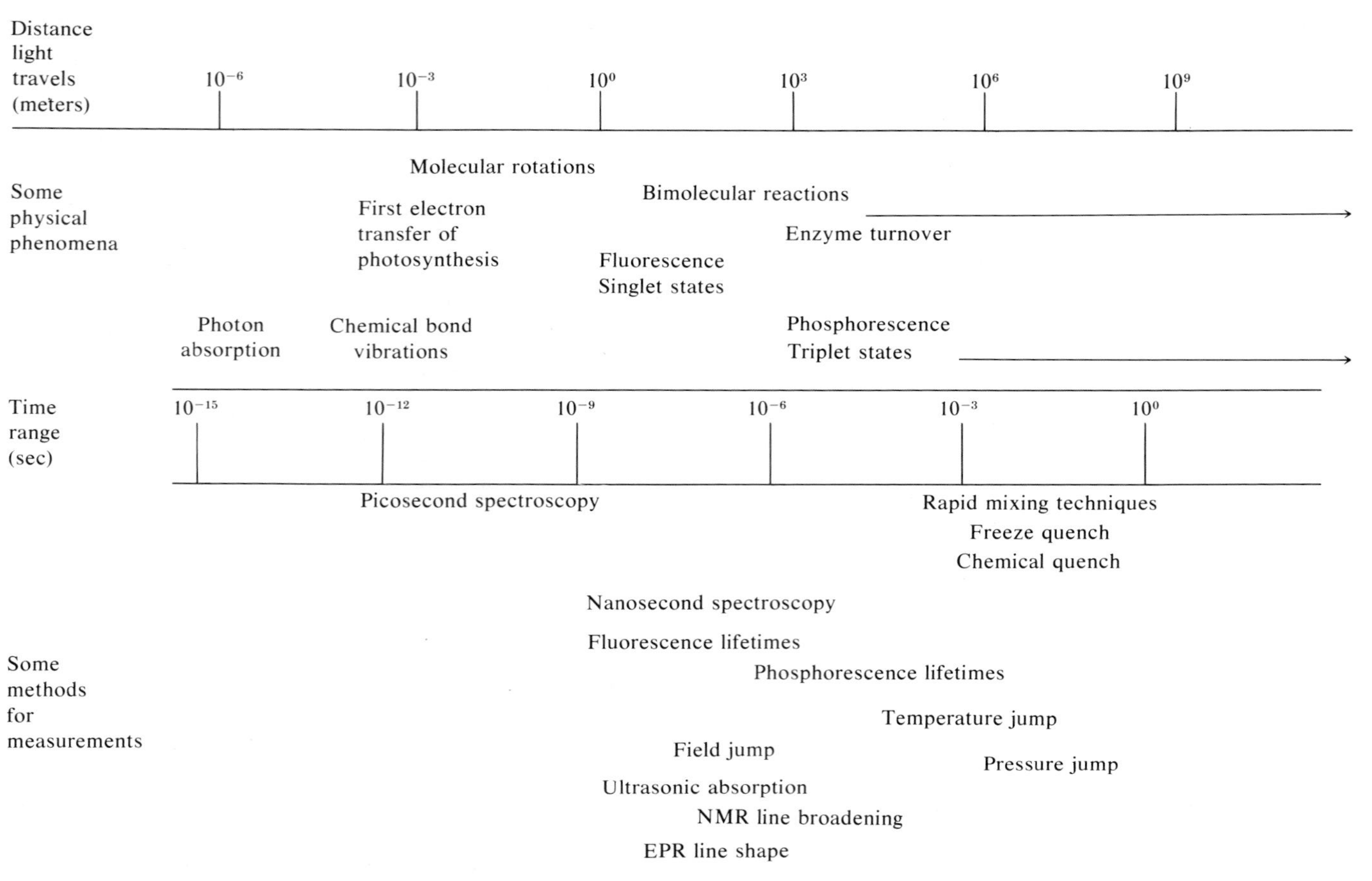

FIG. 1. See legend opposite page.

faster reactions. The methods for measuring the faster reaction rates may be called *relaxation methods*; i.e., a system at or near equilibrium is subjected to a rapid displacement from equilibrium (temperature change, light absorption, pressure change, etc.) and the rate of relaxation back to the equilibrium state is measured. An alternate approach to the study of fast reactions is to lower the sample temperature (Chance, Chapter 9; Frauenfelder, Chapter 28, this volume) and, depending on the activation energy or energies, slowing the reaction to where the available methods are applicable.

[2] Picosecond Spectroscopy in Biological Systems

By P. M. Rentzepis

This chapter describes the state of the art of picosecond spectroscopy and its application to biological systems. Special emphasis is given to the technological aspects and means for detection, display, and processing of data in the picosecond region. The identification and kinetics of short-lived intermediates in photosynthetic (bacteriochlorophyll) and visual (rhodopsin) processes are used as illustrations of the potential application of picosecond spectroscopy for the elucidation of the primary processes of importance in biology.

Fig. 1. A summary of some considerations which relate to the selection of methods of kinetic analysis. Light is the fundamental measure of time, traveling 3×10^{10} cm sec^{-1}. Thus time may be measured in the distance traveled by light, and the upper band of the figure lists the distance traveled by light in various time intervals. For example, light travels only 0.3 μm in 10^{-15} sec or 0.3 mm in 10^{-12} sec. The second band of the figure lists some physical phenomena which occur in the indicated time range. Molecular rotations begin with half-times of near 10^{-12} sec for rotation of small molecules, but the time increases with increasing molecular size. Similarly, molecular singlet states are indicated as having half-lives in the general region of 10^{-7}–10^{-8} sec, but this is only an approximation and both shorter and longer half-lives occur. The lowest band of the figure indicates a few of the experimental methods available for kinetic measurements in the indicated time ranges for their optimal utilization. The methods are usually named for the technique used to induce the reaction. The relaxation methods, for example, may be named for the manner in which the reaction is displaced from equilibrium—such as temperature jump, field jump, or pressure jump—but any technique suitable for measuring the reaction rate (dielectric dispersion, optical absorption, etc.) may be used. Other methods such as picosecond spectroscopy (Rentzepis, Chapter 2, this volume) and nanosecond spectroscopy (DeVault, Chapter 3, this volume) are named for the specialized technology required to make absorbance measurements in the indicated time range.

ISBN 0-12-181954-X

Picosecond spectroscopy is now about 10 years old.[1] Its use is growing continuously because one can perform experiments which reveal previously unknown and until now unobservable events of basic importance not only to biology but to chemistry and physics. This growth is also due to the vast number of ultrafast primary processes which are now accessible to experimental measurement, with relatively high sensitivity and reliability, as a result of recent technological developments which provide means not only for increased reliability in the data but also for the simultaneous display of time, spectrum, and intensity. To perform such experiments, one needs essentially the same tools as in classical spectroscopy with the additional constraints of:

1. A well-defined pulse for excitation which is of picosecond duration with well-defined shape and spectral bandwidth.
2. A clock which allows us to measure time in picosecond units.
3. A light continuum of picosecond duration and a wavelength range spanning thousands of wavenumbers to act as the absorption monitoring light.
4. An experimental laser system which can resolve the time, wavelength, and intensity parameters simultaneously while having a sensitivity approaching that of single photon counting and able to average and display the data in any desirable and meaningful manner.

We shall discuss each of these four constraints and describe the means by which each can be produced in the laboratory and then combined into the complete picosecond spectroscopy optical system shown in Fig. 1a.

The Picosecond Pulse and Its Characterization

We will not attempt a discourse on picosecond pulse generation; instead we will briefly describe the method for its generation and the measurement of pulse shape and duration. Several papers devoted to picosecond pulse generation are available, including those referred to here.[2-6]

The Pulses

The generation of the picosecond pulses is achieved in a cavity [Fig. 1a (2)] formed by 100% and ~50% reflecting mirrors. Between these mirrors is placed the oscillating/amplifying medium such as a flashlamp-pumped Nd^{3+}

[1] P. M. Rentzepis, *Chem. Phys. Lett.* **2,** 117 (1968).
[2] A. J. Demaria, W. H. Glenn, M. J. Brienza, and M. E. Mack, *Proc. IEEE* **57,** 2 (1969).
[3] N. Bloembergen, *J. C.* [N.S.] **1,** 37 (1968).
[4] T. L. Netzel, W. S. Struve, and P. M. Rentzepis, *Annu. Rev. Phys. Chem.* **24,** 473 (1973).
[5] M. A. Duguay and J. W. Hansen, *Opt. Commun.* **1,** 254 (1969).
[6] E. G. Arthus, D. J. Bradley, and A. G. Roddie, *Appl. Phys. Lett.* **9,** 480 (1971).

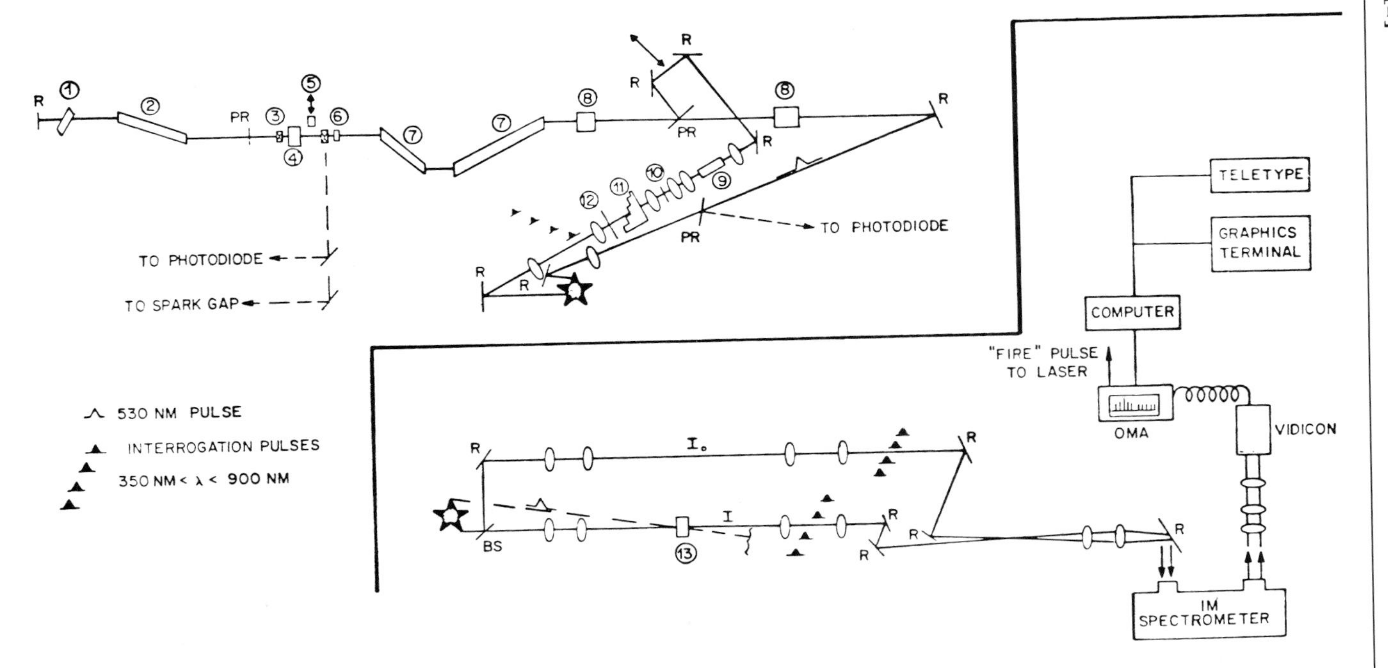

FIG. 1. (a) Double-beam picosecond spectrometer utilizing a silicon vidicon detector. Components: (1) mode-locking dye cell; (2) laser oscillator rod; (3) calcite polarizer; (4) Pockels cell; (5) translatable 90° polarization rotator for 1060 nm radiation; (6) fixed-position 90° polarization rotator; (7) laser amplifier rod; (8) second harmonic (530 nm) generating crystal (KDP); (9) 20-cm octanol cell for generating the interrogation wavelengths; (10) ground-glass diffuser; (11) index matched glass echelon for producing picosecond optical delays between the stacked interrogation pulses; (12) vertical polarizer; (13) sample cell; (R) reflector; (PR) partial reflector; (BS) beam splitter; (OMA) optical multichannel analyzer.

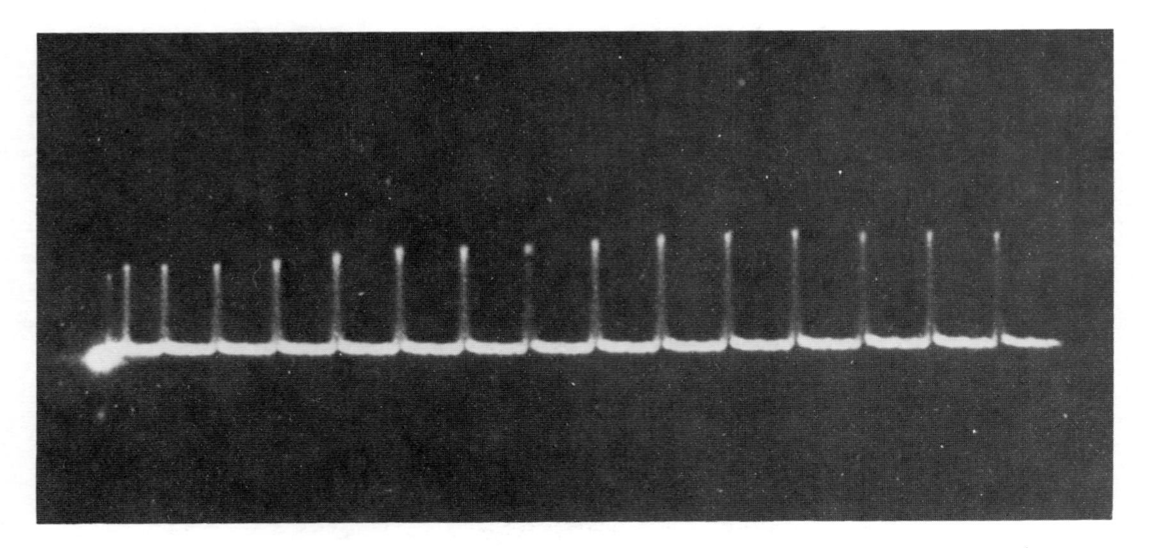

FIG. 1. (b) Oscilloscope (Tektronix 519) trace of the output of a mode-locked laser. The time scale is 20 nsec/div; the separation between pulses corresponds to a time equal to $2l/c$, in this case ~ 7 nsec.

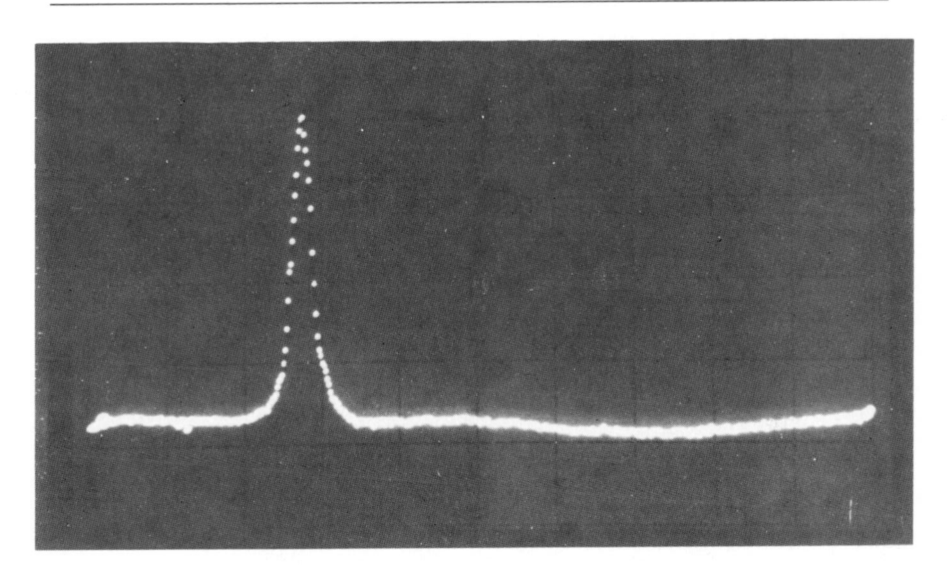

FIG. 1. (c) A display of a single picosecond pulse with a time duration of ~ 6 psec.

glass or ruby rod, preferably with Brewster ends. A 1-cm cell [Fig. 1a, (1)] containing a saturable absorbing dye acts as the mode-locking medium and eliminates satellite pulses. When the radiation within the cavity reaches a certain intensity the dye bleaches and recovers very fast, thus modulating the light and forming periodic pulses with a period of $2l/c$, l being the optical length of the cavity. The output of this laser consists of a train of ~100 pulses (Fig. 1b) separated from each other by $2l/c$ or about 8 nsec for a 1.2-m cavity. When used in biological experiments it is imperative that one avoid excitation by succeeding pulses by using a single pulse (Fig. 1b) which is extracted from the train. This single pulse is attenuated when necessary to prevent saturation, and with appropriate modifications it is used for both excitation and interrogation. Utilization of a single pulse also eliminates the notorious variation in shape and time width within even a single train which frequently gives erroneous results.

A typical single pulse (Fig. 1c) contains an energy of approximately 10 mJ and has a duration of a few picoseconds. Therefore, although the power is of the order of gigawatts, the photon content is low and normally should be amplified to be useful for most experiments other than for a few dyes which have a very high absorption cross-section and fluoresce strongly. Amplification by ~ × 10 is achieved by one or two amplifying rods [Fig. 1a (7)]. Conversion of the 1060-nm light pulse to the more useful second (530 nm) and fourth (265 nm) harmonics is made in phase-matched KDP and

ADP crystals. The energy used per pulse at 530-nm and 2650-nm pulses is about 7 mJ and 2 mJ, respectively. Picosecond pulses by ruby lasers can also be frequency converted by KDP and ADP crystals.[7] The second harmonic of ruby at 347.2 nm and the third harmonic at 231.4 nm have been generated with energy per pulse in the range of a few millijoules.

Since pulse variation is always experienced or expected, one should measure for every experiment the shape and width of the pulse. To this end, an elegant yet simple and inexpensive technique is presented.

The Duration and Shape of Picosecond Pulses

The first method developed which was capable of determining the timewidth of a single picosecond pulse or pulse train is now known widely as the two-photon fluorescence technique (TPF).[8,9] This method uses the simultaneous absorption of two photons via a virtual transition in a dye selected for its high absorption cross-section at the two-photon energy and a subsequent high quantum yield of fluorescence. In practice, the single pulse or pulse train is split into two equal portions. After traversing the paths shown, the two beams are rejoined inside the cell (Fig. 2a) containing a dye such as Rhodamine 6G. At the point where the two pulses overlap the power is sufficient for two-photon absorption to take place although the cross-section is very small ($\sigma \approx 10^{-52}$), and consequently a rather intense fluorescence is observed within the region of overlap (Fig. 2b). The width of the two-photon fluorescence pattern yields a direct measurement of the pulse duration (Fig. 2b). Because this simple and relatively accurate method is inherently incapable of revealing the pulse shape, and because it displays the low 3:1 contrast ratio between the peak intensity of the TPF spot and the background for a perfectly mode-locked pulse, it has been superseded by the three-photon (3PF) method.[10]

This technique offers the advantages of the TPF technique, and in addition to the timewidth measurement it is capable of measuring the shape of the pulse in a fluorescence pattern with a peak-to-background ratio of 10:1.

In principle the three-photon dye should have the same properties as the dyes in the TPF technique, with the obvious variation of the energy gap between the ground and first allowed state being 3 times the laser energy of

[7] See N. Bloembergen, "Optical Masers," pp. 13–22. Polytech Inst., Brooklyn, New York, 1963.

[8] J. A. Giordmaine, P. M. Rentzepis, S. L. Shapiro, and K. W. Wecht, *Appl. Phys. Lett.* **11,** 216 (1967).

[9] P. M. Rentzepis and M. A. Duguay, *Appl. Phys. Lett.* **11,** 218 (1967).

[10] P. M. Rentzepis, C. J. Mitchele, and A. L. Saxman, *Appl. Phys. Lett.* **17,** 122 (1970).

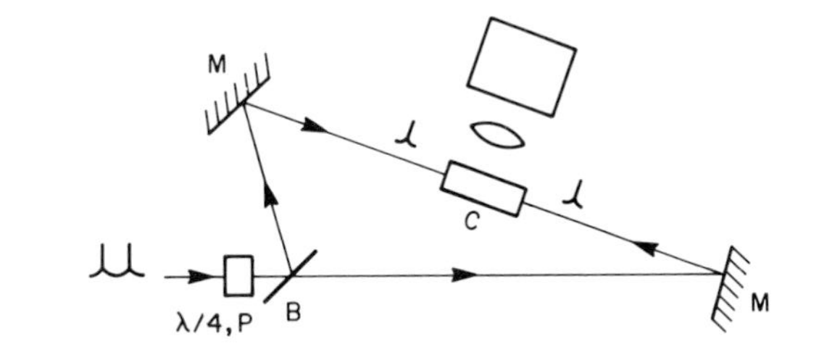

(a)

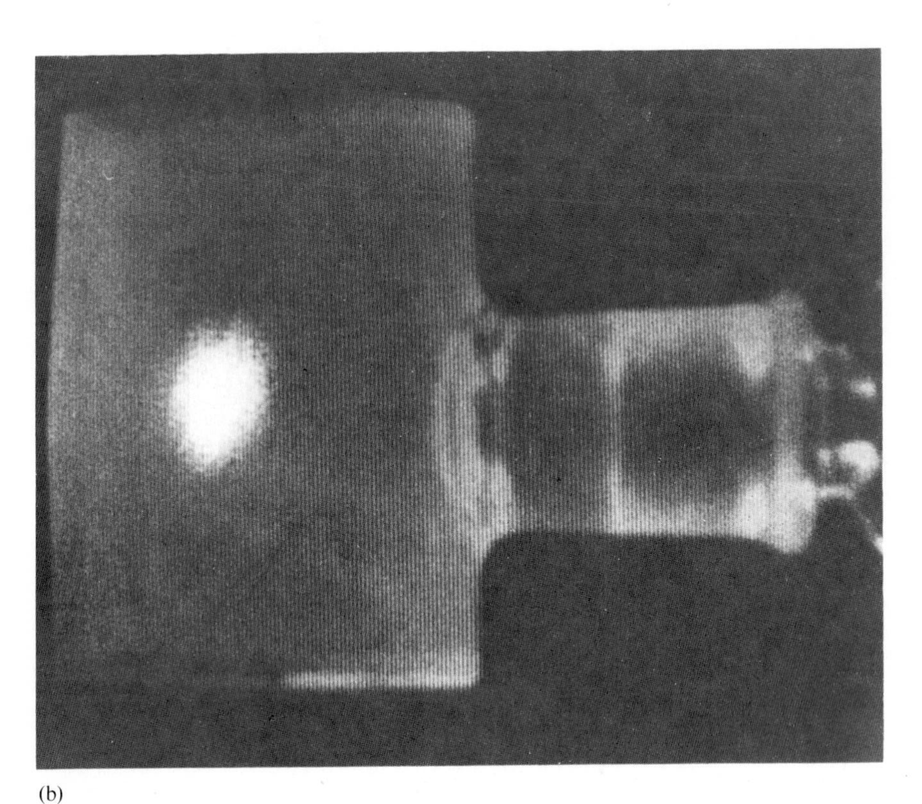

(b)

FIG. 2. (a) Schematic representation of the optical path traversed by the pulse(s). M, 100% reflective mirror; C, cell containing the two-photon fluorescing solution. The quarter-wave plate $\lambda/4$ and polarizer P prevent the light pulse from returning into the laser cavity. (b) Photograph of a picosecond pulse by the TPF method. The pulse has a full width at half maximum of ~4 psec. The peak-to-background ratio is ~2.8 : 1.

the photon. For example, for Nd^{3+} glass laser radiation emitted at 1060 nm (9431 cm^{-1}) the $S_0 \rightarrow S_1$ energy gap should be less than 353.3 nm (28,293 cm^{-1}); this criterion is met by BBOT and dimethyl POPOP in solution. The 3PF experiment of Fig. 3a essentially splits the pulse into three equal parts and directs them to enter the 3PF cell from the three directions X, $-X$, and Y, where they intersect with each other resulting in three-photon absorp-

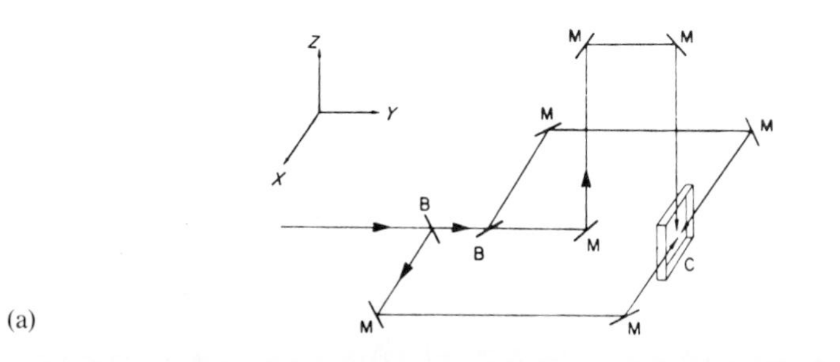

(a)

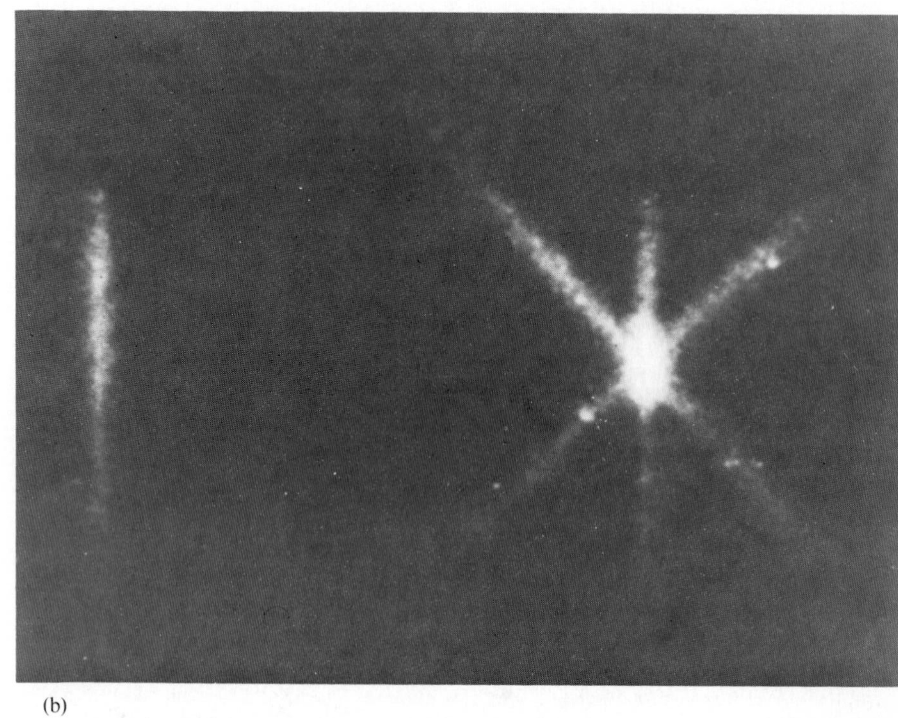

(b)

FIG. 3. (a) Optical arrangement used to obtain two independent delays in 3PF. M is a 100% dielectric mirror, B a beam splitter, and C a cell. (b) 3PF pattern obtained by splitting a single pulse into three equal pulses and overlapping these pulses traveling in directions normal to each of the three lines in the pattern.

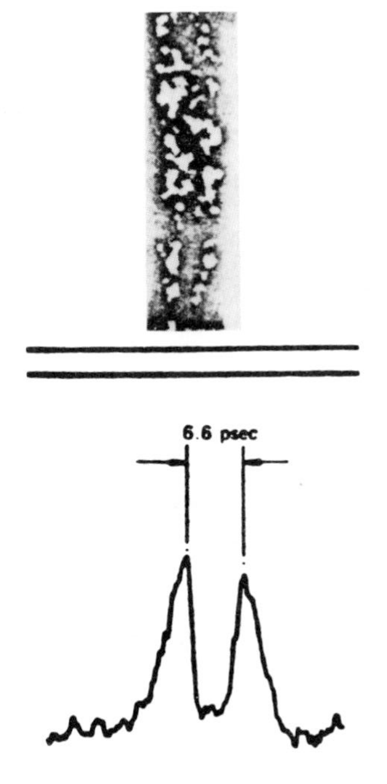

FIG. 4. Streak camera photograph shows a time-resolved pair of picosecond pulses. The microdensitometer trace of the streak camera plot is shown below.

tion followed by fluorescence in the pattern shown in Fig. 3b. The width of each leg is representative of the width of the pulse as in TPF, while the correlation of the patterns formed by each leg provides the data for the determination of the pulsing shape.

Streak Camera

The streak camera[11] provides a unique, but expensive, means for recording picosecond pulses. It has a resolution of 2–10 psec and a high linear dynamic range, and coupled with a silicon image intensifying tube it can display picosecond pulses and fluorescent events in the picosecond scale with high sensitivity and resolution. A picosecond pulse recorded by a streak camera is shown in Fig. 4. The use of the streak camera in the time resolution of chlorophyll is discussed in a following section.

[11] D. J. Bradley and G. H. C. New, *Proc. IEEE* **62,** 313 (1974).

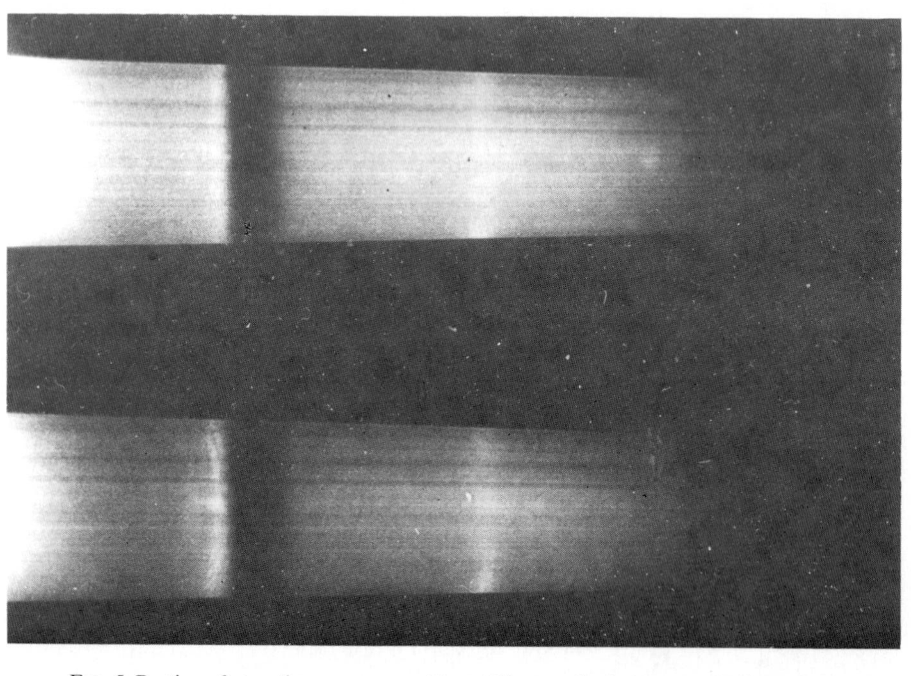

FIG. 5. Portion of a continuum generated by a 530-nm pulse in 20-cm cell of H_2O and D_2O. The spectrum shown spans the region of 530 nm to ~400 nm. The stimulated antistokes and inverse Raman are quite evident on the otherwise smooth spectrum.

The Picosecond Continuum

The identification of an unknown intermediate species which lives as long as several hundred picoseconds or less demands that broad spectral areas be monitored after excitation. This can be achieved by a broad continuum which, unlike the one used in classical spectroscopy, must be of picosecond duration.

Such a continuum has been generated by the interaction of a single picosecond high-intensity pulse with glass[12] or liquids such as H_2O,[13] D_2O,[13] or organic solvents such as hexane, octanol, and many liquid mixtures. We prefer liquids such as H_2O and D_2O because the continuum output has high intensity and picosecond duration and the medium is not damaged as is the case with glasses and other solids. In practice one focuses or collimates the laser pulse into a cell, (~10–20 cm) containing the liquid, e.g., D_2O [Fig. 1a (9)]. The emitted continuum shown in Fig. 5 is the result

[12] R. Alfano and S. L. Shapiro, *Chem. Phys. Lett.* **8**, 631 (1971).

[13] G. E. Busch, R. P. Jones, and P. M. Rentzepis, *Chem. Phys. Lett.* **18**, 178 (1973).

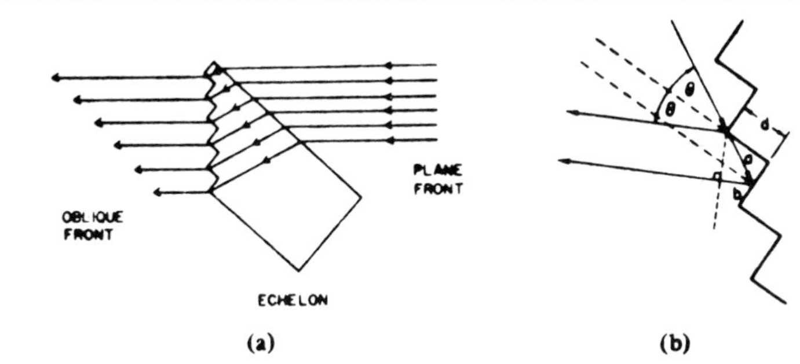

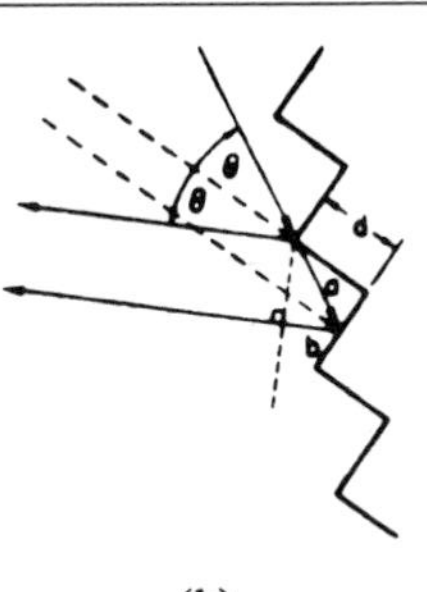

FIG. 6. (a) Transmission echelon for obtaining digital delays. Each segment has a different path length of high index material and thus introduces a different delay to the portion of the wavefront passing through it. (b) Reflection echelon. The intersegment delay is determined by the increment in optical path, $a + b = 2d \cos \theta$.

of the interaction of the 1060 nm and 530 nm in D_2O.[13] Its formation is thought to take place because the high field of the pulse induces a refractive index increase. The liquid now acts as a lens focusing the light into filaments of a few microns in diameter.[14] The power within these filaments is large enough to permit the very low cross-section nonlinear processes such as higher-order stimulated Raman stokes and antistokes scattering and multiphoton processes to become significant. Self-phase modulation builds the intensity of the "side bands," thus generating a smooth continuum of wavelengths spanning several thousand wavenumbers. The continuum generated has been shown to have a timewidth equal to that of the pulse if it is generated carefully.[13] Due to dispersion of the medium and other effects, the appearance time of the continuum at the antistoke side and wavelengths far away from the pumping wavelength can be drastically slower than the generating pulse, resulting in gross errors if not measured.[15] Methods for determining the arrival time of the continuum pulse in the cell and its timewidth have been described earlier by Busch *et al.*[13] This continuum can be used for monitoring optical density changes with high wavelength resolution and picosecond duration. To achieve the last point, however, one needs a device which measures and displays time in picoseconds.

The echelon[16] and the streak camera again can display fluorescent events in time frames of a few picoseconds. Another picosecond clock is shown in Fig. 6. It is simple and inexpensive and can measure and display picosecond absorption and emission events in a digital form. It is known now as the *echelon*. An echelon is a stepped delay formed by a stack of

[14] F. Shimizu, *IBM J. Res. Dev.* **17,** 286 (1973).
[15] C. G. A. O. Varma and P. M. Rentzepis, *Chem. Phys. Lett.* **19,** 162 (1973).
[16] M. R. Topp, P. M. Rentzepis, and R. P. Jones, *J. Appl. Phys.* **42,** 3451 (1971).

plates, fibers, mirrors, cut glass, or polished metal blocks. The spacing, thickness, or length of the steps is normally equal. Thus a single pulse transmitted through an echelon is divided into a number of identical and equally separated pulses, with the separation being determined by the intersegment echelon step thickness. It varies from a picosecond or less to over 100 psec per segment. This device provides a digital clock whose period can easily be changed from a few picoseconds to over 1 nsec by varying the intersegment spacing and number of segments. When a pulse enters the echelon it traverses the various paths defined by the segments. Since the index of refraction of the glass is higher than that of air, the velocity of the pulse is less in glass than air; therefore the part of the pulse passing through the glass will be slowed down relative to its corresponding part in air. The relative delay is a function of the index difference and the echelon thickness. An echelon may have 12 segments each separated by 6 psec and therefore a total span of $\sim$70 psec, while a second might have 5 3-psec segments, 3 100-psec segments, and 3 200-psec segments resulting in a nanosecond span per shot.

Time Wavelength and Intensity Display

If the continuum pulse passed through the echelon, each segment would then contain both time and wavelength information. If we now image this echelon output on the entrance slit of a spectrometer, the resulting histogram is shown in Fig. 7. Notice that a two-dimensional spectrum is displayed. The abscissa shows the typical spectrum of the continuum; however, the ordinate is composed of several lines, each being one echelon segment and separated by 6 psec from the one lying lower. Therefore, if an event (e.g., excitation) takes place coincident with, say, the third segment, any subsequent spectroscopic events can be identified in time by observing the segment where the spectra changes occurred the intersegment time separation. For example, if each segment is separated by 6 psec and the new absorption appears three segments after excitation and decays 10 segments thereafter, we easily deduce that the intermediate is formed within 15 psec and decays within 50 psec.

There are several means for the display of the data. The simplest method is, of course, photographic film. While film is extremely simple to use, it has many disadvantages. These include low sensitivity, difficulty of data reduction, and poor dynamic range and spectral response. The obvious solution is single-photon counting techniques; however, these have to

Fig. 7. Output of the double-beam picosecond spectrometer (Fig. 1). (a) On spectroscopic plate upper I_0, lower I, lines represent echelon segments.

Ultrafast Intermediates in Vision

An illustration of the fascinating, new, important, and previously unattainable data obtained by picosecond spectroscopy is provided by studies of intermediates in the visual process. To the best of our knowledge, the first application of this technique to biology was the measurement of the rate of formation and decay of prelumirhodopsin (bathorhodopsin) at room temperature by Busch *et al.*[22] Previous studies at 77°K had established that the first intermediate, prelumirhodopsin, absorbed in the 550–600 nm region with a maximum near 560 nm. The room-temperature kinetics of this primary intermediate in the visual transduction process could not be observed previously because of its fast formation and decay kinetics.

The experiment of Busch *et al.*[22] was performed with a single 530-nm pulse generated by a picosecond system similar to the one shown schematically in Fig. 1, with the exception that (1) the I_0 was missing and (2) the detector was either a spectroscopic plate or photodiode. The interrogating beam at 561 nm was the Stokes Raman of benzene produced in a 10-cm cell [Fig. 1(9)]. After passing through the echelon forming 10 segments separated by 7 psec, the 561-nm light and the remaining 530-nm light were combined and focused into the cell containing detergent-solubilized bovine rhodopsin. The geometry of the excitation and interrogation spot within the reaction cell was such that the interrogating segment area was much larger than the area of the excitation pulse, thus allowing the outer portion of each echelon segment to probe unexcited rhodopsin solution, while the center portion traversed through the same region as the excitation pulse and monitored the spectral changes induced by the 530-nm pulse. The echelon was then imaged on a spectrometer whose output on a spectroscopic plate is shown in Fig. 9.

In the first set of experiments, an echelon with a 20-psec intersegment time was utilized. The 561-nm prelumirhodopsin absorption was found to occur at the segment coincident with the excitation pulse and to remain for a period of time longer than 400 psec (cumulative timewidth of all segments of the echelon plus the movement of the translation stage). This indicated that prelumirhodopsin was formed within 20 psec and did not decay for at least 400 psec. Attempts to time-resolve the process of formation with a 2 psec/segment echelon showed that the formation time is faster and limited by the width of the excitation pulse. Since the rise time of the prelumirhodopsin is governed by the excitation pulse only, an upper limit ~3 psec can be placed on its rate of formation, and quite possibly it is faster by a factor of 10 or more.

The size of the echelons constructed so far has limited their use to times of only several hundred picoseconds, although in conjunction with inserting a beam splitter and one mirror in the interrogating beam (Fig. 1, insert)

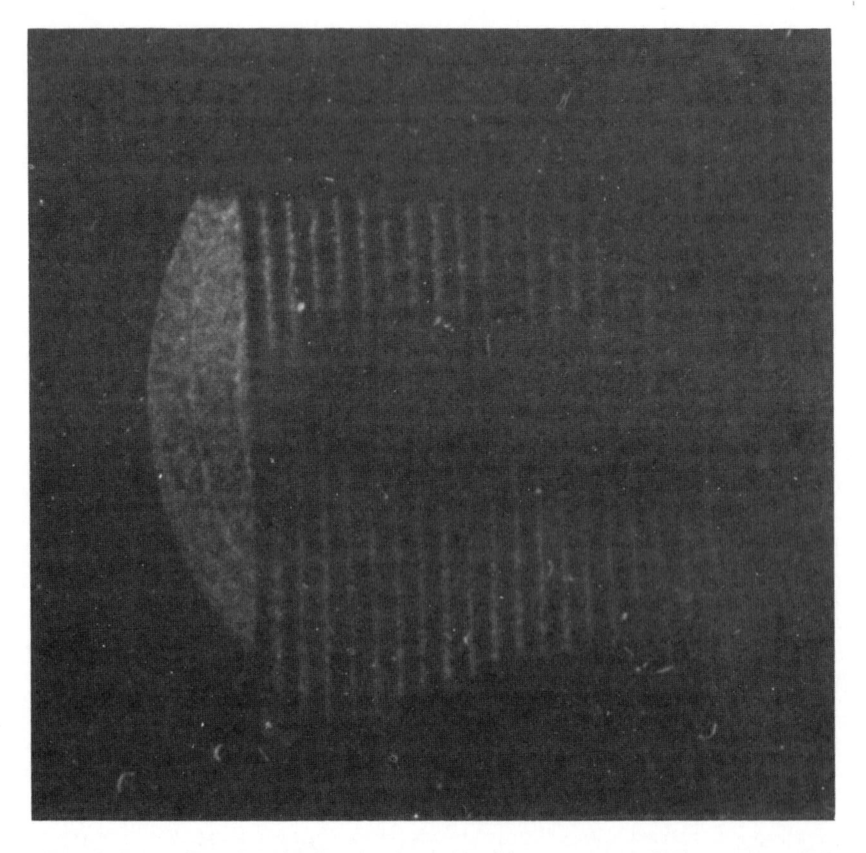

FIG. 9. Formation of prelumirhodopsin obtained by apparatus of Fig. 1. The echelon segments have interrogated the formation of prelumirhodopsin in the sample cell. The time separation between segments is 0.3 psec, and it seems that the absorption (dark area) sets in within 1 segment or 3 psec. The geometry of the overlapping pulses for photographic detection is such that the excitation pulse has a smaller diameter than the interrogating pulse train. The white lines correspond to the part of the interrogating segments not overlapping with the excitation pulse. This area serves as a reference.

one can increase the time to ~10 nsec. To use the same optical system to measure processes which have decay lifetimes of many nanoseconds, such as prelumirhodopsin, one can increase the time-resolving capability of the laser system to more than 100 nsec with the scheme used by Busch *et al.*[22] to measure the decay of prelumirhodopsin at room temperature.

In this case, the pulse train was passed through a Pockel cell where as before the polarization of one pulse was rotated by 180°. Removal of the second polarizer allowed all pulses to travel the same optical path to the KDP crystal rather than be rejected by the second polarizer (Fig. 10). The frequency conversion is polarization dependent; therefore, only the rotated pulse is converted to 530 nm while the rest of the train remains at 1060 nm. A dichroic beam splitter reflects the 530 nm and the appropriate optics guide it to rhodopsin for excitation. The remainder of the train, pulses separated by 7 nsec, are now rotated by $\pi/2$ in an optically active quartz crystal and passed through a second KDP crystal generating 530 nm, then into a Raman cell of benzene generating a set of ~50 pulses of 561-nm light and separation of 7 nsec between pulses. This 561-nm interrogating, I_0, pulse train (Raman shifted) and the 530-nm exciting, I, single pulse were recombined and focused into the cell containing the bovine rhodopsin. Beam splitters before and after the sample cell were set to send the I_0 and I beams into a fast photodiode where the output was fed into an oscilloscope. The decay of prelumirhodopsin could be monitored every 7 nsec (pulse separation) by observing changes in pulse intensity and making the ratio of I/I_0. Experiments were performed at three temperatures, 17.5°C, 22.5°C, and 24.3°C. Although the upper limit of 6 psec was valid at all of these temperatures, the rates of decay were found to be 2.7×10^7, 3.7×10^7, and $4.1 \times 10^7\ sec^{-1}$ for the three respective temperatures. Activation energies and other thermodynamic data have been routinely calculated from this data. The proof that these data are indeed a measure of prelumirhodopsin and a display of its kinetics rests on the identification of the complete absorption band, shape, and isosbestic points.

Recently Sundstrom *et al.*[24] have performed such experiments, and indeed the data confirm that the 561-nm rates do certainly belong to prelumirhodopsin. The Sundstrom *et al.*[24] formation rate at all prelumirhodopsin wavelengths is the same as that found by Busch *et al.* at 561 nm (Fig. 11). The room temperature absorption spectrum of Sundstrom *et al.* corresponds very well with that of Applebury and Yoshizawa at 77°K. However, the 3 : 1 ratio of batho to rhodopsin and the 515-nm isosbestic point measured at 77°K correspond to 1:1 and 525 nm at room temperature. This apparent discrepancy is in fact due to the mode of decay of excited rhodopsin, and recent work by Peters *et al.*[24a] by similar picosecond methods has shown that the ratio is indeed 3 : 1 at 77°K and increases to about 5 : 1 as the temperature is lowered to 4°K. The isosbestic point obviously shifts to lower wavelengths as the intensity increases, and the shape remains the same in such a manner that the isosbestic point occurs at 515 nm at 77°K, in agreement with the work of Applebury[25] and Yoshizawa.[26]

[24] V. Sundstrom, K. Peters, M. A. Applebury, and P. M. Rentzepis, *Nature* **267**, 645(1977).
[24a] K. Peters, M. L. Applebury, and P. M. Rentzepis, *Proc. Natl. Acad. Sci. U.S.A.* **74,** 3119 (1977).
[25] M. L. Applebury, D. Zuckerman, A. A. Lamola, and T. J. Jovin, *Biochemistry* **13,** 3448 (1974).
[26] R. Tukunaga, S. Kawamura, and T. Yoshizawa, *Vision Res.* **16,** 633 (1976).

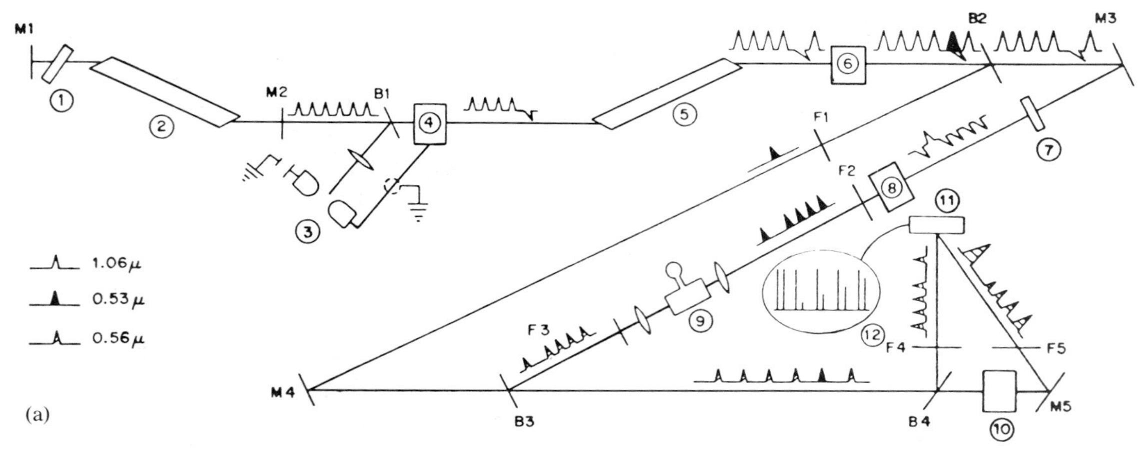

FIG. 10. (a) Diagram of the apparatus used to observe the nanosecond decay of prelumirhodopsin. A pulse early in the mode-locked train of pulses from the Nd^{3+} laser (1.06 μm) has its polarization rotated by the Pockel cell so that it can be doubled (to 0.53 μm) by the KDP crystal. This pulse is then split off, by means of a dielectric reflector, from the remaining pulses in the train and used to excite the rhodopsin in the sample cell. The remaining pulses pass through the 0.53-μm reflector and are used to generate the second harmonic, which is Raman shifted in benzene to 0.56 μm and focused into the sample cell. Appropriate filters are used so that only 0.53-μm light excites the sample and only 0.56-μm light interrogates. Beam splitters placed before and after the sample cell reflect a fraction of the 0.56-μm light to a fast photodiode, providing a measure of I_0 and I, from which absorbance is computed. The optical path difference for the I_0 and I beams and the finite velocity of light allow the same photodiode to resolve the signal into pulse pairs (see Fig. 6) for each pulse in the mode-locked train. Components: (1) cell containing saturable absorbing dye; (2) Nd^{3+} glass-laser rod; (3) spark gap; (4) Pockels cell; (5) Nd^{3+} glass-amplifier rod; (6) KDP SHG crystal; (7) 1.06-μm 90° polarization rotator; (8) KDP SHG crystal; (9) 15-cm cell containing benzene; (10) sample cell; (11) fast photodiode; (12) fast oscilloscope. (b) Oscilloscope trace showing photodiode response to 0.56 μm interrogation. Each pulse in the train (5.5 nsec separation) is split into two parts for the absorption measurements. The more intense pulse in each pair is the I_0 reference, while the lower intensity has monitored the absorbance of rhodopsin: Full scale is ~ 300 nsec. The pulse(s) missing near the beginning of the train were used to generate the 0.53-μm exciting pulse.

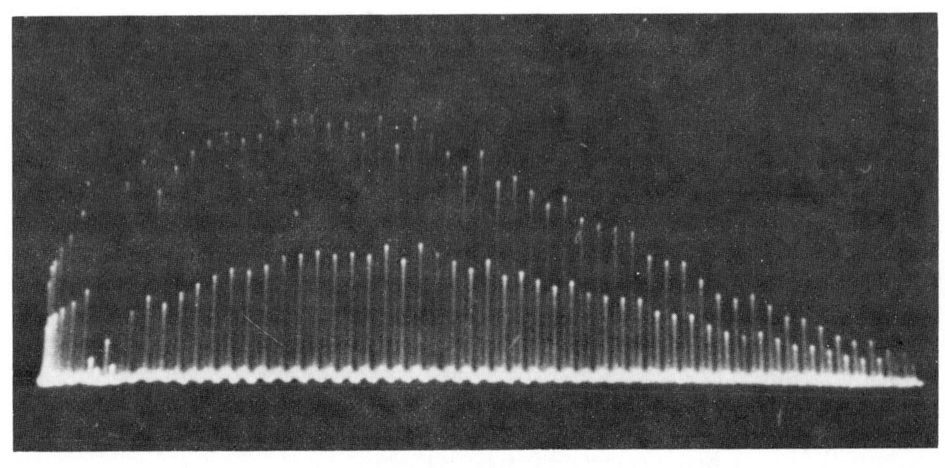

FIG. 10(b).

The picosecond data provide an insight into the primary kinetic processes in vision and possibly pose the question as to what is really the first step in the visual process which has such low activation energy and ultrafast rate even at low temperatures. Should one then expect to find a hindrance in the isomerization process if, in fact, it involves the geometric reorganization of bulky chemical groups? It could, of course, involve only "electron" translocation where the temperature is not as effective. But then why does this change in the intensity ratios with temperature? Of course, tunnelling is a most possible mechanism where the proton associated with the histidine and the protonated Schiff base transfers across the barrier. Although one can propose answers, it is for the experiment to provide unequivocal proof, and for such ultrafast processes, picosecond spectroscopy seems to be ideally suited.

Photosynthesis

Although the photosynthetic cycle is well established, the primary processes in either plant or bacteriochlorophyll have remained quite unknown. Several investigators have made very important contributions, including Witt and his collaborators[27] in plant chlorophyll and Clayton[28] in bacteriochlorophyll. The initial fast events have been studied by fluorescence techniques and absorption spectroscopy. The plant photosynthetic systems have been studied in emission by Kollman *et al.*[29] by time-resolved

[27] H. T. Witt, *Q. Rev. Biophys.* **4,** 365 (1971).

[28] R. K. Clayton and R. J. Wang, Vol. 23A, p. 696.

[29] V. H. Kollman, S. L. Shapiro, and A. V. Campillo, *Biochem. Biophys. Res. Commun.* **63,** 917 (1975).

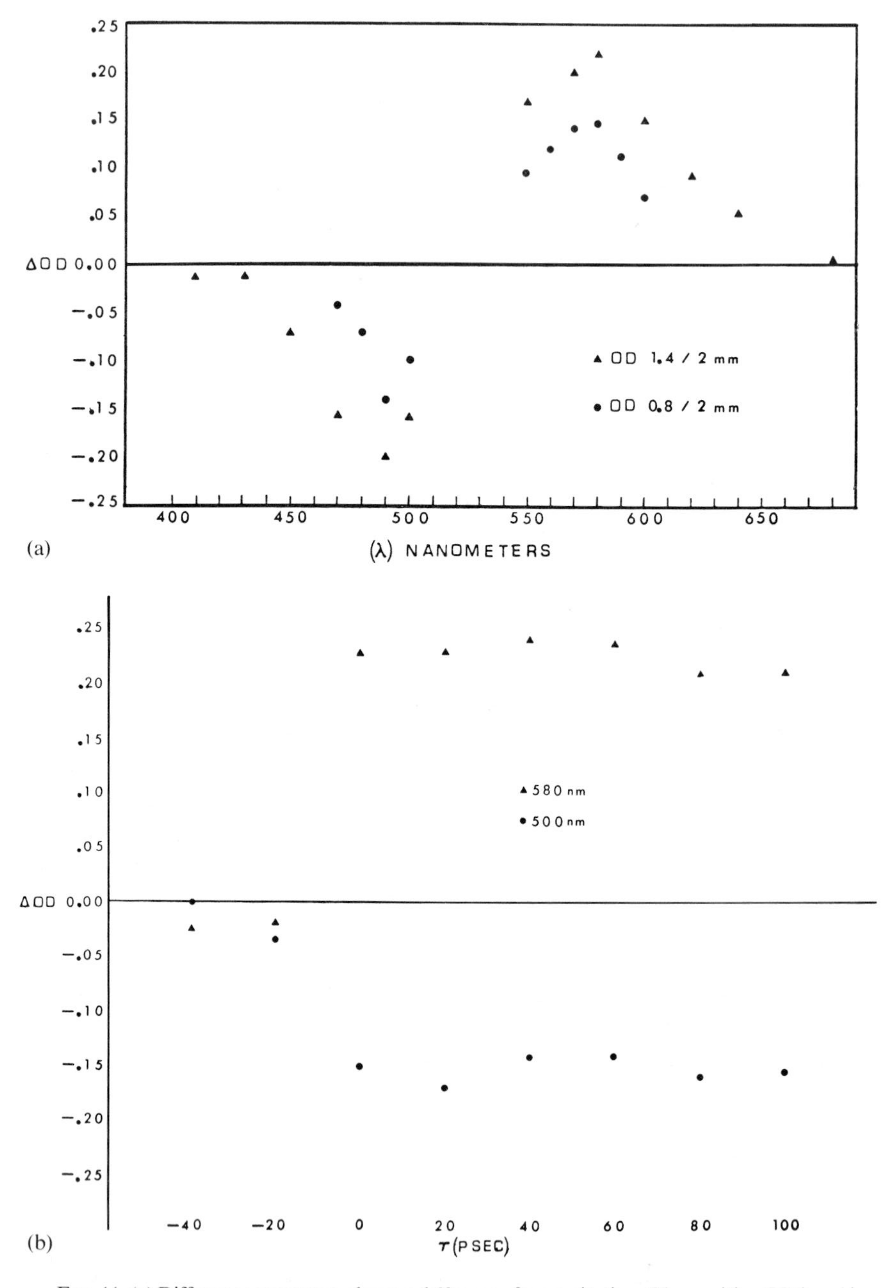

FIG. 11. (a) Difference spectrum observed 60 psec after excitation. The positive OD band is assigned to prelumirhodopsin while the $-\Delta OD$ is due to the bleached rhodopsin. The reproducibility of each point is ± 0.02 OD. (b) (A) Formation of the prelumirhodopsin band with a maximum at 580 nm induced by a 530-nm, 6-psec pulse. (b) Depletion rate of the rhodopsin band at 480 nm after excitation with the same picosecond pulse. Similar kinetics are observed over the entire spectrum. Both figures represent an average of five kinetic records. The variance is given by the error bars.

emission spectroscopy using a streak camera. A streak camera operates on the following principle: light is focused onto a photocathode causing the release of electrons which are then accelerated through the anode. A picosecond pulse from a laser as in Fig. 1 hits a photodiode and generates a voltage pulse which triggers an increasing voltage ramp causing the deflection of the oncoming electrons. The increasing voltage streaks the electrons across the surface of a phosphorescence screen to positions dependent upon the time the photon released the electron. The photographed pattern on the screen gives the lifetime of the event as a function of distance. Bradley and New[11] have written an excellent description of streak camera operation concepts and some of its applications. It is obvious that for fluorescence measurements in the picosecond range, the streak camera offers great advantages over other methods such as the shutter or photodiodes. The dynamic range is large, the reliability greater, and, by the addition of a SIT vidicon, the streak camera offers the additional sensitivity and two-dimensional options. The data can be obtained from a single pulse, eliminating the errors encountered by the dynamic averaging of the whole pulse train and the old but more inaccurate translation stage decay where the lifetime involves several shots, each averaging a complete train with varying timewidth, and intensity per pulse.

The method shown in Fig. 1 and the streak camera can be thought of as quite competitive, with the streak camera having some advantages for fluorescence resolution while our system is probably more versatile and desirable for picosecond absorption spectroscopy. Either system can be used for both emission and absorption with essentially the same sensitivity.

In a series of experiments with streak cameras, several investigators have shown that the fluorescence lifetimes of photosynthetic systems vary from 10 psec to nanosecond. Kollman, Shapiro, and Campillo[29] have shown by means of a streak camera that the lifetime of chlorophyll α is concentration-dependent and varies from 685 $\pm$ 145 psec at 0.00 M to 12 $\pm$ 5 psec at 0.1 M. They found that the fluorescence quenching followed the $1/r^6$ dipole–dipole interaction of the Förster mechanism. The quenching of fluorescence was studied previously by Porter[30] as a function of the quantum yield. Porter and his colleagues[30a] also measured the fluorescence of Photosystem I with a Hadland Photonics streak camera and have come to similar conclusions concerning the lifetime of Photosystem I and its quenching. In addition, Porter has made several very interesting suggestions concerning the energy transfer in chloroplasts. Bacteriochlorophyll is a much simpler system and is therefore amenable to more detailed study

[30] G. Porter, *Proc. R. Inst. G. B.* **47,** 143 (1974).
[30a] G. Porter, E. S. Reid, and C. J. Tredwell, *Chem. Phys. Lett.* **29,** 469 (1974).

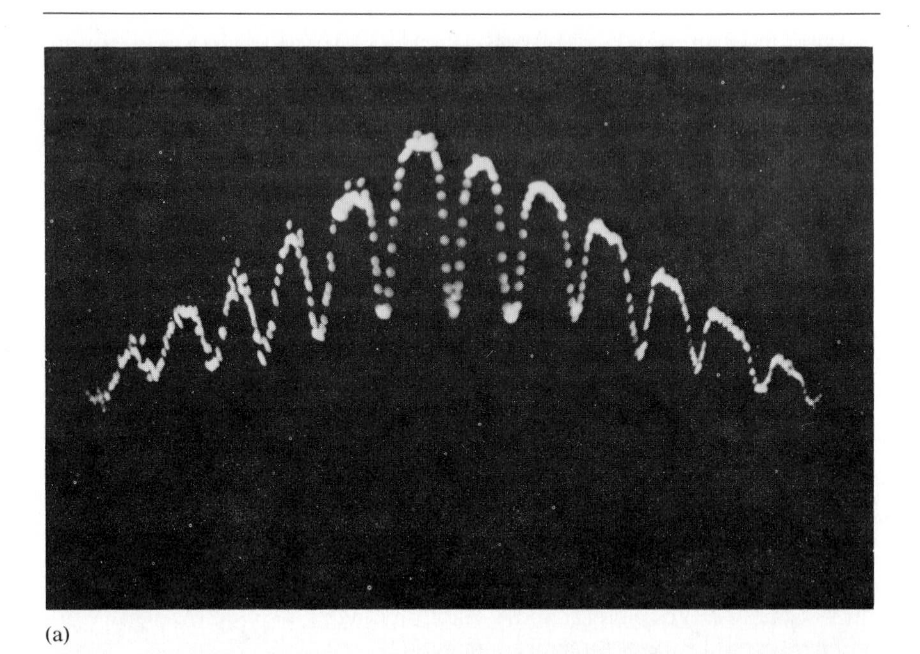

(a)

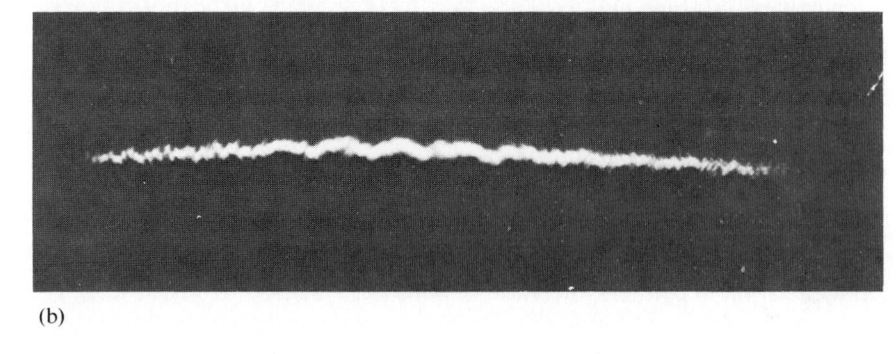

(b)

FIG. 12. Bleaching kinetics of the 860-nm band of bacteriochlorophyll reaction centers. (a) 860-nm light of echelon segments through buffer solution. (b) Same as b absorbed by reaction center 860-nm band—before excitation.

than the chloroplast. This is especially true if one studies the bacteriochlorophyll reaction center rather than the complicated chromatophore.

The first attempt to elucidate the primary photosynthetic intermediates was performed by Netzel *et al.*[31] They excited reaction centers at 530 nm and monitored the absorbance changes at 865 nm. The bleaching of this band was found to occur within 6 psec, as seen in Fig. 12. The changes

[31] T. L. Netzel, J. S. Leigh, and P. M. Rentzepis, *Science* **182,** 238 (1974).

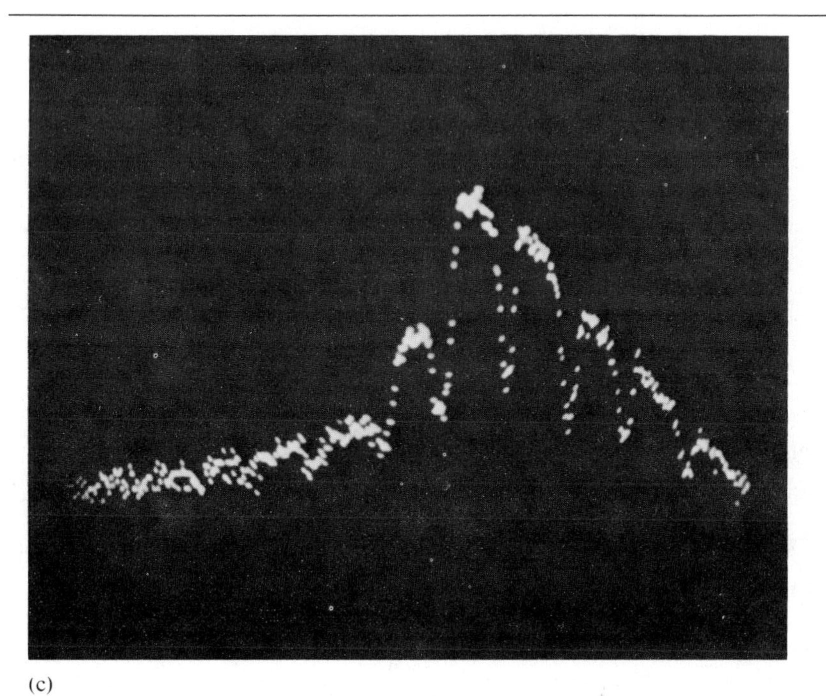

(c)

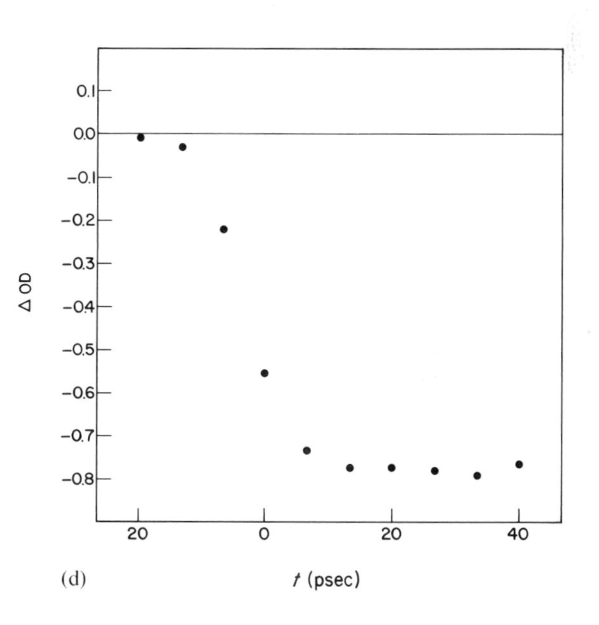

(d)

FIG. 12. (c) Same as b, but with 530-nm excitation coincident with 6th segment from right. (d) Bleaching kinetics of 860-nm band.

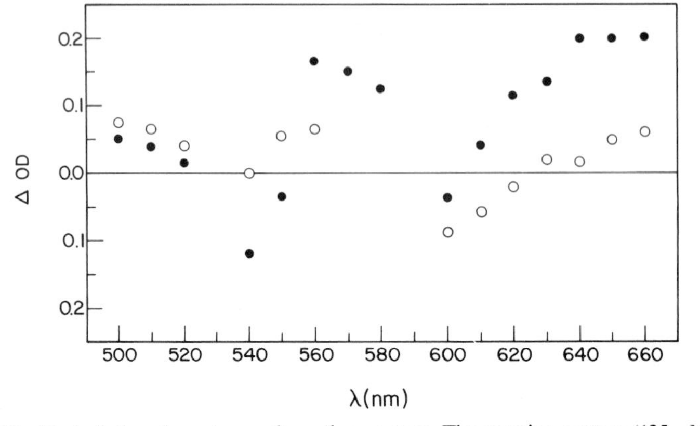

FIG. 13. Photo-induced spectrum of reaction centers. The reaction centers (125 μM) were in 50 mM morpholinopropane sulfonite buffer (pH 7) and 0.1% Triton X-100 detergent and at a redox potential of ~ +200 mV. Absorbance changes were measured in a 2-mm cell, 13 psec (●) and 250 psec (○) after excitation with a 520-nm picosecond light pulse.

throughout the visible spectrum are shown in Fig. 13 as observed by Kaufmann *et al.*,[32] while the kinetics of the predominant band at 540 nm and 640 nm (assigned to pheophytin and bacteriochlorophyll) are shown in Fig. 14.[32] Parsons, Windsor, and their co-workers[23] have obtained similar results for the 640-nm and 540-nm bands. The lifetime of both processes at 540 nm and 640 nm was found to be 120 psec.[32] The correct assignment of this process was made when the kinetics were measured by Kaufmann *et al.*[33] in the case where the primary acceptor, ubiquinone, was part of the reaction centers as the above experiments (Fig. 15) and after the ubiquinone was removed. Figure 15 shows the kinetic behavior of reaction centers when they (a) contain ubiquinone, Q (normal species), (b) are ubiquinone depleted, and (c) have Q added again. This figure shows that ΔOD changes take place immediately after excitation but do return to their original state for longer than 400 psec unless the quinone is present. Therefore, this 8.3×10^9 sec^{-1} rate (120 psec) should correspond to the transfer of charge from the bacteriochlorophyll–pheophytin complex to the quinone and not to the decay of an excited singlet state.

The unequivocal proof that the bacteriochlorophyll is oxidized within 10 psec was provided by the data of Dutton *et al.*[34] which shows that the 1250-nm band (Fig. 16) is formed within 10 psec after excitation. Formation of this absorption band is indicative of oxidized chlorophyll regardless of

[32] K. J. Kaufmann, P. L. Dutton, T. L. Netzel, J. S. Leigh, and P. M. Rentzepis, *Science* **188,** 1301 (1975).

[33] K. J. Kaufmann, K. M. Petty, P. L. Dutton, and P. M. Rentzepis, *Biochem. Biophys. Res. Commun.* **3,** 839 (1976).

[34] P. L. Dutton, K. J. Kaufmann, B. Chance, and P. M. Rentzepis, *FEBS Lett.* **60,** 275 (1975).

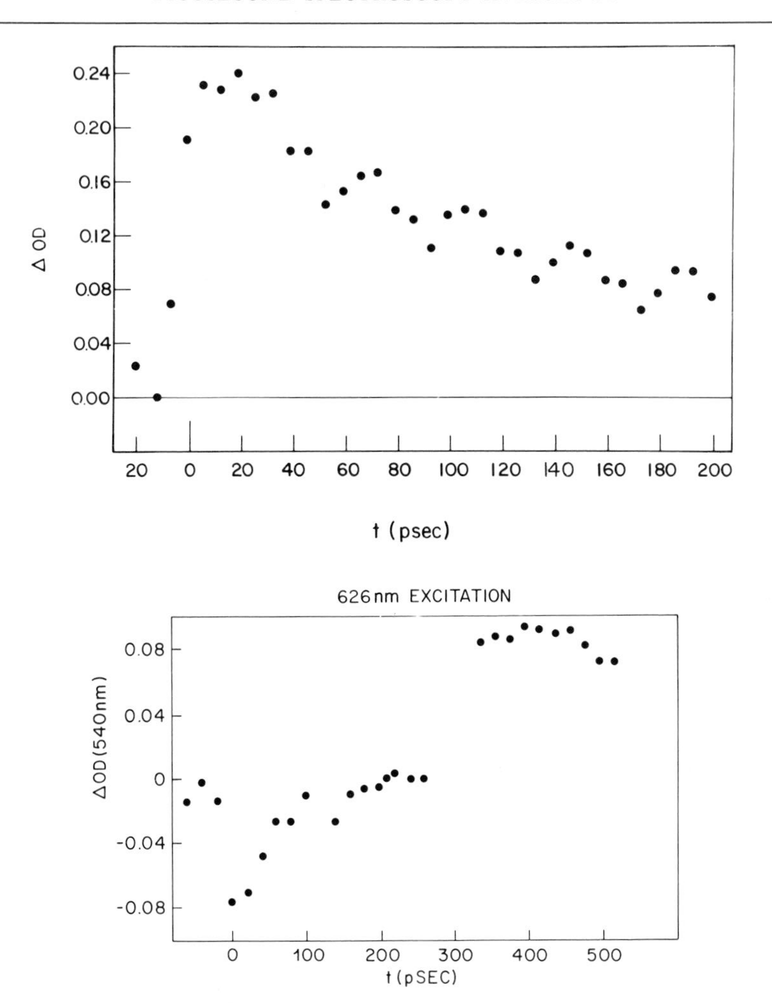

FIG. 14. Optical density changes as a function of time for reaction centers under the conditions of Fig. 13. The 6-psec formation and 120-psec decay of the 640-nm band (upper) correspond quite well with the 6-psec bleaching and 120-psec recovery of the 540-nm band (lower).

the means of oxidation, i.e., photo-induced or chemical. As corroborative evidence of the oxidation kinetics, we monitored the entire 1250-nm band at times varying from a few picoseconds to 1 nsec after excitation. The fact that no change in the kinetics was observed within the first 200 psec after excitation, regardless of whether the reaction center did or did not contain the primary acceptor quinone (Fig. 16), substantiated further our proposal that the oxidation of bacteriochlorophyll takes place within 6 psec with the subsequent charge transfer to Q taking place with a lifetime of 200 psec.

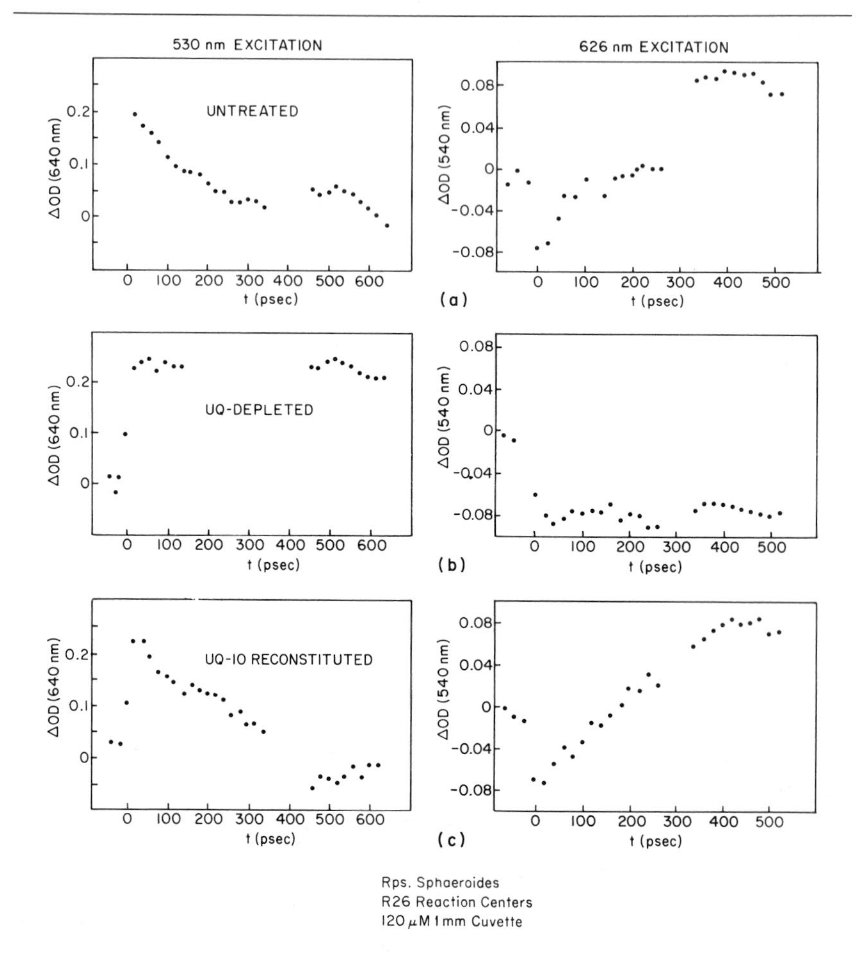

FIG. 15. Reaction center BPh reduction-oxidation in untreated, ubiquinone depleted, and ubiquinone-10 reconstituted reaction centers (c): 120 μM; 1-mm light path measured in each case at 540 nm and 640 nm. Absorbance decrease is a downward movement. In a and c, the absorbance increase at the end of the 540-nm traces is considered to arise from $(BChl)_2^{\pm}$ in the final $[(BChl)_2^{\pm}\ BPh]Q^{\mp}$ state.

This data demanded that a new model be proposed for the photo-induced oxidation of bacteriochlorophyll reaction centers. It was therefore first proposed by Dutton *et al.*[34] that an ultrafast intermediate, *I*, is formed within 6 psec which stabilizes the charge of the bacteriochlorophyll dimer forming a charge transfer complex. The charge is then transferred from this complex to quinone within 120 psec and thereafter the process follows its previously known route:

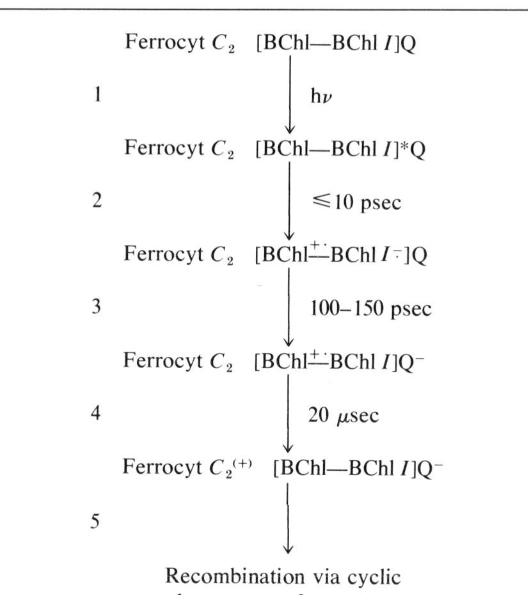

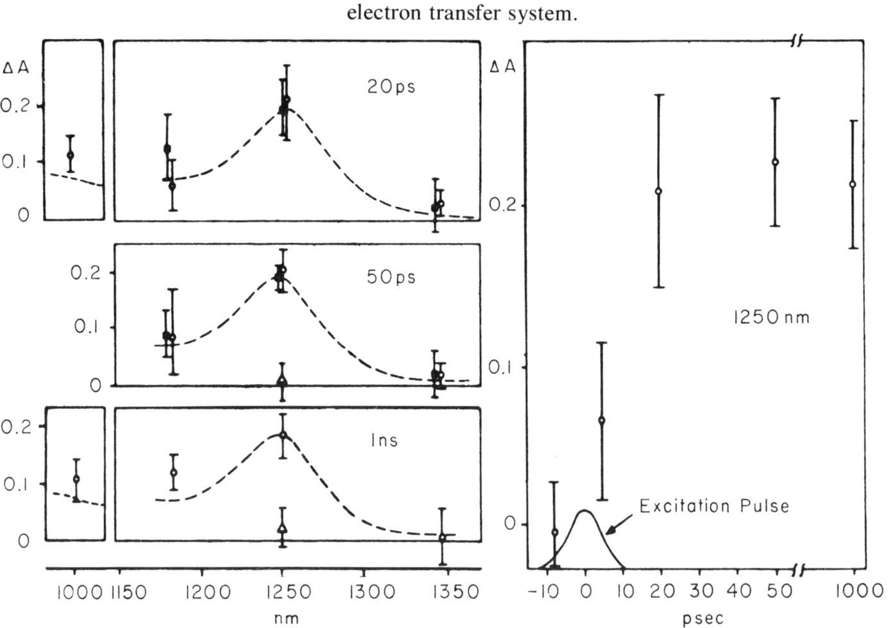

FIG. 16. Picosecond-induced OD changes in the near infrared region of reaction centers. The reaction centers were studied in a 2-mM anaerobic redox cuvette. The reaction centers were poised at different redox states as follows: In the neutral form, BChl—BChl X (○); in the ferricyanide oxidized form, BChl$\overset{+}{—}$BChl X (△); and with the primary electron acceptor reduced, BChl—BChl X^- (■). On the right the figure shows points taken at various times before and after excitation. Shown on the left are points taken various wavelengths and times after excitation. The points are the average of at least four determinations. The error bars represent the mean standard deviation. The laser light used for activation was not saturating. $\dot{+}$ designates a radical ion.

The identity of I can be pheophytin as Fajer[35] suggested or a combination of several species.

These primary processes in rhodopsin and chlorophyll are only an example of the potential uses of picosecond spectroscopy in biology. With extrapolation one can decipher processes with 3×10^{-13} sec lifetimes, thus approaching a vibrational period which is essentially the lower limit for chemical and biological events.

Acknowledgments

The work of Drs. G. E. Busch, D. Huppert, K. J. Kaufmann, T. L. Netzel, W. S. Struve, and V. Sundstrom at Bell Laboratories has formed the basis of this paper. I thank Drs. M. L. Applebury, P. L. Dutton, and K. S. Peters for collaboration and discussions on rhodopsin (MLA and KSP) and bacteriochlorophyll (PLD). I thank Drs. D. C. Douglass, E. O. Degenkolb, and L. J. Noe for continuous discussions and suggestions in the development of this manuscript.

[35] J. Fajer, D. C. Brune, M. S. Davis, A. Forman, and L. D. Spaulding, *Proc. Natl. Acad. Sci. U.S.A.* **72,** 4956 (1975).

[3] Nanosecond Absorbance Spectrophotometry

By DON DEVAULT

It is possible, with a streak camera, to record a whole spectrum in a nanosecond and to record changes in successive nanoseconds. The instruments are expensive ($40,000)[1] and probably do not have the sensitivity required for biological spectrometry[2] which often deals with transmission changes of the order of only 1%. Furthermore when one is dealing with a known reaction it is usually not necessary to obtain a whole spectrum but only the time course of changes at a few key wavelengths.[3] This chapter

[1] Some producers of streak cameras: Hadland Photonics, Ltd., Tel: (0442) 832525, U.K. (or Macro Scientific, Inc., Sunnyvale, Calif., Tel: 408-739-9418); Hamamatsu TV Co., Ltd., Hamamatsu, Japan (or Middlesex, New Jersey 08846); Cordin, Salt Lake City, Utah 84119. See also S. Gordon, K. H. Schmidt, and J. E. Martin, *Rev. Sci. Instrum.* **45,** 552 (1974). The price given is for a Cordin Model 132 + minimum accessories. The Hamamatsu C979 is $47,000 (1977). (Prices approximate only. Dated because of inflation.)

[2] Hamamatsu estimates their dynamic range as possibly 1000 but gives data showing 68. I have no data on the others. However, to make any kind of measurement on a 1% change, the noise should be no more than 0.1%, making imperative a dynamic range of 1000 or a gray-scale resolved to 10 bits.

[3] The relative merits of taking an instantaneous whole spectrum at a given time after the stimulus to the sample *vs.* following the kinetics at a single wavelength is discussed by one of

ISBN 0-12-181954-X

will discuss instrumentation for measuring the time course of small rapid changes of absorbance at a single wavelength.[4] To obtain a spectrum one repeats the experiment at different wavelengths.

According to the title we are discussing instrumentation in which typical or limiting time constants, τ, range from 1000 nsec down to 1 nsec. This corresponds to bandwidths equal to $1/(2\pi\tau)$ or 160 kHz to 160 MHz. Instrument specifications are often in "rise times," T_r, which is defined in several different ways[5] varying from 1.6 τ to 2.3 τ. Sometimes I will call τ the "resolving time." Techniques at the 1000 nsec end of the range are not greatly different from standard spectrophotometry, while those at the 1 nsec end approach the state of the art. How much of the following advice one needs depends upon how far down into the nanosecond range one intends to go.

The parts of a fast-absorbance spectrophotometer are (1) the measuring light source, (2) means of choosing the measuring wavelength if the source is not already sufficiently monochromatic, (3) a cell holding the specimen, (4) a detector to measure, amplify, and record the transmitted (or possibly reflected) measuring light intensity as a function of time, and (5) a means of stimulating or perturbing the sample.

1. The Light Source

The most demanding aspect of fast spectrophotometry is the need for a sufficiently bright measuring light source. The minimum brightness is set by the need to reduce "shot" noise to a suitable level. Shot noise is the noise in the output which results from the corpuscular nature of the measuring light and/or the photoelectric current it produces in the detector. The most critical stage in the measuring light-detector system is that at which the number of such corpuscles is a minimum. In the measuring light beam this will be at the entrance to the detector, after the losses due to absorption in the sample and inefficiencies in the optical parts. In the electrical parts it will be the primary photoelectric current before any amplification. Nor-

the pioneers in the field, G. Porter, *in* "Photochemistry and Reaction Kinetics" (P. G. Ashmore, F. S. Dainton, and T. M Sugden, eds.), pp. 93–111. Cambridge Univ. Press, London and New York, 1967. Note that the streak camera aims to do both.

[4] Another general discussion of fast spectrophotometry is F. E. Lytle, *Anal. Chem.* **46,** 545A and 817A (1974).

[5] T_r is usually defined as the time to rise from 10% to 90% of the ultimate rise. For a sine wave this is $(\sin^{-1} 0.9)/\pi f$, where f is the frequency, taken here to be the bandwidth, or $T_r = 2\tau \sin^{-1} 0.9 = 2.24\tau$. In the case of an exponential approach to a constant displacement, $T_r = \tau \ln 9 = 2.20\,\tau$, where τ is the time constant of the exponential. Another definition used, but not consistently, in the Tektronix catalogue appears to be the time for ¼ of a cycle of sine wave. This give $T_r = \pi\tau/2 = 1.57\tau$.

mally the electric current will have fewer corpuscles (electrons) than there are photons producing them because the quantum efficiency of photoelectron production is usually less than 1. If n photoelectrons are produced in one sample period the noise, i.e., the root mean square of the deviations from sample to sample,[6] will be $\sqrt{n}$. If the noise is required to be less than 0.1% of the signal, then $\sqrt{n}$ must be less than 0.1% of n and n must be more than 10^6. If the sample period is 1 nsec (for a 1-nsec resolution of measurement), the photoelectric current must be more than 10^6 electrons per 10^{-9} sec or more than 160 μamperes.

Suppose, next, that the material in the cuvette transmits only 10% of the measuring light striking it and that the efficiency of the detector in collecting the transmitted photons and converting to photoelectrons is 10%. Then our example tells us that 10^8 photons must strike the cuvette per sample period. This is 10^{14} photons sec^{-1} at $\tau = 1000$ nsec and 10^{17} photons sec^{-1} at $\tau = 1$ nsec. 10^{17} photons of 500 nm wavelength per second is 40 mW of light energy incident on the cuvette. Figure 1 illustrates our experience with various lamps. For specifications other than those assumed in the example, note that noise/signal will be inversely proportional to the square root of the light intensity, and resolving time for a given noise/signal level will be inversely proportional to the first power of light intensity.

It is possible to use smaller measuring light intensity if the experiment can be repeated a large number of times. At 1/100 the light intensity calculated above one could repeat the measurement 100 times and the total amount of light entering during the sum of corresponding sample periods will be the same. A computer would be required to add together the corresponding samples from the different repetitions (Computer of Average Transient). If one uses CAT averaging, some of the following advice needs modification, particularly the needs for measuring-light intensity and for detector capacity.

There are problems with maintaining constancy of light intensity during the measuring interval. Our solution for the boosted incandescent light source, used in the slower measurements, is described elsewhere.[7] The high-pressure mercury arc[8] is constant enough for intervals up to 10 μsec, but it fluctuates over longer intervals. The Xe flash lamp can be given a reasonably flat top for, say 1 μsec, by using a pulse that is considerably broader overall.[9] A common method of getting bright light is to transiently

[6] Poisson statistics.

[7] D. DeVault, *in* "Rapid Mixing and Sampling Techniques in Biochemistry" (B. Chance *et al.*, eds.), pp. 165–174. Academic Press, New York, 1964.

[8] The AHG-1B sold by several lamp companies running at 900 V and 1 A. An example of its use is found in M. Seibert and D. DeVault, *Biochim. Biophys. Acta* **253,** 396 (1971).

[9] We have used the E.G. and G. Co. FX-101 pulsed with 5J. An example of its use is by M. C. Kung and D. DeVault, *Photochem. Photobiol.* **24,** 87 (1976).

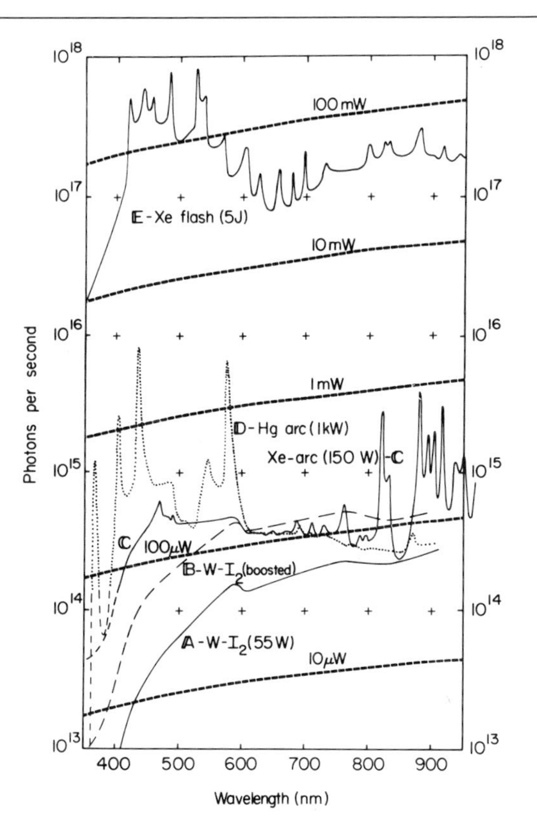

FIG. 1. Measuring light available at the sample from several light sources passed thru a Bausch and Lomb 250 mm, Cat. #33-86-40 monochromator with a blue-blazed, 1200 grooves/mm grating. Slit widths, 2.00 mm, corresponding to 6.7-nm bandwidth. (A) 45 watt W-I_2 lamp (GE, Q6.6AT2½/CL) running at 7.8 V and 7.0 A. (B) The same but boosted for 30 msec to 11.8 V and 9.0 A. (C) 150 W Xe-arc lamp (Hanovia 901C in a Bausch & Lomb housing). (D) 1 kW high-pressure mercury-arc (AH 6-1-B) running at 790 V and 1.36 A. (E) Xe flash lamp (EG&G, FX-101), operated at 14 μF, 845 V giving 5 J per flash. Note that at wavelengths near an intense line as found in the spectra of some sources the wavelength passed by the monochromator set to wide bandwidth can be much distorted. The pulsed Xe arc is not shown. It can be estimated by moving curve C upward by an amount corresponding to the expected enhancement. ×18 to ×1000 are claimed by the papers in footnote 10.

boost an otherwise continuous Xe arc lamp.[10] Taylor *et al.*[10] use a light-sensing feedback control of the arc to produce constant light, and commercial continuous wave (cw) lasers are stabilized by the same method.

[10] Pulsed Xe arcs are described by S. Gordon, K. H. Schmidt, and J. E. Martin, *Rev. Sci. Instrum.* **45,** 552 (1974); W. B. Taylor, J. C. LeBlanc, D. W. Williams, M. A. Herbert, and H. E. Johns, *ibid.* **43,** 1797 (1972); B. W. Hodgson and J. P. Keene, *ibid.* p. 493; S. Martellucci and E. Penco, *ibid.* **37,** 783 (1966).

If a refinement is necessary, one can use a microscope slide to split off a small fraction of the measuring light before it enters the sample and monitor it with a second detector. When the two detectors are balanced against each other their difference, or their ratio without balancing, will measure the changes of absorption in the sample free from fluctuations of the light source.[11]

Note that the use of a bright measuring light in fast measurements does not necessarily mean any greater total exposure of the sample to measuring light. It takes the same total number of photons to make a measurement with a given accuracy whether the measurement takes a long or short time. However, except when using a discontinuous measuring-light source such as a flash lamp, it may be necessary to avoid excessive pre-exposure to the measuring light before the measurement is made. We use a shutter whose opening gives a signal which starts all the other synchronized actions.[7]

2. The Monochromator

In time the tunable, continuously emitting laser[12] can be expected to replace the white light source and monochromator. Present continuous wave (cw) lasers can have trouble with stability, and tuning over more than a narrow range requires change of dyes. (Except in a very specialized case, one needs to be able to vary the wavelength of the measuring light: often to scan a spectrum or to monitor different components of a reaction mixture.) If there is a sufficient selection available, interference filters can replace a monochromator and they have the great advantage, in a light-hungry system, of greater overall transmission.

Because of the need for maximum amount of light one should use the maximum bandwidth of filter or monochromator consistent with the spectral width of the absorption band whose change is being measured. The 6–10 nm bandwidths are common for cytochrome or chlorophyll bands.

Dual-wavelength spectrophotometry—as employed by Chance[13] and widely used to stabilize against slower, nonspecific changes such as of light scattering or of measuring-light intensity—is not needed in fast spectrophotometry. If one arranges to have a short "baseline" on the read-out

[11] A system using a monitoring photomultiplier and difference amplifier is described by J. C. LeBlanc, A. Fenster, D. W. Williams, M. A. Herbert, and H. E. Johns, *Rev. Sci. Instrum.* **44,** 763 (1973).

[12] Tunable cw lasers are produced by Spectra-Physics. Specifications for Model 375 show complete coverage of wavelengths from 430 to 950 nm at 40 mW or more (at least 11 different dyes needed). Amplitude stability = ±.5% with Model 373 stabilizer. $26,425 (1977) with stabilizer, 2 pump lasers, and special circulator.

[13] B. Chance, *Rev. Sci. Instrum.* **22,** 634 (1951).

trace before the stimulus is given to the sample then one has a measurement of the transmission before and after the event in a time too short for slow fluctuations to matter. Of course, if one wants to use two wavelengths simultaneously to extract two separate pieces of information this can be done.[14]

While under certain conditions one can place the monochromator after the sample, there are two reasons for at least sometimes placing it before the sample (between the light source and the sample). If the sample is sensitive to the measuring light, monochromating the light before it reaches the sample can greatly reduce the exposure. In this case, one may also put a shutter in the measuring-light path so that exposure begins only a negligible time before the measurement is to be made.[15] The second reason is that, if the sample is turbid, it would be difficult to find enough light coming from the sample in the proper direction to pass through a monochromator placed after the sample.

If the sample is activated by laser flash, it is normally necessary to use a guard filter between sample and detector which absorbs the flash wavelength and passes the measuring light. Often a colored glass filter will do, but sometimes a monochromator after the sample is necessary.

3. Optics

In arranging the measuring light optics it is helpful to keep in mind the points at which one finds an "aperture" focus and those at which one finds a "slit" focus. These are designated A and S respectively in Fig. 2. The slit foci have the characteristic of narrowness, but may not be uniform. For example, if the source is a tungsten filament, the slit focus will show filament structure. The aperture foci have the characteristic of uniform illumination and thus are good for sample illumination unless the sample is narrow.

4. Sample and Sample Cell

As already mentioned, in contrast to slower measurements, fast measurements tend to be limited by shot noise. The optical density of sample which gives the maximum ratio of signal to shot noise is 0.87, if the total of available measuring light is assumed to be fixed. This value is not critically

[14] P. L. Dutton and J. B. Jackson [*Eur. J. Biochem.* **30,** 495 (1972)] and P. L. Dutton, K. M. Petty, H. S. Bonner, and S. D. Morse [*Biochim. Biophys. Acta* **387,** 536 (1975)] have used 2 wavelengths simultaneously (not time shared) but not below 10 μsec.

[15] A case in which the pre-exposure to the measuring light of a photosynthetic sample was measured is referenced in footnote 8.

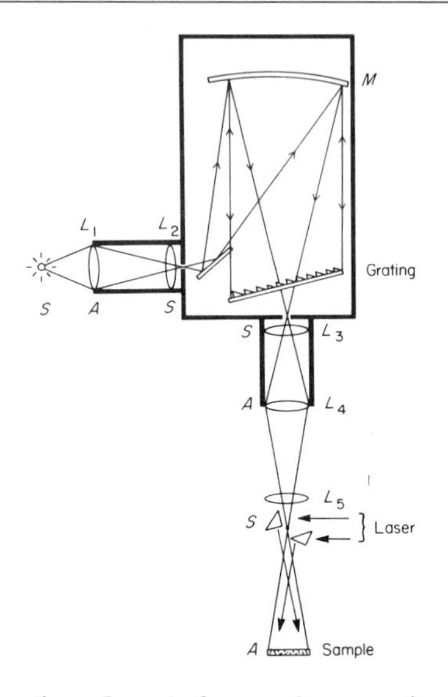

FIG. 2. Optical considerations. Lens L_1 focuses the measuring light source onto entrance slit of the monochromator. The conjugate foci of L_1 are thus "slit foci." Concave mirror M focuses the entrance slit at infinity in the direction of the grating and then focuses the reflections from the grating onto the exit slit. Lens L_4 focuses the exit slit to a slit focus forming a narrow portion of the beam before going to the sample. If the sample were narrow it could be placed at this point. Lens L_2 focuses the aperture at L_1 onto the grating and L_3 focuses it onto the aperture at L_4. L_5 focuses the aperture at L_4 onto the sample. Its position and focal length can be chosen to give an aperture image at the sample whose size matches the sample size. The slit foci are labeled S and the aperture foci, A.

sharp. Any density near this value will do almost as well, but departures by more than a factor of two will noticeably degrade signal-to-noise ratio. A derivation of this number is given in the appendix to this article but it rests upon an optimum between zero optical density which gives no signal, because it contains no sample, and infinite optical density which gives no signal because no light gets through. It is further modified by how the noise varies with the size of the signal. Shot noise is proportional to the square root of the signal and gives $2/\ln 10 = 0.87$ as optimum optical density. If the noise were independent of signal size (as, say, a fixed error in reading the output), the optimum would be $1/\ln 10$ or 0.43. The optical density referred to here is that of everything in the sample including impurities which are proportional to the concentration of the sample and which absorb the measuring light. The assumption is made that the absorbing impurities must go with the sample. If, on the other hand, such impurities could be removed, it would be advantageous to do so.

The optical density of the sample can be adjusted by varying either the concentration or the path length, their product being fixed by the desired optical density.

The area of the cuvette could be important, if either the amount of material available is limited or if the material is sensitive to measuring light. For a fixed optical density, the amount of material used is proportional to the area. If the area of illumination can be varied—as by changing the focal lengths of the lenses, or changing from an aperture focus to a slit focus, or something in between—then the larger the area for a given total amount of measuring light, the smaller will be the exposure of each molecule of material. For this reason we have tended to use cuvettes with a large area (15 mm × 15 mm). However, if the amount of sample material were limited, this would place a limit on the area that could be used.

5. The Detector System

a. Read-out Device. What one uses here may depend upon how much money one can spend. As mentioned in the introduction, we will concentrate on reading out the time course of absorption at a single wavelength. The most versatile instrument for this job is undoubtedly the Tektronix transient digitizer, R7912.[16] This instrument will do in the nanosecond region what the better-known slower transient digitizers[17] can do in the microsecond region. It digitizes the coordinates of a trace. The digital record can then be stored indefinitely, read out immediately, and/or operated on by computer, as, for example, to add together successive traces to get a "CAT" average. The only other method of digitizing nanosecond traces at present is by reading photographs of the traces.[18]

The next most expensive read-out system is a fast-storage oscilloscope newly developed by Tektronix, the Model 7834.[19] At 2.5 cm/nsec writing rate it can store transients 2.5 cm high if they take 1 nsec to rise. The Tektronix 7633[20] stores at 1 cm/nsec. Other storage scopes known to the writer are not fast enough for nanosecond work. The main advantage of storage oscilloscopes is that they simplify the photography of the trace and avoid wasting film on bad traces. They may be a little tricky to adjust to the correct intensity and persistence to get a good stored trace.

[16] $20,000 and up. This is sold as part of several systems designated the WP2000 series.

[17] Biomation, Nicolet, Erdac, Physical Data, and Princeton Applied Research. The Biomation 8100 resolves to about 10 nsec [$9850 (1975)] but none of the others go below 0.5 μsec.

[18] A home-made transient digitizer with a resolution of 2 nsec is described by H. A. Baldis and J. Aazam-Zanganek, *Rev. Sci. Instrum.* **44**, 712 (1973).

[19] $10,000 (1977) with one 7A19, 7B80, and 7B85 plug-ins. Add another $1000 for another 7A19 if differential input is desired.

[20] $6,000 (1977) with 7A18 and 7B50A plug-ins.

Nonstorage oscilloscopes should be chosen with camera and phosphor adapted for high photographic writing rate. A 15 cm/nsec writing speed is obtained with either Tektronix 7904[21] or 7704A.[22] The camera[23] is another $1200 to $1600. The 7904 has a rise time of 0.8 nsec and the 7704A, 4.7 nsec.

b. Detectors. The detector should have a large sensitive area (comparable to the cuvette area or more) to collect as much measuring light as possible from the cuvette. An exception might be a case in which the sample is clear and the transmitted light is still well enough collimated to be focusable to a small area. Usual detectors are photomultipliers, p-i-n type silicon diodes, or vacuum diodes.

Photomultipliers have the great advantage of built-in amplification and variable gain. They are good in the upper half (logarithmic) of the nanosecond region (30–1000 nsec). At 30 nsec the minimum cathode current for 0.1% shot noise would be 5 μA. It is difficult to find a photomultiplier with specifications allowing a larger cathode current than this as would be necessary below 30 nsec.[24] Another problem is that the rise time of many photomultipliers is of the order of 20 nsec due to statistical spread in the transit time of the electrons through the dynodes, although there are ways to solve this problem.[25,26] The dynode voltage supply circuit must provide adequate current to meet the demands of high light levels without too much variation of voltage and the last several dynode stages must be bypassed with capacitors. Fatigue effects[27] encountered on a millisecond time scale in photomultipliers are not important at nanosecond speeds, but linearity should be watched.

Aside from the expensive gated photomultipliers[24] photodiodes are required below 30 nsec. Fortunately the higher light intensities involved reduce the requirements for amplification and external amplifiers of sufficient gain are feasible. Biplanar vacuum photodiodes[28] have very good rise times. However, the greater quantum efficiency of silicon photodiodes[29] helps to combat shot noise by giving a larger primary photocurrent

[21] $7,200 (1977) with one 7A19, 7B80, and 7B85 plug-ins. Add $1000 for a second 7A19 if differential input is desired.

[22] $4,600 (1977) with 7A18 and 7B53A plug-ins.

[23] Tektronix model C-51 with writing speed enhancer ($1570) or model C-27, option 04, with writing speed enhancer ($1160).

[24] Photomultipliers with cathode current ratings higher than 5 μA: I.T.T. Co. F4084 (gated), 50 μA, $3671 (1971); E.G.&G. Co. GPM-50M, gated, 100 μA, $1690 (1971).

[25] Examples of fast photomultipliers: EMI 9810 to 9818 (2 nsec); Amperex XP1020 ($<$2 nsec).

[26] Use of optimum voltages and fewer dynodes: A. Fenster, J. C. LeBlanc, W. B. Taylor, and H. E. Johns, *Rev. Sci. Instrum.* **44,** 689 (1973); G. Beck, *ibid.* **47,** 537 (1976).

[27] See DeVault[7] and Fenster *et al.*[26].

[28] Biplanar diodes: I.T.T. Co. FW114; Hamamatsu Co. R617; Instrument Technology Ltd, Hastings, England, TF50.

[29] Fast silicon photodiodes of large area: E.G.&G. Co. SGD-444 (100 mm^2).

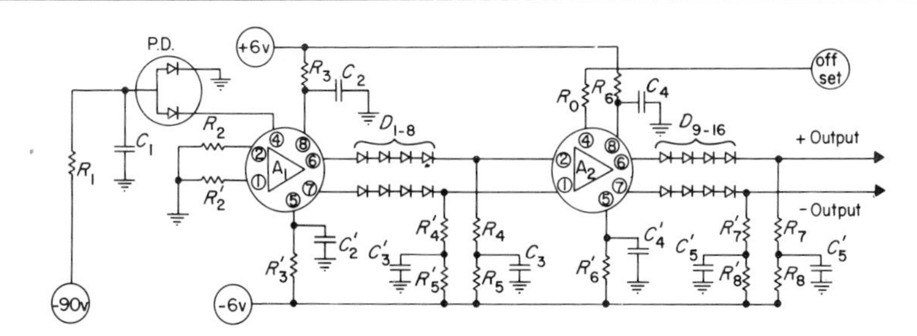

FIG. 3. Diode detector and current-to-voltage signal amplifier. P.D. = SGD-444 photodiode (E.G.&G. Co.); $A_1 = A_2 = \mu A733$ amplifier (Fairchild Co.); D_1–D_{16} = IN4148; $R_1 = R_7 = R'_7 = R_4 = R'_4 = 1\ k\Omega$; $R_2 = R'_2 = 50\ \Omega$; $R_3 = R'_3 = R_6 = R'_6 = 20\ \Omega$; $R_5 = R'_5 = R_8 = R'_8 = 200\ \Omega$; $C_1 = 0.01\ \mu F$; $C_2 = C'_2 = C_3 = C'_3 = C_5 = C'_5 = 0.001\ \mu F$; $C_4 = C'_4 = 0.022\ \mu F$. R_0 = Adjusted to give desired ratio between voltage applied at "offset" and the offset produced in output. One of the outputs goes directly to the input of the amplifier shown in Fig. 3. For further gain at slower speeds the two outputs can go to the inputs of, say, a μA 715 (Fairchild) integrated circuit amplifier. A distinctive feature of this circuit is the use of pin #4 (emitter of one of the input transistors) of the μA733 as a current-input terminal. (Normal input is voltage applied to pins 1 and 2.) Shunt feedback is provided internally in the μA733 by 7 kΩ resistors between output and input on each side of the amplifier. Both share in offsetting the input current so that the output voltage on either side is then $\frac{1}{2} \times 7\ k\Omega \times$ (input current), and the effective input impedance, which multiplied by photodiode capacitance gives the time constant, is very small. Details of this design were worked out by Mr. Drew Henderson. The two amplifier stages apparently have a flat response from 0 to 30 MHz with overall transfer impedance (voltage out/current in) of 35 kΩ on each output.

for a given light intensity. Figure 3 shows a detector-amplifier system, costing not much more than the price of the photodiode ($225) plus labor, which is good to 10 nsec resolving time.

c. Matching Detector to Oscilloscope. Normally one will connect detector to oscilloscope with a length of coaxial cable. At the smaller resolving times it is essential to terminate the cable with its characteristic impedence (usually 50–75 Ω) to avoid reflections of signal in the cable. This termination becomes the output load resistor of the detector and whatever amplifier may be attached to it. If one uses a very high oscilloscope sensitivity, such as 10 m V/cm, the current required through a 50-Ω resistor to get a 2-cm deflection would be 0.4 mA. If the 0.4 mA corresponds to the 1% change in transmission postulated in our earlier example, the total signal would correspond to 40 mA. A current of 40 mA is beyond the output capability of most photomultipliers, and only an extremely bright light source would give this much primary photocurrent from a photodiode. Therefore, an amplifier is ordinarily needed to couple either photomultiplier or photodiode to the connecting cable. By introducing offset (see § *d*) to the amplifier at an early stage, one makes it unnecessary for the amplifier to handle the whole signal and it need amplify only the changes in signal. However, it is well to amplify beyond the 0.4 mA, postulated above, in order to use a less sensitive scale on the oscilloscope.

Photodiodes and photomultipliers are both high-impedance devices putting out a current proportional to the light. The load resistance should be as small as possible to keep the time constant (load resistance × internal capacity) small for fast response. This is accomplished by using shunt feedback in the amplifier.

The amplifier in Fig. 3 is able to put out a signal with up to 4 V peak-to-peak. However, it is limited to 10 mA output, and so could not drive 50 or 75 Ω the full 4 V. It can, however, if the amplifier of Fig. 4 is added as an output stage.[30]

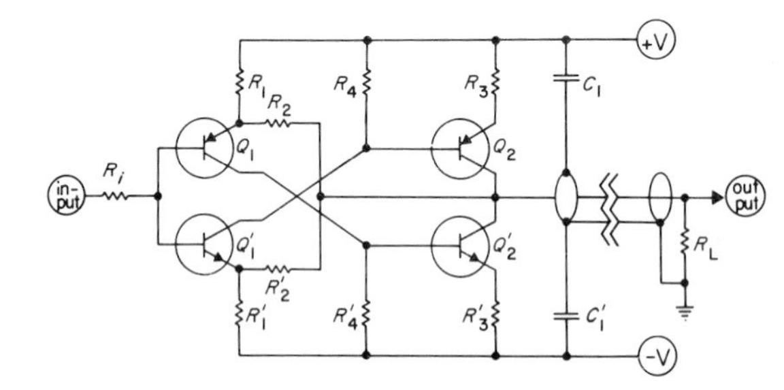

FIG. 4. Output (current) amplifier. $R_i = 270\Omega$; $R_1 = R_1' = 750\Omega$; $R_2 = R_2' = 110\Omega$; $R_3 = R_3' = 91\Omega$; $R_4 = R_4' = 510\Omega$; $R_L = 75\Omega$ (to match RG-59U cable); $C_1 = C_1' = 0.1\ \mu F$; $Q_1 = Q_2 =$ 2N3906; $Q_1' = Q_2' =$ 2N3904. Q_2 and Q_2' should be mounted in good thermal contact with a heat sink. Power supply requirements: $\pm V = \pm 9$ V at 130 mA. Input impedance = 100 kΩ at DC; 1 kΩ at 16 MHz. Another set of values which have been found useful are: $R_i = 200\Omega$; $R_1 = R_1' = 510\Omega$; $R_2 = R_2' = 100\Omega$; $R_3 = R_3' = 51\Omega$; $R_4 = R_4' = 330\Omega$; $C_1 = C_1' = 0.001\ \mu F$; $\pm V = \pm 5$ V at 130 mA. In contrast to the usual emitter-follower output amplifier this circuit is "push-pull"—equal drive for both rising and falling signals. Its rise time appears to be less than 1 nsec.

d. Offset Considerations. When measuring small changes, ΔS, in a large signal S one must offset most of the signal so that the small change will deflect the trace an easily discernable amount. One must also know what the amount of offset is so that one can calculate $\Delta S/(2.3S)$, the change of optical density. We have developed an automatic offset circuit that is very useful.[31] We are trying to publish the circuit elsewhere, but the principle is

[30] Other preamplifiers: J. P. Keene, E. D. Black, and E. Hayon, *Rev. Sci. Instrum.* **40,** 1199 (1969); G. Diebold and R. Santaro, *ibid.* **45,** 773 (1974); R. H. Hamstra, Jr. and P. Wendland, *Appl. Opt.* **11,** 1539 (1972). "Pulse amplifiers" with 1 nsec rise times and band pass limited to frequencies above some given value are fairly common and can be used for fast measurements if the DC light-intensity data can be obtained by parallel use of a slower DC amplifier. See, for example, J. K. Millard, *Rev. Sci. Instrum.* **38,** 169 (1967).

[31] Other offset circuits: S. L. Olsen, L. P. Holmes, and E. M. Eyring, *Rev. Sci. Instrum.* **45,** 859 (1974); D. A. Whyte, *ibid.* **47,** 379 (1976).

as follows: The output from the signal amplifier goes to both the oscilloscope and also a special amplifier that generates an offset current, a current which is fed to an early stage of the signal amplifier in parallel but opposite in sign to the photo-signal current. In Fig. 3 the entrance point of this offset current is indicated. This current counteracts or offsets the photo-signal. It is generated by only a small deviation in the input to the oscilloscope and to the offset amplifier and is just enough to keep the deviation small. Thus the signal-amplifier and the offset-amplifier form a servoloop that keeps the input to the oscilloscope fixed close to some desired value regardless of the size of the photo-signal.

At the time the oscilloscope sweep starts, a signal from the oscilloscope triggers a gate that opens the servoloop and holds the offset constant thereafter at whatever value was found at the moment of opening. Further variations in the photo-signal then record fully on the oscilloscope screen. Before the loop is again closed to make the next measurement the actual offset is read from a meter attached to the output of the offset amplifier. This arrangement assures that the oscilloscope trace will start at the desired height on the screen. If the offset adjustment were not automatic, slow unknown variations of light intensity or sample density could throw the trace completely off the screen before it starts.

e. Bandwidth Variation. For maximum signal-to-noise ratio it is necessary to match the bandwidth of the amplifier to the speed of the measurement. The spectrophotometer will often be used at rates slower than the design maximum. In such a case one should be able to decrease the bandwidth accordingly. We do this with a simple series resistor followed by a capacitor to ground in a box containing a choice of resistors and capacitors.[32] The usual choice is to make $\tau \leqslant 1/200$ of the time for one oscilloscope sweep, so that the trace will correspond to at least 200 resolved points. The bandwidth limiter should be applied at a stage before the noise has been amplified to a point where it is clipped by the amplifier swing limits. The suggestion of footnote 32 is excellent in this respect.

6. The Stimulus

The perturbation administered to the sample could be anything, but it must be fast if it is to be followed by a fast measurement. The experience of this author is entirely with light flashes,[33] but radiolysis pulses,[34] tempera-

[32] Another method is to put a capacitor in parallel with the shunt feedback of the signal amplifier. The capacity times the feedback resistance then gives the "integration time constant."

[33] Pioneering in nanosecond spectroscopy is the group under H. T. Witt. See C. Wolff, H. E. Buchwald, H. Rüppel, K. Witt, and H. T. Witt, *Z. Naturforsch., Teil* **24b,** 1038 (1969).

[34] J. P. Keene, *Nature (London)* **188,** 843 (1960).

ture jumps,[35] shock waves, and electric and magnetic field jumps are also possibilities. Rapid mixing is not yet below the microsecond range but one can think of using flash photolysis to generate a reagent *in situ* to react with a system of interest not itself sensitive to light. The following will consider flash activation.

The pulsed laser is certainly the choice flash generator. Q-switching (control of laser-cavity gain to shorten the pulse) or other gating is necessary to get into the nanosecond range. The coherence (spacial) property of the laser is useful in allowing the laser to be put at some distance (we use 6 m) from the spectrophotometer and readout amplifiers. This helps greatly in reducing electrical interference from the flash-lamp and possibly gating circuits of the laser. The lack of afterglow from the laser (compared to a flash-lamp) is also important because the afterglow can interfere with the spectrophotometric measurements, especially if the sample is turbid.

There are at least three common ways of Q-switching the laser. It can be done with (1) a passive, saturatable dye, (2) a rotating prism, or (3) a Pockels cell. The first has the disadvantage that it is hard to synchronize with the scope trigger so as to obtain a reliable amount of base line before the pulse stimulates the sample. However, this objection only applies at the slower end of the nanosecond range. When the base line need be only 100 nsec or less, the signal delay built into the oscilloscope may be sufficient for generating a base line. In this case the oscilloscope may be triggered by a signal from a small photodiode monitoring the laser flash itself. The Pockels cell has the disadvantage that it requires a heavy electrical pulse to activate it just at the moment one is trying to make the spectrophotometric measurement. The electrical isolation must be extra good. The rotating prism is a simple mechanical device and has served well in the author's ruby laser[36] for 14 years. A magnetic pick-up on it gives a signal a fixed number of microseconds before the prism will be in position to cause lasing, and this signal is used to synchronize the oscilloscope in the upper end of the nanosecond range.

The "delayed sweep" capability of the oscilloscope is very useful for adjusting base lines. Small delays can be introduced into one signal line or another by the simple use of extra lengths of coaxial cable (about 5 nsec/m).[37] Delay lines designed for the purpose can also be obtained.

[35] By electric discharge through the sample microwave heating, etc. [M. Eigen, *Disc. Faraday Soc.* **17,** 194 (1954); M. Eigen and L. De Maeyer *in* "Technique of Organic Chemistry" (A. Weissberger, ed.), Vol. VIII, part II, p. 895. Wiley, New York, 1963; M. Eigen *in* "Nobel Symposium 5" (S. Claesson, ed.), p. 333. Almqvist and Wiksell, Stockholm, 1967]. Also by light flash absorbed in solvent [J. V. Beitz, G. W. Flynn, D. H. Turner, and N. Sutin, *J. Am. Chem. Soc.* **92,** 4130 (1970)].

[36] TRG-Hadron Model 104.

[37] This was done in M. C. Kung and D. DeVault, *Biochim. Biophys. Acta* **501,** 217 (1978). Lytle[4] discusses limitations.

Unless one wants to run out an action spectrum the limited number of wavelengths that may be easily available from Q-switched lasers is not a great disadvantage. The ruby laser's primary output is 694 nm, very good and much more powerful than necessary for stimulating photosynthetic systems. Frequency-doubling gives 347 nm and stimulated Raman effect in hydrogen gas gives 539 nm as the first anti-Stokes line.[38] The 694 nm easily pumps dye lasers to give longer wavelengths.[39] Neodymium lasers give 1.06 μm direct, 530 nm doubled, and 265 nm quadrupled.

If one uses a cuvette whose horizontal cross-section is square one can introduce the laser light at right angles to the measuring light. However, in this case it is particularly necessary to insure sufficiently uniform activation of sample across the face of the measuring beam to avoid errors in the measurement. If the cuvette has large area presented to measuring-light and short measuring-light path length it will be necessary to introduce the laser light in parallel with the measuring light. This can also be more uniform. The arrangement shown in Fig. 2 is particularly suitable for getting both beams through a narrow aperture as in the side of a cryostat holding the cuvette in its interior.

Acknowledgments

The author received valuable advice from Drs. Henry Linschitz and William Parson. The writing of this section was supported by National Science Foundation grants PCM76-15724, PCM76-23744, and PCM77-22086.

Appendix: Derivation of Optimum Sample Density

Let $Y = 2.3\epsilon cl$ where ϵ = extinction coefficient, c = concentration, and l = path length thru the cuvette. Then:

$$S = S_0 e^{-Y} \tag{1}$$

where S_0 = photodetector current with empty cuvette and S with sample in place. For generality we let the noise, N, in the photodetector current be proportional to arbitrary power, n, of the signal; thus:

$$N = k_1 S^n \quad \text{and} \quad N_0 = k_1 S_0^n \tag{2}$$

We also consider two types of spectrophotometric measurement: In case I the measurement is of the total optical density, $Y/2.3$. In this case:

$$Y = \ln S_0 - \ln S \tag{3}$$

In case II the measurement is of a small change in optical density, $\Delta Y/2.3$.

[38] R. W. Minck, R. W. Terhune, and W. G. Rado, *Appl. Phys. Lett.* **3**, 181 (1963).
[39] B. Chance, J. A. McCray, and J. Bunkenburg, *Nature (London)* **225**, 705 (1970).

In this case:

$$\Delta Y = \ln S - \ln (S + \Delta S) \cong \frac{\Delta S}{S} \tag{4}$$

We will assume that ΔY remains a fixed fraction of Y during any manipulation of c or l:

$$\Delta Y = k_2 Y \tag{5}$$

The noise in $\ln S_0$ is:

$$\frac{N_0 \, \partial \ln S_0}{\partial S_0} = \frac{N_0}{S_0} = k_1 S_0^{\,n-1} \tag{6}$$

The noise in $\ln S$ is:

$$\frac{N \, \partial \ln S}{\partial S} = \frac{N}{S} = k_1 S_0^{\,n-1} e^{(1-n)Y} \tag{7}$$

and this is practically also the noise in $\ln (S + \Delta S)$.

The noise in the overall measurements indicated in (3) and (4) will be the square root of the sums of the squares of the noise in the individual terms. Thus the squares of the signal-to-noise ratios are:

Case I:
$$\frac{Y^2}{k_1^{\,2} S_0^{\,2n-2}(1 + e^{2(1-n)Y})} \tag{8}$$

Case II:
$$\frac{k_2^2 Y^2}{2k_1^{\,2} S_0^{\,2n-2} e^{2(1-n)Y}} \tag{9}$$

We find the value of Y which maximizes these expressions by differentiating with respect to Y and setting equal to zero. The result is:

Case I:
$$Y = \frac{1 + e^{2(n-1)Y}}{1 - n} \tag{10}$$

Case II:
$$Y = \frac{1}{1 - n} \tag{11}$$

Case II is assumed in the body of this paper. If the source of noise were the reading error, $n = 0$ and the optimum value of Y would be 1. With shot noise, $n = 1/2$ and $k_1 = \sqrt{q/\tau}$ where q is the charge of an electron. In this case the optimum value of Y is 2 and the corresponding optical density is 2/2.3.

[4] Measurement of Protein Rotational Diffusion in Membranes by Flash Photolysis

By RICHARD J. CHERRY

Introduction

The technique of fluorescence polarization has been used for many years to investigate the rotational diffusion of macromolecules in aqueous solution (for reviews, see Weber[1] and Yguerabide[2]). The method is successful because fluorescence lifetimes, which are typically $\sim 10^{-8}$ sec, are not too different from the rotational relaxation times to be measured. When the relaxation time is longer than about 10^{-6} sec, however, the method fails because fluorescent emission decays before any detectable rotation can occur. Consideration of membrane viscosity leads one to anticipate that the relaxation times of many membrane proteins will be in the microsecond time range, or longer. In order to measure such slow rotation by optical methods, it is necessary to use a spectroscopic state which has a comparably long lifetime. Recently, methods of investigating slow rotational diffusion have been developed which exploit the long lifetime of the triplet state of probe molecules.[3-6] These methods use a flash photolysis apparatus to detect the dichroism of absorption transients induced by a brief pulse of linearly polarized light. A similar approach may also be used to investigate rotation of proteins with intrinsic chromophores, which have long-lived photoproducts. In this article, the methods of making these measurements are described and present applications with membrane proteins briefly summarized.

Principles of Rotational Diffusion Measurements with Triplet Probes

Spectroscopic methods of measuring rotation depend on *photoselection* whereby an oriented population of excited molecules is optically selected from an initial random distribution. This is achieved by excitation with plane-polarized light, so that those molecules whose transition mo-

[1] G. Weber, *in* "Fluorescence Techniques in Cell Biology" (A. A. Thaer and M. Sernetz, eds.), p. 5, Springer-Verlag, Berlin and New York, 1973.

[2] J. Yguerabide, this series, Vol. **26** [24].

[3] K. Razi Naqvi, J. Gonzalez-Rodriguez, R. J. Cherry, and D. Chapman, *Nature (London), New Biol.* **245,** 249 (1973).

[4] R. J. Cherry, A. Cogoli, M. Oppliger, G. Schneider, and G. Semenza, *Biochemistry* **15,** 3653 (1976).

[5] R. J. Cherry and G. Schneider, *Biochemistry* **15,** 3657 (1976).

[6] D. Lavalette, B. Amand, and F. Pochon, *Proc. Natl. Acad. Sci. U.S.A.* **74,** 1407 (1977).

ISBN 0-12-181954-X

ment for absorption is parallel or at a small angle to the electric vector of the incident light are preferentially excited. Signals arising from the excited molecules in general reflect their anisotropic distribution so that emission signals are polarized and absorption signals are dichroic. When excitation is by a brief pulse of light, the initial emission or absorption anisotropy decays as the molecules again become randomized by Brownian rotation. From the rate of decay, rotational relaxation times may be determined.

Figure 1 shows the lower electronic energy levels of an organic molecule with an even number of electrons and illustrates some of the principal optical transitions which occur. Triplet states cannot normally be populated to any appreciable extent by direct absorption, and the absorption spectrum arises from the singlet–singlet transitions S_0–S_1, S_0–S_2, etc. Usually higher singlet states degrade rapidly and nonradiatively to the lowest excited singlet state S_1. The subsequent return to the ground state may occur either directly or via the triplet state. In selected molecules, the S_1-T_1 transition (intersystem crossing) effectively competes with the S_1-S_0 transition. Because the T_1-S_0 transition is spin forbidden, the lifetime of the lowest triplet state (typically $> 10^{-3}$ sec) is much longer than that of the S_1 state (typically $10^{-8}-10^{-9}$ sec). Quenching, especially by oxygen, may considerably shorten the observed triplet lifetime.

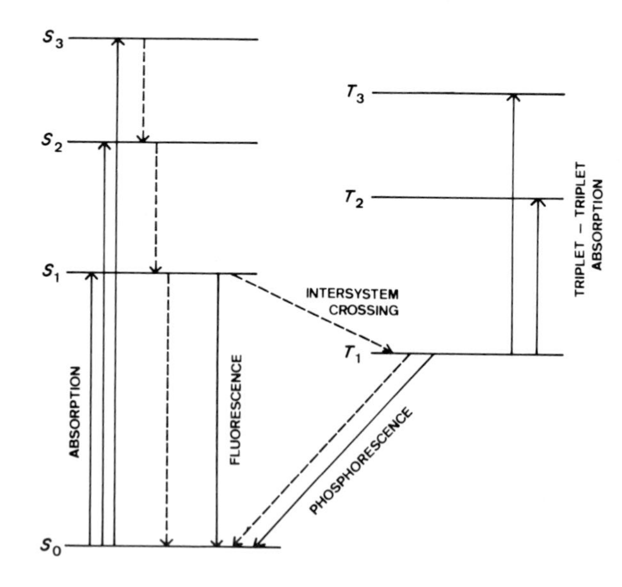

FIG. 1. Electronic energy level scheme. Solid arrows represent radiative transitions, and dashed arrows nonradiative transitions (vibrational states are omitted).

In principle, triplet states may be detected by observing the phosphorescent emission arising from radiative T_1–S_0 transitions. In practice, however, this is difficult in fluid solutions at room temperature, where the nonradiative transition has a much higher probability. Under these conditions, it is much easier to detect triplet states from absorbance changes in the sample. At appropriate wavelengths, an absorbance increase occurs due to the transition from the lowest triplet state to a higher triplet state. Alternatively, a ground-state depletion signal may be observed. This is a decrease in absorbance in the singlet–singlet absorption band due to removal of molecules to the triplet state.

When the triplet state is populated following flash excitation of the S_0–S_1 transition with plane-polarized light, the resulting transient absorbance changes are dichroic due to photoselection. The time dependence of dichroism enables rotational motion to be investigated. Because of the long lifetime of the triplet state, it is possible to detect rotation times as slow as milliseconds, whereas fluorescence methods are confined to times shorter than ~ 1 μsec. It should be pointed out that not only triplet states, but any photoproduct with a suitably long lifetime, can be used in a similar way to measure slow rotational motion.

Experimental Methods

Probes

To make measurements as outlined above, one must have a molecule with suitable spectroscopic properties. While a few membrane proteins have a suitable intrinsic chromophore, it is generally necessary to attach an artificial chromophore or probe. In preliminary experiments, eosin (2,4,5,7-tetrabromofluorescein) was demonstrated to have favorable properties.[3] This has been followed by the development of two reactive derivatives of eosin, eosin isothiocyanate (eosin-NCS) and iodoacetamidoeosin (IA-eosin), which enable the dye to be covalently coupled to proteins (Fig. 2). The isothiocyanate reacts principally with amino residues and the iodoacetamido derivative preferentially with sulphydryl groups.

Eosin-NCS may be prepared by the following procedure. First 50 mg fluorescein isothiocyanate (Serva, isomer I) are suspended in 0.5 ml ethanol. Then 164 mg bromine are added drop by drop, with the suspension being thoroughly stirred throughout. A clear solution is obtained on formation of the dibromo product; as the reaction proceeds further a precipitate of the insoluble tetrabromofluorescein isothiocyanate is formed. After standing for 2 hr at room temperature, the mixture is filtered and the insoluble material washed with 5–10 ml ethanol. The product is solubilized

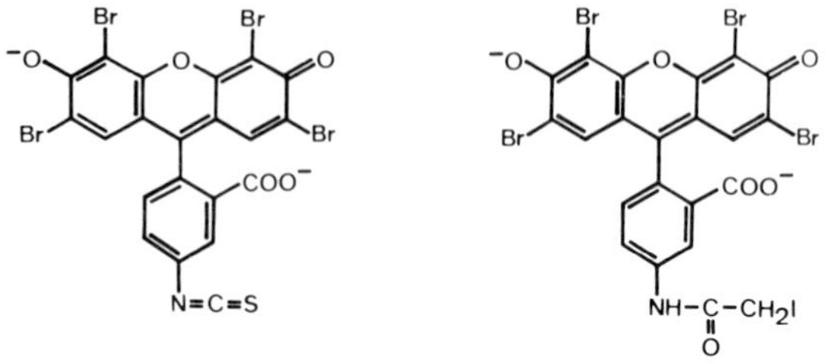

FIG. 2. Triplet probes. (I) eosin-5-isothiocyanate; (II) 5-iodoacetamidoeosin.

in 10 m*M* phosphate buffer, pH 7.4, and precipitated at 4°C with 1 *N* phosphoric acid. The precipitate is collected by centrifugation and washed 3 times with double-distilled water. After lyophilization, the product is stored in the dark at −20°C until used. When examined by thin-layer chromatography (Merck DC Fertigplatten Kieselgel F 254 eluted with benzene-methanol, 2 : 1) the product exhibits a strong main spot together with a very faint second spot.

Eosin-NCS has also recently become commercially available from Molecular Probes Inc., Roseville, Minnesota. IA-eosin may be prepared in a manner similar to the above by bromination of iodoacetamidofluorescein, which is also available from Molecular Probes Inc.

Conjugation of Eosin to Membrane Proteins

The optimal conditions for labeling a given membrane with eosin must be determined for each individual case. The number of eosin molecules bound to an individual protein should ideally be about unity. Less labeling gives weaker signals, while more labeling carries the danger of loss of dichroism due to energy transfer between eosins. Here, two examples are given which should serve as a useful guide for other systems. However, it should be noted that while a rather selective labeling of a particular membrane component is achieved in these two cases, this is not likely to generally apply to other, more complex, systems. Substantial labeling of several membrane components would make the interpretation of the data highly complex and hazardous. One can envisage that in such cases, it will be necessary to develop more selective probes in order to investigate individual proteins.

1. Human Erythrocyte Membrane. The procedure is the same for either of the two eosin derivatives. Red blood cells obtained from fresh or recently outdated blood are washed 3 times with 5 m*M* sodium phosphate buffer (pH 7.4) containing 150 m*M* NaCl. Then 1 mg of the eosin probe (dissolved in isotonic buffer at a concentration of 0.5 mg/ml) is added per 5 mg of packed cells and incubated for 3 hr at room temperature. The cells are then washed twice more with isotonic buffer to remove any unreacted label and subsequently hemolyzed in 40–50 vol of 5 m*M* sodium phosphate buffer, pH 7.4. The ghosts are sedimented by centrifugation for 20 min at 20,000 *g* and washed 3–4 times with hypotonic buffer. All operations except incubation with the probe are performed at 0–4°C. The amount of bound eosin is typically about 2 μg/mg of membrane protein (see next section for method of determination). Using sodium dodecyl sulphate (SDS) gel electrophoresis and selective extraction procedures, it may be shown that most of the label is attached to band 3.[7]

2. *Sarcoplasmic Reticulum Vesicles.*[8] Sarcoplasmic reticulum vesicles are prepared from rabbit skeletal muscle according to published procedures[9,10] and suspended in 50 m*M* potassium phosphate buffer (pH 8.0) containing 1 *M* KCl and 250 m*M* sucrose (buffer 1). The proteolysis inhibitor phenylmethylsulphonyl fluoride (PMSF) is solubilized in ethanol (125 m*M*) and the solution diluted 100 times with water. A small quantity is added to the vesicle suspension to give a final concentration of 5 μ*M* PMSF. The vesicles are incubated for 10 min in the presence of 5 m*M* Mg-ATP and 4 m*M* $CaCl_2$ (protein concentration 15 mg/ml) prior to addition of the probe. IA-eosin is solubilized in buffer 1 (4 mg/ml) and added to the vesicle suspension in the requisite quantity to give 30 μg IA-eosin/mg protein. After incubating for 90 min at room temperature, free label is separated from the vesicles using a Sephadex G-25 column (l-cm diameter, 20 mg dry Sephadex G-25/μg IA-eosin, equilibrated and eluted with buffer 1 plus 5 μ*M* PMSF). The preparation is maintained at 0°C throughout except during the reaction with IA-eosin. Also, all steps are performed in the dark where possible or otherwise under dim red illumination. This precaution is necessary to prevent loss of ATPase activity due to eosin-sensitized photooxidation.[8] The above procedure typically yields ~12 μg eosin/mg membrane protein. SDS gel electrophoresis shows that most of the probe is bound to the (Ca^{2+} + Mg^{2+})ATPase. Eosin-NCS is unsuitable for labeling this system, since in addition to the ATPase, there is heavy labeling of the lipids.

[7] R. J. Cherry, A. Bürkli, M. Busslinger, and G. Schneider, *in* "Biochemistry of Membrane Transport," FEBS Symposium No. 42 (G. Semenza and E. Carafoli, eds.), p. 86. Springer-Verlag, Berlin, 1977.

[8] A Bürkli and R. J. Cherry, unpublished results.

[9] J. D. Robinson, N. J. M. Birdsall, A. G. Lee, and J. C. Metcalfe, *Biochemistry* **11,** 2903 (1972).

[10] G. B. Warren, P. A. Toon, N. J. M. Birdsall, A. G. Lee, and J. C. Metcalfe, *Proc. Natl. Acad. Sci. U.S.A.* **71,** 622 (1974).

Determination of Bound Eosin

The amount of eosin in a given sample may be determined spectrophotometrically. Both eosin derivatives have their absorption maximum at 522 nm in 5 mM phosphate buffer, pH 7.4. The extinction coefficient of probes prepared as described above is typically $8 \times 10^4 - 9 \times 10^4 M^{-1} cm^{-1}$. The change in extinction coefficient on conjugation to protein is less than 5%,[4] although the absorption maximum shifts to longer wavelengths (typically 525–532 nm). Thus the amount of bound eosin may be determined by comparing the maximum absorbance of the sample with that of a known concentration of the free probe in buffer. With membrane samples, the accuracy of the measurement is improved if light scattering is reduced by solubilization in SDS.

Oxygen Removal

Since triplet states are efficiently quenched by oxygen, lifetimes will be short unless oxygen is displaced from the sample. For most practical purposes, it is sufficient to displace oxygen with a stream of argon containing less than 5 ppm oxygen. A slight improvement is obtained with some samples if the argon stream is passed through an Oxysorb (Messer Griesheim, Düsseldorf) to further decrease the oxygen content. To avoid foaming of membrane samples, the argon stream should be directed into the surface of the solution rather than bubbled through it. The flow of argon should be sufficiently vigorous to produce a visible cavity in the solution surface. The sample is contained in a fluorimeter cell which is terminated by about 7 cm of glass tubing. The cuvette is sealed with a tight-fitting rubber stopper. The argon cylinder is connected to the sample via metal tubing terminated by a long needle which penetrates the rubber seal. The seal should be penetrated by a second needle to provide a gas outlet. Normally, blowing argon onto the sample for 10–15 min is sufficient to achieve close to the maximum lifetime.

An additional advantage of displacing oxygen from the samples is that they are protected from photooxidation during the flash photolysis experiment. The enzymic activity of eosin-labeled (Ca^{2+} + Mg^{2+})ATPase from sarcoplasmic reticulum is rapidly lost on exposure to light in air-equilibrated solutions. The above procedure for displacing oxygen affords complete protection against this reaction.[8] The same is true for acetylcholinesterase activity in eosin-labeled erythrocyte ghosts.[11]

[11] E. Nigg and R. J. Cherry, unpublished results.

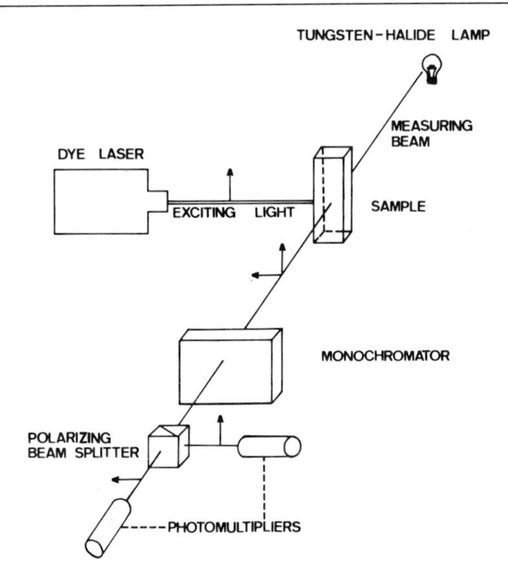

FIG. 3. Schematic diagram of flash photolysis apparatus.

Flash Photolysis Apparatus

A flash photolysis apparatus designed for measuring dichroism of transient absorption signals is illustrated in Fig. 3.[12] The exciting source is an Electro-Photonics Model 43 dye laser which emits a light pulse of up to 200 mJ and 1–2 μsec duration. The dye coumarin 6 gives an untuned emission at 540 nm (10^{-4} M in methanol) which lies within the eosin absorption band. The exciting light is vertically polarized with a Glan-Taylor prism. The sample is contained in a fluorimeter cuvette, either 10 × 10 mm or 10 × 5 mm. With the smaller cuvette, 300 μl of sample are required. The cuvette is inserted into a sample holder which masks the measuring beam down to a vertical slit 4 × 1 mm. The slit should be as close as possible to the side of the cuvette facing the flash. The sample concentration is typically 10^{-6}–10^{-5} M with respect to eosin. With membrane preparations, the concentration is generally limited by the amount of light scattering which can be tolerated.

Absorption changes in the sample following flash excitation are measured with a continuous 100 W tungsten-halide lamp (Oriel Corporation). The wavelength is selected by a Leitz in-line mirror monochromator. Usually, ground-state depletion signals are larger than triplet–triplet absorp-

[12] The instrument described here incorporates features of a prototype built by K. Razi Naqvi and D. Bull at the University of Sheffield.

tion, due to the higher extinction coefficient of the singlet–singlet transition. With eosin, ground-state depletion signals are conveniently measured at 520 nm and triplet–triplet absorption at 610–650 nm. Care must be taken to ensure that artifacts are not produced by the intense laser flash.

After passing through the sample and monochromator, the measuring beam is split into horizontally and vertically polarized components with a Barr and Stroud BC6 polarizing beam splitter. The absorption changes for the two components are then measured simultaneously using two photomultipliers (EMI 9683QB). The high-voltage supply to the photomultipliers is a Brandenburg 475R connected in parallel to the two dynode chains and operated at 500 V.

Before measuring, the steady signals (I_0) from the two photomultipliers are balanced by adjusting the overall voltage to one of them. This is achieved by the variable resistance connected in series with the dynode chain. This procedure compensates for polarizations arising from the optical components. I_0 is typically ~150 mV. When the flash fires, transient intensity changes $I_{\parallel}(t)$, $I_{\perp}(t)$ are recorded by the photomultiplers, where the subscripts $_{\parallel}$ and $\perp$ indicate parallel and perpendicular polarization with respect to that of the exciting flash and t is the time after the flash. The signal $I_{\perp}(t)$ and the difference signal $\Delta(t) = I_{\parallel}(t) - I_{\perp}(t)$ are then displayed on a Tektronix 5103N dual-beam storage oscilloscope equipped with 5A20N plug-in differential amplifiers. In favorable cases, the signals are sufficiently strong for single-shot traces to be evaluated directly from photographs of the oscilloscope display. Better sensitivity and accuracy are obtained with some form of signal averaging. In the present apparatus the signals are amplified by Tektronix AM502 differential amplifiers and fed into a Datalab DL 102A signal averager. Data analysis is accomplished by a Hewlett Packard HP9825A desk-top computer which is interfaced to the signal averager. The results are plotted with a Hewlett Packard HP 9862A plotter.

The apparatus described above has a time resolution of a few microseconds. Faster relaxation times could be measured by using a laser with a shorter pulse width together with faster electronics. The slowest relaxation time which can be measured is limited by the lifetime of the spectroscopic state. The triplet state of eosin permits relaxation times up to about 10 msec to be measured.

Analysis of the Data

The intensity changes measured by the photomultipliers must first be converted to absorbance changes, $A_{\parallel}(t)$ and $A_{\perp}(t)$, using the expressions:

$$A_{\perp}(t) = -\log\left(1 + \frac{I_{\perp}(t)}{I_0}\right) \tag{1}$$

$$A_{\parallel}(t) = -\log\left(1 + \frac{I_{\perp}(t) + \Delta(t)}{I_0}\right) \tag{2}$$

The absorption anisotropy $r(t)$ is given by[13]

$$r(t) = \frac{A_{\parallel}(t) - A_{\perp}(t)}{A_{\parallel}(t) + 2A_{\perp}(t)} \tag{3}$$

where $r(t)$ is independent of the signal lifetime and depends only on rotational motion when the absorption transient exhibits a single exponential decay. Complications can arise when the absorption transient contains more than one component, for example from probes with different binding sites.[14]

The theoretical relationship between $r(t)$ and rotational diffusion has been treated by a number of authors, although the correct solution for the general case of an anisotropic particle (in isotropic solution) was only obtained comparatively recently (for review, see Ref. 14). In the general case, $r(t)$ is the sum of five exponentials. For the special case of spherical particles

$$r(t) = r_0 \exp(-6D_R t) \tag{4}$$

where D_R is the rotational diffusion coefficient and r_0 is a constant which depends on the angle between the transition dipole moments for excitation and measurement and, in practice, on instrumental factors.

For the case of proteins in membranes, it is necessary to consider the anisotropy of the membrane structure. For practical purposes it may be a reasonable approximation to suppose that the anisotropic rotation of the protein is in fact largely determined by the anisotropy of the membrane. A possible model for analyzing the data is shown in Fig. 4, in which it is assumed that rotational motion may be characterized by two diffusion coefficients, $D_{\parallel}$ for rotation about an axis normal to the plane of the membrane and $D_{\perp}$ for rotation about axes lying in the plane of the membrane. Essentially this treatment neglects any asymmetry in the cross-sectional shape of the protein in the plane of the membrane. On this basis, the problem formally is equivalent to that of a body with an axis of symmetry immersed in an isotropic solution. Existing theoretical treatments then give the following expression for the anisotropy of the flash-induced dichroism:[14]

[13] A. Jablonski, *Z. Phys.* **16a,** 1 (1961).

[14] R. Rigler and M. Ehrenberg, *Q. Rev. Biophys.* **6,** 139 (1973).

$$r(t) = \sum_{i=1}^{3} A_i \exp(-E_i t) \tag{5}$$

where $A_1 = (6/5)(\sin^2\theta\cos^2\theta)$; $A_2 = (3/10)(\sin^4\theta)$; $A_3 = (1/10)(3\cos^2\theta - 1)^2$; $E_1 = 5D_\perp + D_\parallel$; $E_2 = 2D_\perp + 4D_\parallel$; $E_3 = 6D_\perp$. θ is the angle between the transition moment for absorption and the normal to the plane of the membrane.

A further simplifying assumption contained in the above equation is that the transition moments for excitation and measurement are parallel. If the transition moments are not parallel, then both orientations appear in the coefficients of the three exponentials.

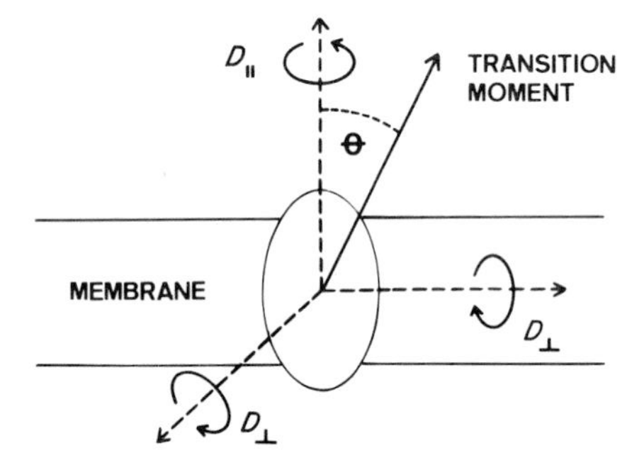

FIG. 4. Model used for interpretation of transient dichroism measurements. $D_\parallel$ and $D_\perp$ are the diffusion coefficients for rotation about axes parallel and perpendicular to the normal to the plane of the membrane. θ defines the direction of the absorption transition dipole moment.

There are a number of indications, for example from chemical labeling experiments, that many integral membrane proteins maintain a constant orientation across the membrane,[15,16] implying that $D_\perp$ is negligibly small. Putting $D_\perp = 0$ in Eq. (5) gives

$$r(t) = A_1 \exp(-D_\parallel t) + A_2 \exp(-4D_\parallel t) + A_3 \tag{6}$$

The above equation predicts that $r(t)$ does not fall to zero but decays to a time-independent value determined by the coefficient A_3.

In principle, the diffusion coefficient $D_\parallel$ is related to the size of the protein and the membrane viscosity. The Stokes-Einstein equation for isotropic rotation is of course not applicable. Treating the protein as a cylin-

[15] M. S. Bretscher, *Science* **181,** 622 (1973).
[16] S. J. Singer, *Annu. Rev. Biochem.* **43,** 805 (1974).

der, with its axis of symmetry normal to the plane of the membrane, leads to the equation[17]

$$D_{\parallel} = \frac{kT}{4\pi a^2 h \eta} \tag{7}$$

where a is the radius of the cylinder, h the length immersed in the membrane, and η the membrane viscosity. The derivation of this equation is based on a hydrodynamic approach which treats the lipid bilayer as a continuum. Its validity has yet to be critically tested.

In the above analysis, θ is regarded as a well-defined angle. This is probably true in one particular case which is briefly discussed later, namely the retinal chromophore of bacteriorhodopsin. Eosin probes, however, usually exhibit independent motion when conjugated to proteins.[5] This motion is rapid and restricted; its effect is that measured r values are smaller than expected since part of the anisotropy is lost in a time shorter than the resolving time of the apparatus. (Note that instrumental factors also reduce the measured anisotropy.) Moreover, the observed plots of r versus t will represent an average of curves, since θ can now take any one of a range of values. A further averaging results when there is more than one binding site for the probe on the same protein. These complexities must be remembered when any detailed interpretation of the data is attempted.

Intrinsic Chromophores

Triplet–triplet absorption arising from the aromatic amino acids tryptophan and tyrosine is readily observed,[18] but dichroism has so far not been detected. It has been proposed that phosphorescence polarization of tryptophan residues may be used to measure slow rotational motion of proteins.[19] Analysis of data from membranes would, however, be complex due to the large number of tryptophans which are generally present.

The general principles of measuring rotational diffusion using flash photolysis of course apply not only to triplet states but to any long-lived photoproduct. The first measurement of protein rotation in membranes[20] exploited the spectroscopic properties of the visual protein rhodopsin, which contains the intrinsic chromophore 11-*cis* retinal. Excitation of rhodopsin produces spectroscopic intermediates which are sufficiently long-lived to permit rotational diffusion measurements in the microsecond time range. More recently, measurements have been made with the related

[17] G. Saffman and M. Delbrück, *Proc. Natl. Acad. Sci. U.S.A.* **72,** 3111 (1975).
[18] S. L. Aksentev, Y. A. Vladimirov, V. I. Olenov, and Y. Y. Fesenko, *Biofizika* **12,** 63 (1967).
[19] G. B. Strambini and W. C. Galley, *Nature (London)* **260,** 554 (1976).
[20] R. A. Cone, *Nature (London), New Biol.* **236,** 39 (1972).

protein bacteriorhodopsin from the purple membrane of *Halobacterium halobium*.[3,21-25] Again this is possible because excitation of the intrinsic chromophore (all-*trans* or 13-*cis* retinal) produces long-lived spectroscopic intermediates. Another system which has been investigated is cytochrome oxidase, where the photoreaction is the dissociation of the cytochrome oxidase–CO complex.[26]

Summary of Applications

The principal application of eosin probes has so far been to study the rotation of band 3 in the human erythrocyte membrane.[7,27] Figure 5A shows an example of the flash photolysis signals obtained. The anisotropy (Fig. 5B) first decreases with time and thereafter appears to be constant. This is the result predicted by Eq. (6) for a protein rotating around a single axis. The diffusion coefficient $D_{\parallel}$ is estimated to be in the order of 1000 sec^{-1}. Protein–protein interaction in the erythrocyte membrane was investigated by observing the effects of spectrin removal on band 3 rotation. It was concluded that spectrin does not inhibit the rotational motion of band 3. However, band 3 is immobilized by conditions which induce aggregation of the particles seen in freeze–fracture electron microscopy.

More recently, we have applied the technique to investigate rotation of the (Ca^{2+} + Mg^{2+})ATPase in sarcoplasmic reticulum from rabbit skeletal muscle. Preliminary measurements indicate a marked tendency towards self-aggregation of the ATPase molcules in the membrane.[8]

Various measurements have been made with bacteriorhodopsin, both in the natural membrane and in reconstituted systems. In the purple membrane the protein is immobilized[3] as anticipated from the existence of a crystalline lattice.[28,29] However, rotational motion of bacteriorhodopsin is observed in the plasma membrane of cells grown in the presence of nicotine.[21] In these cells, bacteriorhodopsin is formed by artificially adding retinal, since nicotine blocks retinal synthesis.[30] It appears that the normal purple membrane patches are not formed under these conditions.

[21] R. J. Cherry, M. P. Heyn, and D. Oesterhelt, *FEBS Lett.* **78,** 25 (1977).

[22] R. J. Cherry, U. Müller, and G. Schneider, *FEBS Lett.* **80,** 465 (1977).

[23] M. P. Heyn, R. J. Cherry, and U. Müller, *J. Mol. Biol.* **117,** 607 (1977).

[24] R. J. Cherry, U. Müller, R. Henderson, and M. P. Heyn, *J. Mol. Biol.* (in press).

[25] M. P. Heyn, P. J. Bauer, and N. A. Dencher, *in* "Biochemistry of Membrane Transport," FEBS Symposium No. 42 (G. Semenza and E. Carafoli, eds.), p. 96. Springer-Verlag, Berlin, 1977.

[26] W. Junge and D. DeVault, *Biochim. Biophys. Acta* **408,** 200 (1975).

[27] R. J. Cherry, A. Bürkli, M. Busslinger, G. Schneider, and G. R. Parish, *Nature (London)* **263,** 389 (1976).

[28] A. E. Blaurock and W. Stoeckenius, *Nature (London), New Biol.* **233,** 152 (1971).

[29] R. Henderson, *Annu. Rev. Biophys. Bioeng.* **6,** 87 (1977).

[30] M. Sumper, H. Reitmeier, and D. Oesterhelt, *Angew. Chem., Int. Ed. Engl.* **15,** 187 (1976).

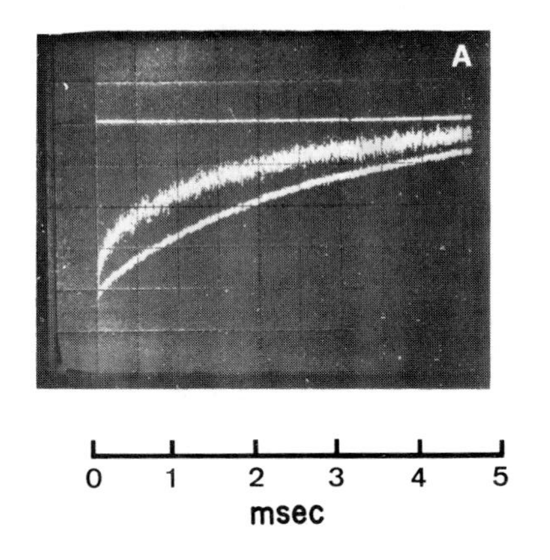

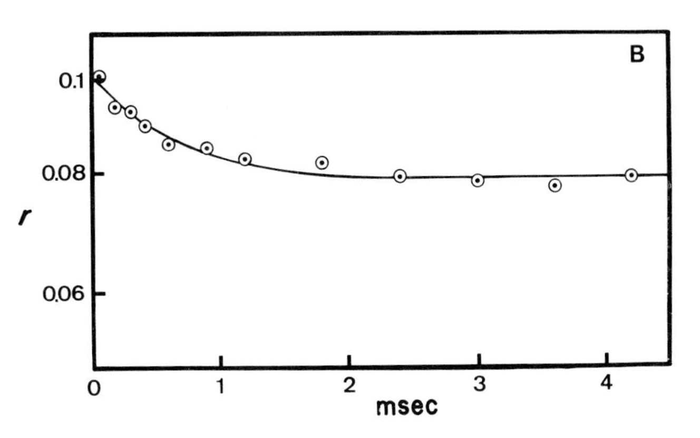

FIG. 5. Results of flash photolysis experiments on human erythrocyte membranes labeled with iodoacetamidoeosin (2 μg eosin/mg membrane protein, 5 m*M* phosphate buffer pH 8.0, 22°C). Most of the label is bound to band 3. (A) Ground-state depletion signals measured at 520 nm. Stronger signal $I_{\perp}$, sensitivity 10 mV/division. Weaker (noisier) signal $I_{\parallel} - I_{\perp}$, sensitivity 5 mV/division. Horizontal trace is base line. Traces are from a single-shot experiment. (B) Time dependence of absorption anisotropy calculated from the above data.

Bacteriorhodopsin has been incorporated into artificial dimyristoyl- and dipalmitoylphosphatidylcholine vesicles.[22] Rotational diffusion measurements have been used in conjunction with other physical techniques (circular dichroism, X-ray diffraction, and electron microscopy) to demonstrate a reversible crystallization of the protein on cooling through the lipid gel to liquid–crystalline phase transition.[24] Of particular interest is the time dependence of r (obtained above the lipid phase transition), which is illustrated in Fig. 6. These results were obtained with light-adapted samples

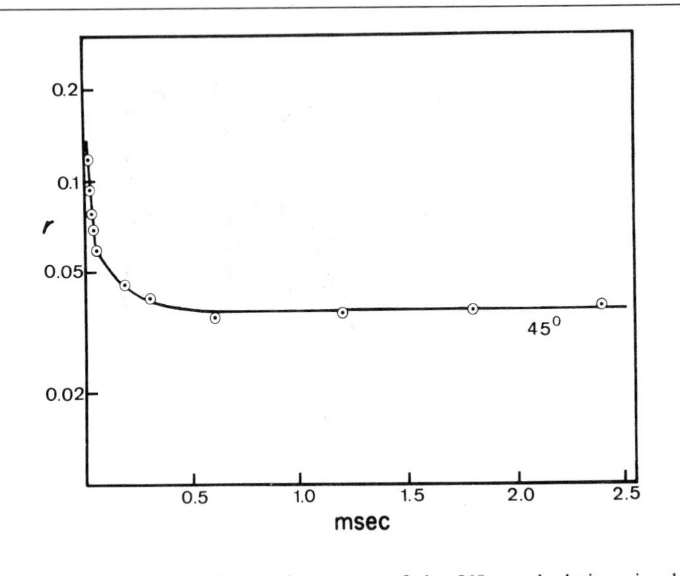

FIG. 6. Time dependence of the anisotropy r of the 568-nm depletion signal for bacteriorhodopsin incorporated in dipalmitoylphosphatidylcholine vesicles measured at 45°C (protein:phospholipid ratio 0.81 w/w, sodium acetate buffer pH 5.0).

in which the retinal chromophore is in the all-*trans* configuration. The absorption anisotropy was calculated from the dichroism of the ground-state depletion signal measured at 570 nm. It can be seen that the curve again has the general form predicted by Eq. (6) in that r does not decay to zero but reaches a time-independent value. By measuring the relative magnitude of the time-independent component, we can determine the angle θ between the transition dipole moment of the 568-nm absorption band (which lies along the long axis of the polyene chain) and the normal to the plane of the membrane. As can be seen from Eq. (5) and (6), this relative magnitude is a function only of θ. On this basis, a value of $\theta = 78° \pm 3°$ is obtained.[23] This agrees well with the value of $71° \pm 4°$ obtained completely independently from the static linear dichroism of oriented purple membranes.[23] This agreement provides support for the validity of Eq. (6) in interpreting transient dichroism measurements with membrane proteins. In addition, this novel method for determining molecular orientation, which does not require oriented membranes, may find application to other systems.

A further interesting observation is that the rotation of bacteriorhodopsin in these reconstituted systems becomes markedly slower as the lipid:protein ratio is decreased.[22] Circular dichroism measurements indicate that this effect is not due to the formation of bacteriorhodopsin aggregates (except at very low lipid:protein ratio).[24] Hence it appears that the membrane viscosity is a strong function of the protein concentration, which is of course well known to be the case for aqueous protein solutions.

In conclusion, the data obtained so far amply demonstrate the usefulness of the flash photolysis technique for studying the rotational motion of membrane proteins. Future investigations may reasonably be expected to yield considerable insight into structural features which affect the rotational motion of membrane proteins, such as their rigidity, their self-association into specific aggregates (dimers, trimers, etc.), and their interactions with other components of the membrane or of the cell.

Acknowledgments

I am indebted to Dr. M. P. Heyn and Dr. N. A. Dencher for critically reading the manuscript, to my colleagues for their contributions to the work described here, and to Professor G. Semenza for his continual support and encouragement. I also wish to thank the Swiss National Science Foundation for financial support.

[5] Measurement of the Change of Electrical Potentials and Currents across Biomembranes in the Range of Nanoseconds to Seconds

By H. T. WITT

First experimental evidence for an electrical field generation across membranes of bioenergetic systems came in 1967 from the observation of special absorption changes of the pigments located in the photosynthetic membrane[1,2] (reported first in footnote 2). These changes are called "field indicating absorption changes." It has been shown thoroughly by kinetic,[1] spectroscopic,[3,4] and electric[5,6] experiments that these changes are indeed caused by the generation of electrical fields in the functional membrane of photosynthesis.

[1] W. Junge and H. T. Witt, *Z. Naturforsch. Teil B*, **23**, 244 (1968).

[2] H. T. Witt, *Fast React. Primary Processes Chem. Kinet., Proc. Nobel Symp., 5th, 1967* p. 81, 261 (1967).

[3] S. Schmidt, R. Reich, and H. T. Witt, *Naturwissenschaften* **58**, 414 (1971).

[4] S. Schmidt, R. Reich, and H. T. Witt, *Proc. Int. Congr. Photosynth. Res., 2nd, 1971* p. 1087 (1972).

[5] H. T. Witt and A. Zickler, *FEBS Lett.* **37**, 307 (1973).

[6] H. T. Witt and A. Zickler, *FEBS Lett.* **29**, 205 (1974).

ISBN 0-12-181954-X

It was shown, furthermore, that these optical changes are caused by electrochromism,[3,4] i.e., by absorption changes, induced mainly through the shift of the absorption bands of the pigments in an electrical field. In green plants the spectrum of these absorption changes in the visible wavelength range is due to the shift of the absorption band of all bulk pigment located in the membrane, i.e., chlorophyll a, chlorophyll b, and carotenoids.[7]

The measurement of electric potentials by electrochromism has been extended to photosynthetic bacteria.[8] In this case the spectrum in the visible range is due to the shift of carotenoids. (The bacteria chlorophylls absorb in the UV and IR.)

The field-indicating absorption change, ΔA, represents an intrinsic molecular voltmeter and ammeter with the following properties:

1. The absorption changes, ΔA, response "promptly."[2,9]
2. They indicate an electric potential difference, $\Delta\phi$, across the membrane.[5]
3. They indicate the potential linearly:[10,11]

$$\Delta \phi (\mathrm{t}) = a \cdot \Delta A(t)$$

The derivative of the change with respect to time is therefore proportional to the flux, i, of charges across the membrane area S:

$$i(t) = C \cdot a \cdot d(\Delta A)/dt$$

C = membrane capacity per area S.

4. The absorption changes, ΔA, have been calibrated in absolute values.[8,10,12]

In respect to the response, to date the absorption changes have been measured in the range of seconds to 20 nsec. The lower limit is caused by the duration of the repetitive laser pulse which was available to us at that time (1967).[2,9] In principle, the technique is applicable to the picosecond range if a corresponding short light pulse is used for excitation of photosynthesis. (The large absorption change at 510 nm measured in chromatophores of bacteria in the picosecond range[13] is probably not

[7] H. M. Emrich, W. Junge, and H. T. Witt, *Z. Naturforsch., Teil B* **24,** 1144 (1969).
[8] J. B. Jackson and A. R. Crofts, *FEBS Lett.* **4,** 185 (1969).
[9] Ch. Wolff, H.-E. Buchwald, H. Rüppel, K. Witt, and H. T. Witt, *Z. Naturforsch., Teil B,* **24,** 1038 (1969).
[10] H. Schliephake, W. Junge, and H. T. Witt, *Z. Naturforsch, Teil B* **23,** 1571 (1968).
[11] H. T. Witt, *in* "Excited States of Biological Molecules" (J. B. Birks, ed.), p. 245. Wiley (Interscience), New York, 1976.
[12] A. Zickler, H. T. Witt, and G. Boheim, *FEBS Lett.* **66,** 142 (1976).
[13] J. S. Leigh, T. L. Netzel, P. L. Dutton, and P. M. Rentzepis, *FEBS Lett.* **48,** 136 (1974).

caused by the electrochromic effect but by the formation of the carotenoid triplet state.)[14,15]

For registration of the signals the repetitive pulse spectroscopic method is most suitable. The extremely high sensitivity of this technique enables the detection of very small signals and thus allows a corresponding high time resolution. Such techniques have already been described in detail in this series.[16]

Indication of electrical potential differences was derived later, also from delayed light emission,[17] salt distribution,[18] electrophoresis of synthetic ions,[19] and microelectrode measurements.[20] In comparison with the electrochromic technique these methods are, however, very slow.

The observed linear indication is expected in the wavelength range in which chlorophyll a and chlorophyll b (permanent dipoles) are responsible for the effect but not in the range where carotenoids (no permanent dipoles) are responsible. The linear response of the carotenoids can be quantitatively explained, however, if the carotenoids are exposed in the membrane to a permanent local field.[4] It was shown recently which arrangement in the membrane is most probably responsible for such a permanent field.[21] It was demonstrated that a special fraction of the carotenoids form asymetrical complexes with the chlorophylls which give rise to a local molecular field.[21,22]

The calibrations were carried out by the following methods.

In thylakoids of green plants:

(a) By measuring the translocated charges.[10]

(b) By using voltage-dependent ionophores.[12]

In chromatophores of bacteria:

By setting diffusion potentials across the membrane.[8] In general it is:

$$\Delta\phi = \frac{\Delta\phi_1 \cdot \Delta A}{\Delta A_1}$$

$\Delta\phi_1$ and ΔA_1 are the values in a single turnover, ΔA arbitrary values.

$$i = C \cdot \frac{\Delta\phi_1}{\Delta A_1} \quad \frac{d(\Delta A)}{dt}$$

[14] Ch. Wolff and H. T. Witt, *Z. Naturforsch.* **24b,** 1031 (1969).

[15] G. Renger and Ch. Wolff, *Biochim. Biophys. Acta* **460,** 47 (1977).

[16] H. Rüppel and H. T. Witt, Vol. 16, p. 316.

[17] J. Barber and G. P. B. Kraan, *Biochem. Biophys. Acta* **197,** 49 (1970).

[18] P. Mitchell and J. Moyle, *Eur. J. Biochem.* **7,** 471 (1969).

[19] L. E. Bakeeva, L. L. Grinius, A. A. Jasaitis, V. V. Kuliene, D. O. Levitsky, E. A. Liberman, L. I. Severina, and V. P. Skulachev, *Biochim. Biophys. Acta* **216,** 13 (1970).

[20] W. I. Vredenberg and W. I. M. Tonk, *FEBS Lett.* **42,** 236 (1974).

[21] R. Reich and K.-U. Sewe, *Photochem. Photobiol.* **26,** 11 (1977).

[22] K.-U. Sewe and R. Reich, *FEBS Lett.* **80,** 30 (1977).

$C \approx 1 \mu F/cm^2$. The value of $\Delta\phi_1$ is 50–100 mV in thylakoids of spinach.[10,12] In chromatophores of bacteria the value during the steady state is 190 mV.[8]

With this technique several valuable information has been obtained:

1. on charge separations and vectorial pathways of charges within the membrane;
2. on the topological arrangement of the reactants in the membrane;
3. on the mechanism of phosphorylation; and
4. on special properties of the membrane, e.g., (a) phase transitions and (b) conformational changes, etc.

Details are reviewed, elsewhere.[11,23,24,25]

The electrochromic method is, furthermore, useful for the analysis of electric phenomena in other biological membranes or interfaces in general. If these are not pigmented, they can be stained with dyes.[26] The method has been extended to bacteria[8] and mitochondria[27] and would be useful for electric measurements at vesicles from nerve membranes.

[23] H. T. Witt, *Qt. Rev. Biophys.* **4,** 365 (1971).
[24] H. T. Witt, *in* "Bioenergetics of Photosynthesis" (Govindjee, ed.), p. 493. Academic Press, New York, 1975.
[25] H. T. Witt, *Biochim. Biophys. Acta* (in press).
[26] H. M. Emrich, W. Junge, and H. T. Witt, *Z. Naturforsch., Teil B* **24,** 1144 (1969).
[27] K. E. Akerman and M. K. F. Wikström, *FEBS Lett.* **68,** 191 (1976).

[6] Relaxation Kinetics of Heme Proteins

By MAURIZIO BRUNORI

1. Introduction

The application of chemical relaxation techniques to the study of the reactions of hemoproteins in solution has been steadily increasing over the last 10 years. The special features of the perturbation methods which have encouraged their application to hemoprotein kinetics may be summarized as: (1) the very high time resolution achieved by the available techniques, which allows coverage of a very wide time range (down to micro- and nanoseconds); (2) the opportunity to investigate near-equilibrium kinetics which provides a means of identifying the principal pathways of the temporal response to a perturbation; (3) the possibility of describing the dynamics of re-equilibration by linear differential equations leading, in principle, to the identification of reaction steps, equilibrium intermediates, and time constants for the system under investigation.

ISBN 0-12-181954-X

In all relaxation methods the reaction is initiated by a change in one of the thermodynamic variables which govern equilibrium. The perturbation applied can be single (step-by-step perturbation) or cyclic (multiple perturbation), but in either case it should respond to the following minimum prerequisites: (1) to be fast compared to the response of the system under investigation; (2) to produce a significant change in the concentration(s) of the reactants.

The general theoretical framework of relaxation spectrophotometry, as well as the methodologies and the technical aspects of these rapid-reaction methods, have been dealt with in great detail in the recent literature.[1-5]

Although in principle all the methods are applicable to the study of hemoprotein kinetics, in practice most of the work carried out to date has taken advantage of the temperature-jump relaxation method.[5] This has been applied to hemoprotein kinetics mainly to investigate (1) ligand-binding reactions, (2) electron-transfer reactions, and (3) conformational isomerizations linked to (1) and/or (2).

2. Treatment of Relaxation Kinetics

2.1. The Relaxation Spectrum

The relaxation spectrum of a system near thermodynamic equilibrium contains information on (1) the time constants (*relaxation times*) and (2) the thermodynamic properties (*relaxation amplitudes*) of the elementary steps of the mechanism in question.

Failure to obtain direct information may similarly depend on (1) a time factor, if a reaction step relaxes too rapidly to be time-resolved with the available techniques, or (2) an amplitude factor, if a reaction step is not observable (due to the absence of a detectable optical signal or of a suitable change in the thermodynamic variables employed as forcing functions).

The theoretical basis of chemical relaxation spectrometry has been dealt with extensively by Eigen and de Maeyer.[2] Here only a minimum of theoretical considerations are presented to provide the basic methodology for calculating the relaxation spectrum of the system under investigation. Moreover we shall deal exclusively with the analysis of single-step perturbation methods.

[1] M. Eigen and L. de Maeyer, *Tech. Org. Chem.* **8,** Part II, 895 (1963).

[2] M. Eigen and L. de Maeyer, *Tech. Chem. (N.Y.)* **6,** Part II, 63 (1974).

[3] K. Justin, ed., this series, Vol. **16** [3].

[4] G. Czerlinski, "Chemical Relaxation." Dekker, New York, 1966.

[5] G. G. Hammes, *Tech. Chem. (N.Y.)* **6,** Part II, 147 (1974).

2.2. *Relaxation Times*

For a system in chemical equilibrium the ratio of products-to-reactants concentration remains constant at constant temperature, pressure, electric field, etc. When a sudden temperature increase is imposed on the system, the equilibrium position will be perturbed if the enthalpy change for the reaction is different from zero (the direction of the shift will depend on the sign of the enthalpy change). The instantaneous concentration of species i, C_i, may follow the temperature change with a finite lag, which is a function of the rate constants and equilibrium concentrations of the species involved.

The time dependence of the concentration of one of the reagents (or products) reveals an exponential approach to the new equilibrium. The observed relaxation spectrum refers, of course, to the conditions prevailing immediately after the onset of the perturbation (i.e., at the higher temperature). The symbols used are clarified in Fig. 1.

If the perturbation is kept small enough, the rate equation which governs the approach to the new equilibrium can always be linearized with respect to the time-dependent concentration variable (*small-perturbation requirement*). Thus for a single-step reaction the rate equation can be reduced to a linear form of the type:

$$\frac{dX_i}{dt} + \left(\frac{1}{\tau}\right)X_i = \left(\frac{1}{\tau}\right)\bar{X}_i \quad \text{or} \quad \frac{dX_i}{dt} = \frac{1}{\tau}(\bar{X}_i - X_i) \qquad (2.1)$$

if $(\bar{X}_i - X_i) \ll C_i$.

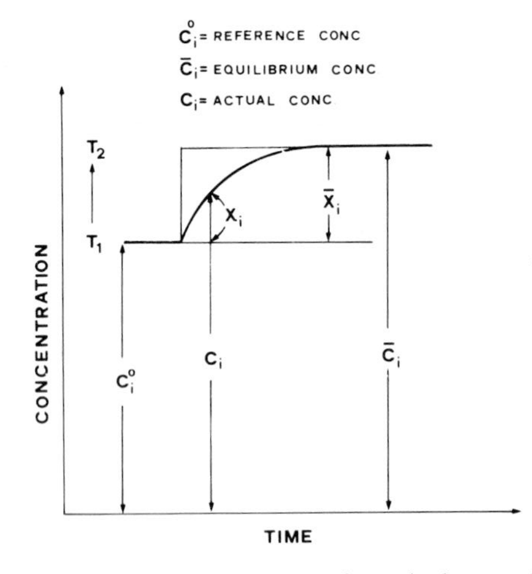

FIG. 1. Time course of concentration change in a single-step (temperature $T_1 \rightarrow T_2$) perturbation experiment.

The relaxation time, τ, is the reciprocal of the first-order decay constant for the approach to the new equilibrium, and can be evaluated from the equation:

$$X(t) - \bar{X} = (X^\circ - \bar{X}) \exp \frac{-t}{\tau} \qquad (2.2)$$

EXAMPLE 2.2.1

Scheme:

$$A + B \underset{k_{21}}{\overset{k_{12}}{\rightleftharpoons}} C$$

Differential equations:

$$-\frac{dC_A}{dt} = -\frac{dC_B}{dt} = \frac{dC_C}{dt} = k_{12} \cdot C_A \cdot C_B - k_{21} \cdot C_C$$

Definitions:

$$C_A = \bar{C}_A - \bar{X}_A + X_A$$
$$C_B = \bar{C}_B - \bar{X}_B + X_B$$
$$C_C = \bar{C}_C - \bar{X}_C + X_C$$

Linearization:

$$-\frac{dC_A}{dt} = -\frac{dX_A}{dt} = k_{12}(\bar{C}_A - \bar{X}_A + X_A)(\bar{C}_B - \bar{X}_B + X_B) - k_{21}(\bar{C}_C - \bar{X}_C + X_C)$$
$$= \left[k_{12}(\bar{C}_A + \bar{C}_B) + k_{21}\right](\bar{X}_A - X_A) = \frac{1}{\tau}(\bar{X}_A - X_A)$$

which can be obtained by (1) substituting according to the definitions, (2) dropping all the terms containing X^2 (because of the small perturbation), and (3) applying the equilibrium condition $k_{12} \cdot \bar{C}_A \cdot \bar{C}_B = k_{21} \cdot \bar{C}_C$.

In a similar manner it is possible to calculate the relaxation time for different simple (single-step) reactions. These are given in Table I.

2.3. *Relaxation Amplitudes*[2,6,7]

Concentration changes are followed using appropriate optical detection techniques, such as optical absorption, fluorescence, or light scattering. The relaxation amplitude is therefore dependent on the particular experimental technique employed, while the relaxation time is not. If we limit ourselves to considerations concerning temperature-jump experiments, we

[6] D. Thusius, *Biochimie* **55**, 277 (1973).
[7] G. Ilgenfritz, personal communication (1976).

TABLE I
RELAXATION TIMES AND AMPLITUDES FOR SIMPLE REACTION SCHEMES

Reaction scheme	τ^{-1}	Γ^{-1}
$A \underset{k_{21}}{\overset{k_{12}}{\rightleftharpoons}} B$	$k_{12} + k_{21}$	$1/\bar{C}_A + 1/\bar{C}_B$
$A + B \rightleftharpoons C$	$k_{12}(\bar{C}_A + \bar{C}_B) + k_{21}$	$1/\bar{C}_A + 1/\bar{C}_B + 1/\bar{C}_C$
$2A \rightleftharpoons A_2$	$4k_{12} \cdot \bar{C}_A + k_{21}$	$4/\bar{C}_A + 1/\bar{C}_{A_2}$
$A + B \rightleftharpoons C + D$	$k_{12}(\bar{C}_A + \bar{C}_B) + k_{21}(\bar{C}_C + \bar{C}_D)$	$1/\bar{C}_A + 1/\bar{C}_B + 1/\bar{C}_C + 1/\bar{C}_D$
$A + 2B \rightleftharpoons C$	$k_{12}(4\bar{C}_A\bar{C}_B + \bar{C}_B^2) + k_{21}$	$1/\bar{C}_A + 4/\bar{C}_B + 1/\bar{C}_C$

can easily see that the amplitude of the effect depends on several terms: (1) the difference in optical properties between the various species (and thus specific molecular parameters, species concentrations, light path, and other geometrical parameters); (2) the enthalpy change for the reaction and the absolute temperature shift imposed on the system; (3) the equilibrium position, i.e., the ratio of concentrations of the various species.

If we limit ourselves to optical absorption methods, and assuming a 1-cm light path, we have the following.

EXAMPLE 2.3.1

Scheme: $A + B \rightleftharpoons C$

Relaxation Time:

$$C(t) = C^0 \exp \frac{-t}{\tau}$$

$$\delta\, OD(t) = \delta\, OD^0 \exp \frac{-t}{\tau}$$

Relaxation amplitude:

$$\delta\, OD^0 = \Delta\epsilon \cdot \frac{\delta K}{K} \cdot \Gamma$$

where $\Delta\epsilon = \epsilon_C - (\epsilon_A + \epsilon_B)$ is a constant term depending on the molecular properties of the species (ϵ being the molar decadic extinction coefficient at a given λ); and

$$\frac{\delta K}{K} = \frac{\Delta H}{RT} \cdot \frac{\delta T}{T}$$

is also a constant term, and is proportional to the enthalpy change for the process in a temperature-jump experiment; and

$$\Gamma = \left[\frac{1}{\bar{C}_A} + \frac{1}{\bar{C}_B} + \frac{1}{\bar{C}_C}\right]^{-1} = \frac{\bar{C}_A \cdot \bar{C}_B}{\bar{C}_A + \bar{C}_B + K^{-1}}$$

expresses the dependence of amplitude on the equilibrium concentration of the species.

Table I reports the Γ factors for several single-step reactions. The linear relationship between the relaxation amplitude (δOD^0) and the concentration variable (Γ) allows one to obtain very easily the product of the constant terms ($\Delta\epsilon \cdot \delta K/K$). Amplitude analysis therefore (1) consists of measurements of amplitudes of relaxation processes as a function of reactant concentration, keeping other variables (such as δT, light path, and observation wavelength) constant, and (2) allows one to determine experimentally either ΔH, if the change in extinction coefficient is known from static spectroscopy, or $\Delta\epsilon$, if the enthalpy change for the process has been determined independently.[5]

Note that this approach may be useful to determine the equilibrium constant of a binding reaction, even for very large values of the affinity constant if enough sensitivity is available.[7,8]

2.4. *Multiple-step Reactions*[2,7,9]

For a system consisting of several elementary steps, the kinetics of establishment of equilibrium can be described by a *set of linear differential equations*, if the small perturbation requirement is fulfilled.

$$X_{\text{tot}} = X_1^0 \exp \frac{-t}{\tau_1} + X_2^0 \exp \frac{-t}{\tau_2} + \ldots + X_n^0 \exp \frac{-t}{\tau_n} \tag{2.3}$$

The existence of coupling between the different elementary steps implies that the temporal response of the concentration of component i is proportional to the concentration of all the other components.

If n independent concentration parameters are present in the coupled system, the set of linearized rate equations reads:

$$\begin{aligned} \dot{X}_1 &= a_{11}X_1 + a_{12}X_2 + \ldots + a_{1n}X_n \\ \dot{X}_2 &= a_{21}X_1 + a_{22}X_2 + \ldots + a_{2n}X_n \\ &\vdots \\ \dot{X}_n &= a_{n1}X_1 + a_{n2}X_2 + \ldots + a_{nn}X_n \end{aligned} \tag{2.4}$$

where the X_i terms are the concentration changes and the a_{rk} terms are the rate coefficients containing equilibrium concentrations and rate constants.

[8] M. Eigen and R. Winkler, *in* "The Neurosciences: Second Study Program" (F. O. Schmitt, ed.), p. 685. Rockfeller Univ. Press, New York, 1970.

[9] M. Eigen, *Q. Rev. Biophys.* **1**, 3 (1968).

If only the kinetic parameters are of interest, the solutions of the characteristic equation yields the relaxation times $(1/\tau_i)$, each one being dependent on all the concentration terms;

$$\begin{vmatrix} (a_{11} + 1/\tau) & a_{12} \cdots & & a_{1n} \\ a_{21} & (a_{22} + 1/\tau) \cdots & & a_{2n} \\ \vdots & \vdots & & \\ a_{n1} & \cdots a_{n2} \cdots & & (a_{nn} + 1/\tau) \end{vmatrix} = 0 \qquad (2.5)$$

The solution of this determinantal equation leads to a polynomial of nth order.[2,9] This procedure is illustrated below for a two-step reaction scheme.

Evaluation of the relaxation times may be numerically very involved when more than two well-separated events are present, and it may not be meaningful or feasible if very strong coupling of the relaxation effects is present. In such cases one may resort to the determination of average relaxation times.[2,10] A nonlinear least-squares method for the numerical analysis of relaxation spectra has been reported by Johnson and Schuster.[11]

EXAMPLE 2.4.1

Scheme:

$$A + B \underset{k_{21}}{\overset{k_{12}}{\rightleftharpoons}} AB \underset{k_{32}}{\overset{k_{23}}{\rightleftharpoons}} C$$

The four concentration variables can be expressed in terms of only two time-dependent variables:

$$\dot{X}_1 = \frac{dX_1}{dt} = \frac{dX_A}{dt} = \frac{dX_B}{dt}$$

$$\dot{X}_2 = \frac{dX_2}{dt} = \frac{dX_C}{dt}$$

since $X_{AB} = -(X_1 + X_2)$. The rate equations can be *linearized* to assume the general form

$$\begin{aligned} \dot{X}_1 &= a_{11}X_1 + a_{12}X_2 \\ \dot{X}_2 &= a_{21}X_1 + a_{22}X_2 \end{aligned} \quad \left\{ \begin{aligned} a_{11} &= -k_{12}(C_A + C_B) \\ a_{12} &= k_{21} \\ a_{21} &= k_{12}(C_A + C_B) - k_{32} \\ a_{22} &= -(k_{21} + k_{23} + k_{32}) \end{aligned} \right.$$

[10] G. Schwarz, *Biopolymers* **6**, 873 (1968).

[11] M. L. Johnson and T. M. Schuster, *Biophys. Chem.* **2**, 32 (1974).

Characteristic equation:

$$\begin{vmatrix} (a_{11} + 1/\tau) & a_{12} \\ a_{21} & (a_{22} + 1/\tau) \end{vmatrix} = 0$$

The solution of the determinant assumes the form of a quadratic:

$$\left(\frac{1}{\tau}\right)^2 + (a_{11} + a_{22}) \frac{1}{\tau} + (a_{11}a_{22} - a_{12}a_{21}) = 0$$

with the solution

$$\frac{1}{\tau_{\mathrm{I,II}}} = -\frac{(a_{11} + a_{22})}{2} \left\{ 1 \pm \sqrt{\left[1 - \frac{4(a_{11}a_{22} - a_{12}a_{21})}{(a_{11} + a_{22})^2} \right]} \right\}$$

Consider:

$$\frac{1}{\tau_{\mathrm{I}}} + \frac{1}{\tau_{\mathrm{II}}} = k_{12}\,(\bar{C}_A + \bar{C}_B) + (k_{21} + k_{23} + k_{32}) = P$$

and

$$\frac{1}{\tau_{\mathrm{I}}} \cdot \frac{1}{\tau_{\mathrm{II}}} = k_{12}(k_{23} + k_{32})\,(\bar{C}_A + \bar{C}_B) + k_{21}k_{32} = Q$$

Plots P and Q vs. $(\bar{C}_A + \bar{C}_B)$ yield two straight lines which allow evaluation of all four rate constants:

$$k_{12} = \text{slope of } P$$
$$k_{21} = \text{intercept of } P - \frac{(\text{slope of } Q)}{(\text{slope of } P)}$$
$$k_{32} = \text{intercept of } \frac{Q}{k_{21}}$$
$$k_{23} = \text{intercept of } P - (k_{21} + k_{32})$$

Approximations. Treatment of relaxation spectra can be considerably simplified if special conditions, which allow applicability of some approximations, are met experimentally:

1. When in a coupled set of reactions *some steps are much faster than others*, this allows effective decoupling of the faster steps, and thus simplifies their expressions.

2. If one component is in great excess over the other(s) its concentration can be treated as constant during the time course of the reaction (*quasi-buffering*). This effectively decouples two steps which are coupled via a common reagent.

3. If some intermediate is in very small concentration at equilibrium, a

steady-state assumption can be introduced whereby the number of observable relaxation times is reduced by one.

The latter point illustrates one of the features of relaxation kinetics and emphasizes the differences between near equilibrium and transient kinetics (typically rapid-mixing kinetics). With the latter approach it may be possible to build up *transiently* large concentrations of an unstable intermediate, and thus analyze its dynamic (or spectroscopic) properties, a goal which cannot be achieved by relaxation kinetics.

The evaluation of the *relaxation amplitudes* for a multiple-step reaction system involves finding expressions which contain the contribution of all elementary steps to the overall amplitude, and thus relate enthalpy and optical signal changes for the elementary steps involved.

Thus for a general case of n independent elementary steps, the relaxation spectrum, expressed in terms of observed optical density changes, is a sum of exponentials of the type:

$$\delta OD_{\text{total}} = \delta OD_1^0\, e^{-t/\tau_1} + \delta OD_2^0\, e^{-t/\tau_2} + \ldots + \delta OD_n^0\, e^{-t/\tau_n} \quad (2.6)$$

A useful procedure for the analysis of individual relaxation amplitudes has been reported by Thusius[6,12] and by Ilgenfritz[7,13] and will not be considered in any more detail in this article.

3. Heme Protein Relaxation Kinetics: Some Examples

Applications of chemical relaxation kinetics to the study of heme protein reactions are presented here in order of increasing complexity. This should not be considered an attempt to provide a review of the literature on the subject, but rather as a selection of a few examples which were of special interest to the author.

3.1. Single-step Reactions

Reactions involving ligand binding to monomeric hemoprotein conform to a single-step process, as shown, for example, by the O_2 binding kinetics of ferrous myoglobin[14] and by the azide binding kinetics of ferric myoglobin.[15] Figure 2 shows that the dependence of the reciprocal relaxation time on the equilibrium concentration of the reactants is linear, according to expectations for a simple bimolecular process. The data refer to (1) Mb

[12] D. Thusius, *J. Am. Chem. Soc.* **94,** 356 (1972).
[13] G. Ilgenfritz and T. M. Schuster, *J. Biol. Chem.* **249,** 2959 (1974).
[14] M. Brunori and T. M. Schuster, *J. Biol. Chem.* **244,** 4046 (1969).
[15] D. E. Goldsack, W. S. Eberlein, and R. A. Alberty, *J. Biol. Chem.* **240,** 4312 (1965).

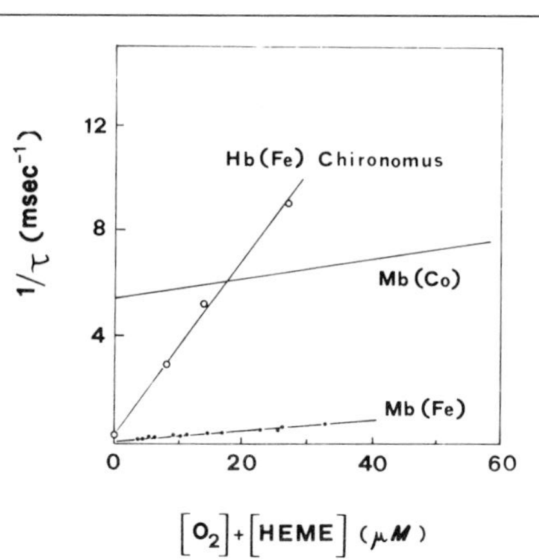

FIG. 2. Dependence of reciprocal relaxation time on the concentration of reactants (at equilibrium). Data for sperm whale Fe myoglobin[14] and *Chironomus thummi* Fe hemoglobin[16] in 0.1 *M* phosphate buffer, pH 7, ≃25°C (after jump); results for sperm whale Co myoglobin,[17] shown as a continuous line, obtained in 0.1 *M* phosphate buffer, pH 7, 10.5°C (after jump).

(from sperm whale),[14] (2) Hb (from *Chironomus thummi*),[16] and (3) Mb (from sperm whale reconstituted with cobalt (Co) containing protoheme[17]). The example clearly shows that information on the relevant rate constants, which could be evaluated from the slope (k^R) and the intercept (k^D), would be very difficult to obtain with other kinetic techniques in view of the very high rates involved.

The particular case of a polymeric protein containing equivalent and independent (noninteracting) sites is of general interest. This is illustrated here by the binding of a ligand (X) to the isolated β chains from human hemoglobins, which are tetrameric (β_4) at high protein concentration.[18] The stoichiometry of binding implies a four-step reaction according to:

$$\beta_4 + 4X \rightleftharpoons \beta_4X + 3X \rightleftharpoons \beta_4X_2 + 2X \rightleftharpoons \beta_4X_3 + X \rightleftharpoons \beta_4X_4 \quad (3.1)$$

For such a system the microscopic equilibrium constants are related by appropriate "statistical factors." Temperature-jump experiments showed that only *one* relaxation time is observable.[14] At constant ligand concentration and partial saturation, the system will continuously undergo "exchange" reactions such as:

[16] G. Amiconi, E. Antonini, M. Brunori, H. Formanek, and R. Huber, *Eur. J. Biochem.* **31**, 52 (1972).

[17] H. Yamamoto, F. J. Kayne, and T. Yonetani, *J. Biol. Chem.* **249**, 691 (1974).

[18] E. Antonini and M. Brunori, "Hemoglobin and Myoglobin in their Reactions with Ligands." North-Holland Publ., Amsterdam, 1971.

$$\beta_4 + \beta_4 X_2 \rightleftharpoons 2\beta_4 X \quad \text{or} \quad \beta_4 X + \beta_4 X_3 \rightleftharpoons 2\beta_4 X_2 \tag{3.2}$$

which are not directly observable if the change in the optical properties per site is independent of the number of sites already occupied in a tetramer. In both examples (3.2) the reactants and the products do not involve a change in C_X, and X is transferred from one species to another via the free ligand. Such an "exchange" process between identical binding sites is associated with a zero enthalpy change, and thus is not observable.

3.2. Two-step Reactions

These have been frequently reported in the course of kinetic investigations of ligand binding to biological macromolecules, which undergo conformational isomerization(s). For clarity they are presented here under two subheadings (A and B), which are special cases of a simple thermodynamic square

$$\begin{array}{ccccc} & M & \rightleftharpoons & M' & \\ & & ④ & & \\ X + & \downarrow\uparrow ① & & ③ \downarrow\uparrow & + X \\ & & ② & & \\ & MX & \rightleftharpoons & M'X & \end{array} \tag{3.3}$$

where M and M' are the two conformers of the ligand-free macromolecule, MX and $M'X$ are the two conformers of the ligand-bound form, and X is the ligand (one binding site per monomer).

3.2.*A*

A reaction which proceeds along the coordinates ① and ② in scheme (3.3) represents the case in which the conformational change occurs in the ligand-bound macromolecule.

Scheme:

$$M + X \underset{k_{21}}{\overset{k_{12}}{\rightleftharpoons}} MX \underset{k_{32}}{\overset{k_{23}}{\rightleftharpoons}} M'X$$

Definitions (see Fig. 1 for reference):

$$X_1 = C_M - \bar{C}_M = C_X - \bar{C}_X$$

$$X_2 = C_{MX} - \bar{C}_{MX}$$

$$X_3 = C_{M'X} - \bar{C}_{M'X}$$

$$X_1 + X_2 + X_3 = 0$$

Linearized differential equations (only two variables needed):

$$\dot{X}_1 = -[k_{21} + k_{12}(\bar{C}_M + \bar{C}_X)]\, X_1 - k_{21} \cdot X_3 = a_{11} \cdot X_1 + a_{12} \cdot X_3$$
$$\dot{X}_3 = -\, k_{23} \cdot X_1 - (k_{23} + k_{32}) X_3 = a_{21} \cdot X_1 + a_{22} \cdot X_3$$

Reciprocal relaxation times:

$$\frac{1}{\tau_I} = -(a_{11} + a_{22}) = k_{21} + k_{12}(\bar{C}_M + \bar{C}_X) + k_{23} + k_{32}$$

$$\frac{1}{\tau_{II}} = -\left(\frac{a_{11}a_{22} - a_{12}a_{21}}{a_{11} - a_{22}}\right) = \frac{[k_{21}k_{32} + k_{12}(\bar{C}_M + \bar{C}_X)]\,(k_{23} + k_{32})}{k_{12}(\bar{C}_M + \bar{C}_X) + k_{21} + k_{23} + k_{32}}$$

Approximations:

1. Bimolecular process faster than the monomolecular—$k_{12}(\bar{C}_M + \bar{C}_X), k_{21} \gg k_{23}, k_{32}$:

$$\frac{1}{\tau_I} = k_{21} + k_{12}(\bar{C}_M + \bar{C}_X)$$

$$\frac{1}{\tau_{II}} = k_{32} + k_{23}\frac{k_{12}(\bar{C}_M + \bar{C}_X)}{k_{12}(\bar{C}_M + \bar{C}_X) + k_{21}}$$

Under these conditions the faster relaxation time follows a simple bimolecular behavior, and the slower one displays a concentration-independent behavior only at high enough concentrations of reactants [when $k_{12}(\bar{C}_M + \bar{C}_X) \gg k_{21}$]. This situation has been frequently encountered in the relaxation kinetics of enzyme substrate or enzyme inhibitor binding.[9,19]

2. Bimolecular process slower than monomolecular one—$k_{12}(\bar{C}_M + \bar{C}_X), k_{21} \ll k_{23}, k_{32}$:

$$\frac{1}{\tau_I} = k_{23} + k_{32}$$

$$\frac{1}{\tau_{II}} = k_{12}(\bar{C}_M + \bar{C}_X) + k_{21}\frac{k_{32}}{k_{32} + k_{23}}$$

3. Intermediate at very small concentration ($C_{MX} \longrightarrow 0$), which allows one to apply the steady-state approximation: $dC_{MX}/dt = 0$. Under these conditions: $X_3 = -X_1$; therefore only one relaxation time is observed.

$$\dot{X}_1 = -\frac{[k_{21}k_{32} + k_{12}k_{23}(\bar{C}_M + \bar{C}_X)]}{k_{21} + k_{23}}X_1$$

$$\frac{1}{\tau} = k_D + k_R(\bar{C}_M + \bar{C}_X)$$

where

$$k_D = \frac{k_{21}k_{32}}{k_{21} + k_{23}} \quad \text{and} \quad k_R = \frac{k_{12}k_{23}}{k_{21} + k_{23}}$$

[19] B. H. Havsteen, *J. Biol. Chem.* **242,** 769 (1967).

EXAMPLE 3.2.1. CYTOCHROME *c* OXIDASE

The temperature-jump relaxation experiments of the kinetics of electron transfer between (mammalian) cytochrome *c* oxidase and cytochrome *c* were found to be consistent with the scheme discussed above. To circumvent the experimental problem represented by small contaminations of O_2, in view of the very high affinity of the oxidase for this ligand, the experiments were performed under pure CO which blocks cytochrome a_3 (and possibly also invisible copper Cu_i) in the reduced state.[20,21]

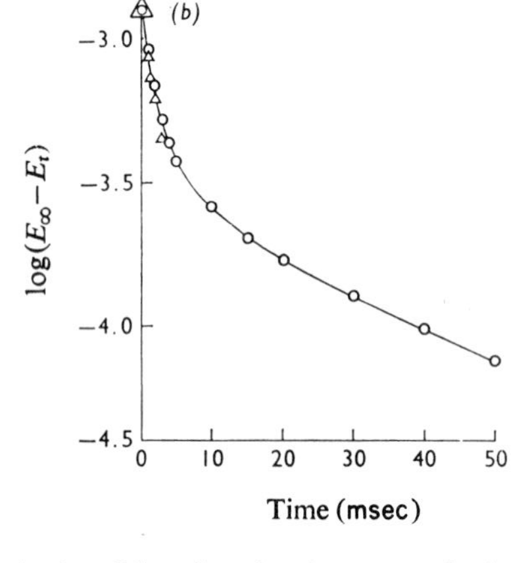

FIG. 3. Semilogarithmic plot of the relaxation time course in the reaction of cytochrome *c* oxidase (as the CO complex of the mixed-valence intermediate) and cytochrome *c*. Conditions: 0.1 *M* phosphate buffer (pH 7.4) + 1% Tween 80 and 25°C (after jump); observation wavelength 547 nm.[22]

The relaxation kinetics of an equilibrium mixture of cytochrome *c* and its oxidase followed at 547 nm (cytochrome *c* absorbance) are shown in Fig. 3. The time course of the approach to equilibrium is fitted with two exponentials, and from experiments performed at different concentrations of reagents the following scheme was proposed:[22]

$$c_r + a_o \underset{k_{21}}{\overset{k_{12}}{\rightleftharpoons}} c_o + a_r \qquad a_r \underset{k_{32}}{\overset{k_{23}}{\rightleftharpoons}} a_o' \tag{3.4}$$

[20] P. Nicholls and B. Chance, *in* "Molecular Mechanisms of Oxygen Activation," (O. Hayaishi, ed.), p. 479. Academic Press, New York, 1974.
[21] J. G. Lindsay, C. S. Owen, and D. F. Wilson, *Arch. Biochem. Biophys.* **169,** 492 (1975).
[22] C. Greenwood, T. Brittain, M. T. Wilson, and M. Brunori, *Biochem. J.* **157,** 591 (1976).

where c_o and c_r indicate oxidized and reduced cytochrome c, a_o and a_r indicate oxidized and reduced cytochrome a, and a_o' indicates a species in which cytochrome a has transferred its electron to a second site (visible copper Cu_v) intramolecularly.

The reciprocal relaxation time for the faster process, which was found to be concentration dependent, could be expressed as:

$$\frac{1}{\tau_f} = k_{12}([\bar{a}_o] + [\bar{c}_r]) + k_{21}([\bar{a}_r] + [\bar{c}_o]) + k_{23} + k_{32} \tag{3.5}$$

The equation can be rearranged into

$$\frac{\tau_f^{-1}}{[\bar{a}_o] + [\bar{c}_r]} = k_{12} + k_{21}\frac{[\bar{a}_r] + [\bar{c}_o]}{[\bar{a}_o] + [\bar{c}_r]} + \frac{k_{23} + k_{32}}{[\bar{a}_o] + [\bar{c}_r]} \tag{3.6}$$

If the third term in Eq. (3.6) is small compared to the others (i.e., applying the approximation of effective decoupling of the first step), then the data could be treated to obtain both the forward and backward rate constants for the c to a electron transfer (see legend to Fig. 4). The second relaxation time (τ_s) was found to increase with total reactant concentrations from approximately 40 sec^{-1} to approx. 100 sec^{-1}, in qualitative agreement with scheme (3.4) (which predicts a concentration-independent relaxation time for the intramolecular step only at high reactant concentrations, where τ_s^{-1} would be equal to $k_{23} + k_{32}$).

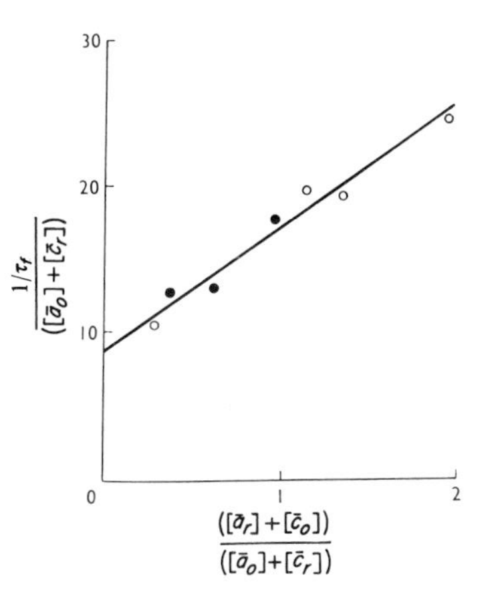

FIG. 4. Kinetics of electron transfer between cytochrome c oxidase and cytochrome c. Temperature-jump data for the faster relaxation time (obtained from relaxation kinetics as in Fig. 3), plotted according to Eq. (3.6). The rate constants for electron transfer between cytochrome c and cytochrome a are obtained from the intercept and the slope: $k_{12} = 9 \times 10^6$ M^{-1} sec^{-1} and $k_{21} = 8.5 \times 10^6$ M^{-1} sec^{-1} (see scheme 3.4 for definition of these rate constants).[22]

EXAMPLE 3.2.2. CYTOCHROME c_{551}-AZURIN

The electron transfer between cytochrome c_{551} (a hemoprotein containing 1 heme/9000 daltons) and azurin (a copper protein containing 1 Cu/16,000 daltons) from *Pseudomonas aeruginosa* has been investigated by temperature jump.[23-25]

The relaxation experiments showed the following features: (1) two clearly resolved relaxation times, with optical density changes in opposite directions, were observed; (2) the two processes have kinetic difference spectra of opposite sign, but identical shape; (3) their reciprocal relaxation times (τ_f^{-1} and τ_s^{-1}) depend on total azurin concentration as shown in Fig. 5.

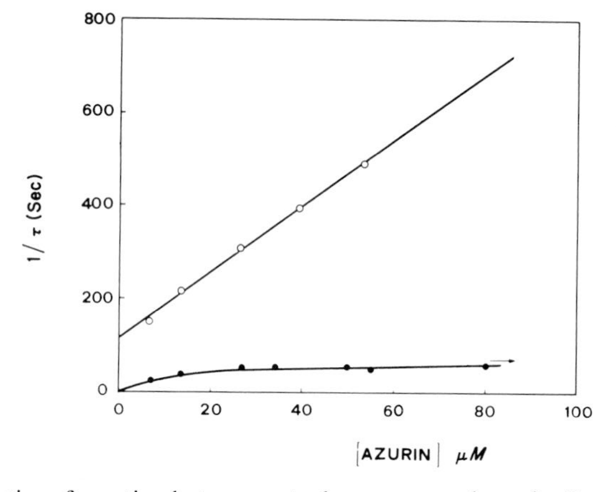

FIG. 5. Kinetics of reaction between cytochrome c_{551} and azurin. Dependence on total azurin concentration of the reciprocal relaxation time for the fast (○) and slow (●) processes. Conditions: 0.1 *M* phosphate buffer, pH 7.0, and 25°C (after jump). Solid lines are calculated (see text).[24]

The scheme which was found consistent with these results (as well as with parallel rapid-mixing experiments) is:[24]

$$
\begin{array}{ccc}
c_r + A_o \underset{k_{21}}{\overset{k_{12}}{\rightleftharpoons}} c_o + & A_r & \\
k_{23} & \downarrow\uparrow & k_{32} \\
 & A_r' &
\end{array}
\qquad (3.7)
$$

where c_r and c_o are (reduced and oxidized) cytochrome c_{551}; A_r and A_o are (reduced and oxidized) azurin; and A_r' is an inactive conformer of reduced azurin.

The reciprocal relaxation times for scheme (3.7) are:

[23] M. Brunori, C. Greenwood, and M. T. Wilson, *Biochem. J.* **137,** 113 (1974).
[24] M. T. Wilson, C. Greenwood, M. Brunori, and E. Antonini, *Biochem. J.* **145,** 449 (1975).
[25] P. Rosen and I. Pecht, *Biochemistry* **15,** 775 (1976).

$$\frac{1}{\tau_f} = k_{12}(\bar{c}_r + \bar{A}_o) + k_{21}(\bar{A}_r + \bar{c}_o) + k_{23} + k_{32}$$

$$\frac{1}{\tau_s} = \frac{\{[k_{12}(\bar{c}_r + \bar{A}_o) + k_{21}(\bar{A}_r + \bar{c}_o)]\,(k_{23} + k_{32})\} - k_{23}k_{21} \cdot \bar{c}_o}{k_{12}(\bar{c}_r + \bar{A}_o) + k_{21}(\bar{A}_r + \bar{c}_o) + k_{23} + k_{32}} \quad (3.8)$$

Observation (2) showed that only one chromophore, that corresponding to the $c_o \longleftrightarrow c_r$ transition, was involved in both relaxation phenomena (over the λ range examined). Thus the faster (concentration-dependent) process (τ_f) acts as an indicator for the slower, spectrally silent, isomerization (τ_s).

Due to the special conditions chosen for the experiment, in which A_r was added progressively (in a titration experiment) to c_o, it was possible to simplify the expressions for the two relaxation times, and fit quantitatively the relaxation data (as shown in Fig. 5), with: $k_{12} = 6.2 \times 10^6\ M^{-1}\ \text{sec}^{-1}$, $k_{21} = 3.4 \times 10^6\ M^{-1}\ \text{sec}^{-1}$, $k_{23} = k_{32} = 40\ \text{sec}^{-1}$, at 25°C and pH 7.

3.2.B.

This case involves a reaction which proceeds along the coordinates ③ and ④ in scheme (3.3), i.e., it describes the situation where the conformational change occurs in the ligand-free macromolecule.

Scheme:

$$M'X \underset{k_{21}}{\overset{k_{12}}{\rightleftharpoons}} M' + X, \qquad M' \underset{k_{32}}{\overset{k_{23}}{\rightleftharpoons}} M$$

Definitions:

$$X_1 = C_{M'X} - \bar{C}_{M'X} = -(C_X - \bar{C}_X)$$

$$X_2 = C_{M'} - \bar{C}_{M'}$$

$$X_3 = C_M - \bar{C}_M$$

Linearized differential equations (only two variables needed):

$$\dot{X}_1 = -[k_{21}(\bar{C}_{M'} + \bar{C}_X) + k_{12}]\,X_1 - k_{21} \cdot \bar{C}_X \cdot X_3 = a_{11}X_1 + a_{12}X_3$$

$$\dot{X}_3 = -k_{23} \cdot X_1 - (k_{23} + k_{32})X_3 = a_{21}X_1 + a_{22}X_3$$

Reciprocal relaxation times:

$$\frac{1}{\tau_{\text{I}}} = -(a_{11} + a_{22}) = k_{21}(\bar{C}_{M'} + \bar{C}_X) + k_{12} + k_{23} + k_{32}$$

$$\frac{1}{\tau_{\text{II}}} = -\left(\frac{a_{11}a_{22} - a_{12}a_{21}}{a_{11} + a_{22}}\right)$$

$$= \frac{k_{23}k_{21}(\bar{C}_{M'}) + k_{32}k_{21}(\bar{C}_{M'} + \bar{C}_X) + k_{23}k_{12} + k_{32}k_{12}}{k_{21}(\bar{C}_{M'} + \bar{C}_X) + k_{12} + k_{23} + k_{32}}$$

Approximations:

1. Bimolecular process faster than monomolecular—$k_{12}, k_{21}(\bar{C}_{M'} + \bar{C}_X) \gg k_{23}, k_{32}$:

$$\frac{1}{\tau_I} = k_{21}(\bar{C}_{M'} + \bar{C}_X) + k_{12}$$

$$\frac{1}{\tau_{II}} = k_{32} + k_{23}\frac{k_{12} + k_{21} \cdot \bar{C}_{M'}}{k_{12} + k_{21}(\bar{C}_{M'} + \bar{C}_X)}$$

When $C_X \to 0$, $\frac{1}{\tau_{II}} = k_{32} + k_{23}$.

2. Bimolecular slower than monomolecular—$k_{12}, k_{21}(\bar{C}_{M'} + \bar{C}_X) \ll k_{23}, k_{32}$:

$$\frac{1}{\tau_I} = k_{23} + k_{32}$$

$$\frac{1}{\tau_{II}} = k_{12} + k_{21}\frac{k_{23}\bar{C}_{M'} + k_{32}(\bar{C}_{M'} + \bar{C}_X)}{k_{23} + k_{32}}$$

EXAMPLE 3.2.3. CARBOXYMETHYL CYTOCHROME *c*

The kinetics of the proton-induced transition of a chemically modified form of cytochrome c^{2+} [i.e., carboxymethyl (CM) cyt *c*, in which met 80 is carboxymethylated and thus incapable of providing the sixth coordination bond to the heme iron] have been investigated by temperature jump.[26]

Only one pH-dependent relaxation time was observed over the range 5.8–8.4 (pK = 7.16). The data were found to conform to the following scheme:

$$\begin{gathered} HP \underset{k_{21}}{\overset{k_{12}}{\rightleftharpoons}} H + P \\ P \underset{k_{32}}{\overset{k_{23}}{\rightleftharpoons}} P' \end{gathered} \tag{3.9}$$

where HP = protonated (high-spin) CM cytochrome *c*, P = deprotonated (high-spin) CM cytochrome *c*, P' = deprotonated (low-spin) CM cytochrome *c*, and H = proton.

The proton-binding step was not observed (zero amplitude) because it is spectrally silent, and is very fast compared to the isomerization (spectrally active). With the approximation that H is buffered, the relaxation time is given by:

$$\frac{1}{\tau} = k_{32} + k_{23}\frac{k_{12}}{k_{12} + k_{21}\bar{C}_H} = k_{32} + k_{23}\frac{K_{12}}{K_{12} + \bar{C}_H} \tag{3.10}$$

[26] M. Brunori, M. T. Wilson, and E. Antonini, *J. Biol. Chem.* **247**, 6076 (1972).

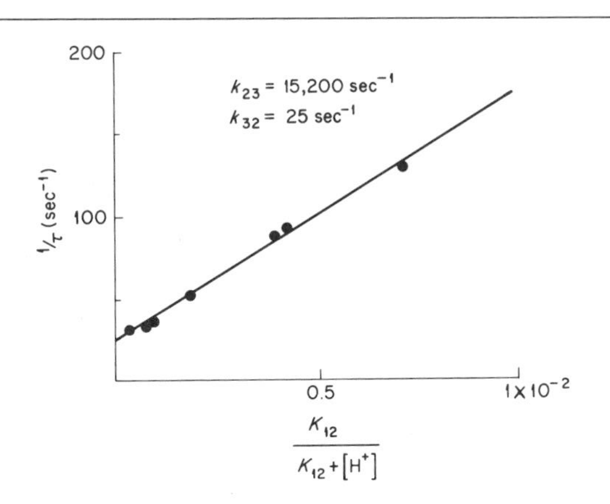

FIG. 6. Kinetics of the pH-dependent transition in carboxymethyl cytochrome *c*. The reciprocal relaxation time is reported as a function of $K_{12}/(K_{12} + H^+)$ according to Eq. (3.10), where $K_{12} = 10^{-10}$ *M*. Conditions: 0.1 *M* phosphate or 2% borate buffers and 25°C.[26] For the definition of the rate constants (k_{23} and k_{32}) see text.

The data in Fig. 6 are plotted according to this equation. They provided estimates of the rate constants for the isomerization (given in the figure) and of the intrinsic acid–base dissociation constant for the relevant group (possibly the amino group of Lys 79).

3.3. More Complex Reactions

It is difficult to provide simple and general rules to deal with the quantitative treatment of the relaxation spectrum of complex mechanisms, although for several possible combinations the expressions for the relaxation spectrum have been calculated explicitly.[4,13,27,28]

Some useful guidelines may be obtained, however, from consideration of the example reported below. From this it appears that (1) the main effort in designing the experiments should be on the selection of proper conditions to isolate, as far as possible, some individual steps, i.e., to reach limiting conditions; (2) the main problem in interpreting the data will be in the selection of the proper approximations.

Allosteric systems. Any general scheme for ligand binding to an allosteric system, such as hemoglobin, involves a number of discrete elementary steps which may be classified as: (1) binding steps (formation of a complex between ligand *X* and binding site on the macromolecule) and (2) ligand-linked conformational isomerizations (structural changes involving

[27] M. Eigen, G. Ilgenfritz, and K. Kirschner, quoted in Ref. 9.

[28] K. Kirschner, M. Eigen, R. Bittman, and B. Voigt, *Proc. Natl. Acad. Sci. U.S.A.* **56,** 1661 (1966).

the quaternary structure of the macromolecule).[9,18,29] For a protein containing *two* identical oligomers and existing only in *two* conformational states (R and T according to the nomenclature of Monod *et al.*[30]) the mechanism of ligand binding may be described by five differential equations (since there are seven concentration variables—three for R, three for T, and X—and two conservation equations).

The relaxation behavior of such a system is considerably simplified, however, by considering that "exchange" reactions between identical binding sites lead to a reduction of the observable relaxation times from 5 to 3. (For a tetrameric protein this degeneracy reduces the number from 9 to 3.)[28]

The relaxation spectrum of an allosteric enzyme with a conformational isomerization *slow* compared to substrate binding has been calculated exactly.[27] This theoretical treatment has been successfully applied by Kirschner *et al.*[28] to the description of the relaxation kinetics of NAD binding to D-glyceraldehyde-3-phosphate dehydrogenase, which appears to conform to a classical Monod's model.

In the case of the reaction of hemoglobin with ligands, however, the ligand-linked quaternary conformational change is known to be *fast* compared to binding.[18] In addition the relaxation kinetic data cannot be fitted with a simple two-states model.[13]

A summary of the relevant data for the reaction of mammalian hemoglobin (HbA) with O_2 is given here.

EXAMPLE 3.3.1. HEMOGLOBIN

Available experimental evidence indicates that every step in the sequence of binding reactions is associated in hemoglobin with an optical density change.[18] Ignoring effects due to ligand-linked dissociation into dimers, the essential findings obtained by temperature jump[13,14,31,32] may be summarized as follows.

1. Under all experimental conditions explored only two relaxation phases were observed, which indicates a high degree of degeneracy of the system and can be (partly) understood on the basis of the arguments given above.

2. The reciprocal relaxation time for the slow phase was found to be linearly dependent on reactant concentrations. Although its interpretation

[29] J. Wyman, *Q. Rev. Biophys.* **1,** 35 (1968).

[30] J. Monod, J. Wyman, and J. P. Changeux, *J. Mol. Biol.* **12,** 88 (1965).

[31] T. M. Schuster and G. Ilgenfritz, *Symmetry Func. Biol. Syst. Macromol. Level, Proc. Nobel Symp., 11th, 1968* p. 181 (1969).

[32] M. Brunori and E. Antonini, *J. Biol. Chem.* **247,** 4305 (1972).

is difficult, because it is the slowest step in a coupled system, this finding proved that monomolecular step(s) related to conformational transitions must be fast compared to binding,[14] in agreement with independent data on the system.[18]

3. At very high fractional saturation the reciprocal relaxation time for the faster phase increases linearly with ligand concentration ($[O_2] \gg$ [heme]), indicating that it reflects an oxygen-binding step.[13,31] Temperature-jump experiments were performed on equilibrium mixtures of HbO_2 + HbCO. Taking advantage of the very high affinity of hemoglobin for CO, the experimental conditions could be adjusted to observe essentially one relaxation time, which was assigned to the reaction:

$$Hb_4(CO)_3 + O_2 \rightleftharpoons Hb_4(CO)_3(O_2) \tag{3.11}$$

Figure 7 depicts the O_2 concentration dependence of the reciprocal relaxation time for this process.[32]

Observation (3) allows one to conclude that it is possible to select experimental conditions in which the relaxation of the last binding step is observed in isolation, and to determine the relevant rate constants from the expression:

$$\frac{1}{\tau} = k_4^D + k_4^R\,(\overline{[Hb_4X_3]} + [\bar{X}]) \simeq k_4^R \cdot [\bar{X}] \tag{3.12}$$

where k_4^D and k_4^R are the dissociation and recombination rate constants for the fourth step, and $k_4^R = 4 \times 10^7\, M^{-1}\, sec^{-1}$ or $4.8 \times 10^7\, M^{-1}\, sec^{-1}$ at 25°.[13,32]

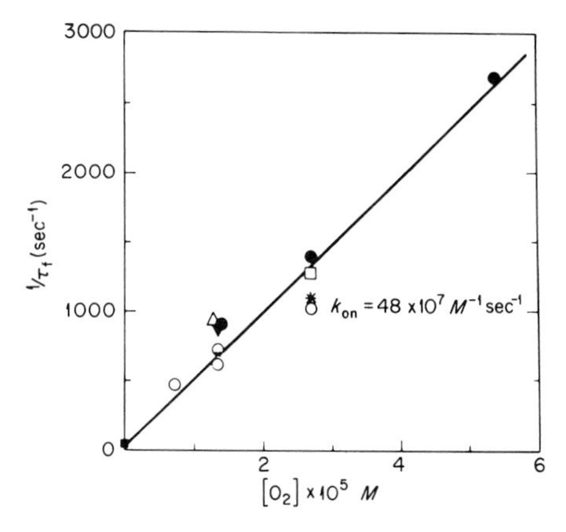

FIG. 7. Kinetics of the reaction of (HbO_2 + HbCO) mixtures with oxygen. Dependence of the reciprocal relaxation time on O_2 concentration (with $[O_2] \gg$ [free sites]). Different symbols indicate different total protein concentrations (from 1.5–124 μM) and different percentages of HbO_2 in the mixture (from 10–30%). Conditions: 0.2 M phosphate buffer, pH7, and 25°C (after jump) (from Ref. 32).

4. At very low fractional saturation (below 5%), the relative amplitude of the faster phase becomes larger, and its reciprocal relaxation time increases linearly with total protein concentration.[13] This shows that under these conditions the first step may be observed in isolation:

$$\mathrm{Hb}_4 + X \rightleftharpoons \mathrm{Hb}_4X_1 \tag{3.12}$$

and, being taken as uncoupled, its reciprocal relaxation time is given by:

$$\frac{1}{\tau} = k_1^D + k_1^R\,([\overline{\mathrm{Hb}}_4] + [\bar{X}]) \simeq k_1^D + \mathrm{k}_1^R\,[\overline{\mathrm{Hb}}_4] \tag{3.13}$$

The corresponding values are: $k_4^D \simeq 700\ \mathrm{sec}^{-1}$ and $k_4^R = 4 \times 10^7 M^{-1}\,\mathrm{sec}^{-1}$.

The results show that three kinetically distinct processes, all reflecting O_2 binding, can be distinguished. Over the range of ligand saturation and protein concentrations explored with the temperature-jump method, relaxation times from less than 100 sec^{-1} to more than 10,000 sec^{-1} were observed.

In their complete analysis of this system Ilgenfritz and Schuster[13] indicate the methodology to deal with different possible allosteric mechanisms in hemoglobin–O_2 reaction. The scheme they propose as consistent with the relaxation kinetics may be summarized as follows:

$$\mathrm{Hb}_4 + 4X \underset{\text{fast}}{\rightleftharpoons} \mathrm{Hb}_4X_1 + 3X \underset{\text{slow}}{\rightleftharpoons} \mathrm{Hb}_4X_3 + X \underset{\text{fast}}{\rightleftharpoons} \mathrm{Hb}_4X_4 \tag{3.14}$$

The slow relaxation time reflects O_2 binding to the second and third site in a tetramer, the second-order rate constant being rate limiting under these conditions. The value which could be obtained from the relaxation kinetics is very close to the overall second-order rate constant estimated by flow.[18,33]

The lesson to be learned from the available relaxation experiments on hemoglobin is that adjustment of the experimental conditions to isolate individual steps is an essential element in any attempt to arrive at a description of the reaction mechanism for a complex allosteric protein.

[33] R. L. Berger, E. Antonini, M. Brunori, J. Wyman, and A. Rossi Fanelli, *J. Biol. Chem.* **242,** 4841 (1967).

[7] Freeze-Quench and Chemical-Quench Techniques

By DAVID P. BALLOU

Introduction

Rapid reaction techniques are essential to the study of chemical mechanisms of enzyme reactions. The application of electron paramagnetic resonance (EPR) is particularly useful in studies of enzymes involved in oxidations since information about oxidation states and environmental features of paramagnetic species can be obtained. The combined use of rapid-reaction and EPR techniques can therefore be extremely valuable to the determination of both the kinetics and the nature of intermediates involved in oxidation-reduction reactions. However, EPR techniques have special requirements: (1) detection with practical sensitivity usually requires relatively long times of observation and possibly repetitive scanning techniques; (2) many species can only be measured at low temperatures; (3) often the spectra should be examined at several incident power levels, modulation widths, and temperatures to fully characterize the species; and (4) the lower sensitivity of EPR relative to optical techniques usually necessitates the use of higher concentrations of reactants making these detailed studies impractical. Thus the usual rapid-flow techniques are not directly applicable.

However, the freeze-quench technique first developed by Bray[1,2] evades these problems. This method utilizes a pulsed-flow version of a Roughton–Millikan continuous-flow system[3] in which a length of tubing of known volume (the reaction tube) couples the mixing chamber to a nozzle which sprays the reaction mixture into a cryogenic liquid (usually isopentane maintained at approximately −140°C). This cold isopentane which is immiscible with the aqueous reactants quenches the reaction rapidly (ca. 5 msec) and leaves a frozen suspension which is then packed into a suitable EPR tube for subsequent measurement. The "reaction age" of the frozen suspension is determined by the flow rate, the volume between the mixer and the nozzle, and the time required for quenching the reaction. The kinetics of a reaction are determined by collecting a series of samples, each produced with a different reaction tube. The freeze-quenched samples can then be measured by all of the low-temperature techniques usually applied to EPR spectroscopy. Such samples may also be used for Mössbauer or reflectance spectroscopy or magnetic susceptibility measurements.

[1] R. C. Bray, *Biochem. J.* **81,** 189 (1961).

[2] R. C. Bray and R. Peterson, *Biochem. J.* **81,** 194 (1961).

[3] F. J. W. Roughton and G. A. Millikan, *Proc. R. Soc. London, A Ser.* **155,** 258 (1936).

ISBN 0-12-181954-X

Design Criteria

Since the important criteria for pulsed-flow and chemical-quench methods have been discussed extensively by Gutfreund in this series,[4] only a few of the important parameters that particularly relate to freeze quenching will be discussed here. The high concentrations of sample used demand the collection of most, if not all, of the mixed sample; to insure that this entire sample is of uniform "reaction age" it must be produced with flow of constant velocity. This implies that the flow system will be accelerated very rapidly to the desired constant velocity, maintained for the required time, and then stopped abruptly.

The low flow rates (ca. 0.61 ml/sec), which are necessary for both conserving sample at longer reaction times and circumventing the problems encountered with sample packing, put certain constraints on the designs of mixers. Although mixing within times shorter than 3 msec can usually be accomplished easily with four-jet tangential mixers, the low flow velocities used in this method demand the use of mixers with very small bore channels to attain the turbulent conditions required for efficient mixing.

The suspension which is produced by freeze quenching must be of such a nature that the particles will sediment to the bottom of the vessel where they can be packed firmly, yet be of such a size that the heat transfer which leads to quenching is as rapid as possible. This involves the careful selection of flow velocity and nozzle diameter.

Apparatus and Methods

Drive System

Three types of apparatus have been used extensively for the freeze-quench technique. All are of similar flow and quench design with the main difference being the method of driving the syringes. The original Bray device[1,2] employs a hydraulic ram. This apparatus has the convenience of being compact, thus making it possible to be used for experiments which must be performed at special locations (e.g., in combination with pulse radiolysis methods[5]). A more commonly used method employs a powerful, precision electric motor controlled by an electronic feedback system. The angular motion of the motor is converted into linear motion for driving the syringes by a screw assembly. This type of device is available commer-

[4] H. Gutfreund, Vol. 16, p. 229.

[5] R. Nilsson, F. M. Pick, R. C. Bray, and M. Fielden, *Acta Chem. Scand.* **23,** 2554–2556 (1969).

cially[6] and although it is the most costly, it is also the most versatile and easiest to use. For example, in the interrupted-flow or "push-push" mode, an initial period of flow mixes the reactants and flows the mixture into a reaction tube. Flow is halted for a predefined delay period and then is re-initiated to spray it into the quenching solution. This mode allows the production of samples which are of relatively long "reaction ages" without the consumption of large volumes of material, yet which are quenched under identical conditions as are the samples of shorter "reaction ages." A third type of drive (Fig. 1) will be described below.

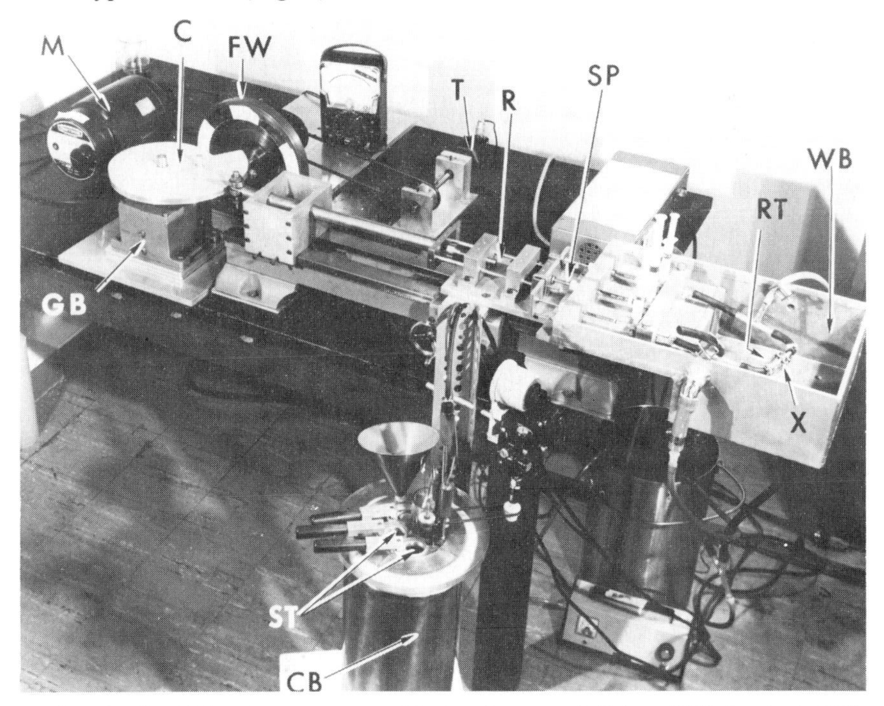

FIG. 1. Rapid-freezing apparatus. (M) electric motor (1/2 hp), (FW) flywheel, (DS) driveshaft, (GB) 100:1 gearbox, (C) linear motion cam, (R) syringe ram, (SP) syringe plungers, (X) mixer, (RT) reaction tubes, (ST) EPR sampling tubes, (CB) cold bath, (WB) aluminum water bath, and (T) tachometer. Reproduced from D. P. Ballou and G. Palmer, *Anal. Chem.* **46,** 1248 (1974), with permission.

An electric motor (M) drives a flywheel (FW) to 500–2400 rpm via a belt and pulley drive. The flywheel is an instant source of a large reservoir of energy and obviates the need for an expensive motor. A gearbox (GB) which reduces the rotational velocity from the flywheel by a factor of 100:1 drives a spiral cam (C), thus converting the angular motion of the motor to a

[6] Update Instrument, Inc., P. O. Box 5177, Madison, Wisconsin 53705.

linear displacement of the syringe ram (R) and the syringe plungers (SP). The reactants are forced through tubing to the mixer (X), the reaction tube (RT), and the nozzle which is mounted in the bottom of the water bath (WB). The mixture is thus ejected as a fine stream into cold isopentane contained in special EPR tubes which are held directly under and within 1 cm of the nozzle. The entire flow system of syringes, connecting hoses, mixers, and reaction tubes is maintained at constant temperature in the aluminum water bath. Details for the construction of this device have been published.[7,8] This driving system is of high performance and is relatively inexpensive to construct, but does not have quite the versatility of the commercial instrument. However, the push-push mode can be achieved by utilizing a cam that is made of two separate, but continuous, equal sections of the spiral. The first section fills the reaction with the volume needed for measurement (ca. 0.3 ml) and the second portion of the cam, which can be offset angularly from the first by an appropriate amount for the desired time delay, drives the sample out and through the nozzles.

Flow System

Three driving syringes are usually included in freeze-quench instruments. This permits one to react two reagents for a prescribed period of time in the reaction tube and then add either a second reactant before freeze quenching or a chemical quenching reagent. Various combinations of Hamilton gas-tight syringes of the size range 0.25–2.5 ml can be used quite successfully over the temperature range of 2–50°C without modification. A recent design change to the 1.0- and 2.5-ml syringes incorporates the combination of an O-ring with a Teflon sleeve. This imparts automatic temperature compensation as well as chemical inertness to the plungers. For EPR work it is often useful to use a 2.5-ml syringe for the protein and 0.5-ml syringe(s) for the other reagent(s). This minimizes dilution of the paramagnetic species which is especially helpful when concentration of the reagents past certain limits is impractical. Hamilton three-way valves are used to introduce reactants from tonometers or storage syringes to the driving syringes.

To minimize contamination from paramagnetic impurities, the tubing used for connecting the mixers, syringes, and nozzles has usually been plastic hose (either polyethylene or nylon). With the advent of HPLC several new types of tubing and connectors have become available. Teflon tubing and fittings such as sold by LDC/Cheminert and Altex are quite

[7] D. P. Ballou, Ph.D. Thesis, University of Michigan, Ann Arbor (1971) (University Microfilms, No. 72-14796. Ann Arbor, Michigan).

[8] D. P. Ballou and G. Palmer, *Anal. Chem.* **46,** 1248 (1974).

suitable and easy to use. Since these types of tubing are permeable to oxygen, anaerobic work requires sheathing this tubing with tubing of larger diameter and passing either anaerobic gas or a solution containing an oxygen scavenger (such as dithionite) through the sheath. However, small-bore stainless-steel and aluminum tubing (1/16 and 1/8-inch OD) which is coated on the inside with either Teflon or glass has recently become available from Alltech Associates. This tubing can be interconnected with standard high-pressure liquid chromatography (HPLC) fittings such as those distributed by Swagelok. Anaerobic experiments should be considerably more tractable with this tubing. Furthermore, the entire fluid system will be noncompressible, yet have sufficient flexibility for easy manipulation (the plastic tubing has a tendency to expand slightly with the fluid pressures which exist with freeze quenching and this is a source of timing uncertainty).

The mixers used most successfully for this technique are four-jet tangential mixers such as described elsewhere[9] or in Fig. 2. The latter mixer has jets which impinge tangentially under turbulent conditions (Reynolds number greater than 2000). An important addition to this design is gasket 2 which has a small hole that produces further mixing. These mixers are easy to construct, have a dead volume of only 1–2 μl, and can be disassembled for cleaning. Tests have shown them to effect more than 95% mixing within 3 msec under a variety of conditions of flow rate, viscosity, and mixing ratios which are appropriate for freeze quenching. In comparison, the commonly employed Gibson mixer[10] (which was not designed for these conditions) is several-fold less efficient at these low flow velocities.

The nozzle is usually made from plexiglass with small holes (1 or 2) approximately 0.2 mm in diameter and 2 mm long.[9] The packing of the sample depends upon the jet size, the flow rate, and even on the properties of the solution. Larger diameter jets yield larger particles which are usually easier to pack, but the freezing time is likely to be longer. Selection of optimum flow rates and jet sizes must be done empirically if the shortest possible quenching times are to be attained. However, a flow rate of approximately 0.7 ml/sec utilizing a jet size of 0.2 mm diameter will usually yield particles which can be packed reasonably well and will have been quenched in approximately 5 msec.

Quenching System

The cooling bath consists of a metal can containing isopentane, placed in a metal dewar, and cooled with liquid nitrogen to about −142°C. A

[9] G. Palmer and H. Beinert, *in* "Rapid Mixing and Sampling Techniques in Biochemistry" (B. Chance *et al.*, eds.), p. 205. Academic Press, New York, 1964.

[10] Q. H. Gibson and L. Milnes, *Biochem. J.* **91,** 161 (1964).

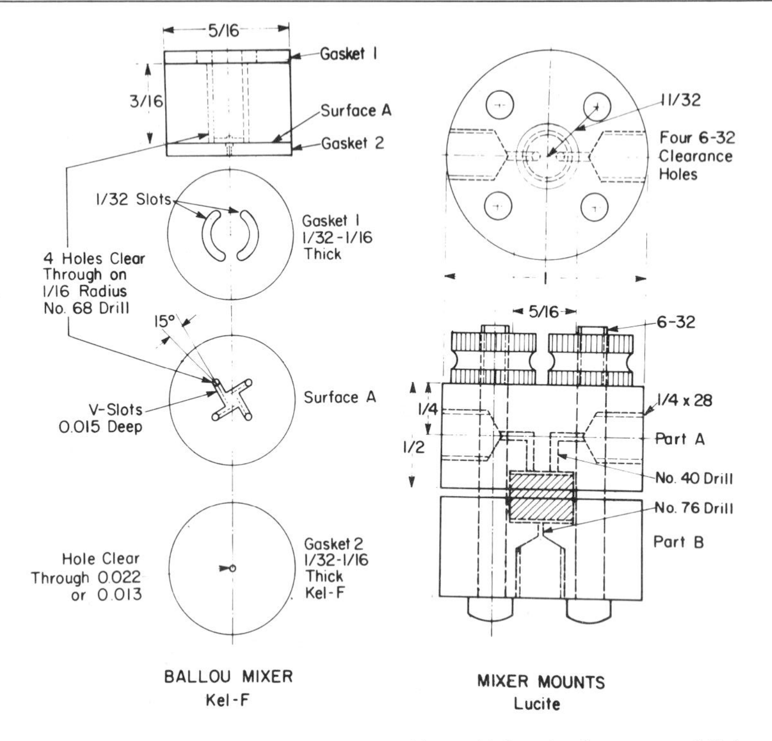

FIG. 2. Rapid-flow mixers and mounts used in rapid-freezing instrument. Mixing takes place on surface A and in gasket 2. Gasket 1 serves to split each of the two solutions into two streams, thus providing for the four-jet tangential mixer. Reproduced from D. P. Ballou and G. Palmer, *Anal. Chem.* **46,** 1248 (1974), with permission.

temperature regulator maintains constant temperature and prevents the isopentane from freezing. A lucite cover holds three sample tubes, provides support for a stirrer, and minimizes the problem of frost contaminating the samples. The sample tubes (Fig. 3) contain cold isopentane for quenching the reaction and are modified from those of Bray and Peterson[2] by fusing two short sections of Vycor tubing (ca. 10 mm diameter) to the top and bottom pieces respectively. The two sections are wetted with glycerol and fastened with a short section of latex tubing and, upon cooling, this assembly becomes rigid. After the sample has been collected and packed at −142°C, the sample is immersed in liquid nitrogen and the isopentane is removed by pouring the main portion off and aspirating the remainder with a syringe needle. The top section is then removed, and the sample can be conveniently stored in liquid nitrogen for subsequent analysis.

The sample is packed firmly into the bottom of the EPR tube with a packer constructed from a stainless-steel rod (ca. 30 cm long, 1.5 mm

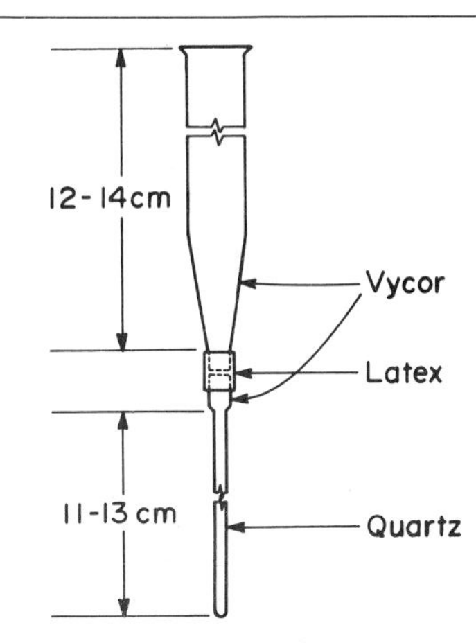

FIG. 3. EPR tubes used for collecting and measuring samples. The bottom section containing packed sample is removed for storing and for determining the EPR spectra. Reproduced from D. P. Ballou and G. Palmer, *Anal. Chem.* **46,** 1248 (1974), with permission.

diameter) with a Teflon tip approximately 0.2 mm smaller than the 3-mm EPR tube. Packing is the most difficult and irreproducible operation of the method and must be practiced for satisfactory results. Even with practice, however, these uncertainties will amount to 5–10%. Another variable is that all samples do not pack to the same density, either because of the intrinsic properties of the samples or because of operator variability. Although the packed density is usually about 60% of the density of normally frozen solutions, this variable must be checked for each type of sample to be measured by measuring a known amount produced by both rapid freezing and normal freezing.

Evaluation

A useful reaction for testing the procedure is that of metmyoglobin with azide since it can be followed by both stopped-flow spectrophotometry and by freeze quenching, the reagents are readily obtainable, and it can be run under pseudo first-order conditions. Figure 4 shows typical experiments at two different azide concentrations. The quenching time can be obtained either by observing the point of intersection of the two lines or from the intersection with the "zero time" shots (i.e., no azide). It can be seen that both methods agree within the uncertainties associated with the method and that the quenching time is approximately 5 msec.

The major problems associated with the freeze-quench technique are: (1) the irreproducibility of packing; (2) uncertainties about artifacts due to

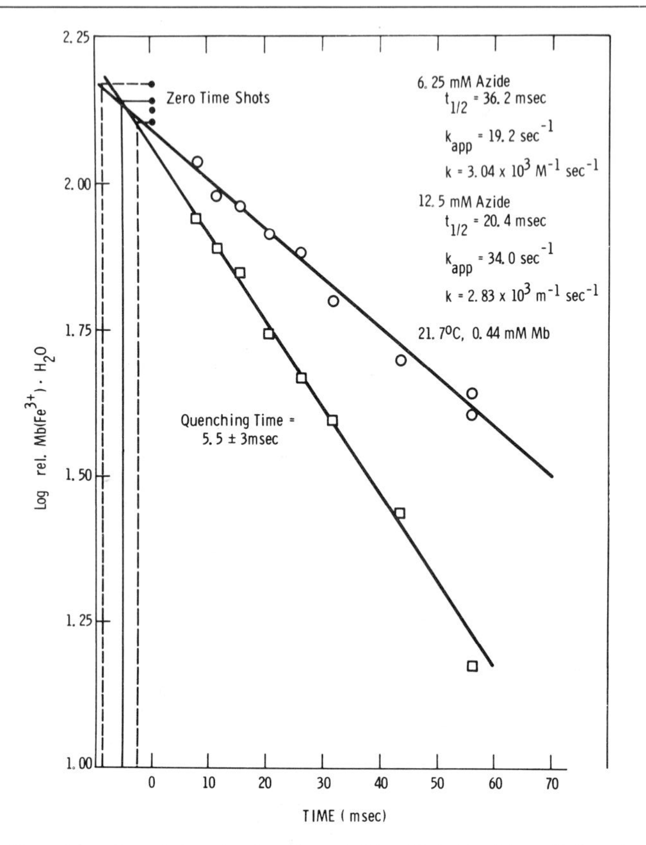

FIG. 4. Quenching-time determination ○ = 6.25 m*M* Azide; □ = 12.5 m*M* Azide. Mb, equine metmyoglobin hydrate. Vertical scale is logarithm of relative size of the high-spin EPR signal elicited by Mb. Horizontal scale is time from mixer to nozzles calculated from flow velocity and reaction-tube volumes. Reproduced from D. P. Ballou and G. Palmer, *Anal. Chem.* **46,** 1248 (1974), with permission.

the freezing process (e.g., some copper-containing samples have slightly different line shapes when frozen normally than when frozen rapidly); (3) uncertainties of whether the reactions may continue to proceed in the frozen state; and (4) obtaining anaerobic conditions during the quenching process. Oxygen is far more soluble in cold isopentane, and its removal in any convenient way is impossible. Attempts have been made with some success by bubbling the isopentane with nitrogen,[11] although even with this method some enzymes such as cytochrome oxidase cannot be maintained fully reduced during the freezing.

[11] R. C. Bray, D. J. Lowe, C. Capeillére-Blandin, and E. M. Fielden, *Biochem. Soc. Trans.* **1,** 1067 (1973).

Nevertheless, the time resolution (5–6 msec) is in many cases sufficient to make the method complementary to the stopped-flow technique. Furthermore, the types of information retrievable from EPR by resolution of the complexities in signal shapes, microwave power saturation behavior, and temperature dependence cannot be obtained by any other means.

Freeze quenching has therefore been most useful in studies of such systems as xanthine oxidase, hemoglobin, nitrogenase, laccase, ferredoxins, superoxide dismutase, and complexes of the respiratory chain. In all cases, to avoid artifacts, interpretation of data must be made with caution and comparisons with stopped-flow and other methodologies are mandatory.

The basic methodology can be used without modification for chemical-quenching studies where a third syringe and second mixer are used to introduce the quenching reagent which is usually acid, base, or reagents such as SDS. This method produces a sample which can be analyzed by a large variety of methods. For example, radioactivity can be measured to analyze incorporation or release of tritium, or product formation can be measured by HPLC methods. The chemical-quenching technique is especially useful in studying reaction mechanisms when the kinetics of an intermediate which have been observed by stopped-flow or freeze-quench methods can be correlated with some particular transformation of a substrate.

[8] Flash Photolysis Techniques

By QUENTIN GIBSON

Introduction

The aim of presenting a working description of apparatus and procedures used for flash photolysis as applied to enzymology appears scarcely feasible. There is no accepted set of equipment which is widely used—rather, each worker has put together an assembly of units as dictated by cost, availability, and the nature of the problem, and no two of these apparatus are alike. In the author's laboratory at least 10 distinctly different apparatus have been set up during the last 20 years with various light sources, geometries, and data-recording systems, each adapted to a specific problem. An attempt is made here to describe the classes of subunits available, and to discuss the principles governing their choice for specific ends.

ISBN 0-12-181954-X

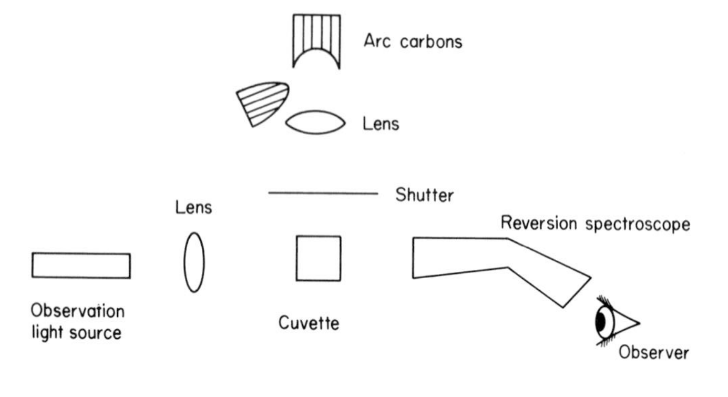

FIG. 1. Schematic drawing of Hartridge and Roughton's flash photolysis experiment. The source of the flash was the negative crater of an arc lamp. Light could be admitted to the cuvette containing the photosensitive system by means of the shutter. In the light the equilibrium $HbCO + O_2 \rightleftarrows HbO_2 + CO$ is displaced in favor of HbO_2. On cutting the photolysis light off suddenly with the shutter the dark equilibrium was reestablished. The reaction was followed visually with the reversion spectroscope, an instrument which presents two spectra to the observer, one above the other, with the order of appearance of the colors reversed. The position of an absorption band could be rather precisely determined by shifting the spectra until a continuous line appeared to cross both of them. The total difference in position was about 6.5 nm. This method of observation, which permitted the experiment, also meant that the method could not be generally used.

General Principle

The essential feature of flash photolysis is that light is used to displace an equilibrium or to perturb a steady state in a time which is short when compared with the dark relaxation time. The invention of the method and its first biochemical application are due to Hartridge and Roughton[1] who studied the replacement of oxygen by carbon monoxide when these ligands are bound to hemoglobin. In their experiments, performed more than 50 years ago, a solution of carbon monoxide hemoglobin in the presence of dissolved oxygen was exposed to the light from a carbon arc lamp (Fig. 1). As the quantum yield for the photodissociation of CO is perhaps 20 times that for O_2, CO is replaced by O_2 in the light, and in the dark the original equilibrium is reestablished in a reaction which could be followed using Hartridge's ingenious reversion spectroscope. Unfortunately, the idea of Hartridge and Roughton was a full generation ahead of the technology needed to give it practical effect. The photolysis light was simply not bright enough, and the visual method of observation was both highly specific and very slow, restricting work to relaxation rates of the order of 0.1/sec. As a

[1] H. Hartridge and F. J. W. Roughton, *Proc. Soc. London, Ser. B* **94,** 336–367 (1923).

result, their work seems to have been without influence on later developments, and no further use of the method was made by the original authors. In the following 30 years the technical base needed to give effect to the original idea appeared in the forms of the electronic flash discharge lamp, the photomultiplier, and the cathode ray oscillograph.

Light Sources

It is convenient to express the brightness of a source in terms of the first-order rate constant for photodissociation of a complex exposed to it, although this is a quantity which depends very much both on the photosensitive system and on the physical layout of the apparatus. The numbers given are therefore almost arbitrary: they refer to carbon monoxide hemoglobin. Conversion to another system would require calculation of the overlap integrals for the flash and the absorption spectra as well as a knowledge of their quantum yields. With these severe restrictions in mind the comparison would run as follows: carbon arc 10/sec, xenon arc 50/sec, medium-voltage (3 kV) photographic flash discharge tube 50,000/sec, high-voltage (20 kV) flash discharge tube 500,000/sec. The enormous advantage of the transient over the continuous sources is obvious. Continuing along the same lines, a dye laser using coumarin 6 G might achieve 10^9/sec for a suitably restricted volume of solution, and a Q-switched frequency-doubled neodymium glass laser a rate of perhaps 10 times more still, though its utility would be restricted by the small numbers of photons available in the pulse.

Although the problem of delivering quanta quickly enough has been solved, in most cases this is not the end of the problem. The light pulse must end abruptly as well if a reversible system is being studied. Suppose, for example, the system COHb $\rightarrow$ CO + Hb is being examined with a concentration of CO sufficient to give a dark recombination rate of 10/sec. With a flash giving a peak dissociation rate of 20,000/sec and assumed exponential, if it is required that the light reaction drop to 1% of the dark recombination rate before measurements of the recombination rate are accepted as reliable, 18 half-lives must elapse, or about 3.5 msec for a flash of 200 μsec half-life. For this reason there was a trend towards the use of high voltages giving shorter flashes with half-lives of 10 μsec or less, permitting observation to start in 0.1 msec or so. The use of voltages above about 5 kV, however, is associated with a high price in terms of loss of reliability, short life of the flash lamps, and increasing difficulty in designing and constructing safe equipment. The difficulty of providing secure containers for high-voltage capacitors while keeping leads short is such that in my laboratory their terminals were fully exposed and the apparatus as a whole was treated as highly dangerous. Each capacitor was *always* discharged individually with a shorting stick immediately before working on any part of the equipment. This rule was followed irrespective of the wiring arrangements, since, though capacitors may have been wired in parallel, it should never be assumed that they are still so wired, since the large currents cause failure at junctions where there is even a trivial contact resistance. The requirement for immediacy is due to the general property of high-voltage capacitors, which have been charged for some time, of reacquiring quite a healthy charge on standing on open circuit after discharge.

A further difficulty with high-voltage systems is that the discharge consists of several exponentials. The initial flash is succeeded by a glow with perhaps 2% of the initial intensity and a decay time of the order of 100 μsec. A very much weaker glow with a decay time of seconds follows, and is due to phosphorescence in the walls of the discharge tube. Phosphorescence will seldom give any trouble; indeed, the author only noticed it when attempting to induce bioluminescence in a bacterial protein system. The problem was overcome only by separating the observation and irradiation volumes by transferring the solution to the observation vessel via a flow system after irradiation.

Quite recently photographic flash apparatus have appeared commercially in which the flash is interrupted by a solid-state switching device—a thyristor. This is applied, typically, to a low-voltage (say 200 V) apparatus and may operate after a specified proportion of the total energy of the discharge has been delivered or by means of an integrating light sensor. If the light available is sufficient such an apparatus is cheap, reliable, and convenient, but the intensity is necessarily low as compared with high-voltage apparatus.

Although most of the components required for a high-voltage system are available cheaply as obsolete radio transmitting equipment, it is recommended as safer not to try to build power supplies unless you are specially experienced.

More recently, lasers have come to the fore as sources of light for flash photolysis; at first these used ruby or neodymium rods giving energy in the red or near infrared respectively. These are wavelengths of little use in many systems because they are not absorbed sufficiently strongly. Frequency doublers were available, but they were, typically, of low efficiency, and the final output energy was insufficient for many purposes. In all other respects they approached the ideal for a flash photolysis source. The pulse was short (tens of nanoseconds), and died away abruptly, while the energy appeared in the form of a slightly divergent beam which could be precisely handled over distances of meters using suitable lenses. These advantages have carried over to the dye laser in which the ruby rod is replaced by a tube containing a solution of a strongly fluorescent dye. This emits laser light at about the wavelength of fluorescence in a pulse typically 0.5 μsec long with a steeper-than-exponential cutoff. The energy output depends on many factors, but may reach a joule—ample for most purposes. It would seem, at first sight, as though many dyes could be used and that the output wavelength could be tailored to the system being examined. In practice, this is not so because of the requirement that the dye be able to cycle repeatedly between the ground and singlet excited states without significant population of other absorbing species. Should other absorbing states be significantly populated the resulting loss of energy will either prevent laser action, or permit a short flash of low energy.

This is clearly a time-dependent effect. If quanta can be delivered rapidly enough to the laser tube so that lasing occurs promptly, there will be less opportunity for population of inhibitory forms. More dyes can be made to lase, and higher efficiencies obtained using the standard dyes. Unhappily, the requirement that the pumping flash start quickly can be met only by using high voltages of 20 kV or more, and the equally important requirement that the flash energy be transferred efficiently to the dye tube requires either a coaxial combined lamp and dye tube, or an elliptical cavity with lamp and tube at the foci. As a result it is scarcely practical to construct an efficient dye laser in a biochemical laboratory, and the successful operation of a commercial instrument requires considerable care. The dangers attending the use of high-power lasers are so well known that it is scarcely necessary to repeat the warnings about the occurrence of retinal damage.

Data Recording

The second main failing of Hartridge and Roughton's apparatus, after the inadequacy of the photolysis light, was the slowness of the method of observation which required visual matching of the position of absorption bands. At that time (1920) it is just possible that a team of physicists and engineers could have been assembled to apply photoelectric cells, vacuum-tube amplifiers, and cathode-ray tubes to follow the reactions. When the problem was again attacked the instruments required were all readily available, though not at first applied by physical chemists, who were interested both in the spectrum of the immediate photoproducts and in events within the lifetime of the photolysis flash. They therefore used a spectrograph and a second smaller electronic flash fired at a known interval after the main flash to record the absorption spectrum of the contents of the photolysis tube.

More recently, a photomultiplier with a cathode follower, or an operational amplifier in the current-to-voltage mode, has been used to drive a suitable cathode-ray oscillograph. The only serious problem is that the photoflash may overload the amplifier, delaying the response until it recovers. Early workers used mechanical shutters, usually rotating disks, but ordinary Compur-type camera shutters could probably be applied. A special problem remains where a slow relaxation of small amplitude is to be recorded, calling for a long time constant in the system. The flash, however, is short, and the recording device will not have time to reach an appropriate zero position before the relaxation process begins. It would be appropriate to use two different time constants switched by suitable logic, but this has not been done (so far as the author knows). Photodiode–amplifier combinations are becoming worth consideration as detectors. They are mechanically rugged and are available in an increasing range of types. So far, their quantum efficiency has been below that of a photomultiplier, with correspondingly bad effects of the signal/noise ratio. If, however, only large absorbance excursions are to be measured, their small size and good mechanical characters might make the photodiode the best choice.

The next unit to be considered in assembling a system is the observing light source. This has usually been a tungsten filament because of simplicity, steadiness, and reliability. The drawback is low surface brightness so that a considerable area (1 cm^2) of cell must be illuminated to get a high enough photon flux for a good signal/noise ratio. Less steady, but considerably brighter sources are zirconium arcs (Sylvania) and xenon arcs. The Zr arc is a circular source which for 100 W is 0.072″ in diameter. It can be imaged on the cuvette, and again at the monochromator slit, so that much smaller areas may be used. With highly photosensitive systems, substantial effects may be produced by the observation beam, corresponding to dissociation velocities of up to 20/sec for the unscreened source; therefore an appropriate glass filter, or better, an interference filter, should be put between the source and the cuvette. If the geometry of the system is appropriate, a tungsten lamp is probably the first choice as a light source.

The geometry of the system is both neglected and important. In the earlier biological experiments an integrating sphere was used to contain the flash tube and cuvette with the idea of providing uniform illumination of all four surfaces of a square vessel. This arrangement is difficult for providing temperature control, which requires Lucite jackets. With rectangular cuvettes using a short light path but large area (as for a tungsten lamp), the light must pass through the temperature-control fluid, an arrangement which works very well if a bubble trap is included in the system. It is then convenient to circulate a fixed volume of liquid through an otherwise closed system including a heat exchanger.

The integrating-sphere system does not lend itself well to the use of complementary filters since the convenient Corning glass series are available as 2″ squares. With a commercial photoflash unit the protective bulb may be enclosed in a small beaker and the space between bulb and beaker filled with a filter solution, the whole going into the integrating sphere as usual. There is a good deal of magic about integrating spheres, and traditional coatings such as magnesium oxide may be applied by burning ribbons of the metal. My impression is that with irregularly shaped vessels, the nature of the surface does not make a great deal of difference, and a lining of aluminum foil is easily replaced and is as good as anything. This would not be true of an elliptical cavity, but apart from dye lasers, these do not seem to have been used. An advantage of the integrating-sphere approach is that any convenient shape of cuvette may be used at will to contain the solution.

If a long (10-cm) path is desirable the cuvette may pass right through the light box and be surrounded by a jacket which will serve at the same time for temperature regulation and to contain a filter solution.

The use of laser light poses an entirely new set of considerations which affect the choice of observing light source and of experimental geometry. The basic difference is that the laser provides less total energy by a factor of perhaps 20, but it is concentrated both in wavelength range and by virtue of its near parallelism. The layout of the apparatus should allow for the illumination of a small area of the cuvette, and since it is only possible to illuminate from one face, the depth must be small also. Indeed this problem, and a need for large light input, has led at least one worker to use two lasers, one on each side of the cell.

With highly photosensitive systems the laser beam may be defocused and sufficient quanta still be delivered to the sample to permit use of a large-area observing beam from a tungsten lamp, but in less favorable cases other geometries than the right-angle between observing and photolysis beams may be considered. Quite short optical path lengths then become necessary ($\simeq$2 mm), and an arc lamp must be used to supply the observing beam. Obviously, the effect of the observation beam on the system must be taken into account, as it may correspond to a dissociation velocity constant of 5–10/sec. Appreciable heating effects from the observing beam may also occur with convective flow. This poses a real stability problem in a vertical cell since there is no counterforce to oppose it and convection will occur with quite small temperature gradients.

The relative ease with which laser light may be focused brings other problems arising from its intensity. It is easy to calculate that an ordinary flash lamp will not heat a solution very much under practical conditions. As already mentioned, the rate of removal of CO from myoglobin may be, say, 60,000/sec. In a solution 10 μM in heme the total broken down would be 0.6

M/sec. At 40 kcal/Einstein the rate of temperature rise would be 13°/sec—but the light intensity is maintained for only 0.2–0.3 msec, for a temperature rise of 0.01°. Note, however, that for 100 μM heme, the rise would be 0.1°, assuming that a correspondingly thinner film of myoglobin is used so that self-absorption is not limiting, and so on. With a conventional flash, really high concentrations are unlikely to be used because the rate of the recombination reaction would often become too competitive with the decay of the flash. With the brief laser flash these limitations largely disappear, and the heating situation must be reassessed. For purposes of calculation it is convenient to take an implausible situation, and to suppose that photolysis is being performed at an isosbestic wavelength where a quarter of the radiation energy is taken up by the system being examined. Then, if a laser pulse of 1 J is focused down to a diameter of 2 mm in a 2-mm-depth cell, 0.25 J will be delivered to a volume of 6 μl, with a 10° temperature jump. Nearly all of this temperature rise is associated with the absorption of quanta by deoxyhemoglobin, in the specific system being considered here, and is due to the use of an excess of energy over that required to give, say, 99% complete photodissociation. With highly concentrated solutions a different limitation appears. If a film of hemoglobin crystals were studied, with a protein concentration of approximately 40 m*M* (heme), a minimal temperature rise of 3.2° would be expected, using a quantum efficiency of 0.5 and assuming 40 kcal/Einstein. In practice, a significantly greater temperature jump must occur, even if a favorable wavelength where Hb absorbs less than COHb can be used.

In the discussion so far it has been assumed, explicitly or implicitly, that a reversible reaction is being studied. This limitation may sometimes be avoided by combining flow and flash techniques. An example is offered by the cytochrome oxidase reaction first studied photochemically by Gibson and Greenwood.[2] This enzyme reacts so rapidly with oxygen as to be effectively outside the range of conventional stopped-flow methods. The reaction is not reversible by light, and so cannot be studied directly by flash photolysis. If, however, the carbon monoxide compound of cytochrome oxidase is mixed with oxygen, oxidation of the enzyme is limited by the rate of dissociation of CO from the complex. The CO complex is photosensitive, and if the mixture is exposed to a flash, the oxygen reaction may be initiated much more rapidly than is possible by flow methods, i.e., within a microsecond or so. This procedure was used quite effectively to follow the oxidation of a_3, and a, and later that of Cu^+, by molecular oxygen. The flash was, however, provided by conventional tubes, so that the reaction with oxygen was going on during the flash. It was, therefore, necessary to

[2] Q. H. Gibson and C. Greenwood, *Biochem. J.* **86,** 541–555 (1963).

assume that any oxygenated compound of cytochrome a_3 was photoinsensitive, an assumption justified by analogy with hemoglobin rather than by experiment. It is quite possible that rather different results would be obtained if a laser were used instead. In general, when flow–flash methods are used with rapidly reacting systems, a case for a laser may be made.

Finally, unlike many other methods of initiating a reaction, flash photolysis can be applied to gases, liquids, or solids, and to intact biological structures as well as to solutions of purified proteins. In particular, low temperature may be used with profit, as for example in the early work of Yonetani[3] with cytochrome oxidase, and in the work of Frauenfelder with hemoglobin.[4]

It will be clear to the reader that flash photolysis cannot be regarded as a standard biochemical procedure, and an experimenter must set up his own apparatus. It is for this reason that this article has dealt in generalities, rather than offering specific descriptions of apparatus. The relatively restricted range of application of flash methods has discouraged manufacturers from offering commercial instruments, and it remains true that the method has not been fully exploited, especially in conjunction with modern flash equipment and in flow–flash systems.

In an attempt to make up partially for the lack of specifics, some references to methods papers are noted.[5-9] Although most of these may be regarded as outdated, the simplest types of equipment deserve to survive because they are easy to operate and reliable. Full attention can then be given to the experiment proper. Hudson[5] is included to emphasize the gap between the techniques actually available (though sometimes at great cost) and those which have been used so far in this area of enzymology.

[3] T. Yonetani, *Oxidases Relat. Redox Syst., Proc. Symp., 1964* Vol. 1, p. 173 (1965).

[4] H. Frauenfelder, this volume [28].

[5] B. S. Hudson, *Annu. Rev. Biophys. Bioeng.* **1,** 135–150 (1977).

[6] R. G. W. Norrish and G. Porter, *Discuss. Faraday Soc.* **17,** 40–46 (1954).

[7] S. Claesson and L. Lindquist, *Ark. Kemi* **11,** 535–561 (1957).

[8] C. Greenwood, *in* "Rapid Mixing and Sampling Techniques" (B. Chance *et al.*, eds.), pp. 157–163. Academic Press, New York, 1964.

[9] G. Porter, *Tech. Org. Chem.* **8,** 1055–1106 (1963).

[9] Cytochrome Kinetics at Low Temperatures: Trapping and Ligand Exchange

By BRITTON CHANCE

The remarkable progress in the study of kinetics at low temperatures had its origins in the development of optical techniques sufficiently sensitive to detect cytochromes in highly opaque freeze-trapped samples. In the first instance, the split-beam wavelength-scanning spectrophotometer[1] and the dual-wavelength spectrophotometer[2] were employed for static spectra, and kinetic changes, respectively. These techniques led first to a quantitation of low-temperature spectroscopy of cytochromes[3] and the discovery of a number of novel kinetic phenomena at low temperature, particularly the oxidation of cytochromes in photosynthetic bacteria at liquid N_2 temperatures[4] and in what is now considered to be the first example of the electron tunneling phenomenon in biological systems.[5] Observations on the kinetics of photolysis in recombination of the carbon monoxide compounds of myoglobin and cytochrome oxidase have demonstrated much greater reactivity of CO toward myoglobin than toward cytochrome oxidase at low temperatures. These methods have led to studies of photolysis and recombination with CO at low temperatures, described on the one hand as a purely kinetic study[6] or as an analytical study by Sato and his colleagues.[7]

The flash-photolysis activated ligand exchange reactions of hemoglobin and cytochrome oxidase stem logically from the pioneering studies of Porter[8] and his colleagues as applied effectively by Greenwood and Gibson,[9] Antonini,[10] and Greenwood *et al.*[11] to problems especially having to do with cytochrome oxidase. Gibson *et al.*,[9] in particular, have used flash

[1] C. C. Yang and V. Legallai, *Rev. Sci. Instrum.* **25,** 801 (1954).

[2] B. Chance, *Rev. Sci. Instrum.* **22,** 619–639 (1951).

[3] R. W. Estabrook, *J. Biol. Chem.* **233,** 781–794 (1956).

[4] B. Chance and M. Nishimura, *Proc. Natl. Acad. Sci. U.S.A.* **46,** 19 (1960).

[5] D. DeVault and B. Chance, *Biophys. J.* **6,** 826–847 (1966).

[6] R. H. Austin, K. W. Beeson, L. Eisenstein, H. Frauenfelder, and I. C. Gunsalus, *Biochemistry* **14,** 5355–5373 (1975).

[7] N. Sato, B. Hagihara, T. Kamada, and H. Abe, *Anal. Biochem.* **74,** 105–117 (1976).

[8] G. Porter, *Fast React. Primary Processes, Chem. Kine., Proc. Nobel Symp., 5th, 1967* pp. 141–164 (1967).

[9] Q. H. Gibson, C. Greenwood, D. C. Wharton, and G. Palmer, *J. Biol. Chem.* **240,** 888–894 (1965).

[10] E. Antonini, M. Brunori, A. Colosimo, C. Greenwood, and M. T. Wilson, *Proc. Natl. Acad. Sci. U.S.A.* (in press).

[11] C. Greenwood, T. Brittain, M. Wilson, and M. Brunori, *Biochem. J.* **157,** 591–598 (1976).

ISBN 0-12-181954-X

photolysis ligand exchange of oxygen for CO in soluble oxidized cytochrome oxidase and have characterized a number of kinetic constants for electron transfer processes.

Design Philosophy

Some important modifications of the design of spectrophotometers are necessary for operation at low temperatures. One special change is the need to gather much of the light from the sample, which is nearly totally scattered when opaque suspensions of organelles are frozen at low temperatures. Thus, detectors which are closely coupled to the sample, either by using small-diameter Dewar flasks or, more recently, light guide coupling, seem highly appropriate. The latter serves optimally to couple the samples, and thus has been used in some of the designs that are discussed below.

For wavelength-scanning spectrophotometers, low-temperature operation requires very wide dynamic range in the detection; thus dynode voltage-control circuits are appropriate to cover the range of sensitivities needed to permit satisfactory operation over the wide range of opacities at various wavelengths. To obtain perfect cancellation dynode regulation circuits may be supplemented by digital memories and correcting circuits. The digital memory may be used exclusively if a reference wavelength is employed and the scanning is essentially in a dual-wavelength mode. For wavelength scans of small magnitude (± 50 nm) the required dynamic range is dramatically diminished by the dual-wavelength method. For scans of larger magnitude (± 200 nm) the dual-wavelength method has little advantage over the single-wavelength scan.

Data acquisition from multiple components may be obtained by a wavelength-scanning device or a multichannel time-sharing system. The choice may be dictated by kinetic requirements, on the one hand, and information-gathering requirements on the other. In the initial phases of a spectroscopic exploration, wavelength scanning is of the utmost value in identifying the relevant peaks and the nearby reference wavelengths which may be used for subsequent dual-wavelength recording of kinetic changes. When the characteristics of the intermediates in the reaction are adequately known, and suitable reference wavelengths for the various intermediates found by wavelength scanning, it is relatively more efficient to time-share wavelengths appropriate to these compounds and to record analog signals with appropriate conversion to digital where necessary. The numerical aperture of a time-sharing system using interference filters is higher than that of a monochromator in the same price class, and information is being gathered nearly simultaneously at the various wavelengths, and thus is optimal for kinetic recordings.

One advantage of operation at low temperatures is that rapid recording may be unnecessary; usually, the temperature at which the reaction is initiated can be lowered to suit the bandwidths of the available instrument in view of the signal-to-noise ratio obtained from the samples. Response times of a few hundred milliseconds in recording cytochrome kinetics can readily be obtained at $-100°$ with 5–10 mg/ml of mitochondrial protein. Reduction of bandwidth is desirable for the slower phases of the kinetics. However, wavelength scanning records different wavelengths at different times, and thus requires interpolation for kinetic study.

Recording on Log Time Base

The article by Frauenfelder[12] describes the advantages of recording on a logarithmic time base so that multiple phases of power-law kinetics can be precisely delineated. In the special systems cited there—myoglobin, hemoglobin, and some model systems—the method appears justifiable. However, in the study of the kinetics of intermediates of enzyme reactions, the system is not appropriate, and could give completely misleading information on late-appearing intermediates, etc., on which detailed kinetic data are required. For this reason initial explorations of reaction kinetics should be made by recording simultaneously on linear time bases of different speeds. Then after the kinetics have been examined, the appropriate time-base algorithms can be employed without undue loss of information.

Types of Time-sharing Equipment

Two types of time-sharing spectrophotometers have been found useful. One operates at low speeds and is synchronous with the line frequency, affording roughly 8 msec/aperture with a maximum number of four apertures. The four wavelengths may be used for two separate dual-wavelength recordings, or for the recording of three signals against a signal reference wavelength.

High-speed time-sharing is obtained from free-running turbines, rotating four filters at speeds up to 500 Hz or 0.5 msec/aperture. The other features of this apparatus are described in a recent publication.[13] Such units are operable over wide frequency ranges, have the same optical aperture as the slower units, and are considered to be useful for a wide variety of applications.[13]

[12] H. Frauenfelder, this volume.

[13] B. Chance, *Anal. Biochem* **66**, 498–514 (1975).

Triple-trapping Method

To obtain maximal advantage from the low-temperature technique, it often is appropriate to trap a state of the system at room temperature, and to activate the system momentarily at another temperature, and finally to trap the partially or fully reacted state at very low temperatures. This is exemplified by studies of intermediates in the cytochrome oxidase reaction.[14-16] To start the reaction with oxygen at low temperature, oxygen must be mixed with the cytochrome oxidase–CO compound sufficiently rapidly and at a sufficiently low temperature so that none of the carbon monoxide is dissociated; hence, the first trapping step is carried out at subzero temperatures. Second, the reaction must be started at such a low temperature that the appropriate intermediate—presumably one of the primary intermediates of cytochrome oxidase with oxygen—reaches its maximal concentration on a time scale appropriate to the freeze-trapping and recording time scales. Thus, when the maximal concentration has been reached, the third trapping step is employed rapidly to cool the system so that no further reaction takes place, and the products of the reaction can be examined at leisure by a variety of "slow" physical methods such as EPR, etc.

Preparations

In these studies any membrane preparation can be used which contains cytochrome oxidase at 0.1 nmole/mg at concentrations of 1–20 mg/ml. The properties should not be significantly altered by the addition of the 30% ethylene glycol required to reach −25°, at which temperature oxygen mixing can occur satisfactorily. An improved solvent mixture appropriate to −22° consists of 15% ethylene glycol, 15% DMSO, and 5% methanol.[17]

Methods of Adding Oxygen

The method of stirring, or of oxygen generation by decomposition of H_2O_2 by catalase, seem less quantitative than the method exemplified by Fig. 3 below, where dilutions of a small amount of the preparation by a large amount of oxygen-saturated buffer have been employed.

[14] B. Chance, C. Saronio, and J. S. Leigh, Jr., *Proc. Natl. Acad. Sci. U.S.A.* **72,** 1635–1640, (1975).

[15] B. Chance, C. Saronio, and J. S. Leigh, Jr., *J. Biol. Chem.* **250,** 9226–9237 (1975).

[16] B. Chance, J. S. Leigh, Jr., and J. I. Ingledew, in preparation.

[17] B. Chance and F. Itshak, in preparation.

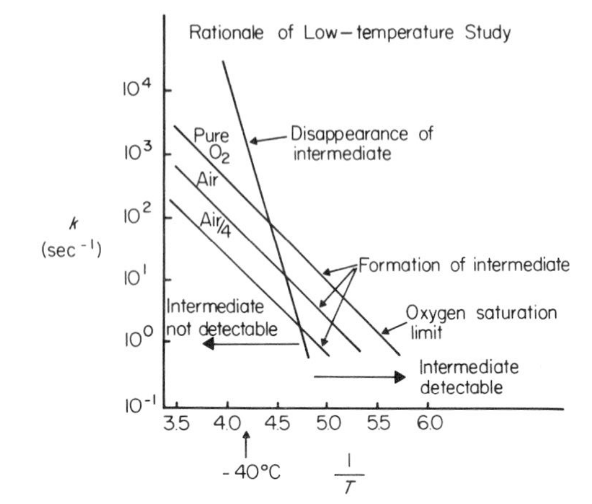

FIG. 1. A schematic diagram of parameters in the choice of temperature regions appropriate to detection of labile intermediates.

Oxygen Solubilities at Low Temperatures

The solubility of oxygen in 30% ethylene glycol at $-25°$ is approximately 2 m*M*, as obtained by addition of aliquots to either the oxygen electrode at 23°, or to an anaerobic DPHN–oxidase system.[17]

Basic Design of the Experiment

Figure 1 illustrates the design philosophy in the trapping method for a hypothetical one-intermediate reaction. Displayed in this Arrhenius graph are the schematic characteristics of the combination of cytochrome oxidase and oxygen at three different oxygen concentrations, each moving the Arrhenius plot in parallel to higher velocities at a given temperature. The intersecting line represents the decomposition of the intermediate, which is drawn here at a significantly higher activation energy than that for the formation of the first intermediate. Therefore, there are intersections of the two lines at which the formation and disappearance of the intermediate have equal rates and the concentration of the intermediate is half maximal. The point of intersection is oxygen-dependent; the rate is highest at the highest temperature and oxygen concentration. The intersections of the lines are indicated here to be at temperatures which are in the subzero region, as has been found to be the case in the cytochrome oxidase–oxygen reaction. Thus, studies at temperatures well below the freezing point are necessary; the requisite temperature for the cytochrome oxidase–oxygen reaction is well below −100°. Thus, no aprotic solvent system for reactions in the liquid state can be considered, and resort to flash photolysis activation of the reaction by ligand exchange of carbon monoxide for oxygen in the solid state is necessary.

Figure 2 illustrates schematically the parameters in the design of a flash photolysis and ligand exchange reaction. In an actual case, this would require oxygen addition. In principle, the method requires that carbon monoxide be added to cytochrome oxidase in a time short compared to the spontaneous dissociation of CO from cytochrome oxidase in the dark. We have arbitrarily taken 10 kcal as the energy of activation to show how this parameter affects the design of the experiment. For these conditions, the graph indicates that at −23°, the half time for the dissociation reaction would be approximately 100 sec for a 10 kcal activation energy. Thus, an oxygen-mixing method that requires approximately 10 sec would give an acceptably small degree of dissociation of the CO compound.

For a temperature of −23°, 30% ethylene glycol gives an adequately low freezing point and an adequately low viscosity so that mixing of oxygenated buffer with the membrane preparation is feasible. Electron transport from cytochrome oxidase to oxygen is inhibited by about 20% and energy coupling by about 50% in 30% ethylene glycol at −20°.[12,17] A

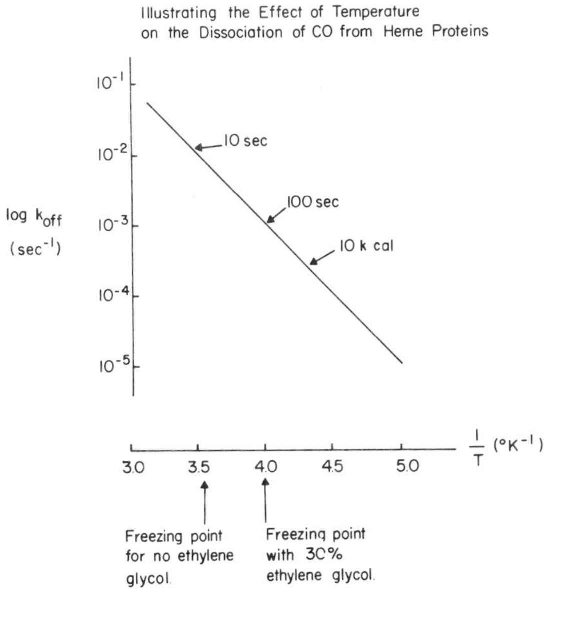

FIG. 2. A schematic diagram of parameters in the choice of oxygenation temperature.

modified solvent that freezes at −22° contains 15% ethylene glycol, 15% DMSO, and 5% methanol[12] and causes negligible uncoupling and inhibition of respiration.[17]

The current procedure for preparing samples for the triple-trapping method is shown in Fig. 3. A small volume of the highly concentrated membrane-bound or soluble oxidase is placed in the bottom of a 1-ml Lucite cuvette of an optical path length of 1–4 mm and an optically active area of 20 × 20 mm. An oxygen-saturated ethylene glycol solution is prepared at −23°, where the solubility is 2 m*M*. The cuvette is then filled with the oxygen-saturated medium, and the two layers are mixed with a wire stirring rod. The sample can be immediately transferred to a freezing bath and stirred during the 5 or 10 sec during which the temperature has dropped sufficiently to start freezing. Thus, the time of exposure of the CO-bound oxidase to oxygen at −23° is no more than a few seconds, and nearly 100% yields of the oxygenated CO compound are obtained.

The mixed valence state of cytochrome oxidase in which heme a and its associated copper become oxidized and heme a_3 and its associated copper remain reduced, due to binding of the heme a_3 to carbon monoxide, is prepared by addition of 2 m*M* ferricyanide to the reduced, CO-saturated oxidase in 30% ethylene glycol at −23°, the reaction being allowed to proceed for 30 sec. The small amount of oxygen added at this time does not react because CO is still tightly bound to the oxidase at this temperature. The conversion to the mixed valence state is more than 95% complete in this time interval. At the end of 30 sec, the oxygen-saturated ethylene glycol is added, and the procedure continues as described above.

Experimental Recordings

The sample prepared as above can be adjusted to a suitable temperature for the formation of the appropriate species of oxygen compound of cytochrome oxidase. Figure 4 displays appropriate traces for the formation of the

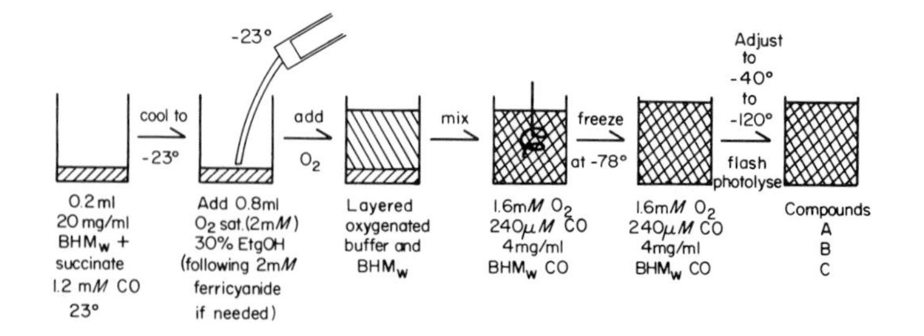

FIG. 3. A sequence of steps in the trapping procedure.

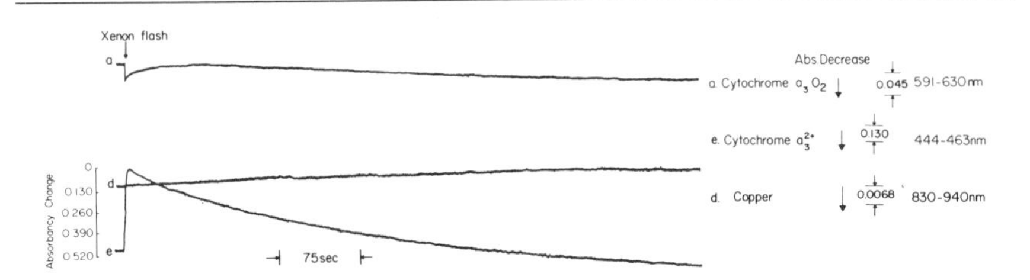

FIG. 4. Typical recordings of compound A and B in the reaction of cytochrome oxidase with O_2.

oxy compound and its conversion to a peroxy compound at three pairs of dual wavelength readouts, e.g., at 591–630 nm for observing the formation and disappearance of compound A (trace *a*). In trace *e*, the kinetics are recorded at 444–463 nm, and in trace *b*, at 830–940 nm. These traces correspond, respectively, to the formation and disappearance of compound A, the disappearance of reduced cytochrome a_3, and the oxidation of copper associated with heme a_3. The sample is equilibrated at the appropriate temperature (−105°) with the measuring light cut off. The sample is then illuminated, the dual-wavelength circuits are balanced, and the sample is flash-photolyzed with either a laser or xenon flashlamp. The reaction kinetics are recorded as shown on a relatively fast time scale and, simultaneously, on a relatively slow time scale.

The dramatic effects of ferricyanide upon the kinetics recorded in the cytochrome a_3 heme region and in the region appropriate to the copper component are illustrated in Fig. 5. In comparing this figure to Fig. 4, note that the temperature difference, in the absence of ferricyanide (ferricyanide zero), causes the formation and disappearance of compound A (trace a) to occur within approximately 10 sec. This demonstrates vividly how the time scale may be altered by the temperature at which the flash photolysis occurs. A similar speeding of the kinetics in the region of the gamma band of cytochrome a_3 is observed where the disappearance of absorption occurs in approximately 10 sec and proceeds to a much greater extent than at the lower temperature of Fig. 4. Trace c at 608 nm shows some oxidation of the cytochrome oxidase heme by its downward deflection. At this temperature the copper absorbancy change springs into action, and instead of the very small show of absorbancy, a large deflection is recorded which appears to be biphasic; i.e., a fast phase completed in approximately 20 sec followed by a rather slower one, each attributed to the oxidation of two forms of copper in cytochrome oxidase.

If the mixed valency state is established now (pretreatment with 400 μM of ferricyanide according to the procedure above), the formation of com-

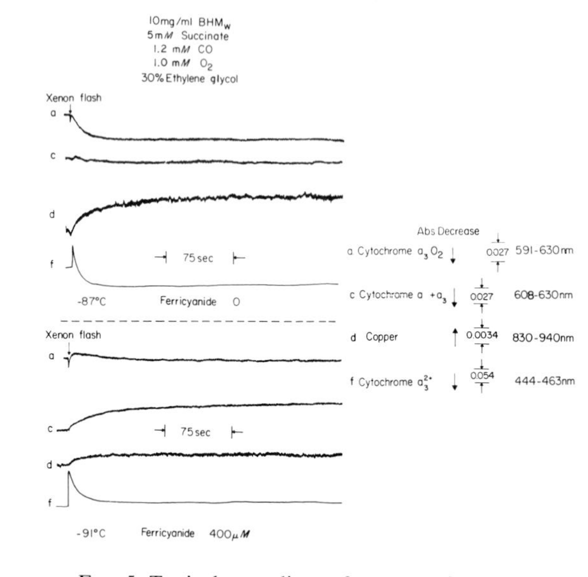

FIG. 5. Typical recordings of compound A and C.

pound A is barely resolved in trace a, but the compound has not disappeared in the same way as it did in the absence of ferricyanide. Similarly, in trace f at 444 nm the extent of reaction is considerably less than that in the absence of ferricyanide because the heme *a* has already been oxidized by prior treatment with ferricyanide and only the heme a_3 can react. Thus, trace f gives, in relation to the size of the absorbancy change on photolysis of the CO compound, a very effective indication of the contribution of heme a_3 at 444 nm. Trace c is different; instead of an absorbancy decrease which might correspond to an oxidation of a heme of cytochrome a_3, there is an abrupt increase of absorbancy which would initially be interpreted as a further reduction of the heme of cytochrome a_3. Since, however, cytochrome a_3 remains reduced, the increase of absorption at 608 nm may be due to charge transfer between iron and copper or oxygen in compound C. Trace d is a much smaller amplitude, partly because the ferricyanide has already oxidized the copper associated with heme *a*. However, observations at other wavelengths indicate clearly that the copper associated with heme a_3 absorbs at different wavelengths from 830 nm and can be observed optimally to be oxidized at 744 nm. Thus, the small change of 830 nm is a portion of the signal ascribable to the oxidation of copper associated with heme a_3.

Summary

The multiwavelength time-sharing systems seem appropriate to recording the complex kinetics of the intermediates in the cytochrome oxidase–oxygen reaction and presumably in many other types of enzymic systems involving multiple intermediates. Furthermore, the great degrees of freedom in choosing the reaction rate and the lifetime of the intermediate underlies the importance of temperature as a parameter in observing the cytochrome oxidase–oxygen kinetics.

Predictions for the Future and Drawbacks of the Present Method

The present method is limited to photolysis-induced ligand exchange and therefore is suitable for CO–oxygen reaction but unsuitable for a variety of reactions with other ligands. The latter are fortunately slower and not expected to produce the multiple intermediates as observed in the oxygen reaction. The reactions in the solid state are not complete in themselves; thus it is appropriate that they be supplemented by low-temperature rapid-flow apparatus at as low temperatures as the membrane-bound system can be stabilized in aprotic solvent. Thus, of greatest interest in research progress at low temperatures are kinetic studies in solution at acceptable values of concentration of aprotic solvent. This would represent an essential companion research to the trapping methods that operate in the frozen state but are limited by flash-activated ligand-exchange reactions.

[10] Special Techniques for the Preparation of Samples for Low-temperature EPR Spectroscopy

By HELMUT BEINERT, WILLIAM H. ORME-JOHNSON, and GRAHAM PALMER

I. Introduction

This volume, as well as other volumes of this series, contains a number of articles concerned with various aspects of electron paramagnetic resonance (EPR) spectroscopy. Most of these contributions deal with theoretical background, scope, applicability, and results obtained in specific areas of EPR spectroscopy. There is, however, a need to discuss the preparations and manipulations preceding spectroscopy proper so that results of

ISBN 0-12-181954-X

the desired precision and decisiveness are obtained. Aspects of this nature will be treated in the present article.

It may be asked why preparation of samples for EPR spectroscopy should differ sufficiently from general practice in other methods used in biochemistry, e.g., spectrophotometry, to make this necessary. The main reason for different requirements stems from the geometry of the sample container, which is dictated by the spectroscopic procedure, and from the relatively low sensitivity of EPR spectroscopy, e.g., as compared to most optical methods. Thus, high concentrations of protein are often required for satisfactory resolution. Since it is undesirable to change this concentration appreciably in the course of an experiment, additions to be made must remain in the range of a few microliters. It will be appreciated that mixing and moving small quantities of viscous solutions, within apparatus with parts of narrow diameter under anaerobic conditions, such that reproducible and quantitative results are obtained, is not simple.

Since most features observable by EPR spectroscopy in the field of oxidoreduction and of electron-transfer systems will require low-temperature spectroscopy—even signals from semiquinones of ubiquinone or flavin usually show a poor signal-to-noise ratio in the liquid state—the discussion below will mainly deal with procedures suitable for low-temperature work.

Other articles in this volume and in previous volumes of *Methods in Enzymology* which are related to this article and should be consulted in appropriate instances are the following: the chapters by J. A. Fee[1] and by L. J. Berliner[2], on more general aspects of EPR spectroscopy; the chapter by H. Beinert[3] on EPR-detectable components of the mitochondrial electron-transfer system; that of N. R. and W. H. Orme-Johnson[4] on quantitation of cytochrome P-450; that of W. H. Orme-Johnson and R. H. Holm[5] on the quantitation of FeS clusters; that of D. P. Ballou[6] on rapid freeze-quenching as required for kinetic studies monitored by low-temperature EPR; and the chapters by G. Palmer on EPR spectroscopy[7] and low-temperature reflectance spectroscopy.[8] The earlier chapters by Palmer contain a number of hints which will not be reiterated here; on the other hand, experience gathered in the 10 years intervening since that first

[1] J. A. Fee, this series, Vol. 49 [20].

[2] L. J. Berliner, this series, Vol. 49 [18].

[3] H. Beinert, this volume [11].

[4] N. R. Orme-Johnson and W. H. Orme-Johnson, this series, Vol. 52 [26].

[5] W. H. Orme-Johnson and R. H. Holm, this series, Vol. 53 [29].

[6] D. P. Ballou, this volume [7].

[7] G. Palmer, this series, Vol. 10 [94].

[8] G. Palmer, this series, Vol. 10 [93].

writing has led to the development of methods which are more suitable for the acquisition of quantitative information. While for exploratory work the procedures outlined 10 years ago may still be preferable, since they require less specialized equipment and considerably less skill, this article will describe or refer to contemporary state-of-the-art techniques with which most of the presently published information in the field has been obtained.

Some of the procedures described below, such as purification of gases, will be more generally applicable, while most were specifically developed for low-temperature EPR spectroscopy.

II. Filling EPR Tubes Aerobically

With most enzymes of oxidoreduction it is necessary to work under exclusion of oxygen. The largest portion of this article will, therefore, be devoted to approaches suitable for this purpose. For simple additions to the relatively narrow EPR tubes the methods and tools mentioned by Palmer[7] can be used, i.e., polyethylene hoses (attached to hypodermic needles) which clear the walls of the tubes sufficiently so that when removing the hose one is not dragging out much of the liquid again by capillary action. Very wide hoses are suitable where pastelike material, such as a concentrated mitochondrial or bacterial suspension, is added. After addition of viscous materials it may help to detach the hose from the needle, centrifuge tube and hose for a few turns, and then withdraw the hose very slowly. When whole tissue is to be examined at low temperature in an EPR tube of a few millimeter's bore it is generally very difficult to stuff the material down the length of the tube. In these cases one may use a tube open at both ends, apply mild suction from an aspirator to one end and draw the tissue up, and then, if necessary, pack it together with a glass rod or applicator stick (or a packer as used in the freeze-quench procedure). Powders of frozen material, such as tissue ground under liquid nitrogen, are also best transferred to an EPR tube by the techniques used in the freeze-quenching method.[6]

When a few microliters of a substrate or inhibitor are to be added to a tube, again a calibrated, narrow, polyethylene hose (e.g., Clay Adams PE-50 attached to a 22-gauge needle) or a micro-glass capillary attached to a plastic hose may be used. It is not advisable to use a rigid glass pipette, since breakage is bound to occur occasionally. It is also good practice, when pipetting small quantities down a tube, to make sure during the process that the liquid is still in the hose or capillary as one moves down to the point of delivery, that it is finally expelled, and that part of it is not dragged up again during removal of the pipette. Obviously, when adding larger volumes, errors arising from capillary forces may be minor.

III. Anaerobic Procedures for Sample Preparation

1. Construction of an Anaerobic Gas-train.

a. General Remarks

Note that exclusion of oxygen in the sense required in some work in physics and chemistry is not intended; thus the much misused term "absolute anaerobicity" is not applicable here. Apparatus, procedures, and time required would be much too costly and probably in most cases unsuitable for handling of protein solutions, which are sensitive to elevated temperatures and processing and have a limited stability. Compromises must therefore be reached, and techniques must be applied which allow reasonably rapid handling of small samples under far-going exclusion of oxygen. Since in most cases relatively concentrated samples are required for EPR spectroscopy proper, minor oxygen contamination may be tolerable. It is good practice to consider what is "tolerable" by some simple calculations based on concentrations, volumes, etc. It is also apparent that it is advantageous to keep gas volumes to a minimum in all apparatus involved. It is a good operational philosophy, however, to employ stringent criteria throughout, so that the more readily preventable sources of contamination are excluded and a minimum of uncertainties remains. In addition, *maintaining* an anaerobic system in top working condition requires constant vigilance and a healthy skepticism toward the reliability of all components of the system being employed.

The construction of a system for the purification and delivery of "oxygen-free" gas requires consideration of three basic components: the source of gas, the method for removal of oxygen, and the means whereby the gas is transported between the various components of the system.

There are many ways to implement each of these considerations, but the success of the final system can be judged only within the requirements of a given application. In the following sections we will enumerate those aspects which need to be considered in any effective anaerobic train, and we will present methods for evaluating a train's performance. Because every laboratory has applications unique to its own area of research, we will not present a specific design in detail. Rather we will stress those requirements which we deem to be essential to any system to be used with solutions of proteins in glass vessels.

b. Choice of Gas

The gases that are most frequently used for inert atmospheres are nitrogen and argon, while helium, hydrogen, and carbon monoxide are of occasional utility. Of these alternatives argon is probably the most suitable vehicle for such experiments. Although nitrogen is less expensive than argon, the additional cost of the latter is for most applications negligible, although in situations where large quantities of gas are consumed, e.g., anaerobic enzyme purification, nitrogen may be preferred.

Argon is biologically inert and denser than air: this latter property can be of value in situations where it is necessary to open up an anaerobic vessel briefly and yet minimize contamination by atmospheric oxygen. This manipulation is often successful if the argon is maintained at a positive pressure and the aperture in the vessel kept very small, e.g., a l-mm capillary tube. (Examples of such manipulations are provided below.) An additional virtue of argon is that its solubility increases with increasing temperature. Consequently, removing a sample from ice and warming it up for the purposes of making an absorbance measurement is not frustrated by the formation of bubbles as the gas comes out of solution.

Nitrogen is a popular alternative to argon but other than the factor of cost it offers to no practical advantage: we might point out that a size A cylinder of gas might well last a year in most applications. A special disadvantage of nitrogen is that it is a substrate for a presently much studied enzyme system, nitrogenase.

Helium has found some use in the provision of inert atmospheres: its only obvious advantage is that it can be purchased (at significant cost) at a purity which renders additional purification unnecessary. However, as we will describe below, the techniques for removal of oxygen are straightforward and readily implemented so that this feature will only rarely be of value.

The two remaining gases, hydrogen and carbon monoxide, have obvious hazards associated with their use, and should be avoided unless specifically required by the experiment in progress. Both of them have the specific advantage that they regenerate one or another of the heterogeneous catalysts employed in scavenging the oxygen. It is important to note that with carbon monoxide the product of regeneration is carbon dioxide which may well lower the pH of the experimental solution unless precautions are taken to remove it from the gas stream: passing the scrubbed gas through a suitably basic solution will both remove the CO_2 and perform the valuable function of humidifying the gas.

c. Gas Delivery

Gases bought in large quantity are delivered in the familiar 5-ft-tall type A cylinder which is provided with CGA female-type fittings for mounting gas regulators. It may not be widely appreciated that standard commercial regulators are not gas tight and a significant source of oxygen contamination arises via inboard leakage at the gas regulator. A mandatory requirement for a high-quality anaerobic system is a (two-state) gas regulator with low inboard leakage: a leakage rate of less than 1×10^{-10} ml helium min^{-1} is acceptable. Significantly better performance can only be obtained at great inconvenience, such as elimination of the pressure gauges needed for efficient operation of the system, and this disadvantage does not compensate for the added performance.

Transport of the gas from the regulator to the rest of the train should be achieved using either metal or glass tubing. For most configurations glass tubing should be kept to a minimum and either thin-wall stainless-steel or copper tubing should be used wherever feasible. The former is rigid and needs a special tool for bending to the shape required. The copper tubing is easily bent but is susceptible to fatigue if it is subjected to many cycles of flexing. This becomes apparent as lateral surface cracks in the metal. The metal tubing should be grease- and oil-free. This requirement is met by commercial air-conditioner tubing. Material from other sources should be flushed thoroughly with a suitable organic solvent and dried with compressed gas.

The connection of the metal tubing to the regulator, and indeed all joints, should be made in a manner known to be leak-free. The most rigorous method is to solder[9] the two components together. However, this requires certain skills, and the resultant assembly is not easily modified. Metal-to-metal compression-type fittings (e.g., Swagelok, Gyrolok) are leak-free when carefully installed and are readily removed in the event that a modification of the system is necessary. The manufacturers' directions regarding reuse of a disassembled fitting should be adhered to. Similar fittings exist for metal-to-glass joints; these employ an internal O-ring to secure the seal to the glass. An alternative procedure is to ensure that one of the two pieces of tubing (metal or glass) will fit inside the other for a reasonable distance, e.g., 75 mm. The inner piece is coated with epoxy over the common length and inserted into the outer component, thus filling the annular space with the epoxy glue. The adhesive is also applied over the outer surface of the joint. Finally, glass-to-glass connections, where they are necessary, should be made using ground-glass joints. The conical variety are generally less prone to leak than are ball joints, while their rigidity is a disadvantage. Some flexibility may be introduced by placing spiral sections of metal tubing in the line. Glass-to-glass joints should be selected so that the ground surface area is as large as possible relative to the

[9] For the nonexpert soft-soldering is recommended. Silver solder and brazing may leave pinhole leaks in stainless steel unless conducted by a specialist in vacuum apparatus construction. In any event joints should be tested for leaks.

bore of the tubing, and, when assembled, the joint should be compressed using either springs or clamps.

d. Removal of Oxygen

The most popular method for removing the oxygen contamination from carrier gases has been to scrub the gas with a solution of a readily autooxidizable reagent—such as alkaline pyrogallol, naphthoquinone-β-sulfonate, or chromous or vanadous sulfate—which is then subsequently rereduced with dithionite or zinc. In reality, the efficiency of these liquid scrubbing reagents is poor and their use cannot be recommended. Past success with these agents is mainly attributable to the fact that the final experiment was performed *in vacuo*.

There are a variety of solid agents which work extremely well and which should be used in place of the liquid reagents. These are (1) metallic copper, (2) BTS catalyst, a copper derivative deposited on an inert support, (3) chromium or platinum agents deposited on molecular sieves, and (4) special alloys. Of these, the ones falling in the second and third categories are to be preferred, simply because they operate at lower temperatures; indeed, some of them operate at room temperature. Metallic copper and the zirconium-titanium-nickel alloys operate at 400–500°C which is an unnecessary inconvenience since one must subsequently cool the gas.

By far and away the most convenient method employs the newly developed molecular sieve catalysts. These can now be purchased from most major gas suppliers. They are supplied in a can and must be installed in a special cylinder also available from the supplier. This cylinder has threaded openings into which components for compression fittings can be mounted, or alternatively they can be installed directly on the regulator exit. This latter alternative is not recommended for it is desirable to install a second molecular sieve between the cylinder and oxygen scavenger: this second sieve serves to remove moisture and oil from the gas and thus enhances the performance of the oxygen removal system. These cartridges have the capacity to purify one large cylinder of gas and when spent must be replaced. A simple arrangement would be to purchase cylinder and cartridge at the same time. Nominally these cartridges reduce the O_2 content of the gas to 0.1 ppm. We have no data on the accuracy of this figure.

The most widely used of the above alternatives is BTS R3-11 catalyst manufactured by the BASF Corporation and available through Kontes Glass Co. A similar preparation, Ridox, is available from the Fisher Scientific Company. This material is a form of activated copper and is supplied as green cylindrical pellets about the size of a pencil eraser. For use the material should be pulverized to 200–400 mesh (care: do not breathe the powder) and poured into a column about 5 cm in diameter, 25 or more cm long. This column should bear a heating element over its whole length (although this material will function at room temperature, its capacity is markedly enhanced at high temperature). The column is installed in the anaerobic apparatus, and when construction is complete the catalyst is activated by passing 5% H_2 in N_2 through the column at 120°C. (The reaction is markedly exothermic and if pure H_2 is used the temperature might rise to a level which will damage the catalyst.) When regeneration is complete, the catalyst will be seen to be black and any moisture condensed on the exit port of the column during regeneration will evaporate. The regenerating gas can now be replaced by the working gas. This is most readily effected by installing the two gas cylinders adjacent to one another and using a gate valve to select one or the other cylinder. A recommended orientation for the "furnace" is vertical, with impure gas entering at the top and exiting at the bottom. With this orientation any "suckbacks" of water from the following humidifying vessel cannot enter the column to any great distance and thus the integrity of the catalyst is maintained.

A metallic copper purifier is similar: it contains copper turnings which have been thoroughly degreased, the regenerating gas can be pure H_2, and the operating temperature is 450°C. When activated the metal is bright and shiny; when ready for regeneration it is dull and red.

For applications where a *continuous* source of high-purity gas is needed (e.g., glove box) a gettering furnace is available (Model 8301 Hydrox Purifier, Matheson Company). This compact unit will service 50–70 standard large gas cylinders and reduce the O_2 content to ca. 0.1 ppm. Note, however, that this unit runs at ca. 500°C and that the gas flow must be at least 10 ml/min when it is at temperature—otherwise it burns out.

The next component in the train is a wash bottle fitted with a porous sintered sparger and containing water or buffer. This is required to restore water vapor to the very dry gas.

e. Gas Manifold

The final component is the system whereby the pure moist gas is delivered to the sample container. In general this is a manifold—a glass tube on which is mounted the desired number of outlet ports. Each of these is a vacuum stopcock of high quality. The tube also bears a three-way stopcock which is used to select the connection of either vacuum or inert gas lines to the manifold. A mercury-in-glass pressure manometer is a useful adjunct: it is particularly valuable for detecting minor leaks in the system. The manifold can be evacuated, the three-way stopcock closed, and the mercury manometer observed over a period of time. A stable reading is evidence for the lack of any leaks. Each outlet tap may be connected to a piece of experimental apparatus. This connection should be made in one of two ways: flexible stainless-steel bellows (e.g., Cajon Company) or small-diameter flexible copper tubing. Any significant length of rubber tubing, even butyl rubber, will lead to contamination. However a small, 1–2″ length of butyl rubber can be used in less critical applications for the final connection of the metal tubing to the experimental apparatus. This is, however, an undesirable compromise, and the length of this piece of tubing should be kept to the absolute minimum.

f. Gas Analysis

The only test of the successful anaerobic train is the level of oxygen present in the final gas. Thus it is important to be able to measure low concentrations of oxygen with reliability. To obtain a measurable value it is necessary to sample a large volume of gas. This can be achieved by modifying a 1-liter round-bottom flask so that a 1-cm spectrophotometric cell is fused to the surface opposite the standard taper joint. The optical cell should be attached in such a way that the whole assembly can be inserted into a spectrophotometer. Six milliliters of a solution containing 0.1 *M* pyrophosphate (pH 8.3), 1 m*M* EDTA, and 10^{-4} *M* lumiflavin-3-acetate (or other flavin) are placed in the flask which is then closed with a vacuum stopcock mounted on a 24/40 male joint. The flask is attached to the train and rendered anaerobic via 6–10 cycles of vacuum and gas, being finally left *in vacuo*. The flavin solution is then reduced by exposure to visible light, e.g., the sun: the illumination is continued until the solution is nonfluorescent and remains nonfluorescent even after vigorous shaking in the dark. All subsequent manipulations are performed in minimum light. The flask is placed in the spectrophotometer and the absorbance measured at 445 nm (flavin maximum) and 520 nm (to check for nonflavin absorbance, since scattering may develop with prolonged shaking). The flask is reattached to the train, the connecting tube thoroughly cycled between vacuum and gas, the flask opened under vacuum, and the gas admitted to the flask until flow has stopped. The flask is returned without shaking to the spectrophotometer, and the absorbances at 445 and 520 nm are recorded again: they should not have changed. The flask is now shaken in the dark with occasional measurement of the absorbance of the solution. If the absorbance changes (A_{445}–A_{520}) are very small (ca. 0.01–0.02) shaking is continued overnight. If the increment is large ($\geqslant$0.1 over a 15-min period) then the train is not functioning correctly and must be overhauled. With an inlet gas pressure of 2–4 psi, an increment of 0.01 absorbance units corresponds to 0.14 ppm of O_2 present in the gas [ΔA_M (445 nm) for oxidation of flavins = 10,000 cm^{-1}]. It may be observed that the absorbance at 445 nm increases slowly but continously over the 24-hr period: most of this increase appears to arise through the formation of a faint turbidity. This is corrected for by the measurements at 520 nm.

A method similar to the above has been described by Sweetser.[10]

2. *Demountable EPR Tubes and Accessories*

In order to maintain a low and stable temperature of the sample in the EPR cavity, it is advantageous to use an insulated container atop the insert dewar within the cavity. This container cannot usually accommodate large pieces of glassware and it is disadvantageous to have to cool large masses of apparatus. Thus, the EPR tube should be made in a form in which it can be reversibly attached by a vacuum-tight connection to other equipment in which sample preparation is effected. A simple method of doing this is shown in Fig. 1 where item A is a standard 5 mm (OD) × 150 mm thin-wall quartz EPR tube attached to a 7/15 outer joint. This joint may be either of quartz (we have on occasion had difficulty in obtaining accurately tapered quartz joints) or borosilicate with a graded seal intervening between joint and EPR tube. The total cost of the unit is about the same with either arrangement.

Under vacuum, most protein samples will foam inconveniently in the narrow confines of a 5-mm EPR tube, so that degassing of samples is done in a separate chamber, followed by transfer to the EPR tube. To this end, the apparatus is left under vacuum after the conclusion of degassing and reactant mixing. The partial vacuum (water vapor is present) allows one to make the fluid leap smartly from one chamber to another by a judicious combination of shaking and application of finger heat to gas spaces behind the sample. Normally, samples are then frozen in the bottom 25 mm of the EPR tube by slow immersion of the tube in liquid nitrogen or more rapidly (ca. 1 sec) by immersion in stirred isopentane at $-140°C$. (The isopentane is cooled in a cup suspended in a dewar containing liquid nitrogen and is rapidly swirled with an overhead stirrer.) If one desires an estimate of the volume of fluid left in the apparatus after evacuation and flushing, one may want to wait some 15 min until the remaining water has distilled down to the frozen sample tip of the EPR tube. Measurement of the height of the ice column (after removing gas bubbles by local thawing, if necessary) affords an estimate of the amount of water present after deoxygenation. The tube is then detached from the rest of the apparatus and stored in a dewar under liquid nitrogen, pending examination by EPR or diffuse-reflectance spectroscopy. Tubes are thawed for cleaning by rapid immersion in water and

[10] P. B. Sweetser, *Anal. Chem.* **39**, 979 (1967).

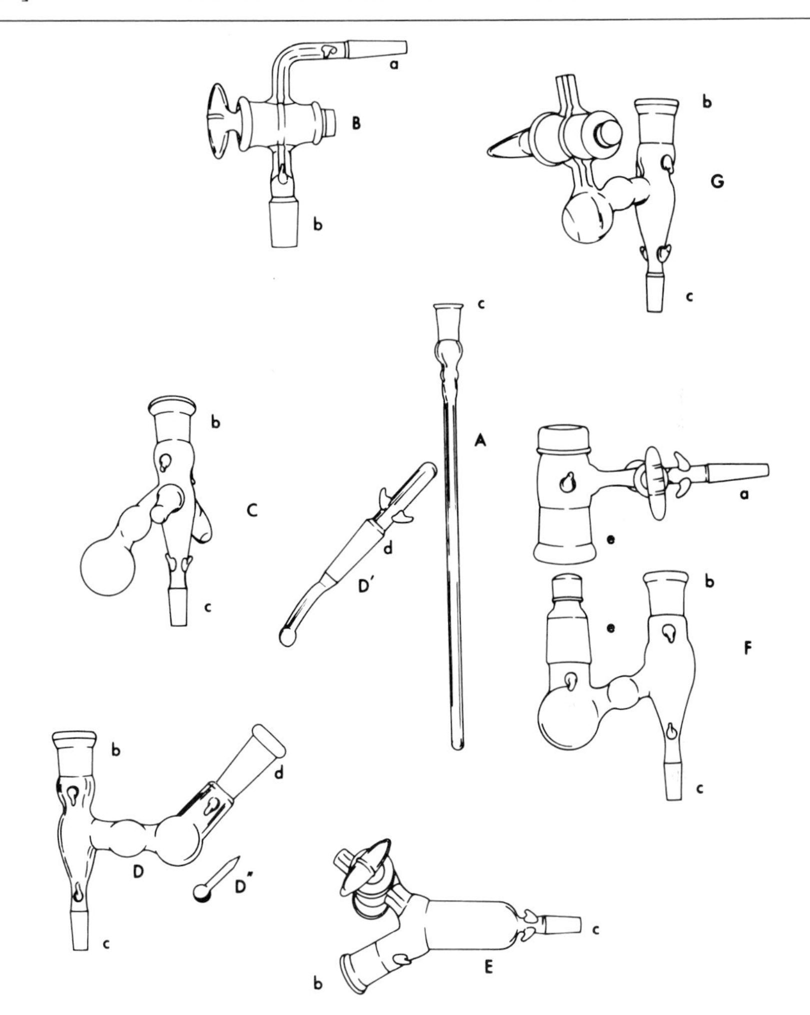

FIG. 1. Demountable glass apparatus for anaerobic sample preparation. (A) EPR tube (quartz) sealed to ground-glass joint; overall length is 19 cm. (B) stopper for connection of apparatus to vacuum line; the stopcock is of the hollow high-vacuum type. (C) Thunberg chamber for mixing of air-stable substances after deoxygenation. (D, D′, D″) Assembly for releasing solid titrant after deoxygenation. (E) chamber for equilibration of solution with gas phase in a thermostat. (F) Double septum seal chamber. The tandem septa define an evacuable intermediate chamber in which the oxygen-contaminated solution in a syringe tip can be deposited. (G) Thunberg chamber with vacuum stopcock fitted to the main sample compartment. Samples can be transferred via syringe through the stopcock. Further details are given in the text. Ground joint (standard taper) sizes: a, 7/25; b, 12/18; c, 7/15; d, 10/30; e, 19/22. The joints are greased with Apiezon N and secured with springs or small rubber bands using the hooks shown.

cleaned in hot 0.1% Haemosol,[11] the Apiezon N grease having been removed from the ground joint with chloroform.

With an EPR tube as well as a stopper with a vacuum stopcock (Fig. 1B) fitted to any of the types of chambers shown in Fig. 1, anaerobic samples may be prepared by cycles of evacuation and flushing with gas deoxygenated as described above. The process is repeated 6–10 times to lower the oxygen content to suitable levels, with most of the time being taken in the vacuum portions of the operation, during which the apparatus is gently shaken to speed the removal of gases. Excessive foaming of protein solutions must be avoided. In case of strong foaming a greater number of brief cycles may be necessary initially. The chambers shown in Fig. 1 are types specifically devised for various manipulations of the samples subsequent to deoxygenation.

An adaptation[12,13] of the classical Thunberg tube is shown in Fig. 1C. This is useful where air-stable materials are to be mixed after deoxygenation. The protein is normally placed in the large side arm, and titrants, cofactors, etc. are placed in the smaller side arms. Worthy of note is the small antechamber between the main bulb and the rest of the apparatus: this serves to help break up any bubbles that may form during evacuation. After deoxygenation, the contents of the chamber are mixed by rolling the apparatus around its long (horizontal) axis, and the solution is then tipped over into the EPR tube joint and moved down into the tube by gentle heating of the upper part and cooling below the joint.

The remainder of the chamber types were devised for applications in which air-sensitive materials must be added after oxygen has been removed.

Apparatus D (Fig. 1) is used for anaerobic additons of a solid reactant.[14] This will be discussed below under anaerobic titrations.

Apparatus E (Fig. 1) was developed[15] to allow equilibration of a protein solution with a gas phase at a fixed temperature, injection of an air-sensitive material, and rapid mixing and freezing of the sample. The apparatus is assembled from units A, B, and E, degassed, and a protein solution is put in the enlarged portion of E while the equipment is held in a horizontal plane. The apparatus is then placed in a reciprocating water bath in the same orientation and shaken at right angles to its main axis, while connected to a gas manifold via B. When the system is at equilibrium, the stopcock on E

[11] There is evidence that glassware cleaned with strongly acidic or alkaline cleaning agents retains gases more tenaciously at its surface.

[12] B. F. van Gelder and H. Beinert, *Biochim. Biophys. Acta* **189,** 1 (1969).

[13] R. E. Hansen, B. F. van Gelder, and H. Beinert, *Anal. Biochem.* **35,** 287–292 (1970).

[14] W. H. Orme-Johnson and H. Beinert, *Anal. Biochem.* **32,** 425 (1969).

[15] L. C. Davis, M. T. Henzl, R. H. Burris, and W. H. Orme-Johnson, *Biochemistry* (in press).

may be opened (gas pressure must be above 1 atm!) and a further reactant may be added. The stopcock is reclosed, the apparatus shaken, and the sample tipped down into the EPR tube and frozen in an isopentane bath. With practice the time from addition of the last reactant to freezing may be as short as 5 sec.

The equipment[16] marked F in Fig. 1 is used where the transfer of oxygen-sensitive substances by syringe is required, and where small amounts of oxygen entering with the needle cannot be tolerated. The important feature is the intermediate chamber formed by joint e, bounded by tandem rubber serum stoppers. In use, the lower portion and the intermediate chamber are evacuated and flushed with gas in order to bring the dissolved oxygen concentration in the lower septum to a low level. Both chambers are then let stand under deoxygenated gas at ca. 1 atm, and a syringe needle with the protein sample or titrant is introduced through the first septum. A small amount of the oxygen-contaminated material in the needle tip is expelled into the intermediate chamber, and the needle is then passed through the lower septum into the large sample compartment, where the solution is delivered. This apparatus is sufficiently leak-tight that 25 μM $[Fe_4S_4(PhS)_4]^{2-}$ can be handled[16] with greater than 95% recovery during a 30-min incubation at room temperature, and 1-mM solutions of the same ion show comparable stability for at least 3 days.

For less-demanding applications the rather convenient equipment of Fig. 1G may suffice.[17] In this apparatus, a vacuum stopcock gives access to the side compartment of a Thunberg chamber. The apparatus is degassed, placed under positive pressure, the stopcock is opened, and the solution of oxygen-sensitive material is introduced through the stopcock. For more critical experiments the exit tube may be capped with a small septum and kept filled with oxygen-free gas. The operation of this apparatus is then rather like that of Fig. 1F. For many applications, however, the use of the septum is superfluous; for example, in many experiments with nitrogenase[17] in which the proteins were kept in 2–10 mM sodium dithionite solution, transfer of solutions directly into the Thunberg chamber was found to give results identical to experiments in which the septa were used.

3. *Anaerobic Oxidoreductive Titrations*

There are two procedures by which most of the recent work in this area has been done, one employing liquid titrant and continuous titration of a

[16] B. A. Averill, J. R. Bale, and W. H. Orme-Johnson, *J. Am. Chem. Soc.* **100,** 3034 (1978). See also W. H. Orme-Johnson and R. H. Holm, this series, Vol. 53 [29].

[17] W. H. Orme-Johnson, W. D. Hamilton, T. Ljones, M.-Y. W. Tso, R. H. Burris, V. K. Shah, and W. J. Brill, *Proc. Natl. Acad. Sci. U.S.A.* **89,** 3142 (1972).

single large sample,[18,19] the other using titration with solid titrant of a number of individual samples, so that every point on a titration curve requires a completely separate sample. In what is called continuous method the final EPR readout is, of course, also done on separate samples that are derived from the larger sample which is being titrated. Both procedures have their advantages and disadvantages and both require considerable practice and skill. The choice will have to be made according to the problem and material at hand and partly also according to available expertise. If quantities of material are limited or the protein is not stable during handling at room temperature for 2 hr or more, discontinuous titration of individual samples is preferable. However, this method is more prone to suffer from errors or shortcomings affecting individual samples in different ways.

a. CONTINUOUS TITRATION

The device and its attachments are made entirely from glass or quartz.[20] The progress of the titration reaction is monitored by spectrophotometry together with discrete sampling of the titration mixture; the individual samples are transferred into EPR sample tubes prior to freezing and examination by low-temperature EPR spectroscopy. This device has been of great utility in reductive titrations of bacterial ferredoxin,[21] xanthine oxidase,[22] cytochrome oxidase,[23] and complex III from yeast mitochondria.[24]

An apparatus which utilizes a principle similar to that described here has recently been developed for electrochemical experiments.[25]

1. Apparatus. The device is depicted schematically in Fig. 2. It consists of a large body, B, to which is fused (1) a 2-mm light path quartz cuvette Q fit with a graded seal (Hellma Cells, Inc., Jamaica, N.Y., Cat. No. 220); (2) a side-arm, A, bearing a capillary 7/25 inner joint, J1 (Kontes Glass Co., Vineland, N.J., Cat. No. K663250-0071); (3) a sampling pipette P made from a 0.5-ml graduated transfer pipette. The transfer pipette penetrates B at its upper margin and extends into a conical extrusion, C, present in the side of the main body. The end of the pipette is sharply bevelled and

[18] B. D. Burleigh, Jr., G. P. Foust, and C. H. Williams, Jr., *Anal. Biochem.* **27,** 536 (1969).
[19] G. P. Foust, B. D. Burleigh, Jr., S. Mayhew, C. H. Williams, Jr., and V. Massey, *Anal. Biochem.* **27,** 530 (1969).
[20] G. Palmer, *Anal. Biochem.* **83,** 597 (1977).
[21] R. Matthews, S. Charlton, R. H. Sands, and G. Palmer, *J. Biol. Chem.* **249,** 4326 (1974).
[22] J. S. Olson, D. P. Ballou, G. Palmer, and V. Massey, *J. Biol. Chem.* **249,** 4363 (1974).
[23] J. Siedow, Power, S., de la Rosa, F. F., and Palmer, G., *J. Biol. Chem.* **253,** 2392 (1978).
[24] Babcock, G., Vickery, L., and Palmer, G., *J. Biol. Chem.* **253,** 2400 (1978).
[25] J. L. Anderson, *Anal. Chem.* **48,** 921 (1976).

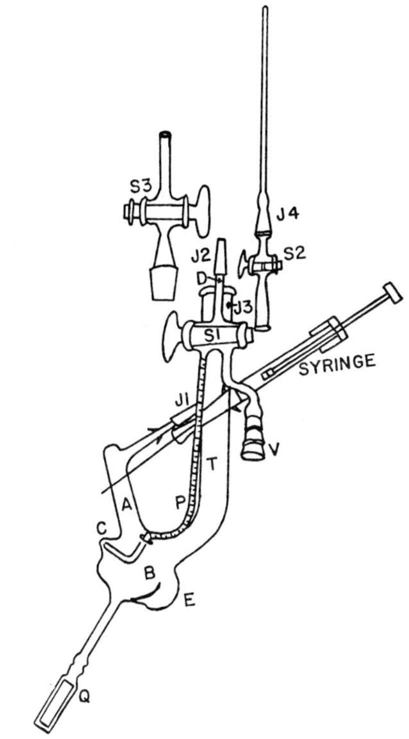

FIG. 2. Schematic outline of the combined optical EPR titrator with accessories. See text for explanation.

reaches to within ca. 1 mm of the tip of the cone. The portion of the transfer pipette external to B is approximately parallel to the large tubing T and terminates in a three-way capillary-bore stopcock, S (Eck & Krebs, Inc., Long Island City, N.Y. 11101, Model 5022, 1-mm bore). The adjacent arm, V, of the stopcock bears a 10/30 outer standard-taper joint for easy connection to a vacuum line, and the distal arm, D, is fused to a second 7/25 inner, capillary, standard-taper joint J2. The length of this latter portion of glass tubing is kept to a minimum. (4) A length (ca. 8″) of 0.75″ diameter glass tubing T terminating in a 19/22 outer joint, J3. All preliminary additions are made through this opening. This tubing also serves to support the transfer pipette via a glass bridge to the arm V of the three-way stopcock.

There is a second bulbous extension E in the main chamber, B; this is used to contain the protein solution during the agitation maneuvers employed while the sample is made anaerobic.

Additional components which are needed during the manipulation of the titrator are:

(1) A number of quartz EPR tubes (ID ca. 3 mm; OD ca. 4 mm) 5″ long, terminating in a quartz 7/25 outer joint, J4 (Westglass 12440 Exline, El Monte, Cal., Cat. No. W-1510-Q).

(2) A two-way stopcock S3 (Eck & Krebs, Model 5004, 1-mm bore), terminating at one end with a 19/22 inner joint.

(3) Caps made from a 7/25 outer joint (Kontes Glass Co., Cat. No. K662500-0725) closed at the tubular end.

(4) A 1-mm capillary two-way stopcock, S2 (Eck & Krebs, Cat. No. 5004) which bears a 7/25 inner joint for attachment to J2 and a 7/25 outer joint for attachment to J4.

(5) A titration syringe and titrant flask essentially similar to those described by Burleigh *et al.*[18] The titrant flask consists of a conical vessel modified by attaching a 9/22 outer joint to the mouth and a straight-hose two-way stopcock near the base. The stopcock is terminated by a 7/25 inner joint, and replaces the gas lock unit at location B9 of the flask described by Burleigh *et al.* (Codes prefaced with the letter B identify components depicted in Fig. 2 of Burleigh *et al.*[18] In our experience the rubber septa present in the gas-lock unit in both the cuvette and flask unit described previously[18] contain a significant quantity of dissolved oxygen which interferes with the quantitative character of the experiment.) All standard-taper joints are secured to one another with short metal springs. The titrant syringe is a Hamilton gas-tight syringe of capacity 0.5–1.0 ml and fitted with a threaded plunger. The syringe has been modified so that the needle and luer base feeds through a 7/25 outer joint which is epoxied to the barrel of the syringe (Fig. 2).

2. Procedures. PREPARATION OF TITRANT AND SOLUTION TO BE TITRATED. In a preliminary operation the titrant solution, usually dithionite, is prepared: this operation parallels that described by Burleigh.[18] About 50 ml of 0.1 *M* pyrophosphate (pH 8.3) together with a magnetic stirring bar are placed in the flask which is closed by means of a 19/22 inner joint bearing a stopcock. The flask is placed on a magnetic stirrer and subjected to several cycles of vacuum and inert gas with rapid stirring. When the solution is judged adequately free of air, the stirrer is turned off and the flask left onto a positive pressure of gas. The solid titrant is weighed out (e.g., 5–50 mg of dithionite), the top of the flask cautiously removed, the solid deposited into the buffer, the top replaced, and the vacuum immediately applied to the flask. The stirrer is restarted, and several additional cycles of anaerobiosis are performed. When this operation is complete, the syringe is filled with the titrant solution as follows: the titrant flask is placed in a horizontal position with the capillary stopcock upwards. With stopcock BB open, a positive pressure of argon is applied to the flask and stopcock B9 is then opened. The syringe, with plunger fully displaced to the base, is mounted on the 7/25 inner joint, locked in position, and the flask rotated back to its normal, upright position. The syringe needle is now submerged beneath the liquid and the syringe filled by withdrawing the plunger. The flask is placed back in the horizontal position, the syringe removed, and stopcock B9 closed. The syringe is inverted, any gas bubbles ejected, and the titrant eliminated into a waste container. The syringe is refilled two more times to ensure that the titrant has not been contaminated with oxygen. At the conclusion, taps BB and B9 are closed and the flask placed aside. (In agreement with the report of Foust *et al.*,[19] we find that solutions of dithionite are stable for up to a week when prepared as described.) The titrant is then standardized spectrophotometrically by following the reduction of lumiflavin 3-acetate as described by Burleigh *et al.*[18] An optical cuvette is modified by addition of a 19/22 outer joint and a sidearm bearing a 7/25 inner capillary joint (cf. Fig. 2). Three milliliters of a 10^{-4} *M* solution of flavin are placed in the cuvette and the two joints closed, the 19/22 joint with a stopcock similar to that employed with the titrant flask and the 7/25 inner joint with a 7/25 outer cap. The solution is made anaerobic, placed under positive pressure and the cap removed and replaced with the syringe as described in detail below for protein solutions. [The cuvette unit employed for the standardization is modified from that described previously (Fig. 1[18]) in that the gas-lock unit has been replaced by a 7/25 inner standard-taper joint and cap: the mounting of the syringe on this joint follows the procedure described below for the combined optical–EPR titrator.]

The protein solution to be titrated is prepared in the titrator. The joints J1 and J2 are closed with caps and the main arm T is closed with the stopcock S3. The apparatus is then connected to the anaerobic train and the gas atmosphere replaced with freshly purified argon, with the titrator in a vertical position. This is achieved by first eliminating the original atmosphere via application of a vacuum and then replacing the vacuum with the fresh inert gas. The titrator is then laid horizontal in a plastic dishpan with the bulbous portion immersed in ice water, and agitated for 2–3 min to equilibrate the protein solution with the gas phase. Occasionally the titrator is returned to the vertical position to ensure thorough mixing with any liquid which may have been retained in the optical chamber. At the end of the equilibration period, the titrator is returned to the vertical position and the argon atmosphere replenished. Note that equilibration is not executed *in vacuo* for this leads to frothing and denaturation of the protein samples.

This equilibration cycle is repeated 6–10 times depending upon the vigor with which the agitation is executed. (During the early cycles of this process, tap S1 is open, thus eliminating air from its bore.)

At this point tap S3 is opened and the titrator is exposed to a positive pressure of argon. The cap on J1 is slowly removed and the inner joint immediately regreased. Some titrant is expelled from the syringe; the needle is almost seated on the joint. The syringe is held in this position for ca. 20 sec to allow the escaping argon to displace any air remaining in the outer joint attached to the syringe. The syringe is then seated in position and secured in position with springs. Tap S3 is closed, and the titrator is removed from the anaerobic train. The optical spectrum is now recorded to establish the magnitude of any evaporative losses during the preceding manipulation.

To calculate the equivalents of reductant added at each point throughout the redox titration, it is necessary to know both the total amount of titrant added at each point and the volume of solution removed from the titrator with each sample, for the total volume in the titrator decreases substantially through a titration, typically from 5 ml to 1 ml; thus additions at later stages of the titration produce much greater amounts of reduction than do the same additions at the beginning of the titration.

CALIBRATION OF THE TITRATOR TRANSFER PIPETTE. To establish the volume removed with each sample, it is necessary to calibrate the transfer pipette. This is achieved by filling the pipette to one of its engraved graduation marks with a solution of 10.0 m*M* ferricyanide in precisely the same way as is employed in obtaining a typical EPR sample (see above). However, at this point the balance of the ferricyanide solution is removed from the titrator which is then rinsed carefully with water, then methanol, and finally dried with a stream of nitrogen gas. Care is taken at this point not to make contact between the tip of the transfer pipette and the rinsing fluid. Then 10.0 ml of water are pipetted into the titrator, and the ferricyanide sample contained in the transfer pipette is readmitted to the vessel by opening tap S2 to air. The pipette is rinsed by sucking the diluted ferricyanide solution up into the transfer pipette several times. The absorbance of this solution is measured at 420 nm (A_{420}), and the volume corresponding to the specific calibration mark (V_c) is calculated from:

$$V_c = \frac{10 \times A_{420}}{2.04 - A_{420}}$$

The apparatus is cleaned and dried and the procedure repeated for each of several appropriate calibration marks on the transfer pipette.

TITRATION AND SAMPLING PROCEDURE. In addition to obtaining an initial optical spectrum, it is routine to remove a sample of reaction mixture prior to commencing the titration. The sampling procedure is as follows. Stopcock S2 is mounted on joint J2 and an EPR tube mounted on S2. The stopcock and EPR tube are then evacuated and filled with argon 6 times via the three-way stopcock and sidearm V; they are finally left under vacuum. The titrator is placed in a horizontal position with the conical extrusion C lowermost so that the top of the sampling pipette is below the surface of the protein solution. Stopcock S1 is now rotated *very* slowly so that the vacuum present in the EPR tube is applied to the liquid. The liquid will rise in the pipette, and when an adequate amount of liquid has entered the pipette, stopcock S1 is closed. The position of the meniscus relative to the graduations on the pipette is noted; this allows subsequent quantitation of the volume of liquid removed. The titrator is now placed in a vertical position, and the protein solution is allowed to drain into the optical chamber and away from the tip of the transfer pipette. Stopcock S1 is once more rotated very slowly allowing the liquid present in the transfer pipette to be transferred into the EPR tube by virtue of the pressure differential in the main titrator and the sampling container (i.e., EPR tube). When transfer is complete, stopcocks S1 and S2 are closed, the EPR tube plus stopcock S2 is detached as one unit, the liquid is shaken to the bottom of the EPR tube, and the sample is then frozen in liquid nitrogen prior to examination by EPR. Stopcock S2 is now dismounted from the EPR tube and, after cleaning, is available for subsequent samples.

In the above procedure, it is important that all transfers be carried out slowly so that surface tension moves the liquid sample as a continuous unit, thus minimizing contamination of subsequent samples by liquid left as a film on the walls of the transfer pipette.

Addition of titrant is achieved by rotating the threaded plunger by controlled amounts: displacements produced by as little as 1/4 of a turn can be achieved reproducibly. The titrator is then tilted to wash the needle tip with enzyme solution, and the mixture of protein plus titrant is agitated to achieve complete mixing. The titrator is then placed in a spectrophotometer and the progress of reaction followed spectrophotometrically. When the reaction is complete, an optical spectrum can be recorded and, if desired, a sample removed for EPR spectroscopy as described above. This procedure is repeated until the titration has been carried to completion.

3. Extensions and Possible Improvements. Several comments can be made which might be of value to anyone considering the construction of this device. First of all, ready and complete transfer of liquid in and out of the optical cuvette can be obtained only with cuvettes of 2-mm path length or greater; it is most difficult to obtain easy transfer with 1-mm light path cuvettes. Furthermore, there should be no constriction at the junction of the cuvette and the attached tubing: this requirement is met by the cuvette recommended.

Second, it may well be the case that reversing the arrangement of inner and outer joint at J1 and mounting an inner joint on the syringe can minimize the hazard of introducing air into the system when the syringe is mounted in place. This alternative configuration has not been tested.

Finally, it should be obvious that the utility of this device is not limited to EPR, but the receiving vessel can in fact be designed to fit the requirements of many spectroscopic methods; NMR amd Mössbauer spectroscopies come immediately to mind.

b. Discontinuous Titration with Solid Titrant

The apparatus has been shown in Fig. 1, where related devices are also seen. In operation a small (1–10 mg) aliquot of solid is weighed into a bulb with a stem (Fig. 1D″) prepared from a melting-point capillary (e.g., Kimax #34505). The bulb is evacuated and the stem is sealed off (a 22-gauge square-cut piece of hypodermic needle stock, on the end of a gas-oxygen torch, provides a small flame for this). Bulb D″ is placed in D, the protein solution is layered in around the bulb, the bulb-crusher arm D′ is inserted and sealed in place with greased joint d, and the protein solution is degassed as above. At the appropriate moment, the bulb D is broken by carefully twisting the outer end of D′ so that the flattened inner end of D′ crushes D″ against the side of the main bulb of D, which should be flattened for this purpose. The apparatus is gently shaken to dissolve the solid titrant, the glass particles (generally somewhat paramagnetic) are allowed to settle, and the solution is decanted into the EPR tube portion of the apparatus as described above.

It is of interest to describe the preparation of diluted solid sodium dithionite, a low-potential reductant of wide applicability in titrations of biological electron carriers. A 500-ml flask and 12[26] Teflon or glass balls (18 mm diameter) are dried for 16 hr at 70° and subsequently cooled *in vacuo*. The dilution is made from solids dried *in vacuo* over P_2O_5 for several days, and a typical formulation consists of 100 mg sodium dithionite (British Drug Houses), 10 gm KCl (Mallinkrodt AR), and 300 mg Tris base (Calbiochem "low heavy metal") weighed into the dry flask which is then evacuated and flushed 4 times with Ar while being rotated slowly on a Buchler evaporator. The apparatus is then sealed (vacuum stopcock) and rotated at 60 rpm for 16 hr, the flask being immersed as ice during this time. The apparatus is warmed to room temperature, and the fine powder is transferred to storage vials and kept in a vacuum desiccator. Such a powder may be tested[14] with flavin mononucleotide (FMN) or cytochrome c and is usually found to have greater than 90% of the titer of $Na_2S_2O_4$ expected from the composition. A few nanomoles of reductant are easily weighed out using a dilution such as is described here.

c. Electrochemically Defined Samples

It often may be desirable to prepare sample solutions in which the oxidation-reduction potential is controlled at known values and samples

[26] The Teflon must be thoroughly freed of oxygen.

are subsequently frozen for EPR spectroscopy. Generally, one may wish to use a series of mediators in solution along with a platinum electrode/reference electrode combination and adjust the potential in a thermostated closed compartment by additions of oxidants or reductants.[27] An automatic potentiostat for doing this has been described,[28] from which samples could be removed to a sample tube like that of Fig. 1A. The titration of a nitrogenase component protein[29] was accomplished with the removal of samples at intervals via gas-tight syringe to serum-capped EPR tubes. Heavy redox buffering was used to overcome the problem of residual oxidants in the EPR tubes. Chambers of the type shown in Fig. 1F and G should also be useful in this connection.

IV. Combination of Low-temperature EPR with Optical Spectroscopy

Apparatus and procedures suitable for this purpose have been described above under anaerobic titrations (III,3,a). Similar apparatus can, of course, be constructed for handling individual samples without the continuous titration features. Sometimes objections may arise to performing EPR and optical spectroscopy on different aliquots of a sample. Procedures have, therefore, been devised which allow both measurements to be made on the same aliquot.

When EPR spectroscopy in the liquid state is feasible, it is relatively simple to design a holder for a quartz flat cell which will allow mounting in a spectrophotometer with a sample compartment of suitable dimensions. For instance, a "tower"-type raised cover for the compartment may be built, with a holder protruding downwards, which can accommodate the flat cell.

When EPR spectroscopy is to be carried out at low temperature, optical measurements may be made through a vertical center section of ordinary EPR tubes, provided the tube is suitably masked and reproducibly positioned. A device suitable for this purpose has been described[13] with tubes of 4-mm inner diameter. At the concentrations usually required for EPR spectroscopy excellent optical spectra can be obtained in the absorbance range of 0–1.0. In addition it is always feasible to obtain low-temperature reflectance spectra[8] on the frozen samples. Usually these spectra do not lend themselves to more than semiquantitative evaluation; however, they may be valuable in ascertaining the presence or absence of certain features (e.g., reduced cytochrome c in cytochrome c–cytochrome c oxidase reac-

[27] P. L. Dutton and D. F. Wilson, *Biochim. Biophys. Acta* **346,** 165.

[28] B. Ke, W. A. Bulen, E. R. Shaw, and R. H. Breeze, *Arch. Biochem. Biophys.* **162,** 301 (1974).

[29] W. G. Zumft, L. E. Mortenson, and G. Palmer, *Eur. J. Biochem.* **46,** 535 (1974).

tion mixtures). Also, since in reflectance spectroscopy long-wavelength spectral features are intensified, a number of bands, barely detectable by absorption spectrophotometry in the liquid state, have been recognized and exploited.

V. Combination of Low-temperature EPR and Mössbauer Spectroscopy

Often when Fe-containing proteins are being investigated, it becomes desirable to combine EPR and Mössbauer[30] studies. The most often adopted approach[31,32] has been to prepare a large volume of Fe^{57}-enriched protein and subsample it for EPR and Mössbauer spectra. Since optimal sample volumes for the two techniques are about 0.3 and 1.0 ml, respectively, and since in any event one may want to make sure that the material in the Mössbauer cell is in a known EPR state, it is convenient to do the two measurements on a single sample. This may, for example, be accomplished by making the sample up in a 12 mm (ID) by 14 mm (OD) by 18 mm (height) nylon cell contained within a chamber to the top of which is affixed (via a large short ground joint) a double-septum seal as in Fig. 1F. (The chamber replaces the large side bulb and is connected to a 12/18 outer joint for evacuation through a stopper as in Fig. 1B.) The apparatus is partially immersed in liquid nitrogen to freeze the sample, and the frozen sample is subsequently removed and stored under liquid nitrogen. For EPR spectroscopy, a large sample-access cylindrical cavity (e.g., the Varian #917810) is employed with a quartz insert dewar (OD 24 mm, ID 16 mm) operated with N_2 or He boil-off gas cooling analogously to the normal rectangular cavity system.[1] The EPR signal from such a combination is quite sensitive to positioning in the cavity and, in general, the precision attainable with the normal system ($\pm 5\%$) is not easily realizable unless an internal paramagnetic standard can be included.

[30] See E. Münck, this volume [20].

[31] E. Münck, H. Rhodes, W. H. Orme-Johnson, L. C. Davis, W. Brill, and V. K. Shah, *Biochim. Biophys. Acta* **400,** 32 (1975).

[32] B. E. Smith and G. Lang, *Biochemistry* **137,** 169 (1974).

[11] EPR Spectroscopy of Components of the Mitochondrial Electron-transfer System

By HELMUT BEINERT

Introduction

General principles and techniques of EPR spectroscopy have been discussed in two articles of this treatise (this series, Vol. 49 [18, 20]) and special techniques particularly suited for work on oxidoreductive enzymes, such as those that form the subject of the present article, have been dealt with in this volume [10]. This article will, therefore, only be concerned with features specific to the EPR of components of electron-transfer systems with emphasis on those of mammalian mitochondria.

EPR-detectable Components and Properties of Their Signals

At the present state of our knowledge and technology, the following components of such systems are amenable to detection by EPR spectroscopy: heme-, copper-, iron-sulfur-, molybdenum-, and flavoproteins and quinones such as ubi- or naphthoquinones. In addition, iron compounds of unknown structure and function (signal at $g = 4.3$) appear to equilibrate with electron-transfer systems over a range of rates; these may be sufficiently rapid so that these electron acceptors cannot be completely discounted, particularly in equilibrium-type experiments.

Table I shows various subclasses of these categories, a number of their EPR features, and conditions for their observation. Specific comments on the numbers given in this table are incorporated into footnotes. Some more general comments, however, are also in order.

The magnetic field regions (on a g-factor scale) within which signals are likely to be found are given so as to include the broadest range of values observed (according to the author's knowledge) in biological materials for the particular paramagnetic species. This does not mean that the values actually observed will be randomly distributed within that range, but they will, in general, be found in a more narrow field region. In Table I this latter region is given in parentheses.

Most signals can be observed over a range of ~30 to several hundred degrees. In some cases the upper limit is not known. Unless special circumstances exist, such as, e.g., a temperature-dependent population of states in spin >1/2 systems, one would always expect to be able to detect a certain signal as the temperature is lowered toward 0°K, since the sensitivity of detection increases with T^{-1}. In practice, however, most instruments (and practically all commercial ones) cannot operate reliably at the low

ISBN 0-12-181954-X

TABLE I

EPR CHARACTERISTICS OF SPECIES EXPECTED IN PREPARATIONS OF ELECTRON-TRANSFER SYSTEMS

Effective spin state	Heme		Copper	Fe-S		Molybdenum	Flavin or ubiquinone semiquinone
	1/2 (low spin)	1/2[a] (high spin)	1/2	1/2		1/2	1/2
Formal ox. state as observed	+3	+3	+2	+1[b]	+3[b]	+5	
	x y z	x y z	x y z	x y z	x y z	x y z	x y z
g value range[c,d]	0–2; 1.5–3; 2–4	1.5–2; 4–6; 6–8	1.9–2.1; ~2 2.1–2.4	1.7–2.0; 1.8–2.1; 2.0–2.2	2.0; 2.0; 2.0–2.1	1.93–1.97; 1.95–1.98; 1.97–2.1	2.00; 2.00; 2.00
	(1.2–1.9; 2.2–2.5; 2.2–3.4)	(1.8–2.0; 5.5–6.0; 6.0–6.5)	(2.0–2.1; 2.0–2.1; 2.1–2.4)	(1.8–1.9; 1.9–2.0; 2.0–2.1)	(as above)	(1.95–1.97; 1.95–1.98; 1.97–2.0)	(as above)
Temp. (K)	<120	<300	<150	<150	<30	<300	<300
Optimum T	10–20	10–20	~80	10–20	4.2	100–150	150–300
Power (μW) P at optimum T	1000	5000	10,000	10–1000	100	1000	<10
Separation T^d	30	4.2	120			150	150
Separation p^d	100	100	1000			1000	1–10

[a] Although the spin state for high-spin iron is 5/2, the effective spin is 1/2 for the single transition that is observed as shown.

[b] The formal oxidation state given here is that applying only to the Fe-S cluster itself, not including the RS-ligands as has often been done in the past. Reduced ferredoxin is then $[2Fe\text{-}2S]^{1+}$ or $[4Fe\text{-}4S]^{1+}$; oxidized ferredoxin $[2Fe\text{-}2S]^{2+}$ or $[4Fe\text{-}2S]^{2+}$ and superoxidized ferredoxin, as it occurs in Hipip. is $[4Fe\text{-}4S]^{3+}$.

[c] The symbols x, y, and z as used here imply the relation $x < y < z$. In cases where an exact assignment of g values has been made or will be made, the values have to be exchanged correspondingly, if this relationship does not hold.

[d] See text.

microwave powers required to avoid saturation at very low temperatures and cannot produce the amplification necessary for operation at such low powers (nano- and picowatt region!). The optimum temperature as given in the table therefore indicates a range in which saturation problems are manageable, sensitivity of detection is approximately optimal for conventional EPR spectrometers, and good resolution is obtained. With signals of relatively rapid spin relaxation rate (i.e., broad lines), such as those of most Fe-S centers, resolution is rapidly lost above ~15°K, whereas for Cu and Mo signals, which are less relaxed, resolution may be better at higher temperatures. The optimum T as given in the tables is meant for the signal under consideration as such and does not take account of interferences by other overlapping signals. Because of possible interference, when materials with a number of paramagnetic components are studied (e.g., mitochondria), the temperature range over which one can study a signal efficiently may be much more limited. In Table I, the lines marked "separation T or P" attempt to specify conditions of temperature and power under which a signal of a certain component can best be separated from or emphasized over other signals usually associated with it in electron-transfer systems. It must be remembered, however, that, depending on what interferences are to be eliminated, the conditions for separation may have to be varied so that the number given can only be a guide for the situation most commonly encountered. It should also be noted that in many instances the "separation" conditions may not be the optimal conditions for quantitative work, since they represent a compromise, where, e.g., one may tolerate some saturation of the signal he wants to study, if at the same power interferring signals are effectively eliminated.

As evident from the foregoing, there is no single condition under which all components of systems with multiple paramagnetic species can be optimally observed; however, this very feature makes it possible to separate to some extent overlapping signals. In the following some special precautions or limits as to optimal conditions of spectroscopy, specifically for components of the mammalian mitochondrial electron-transfer system, will be mentioned and references to the literature where further information may be found will be given.

Cytochrome aa_3 (Cytochrome c Oxidase). The EPR signals and problems in their quantitative evaluation have been described in some detail.[1-4] There are differences in detail between signals observed with different

[1] C. R. Hartzell and H. Beinert, *Biochim. Biophys. Acta* **368,** 318 (1974).

[2] C. R. Hartzell and H. Beinert, *Biochim. Biophys. Acta* **423,** 323 (1976).

[3] H. Beinert, R. E. Hansen, and C. R. Hartzell, *Biochim. Biophys. Acta* **423,** 339 (1976).

[4] R. Aasa, S. P. J. Albracht, K.-E. Falk, B. Lanne, and T. Vänngård, *Biochim. Biophys. Acta* **422,** 260 (1976).

preparations, particularly of the high-spin signals ($g = 6;2$). These signals are very sensitive to temperature. In quantitative comparison, careful control of temperature and signal positioning (which influences temperature in a gas-flow system!) is therefore necessary. The major low-spin signal ($g \sim 3; 2; 1.5$) is usually asymmetric, indicating the presence of more than one species. The shape of this signal is very sensitive to environment (ions, gases, solvent).[1] It is, therefore, very hazardous to compare, for quantitative evaluation, simply signal size after various treatments. In quantitative evaluations of the copper signal, it must be considered that this signal is composed of at least two species. The so-called "inactive" copper, which shows hyperfine structure of the Cu-nucleus with spacings of $\sim$150 G, is readily saturated with microwave power. This can be used to advantage in spectroscopy at $\sim$10°K, where the signal from "inactive" copper can be suppressed. The quantity of "inactive" copper varies widely from preparation to preparation and between types of preparations.

Cytochrome c. The EPR spectrum of cytochrome c has been well investigated under a number of conditions.[5-9] In experiments with both cytochrome c and its oxidase, the overlap of the low field and center resonances complicates quantitative evaluations, particularly of the component present in lower quantity. Computer subtraction of the signals of the pure species is not very helpful in most cases, since small but significant shifts of resonances and changes of shape occur when both proteins are present. A procedure applicable for comparisons of intensities under a number of conditions is to determine the intensity of the low-spin signal of cytochrome c from a point chosen on the envelope of the low-field peak sufficiently far down-field so that the low-field resonance of cytochrome oxidase does not interfere, and to determine the intensity of the low-spin signal of cytochrome c oxidase from the high-field peak at $g = 1.5$. For the latter, signal averaging is often required.

Cytochrome c_l. For this cytochrome only the low-field peak is known with certainty,[10] but the center- and high-field peaks will probably, in any event, not be useful for quantitative work. The low-field peak overlaps with that of a b cytochrome (b_{561}), and the separation of the contributions of the two components is even more difficult than described for the pair cyto-

[5] I. Salmeen and G. Palmer, *J. Chem. Phys.* **48,** 2049 (1968).
[6] C. Mailer and C. P. S. Taylor, *Can J. Biochem.* **49,** 695 (1971).
[7] C. Mailer and C. P. S. Taylor, *Can J. Biochem.* **50,** 1048 (1972).
[8] W. B. Mims and J. Peisach, *Biochemistry* **13,** 3346 (1974).
[9] W. E. Blumberg, J. Peisach, B. Hoffman, E. Stellwagen, E. Margoliash, L. Marchant, J. Tulloss, and B. Feinberg, *Fed. Proc., Fed. Am. Soc. Exp. Biol.* **32,** 469 (1973).
[10] N. R. Orme-Johnson, R. E. Hansen, and H. Beinert, *J. Biol. Chem.* **249,** 1928 (1974).

chrome c and aa_3. Cytochrome c_1 can be differentially reduced with ascorbate.

Cytochrome b. A number of different signals of b cytochromes have been observed. The only signals of mitochondrial b cytochromes which are almost with certainty derived from the native components are those attributed to b_{561} and b_{565}.[10] The signal of the b cytochrome observed in complex II is different from these and has never been reported for mitochondria or submitochondrial particles. For both b_{561} and b_{565}, only the low-field peaks are readily observed, although for b_{561} the center line is almost certainly located at $g \sim 2$. The high-field lines are probably at very high fields, beyond the reach of conventional magnets, and are expected to be rather broad and not useful for quantitative work. The sharpness of the low-field line of b_{565} is surprising. Judging from the shape of this line, it represents more than one component. Since for this cytochrome only one g value is known, its quantity can be estimated from this low-field peak only with certain assumptions. A quantitative comparison of the *b* cytochromes represented in the two signals, which has appeared in the literature,[11] is therefore open to question.

Semiquinones. Both flavin and ubiquinone can be detected by a free-radical type signal at $g = 2.005$ when they are in the semiquinoid state.[12] In many instances, preparations of the electron-transfer system may contain both flavin and ubiquinone, so that an unambiguous identification from the signal alone is not possible at the present state of our knowledge. Inferences from width and shape of the signals of free flavin and ubisemiquinone are not very helpful since the effects of binding to protein or membranes and other influences of the milieu in the preparation studied are not known. Since there are workable procedures for the extraction of ubiquinone from mitochondria or SMP,[13] some reasonable inferences as to the identity of the signal observed can be made, but it is unknown what influence extraction of ubiquinone may have on the appearance of flavin in the semiquinoid form. Rather generally the quantities of either flavin or ubiquinone represented in the radical signals observed with mitochondria or submitochondrial particles are a small fraction of the amount of these substances present. This is different with some purified components of the electron-transfer system.[14]

[11] D. V. Dervartanian, S. P. J. Albracht, J. A. Berden, B. F. Van Gelder, and E. C. Slater, *Biochim. Biophys Acta* **292,** 496 (1973).

[12] D. Bäckström, B. Norling, A. Ehrenberg, and L. Ernster, *Biochim. Biophys. Acta* **197,** 108 (1970).

[13] B. Norling, E. Glazek, B. D. Nelson, S. L. Ernster, *Eur. J. Biochem.* **47,** 475 (1974).

[14] H. Beinert, B. A. C. Ackrell, E. B. Kearney, and T. P. Singer, *Eur. J. Biochem.* **54,** 185 (1975).

It is possible that in the more integrated preparations little semiquinone of either substance is formed, but the possibility must also be considered that the semiquinone(s) is (are) present, and that it is not represented in the usual radical signals because of some interaction with other paramagnetic components in the system. In fact, for ubisemiquinone this latter possibility has been verified. Two fairly strong resonances ($\Delta H \sim 80$ G) centered at $g \sim 2$, which are readily observed at <20°K in whole tissue, mitochondria, and submitochondrial particles over a relatively wide range of partial reduction, are almost certainly due to the interaction of ubisemiquinone and another paramagnetic component, presumably a second ubisemiquinone molecule.[15,16] Since *in vitro* the equilibrium in the system

$$2 \text{ ubisemiquinone} \rightleftharpoons \text{ubiquinone} + \text{ubiquinol}$$

is strongly in favor of disproportionation, it appears that ubisemiquinone is stabilized in this interacting form.

NADH Dehydrogenase. Whenever feasible, the low-field lines of the Fe-S centers of this dehydrogenase (see Table II) should be used for quantitative work, since they are better separated from each other than the center and high-field lines. Particularly the line(s) at $g = 1.95–1.92$ may have many contributions, some from even yet unknown components, so that considerable errors can arise. Center(s) 1 (a,b) is, however, hard to estimate from its low-field line, as this is too close to the semiquinone signal, to those from high potential-type Fe-S proteins, to the low-field line of the Fe-S protein of the cytochrome bc_1 complex (Rieske protein), and to those from succinate dehydrogenase. The size and shape of the line of center 1 at $g \sim 1.94$ depends very much on the contributions from other Fe-S centers and can only be used for quantitative evaluation to a rather limited extent. It is useful to visualize the mutual influences of overlapping signals by computer simulation, and in some limited instances calibration spectra of mixtures may be simulated. Good examples of this exist.[17,18] In complex preparations, it is almost impossible to separate quantitively the contributions of centers 3 and 4, particularly since the assignment of the center and high-field resonances to these centers is not clear at this time.[17–20]

[15] F. J. Ruzicka, H. Beinert, K. L. Schepler, W. R. Dunham, and R. H. Sands, *Proc. Natl. Acad. Sci. U.S.A.* **72,** 2886 (1975).

[16] W. J. Ingledew, J. C. Salerno, and T. Ohnishi, *Arch. Biochem. Biophys.* **177,** 176 (1976).

[17] S. P. J. Albracht and G. Dooijewaard, *in* "Electron Transfer Chains and Oxidative Phosphorylation" (E. Quagliarello *et al.*, eds.) p. 49. North-Holland Publ., Amsterdam, 1975.

[18] S. P. J. Albracht, G. Dooijewaard, F. J. Leeuwerik, and B. Van Swol, *Biochim. Biophys. Acta* **459,** 300 (1977).

[19] T. Ohnishi, *Biochim. Biophys. Acta* **387,** 475 (1975).

[20] H. Beinert, *in* "The Iron-Sulfur Proteins" (W. Lovenberg, ed.), Vol. 3, p. 61. Academic Press, New York, 1976.

Center 1 can be observed at higher temperature than the other signals, but usually, under these conditions, the signals become sufficiently poor that quantitative evaluation is difficult, particularly since broad underlying lines from other centers may still be present. The low-field line of center 3 (and possibly also center 4) is well singled out at low temperature and high microwave power, whereas that of center 2 is readily saturated under these conditions. There is evidence for the heterogeneity of center 1.[17-19,21] The implications of this are not clear at this time. It is likely that in most circumstances only one of the components is being observed.[18,19] The original literature will have to be consulted on this point. Similarly, additional Fe-S centers (5 and 6) that have been reported to be present in NADH-dehydrogenase are poorly characterized, so that they will not be considered here.[17-22]

Succinate Dehydrogenase. Three different signals typical of Fe-S components have been observed in preparations of succinate dehydrogenase at different oxidation states and have also been reported for mitochondria and submitochondrial particles under corresponding conditions.[14,23,24] One of these signals is of the type exhibited by the high-potential Fe-S protein of *Chromatium*,[25,26] which is that of a ferredoxin, higher by one electron in its oxidation state than, e.g., oxidized spinach ferredoxin. This oxidation state is expressed in shorthand notation at the 3^+ state, with oxidized and reduced spinach ferredoxin being in the 2^+ and 1^+ state, respectively. Whenever signals appear from the paramagnetic 1^- state of a Fe-S center, one must ask whether the same Fe-S center may not also show a signal in the 1^+ state (the 2^+ state is diamagnetic). Thus the appearance of three Fe-S type signals in succinate dehydrogenase does not *per se* mean that there are three different Fe-S centers present, although this has become very likely. The center present in the 3^+ state in the oxidized state of mitochondria and SMP apparently becomes very labile on extraction of succinate dehydrogenase. The corresponding signal is therefore readily lost, unless special precautions are taken.[14,24] This Fe-S center has been called center 3 in some publications.[24] Center 1 of succinate dehydrogenase is that observed on reduction with succinate by its typical signal characteristic of reduced ferredoxins. This signal is readily observed at liquid nitrogen temperature.

[21] S. P. J. Albracht, *Biochim. Biophys. Acta* **347**, 183 (1974).

[22] H. Beinert and F. J. Ruzicka, *in* "Electron Transfer Chains and Oxidative Phosphorylation" (E. Quagljarello *et al.*, eds.), p. 37. North-Holland Publ., Amsterdam, 1975.

[23] T. Ohnishi and J. C. Salerno, *J. Biol. Chem.* **251**, 2094 (1976).

[24] T. Ohnishi, J. Lim, D. B. Winter, and T. E. King, *J. Biol. Chem.* **251**, 2105 (1976).

[25] K. Dus, H. De Klerk, K. Sletten, and R. G. Bartsch, *Biochim. Biophys. Acta* **140**, 291 (1967).

[26] T. H. Moss, D. Petering, and G. Palmer, *J. Biol. Chem.* **244**, 2275 (1969).

TABLE II

EPR CHARACTERISTICS OF INDIVIDUAL Fe-S PROTEINS OR CENTERS OF MAMMALIAN MITOCHONDRIA

		DPNH dehydrogenase center				Succinate dehydrogenase center			b–c_1 (Rieske)	ETF-UQ oxido-reductase	Hipip soluble aconitase	Outer membrane protein
		1	2	3	4	1	2	Superoxidized (Hipip)				
Formal ox. state as observed[a]		+1	+1	+1	+1	+1	+1	+3	+1	+1	+3	+1
g value[b]	la;	1.94[c,d]	1.93[d,e,f]	[1.86][e]	[1.86][e]	1.91[g,h]	1.91[g,h]	1.99[i]	1.81[j] (1.78)[k]	1.89[m]	[2.00][n]	1.89[p]
		1.94	1.93	[1.89]	[?]	1.93	1.93	2.01	1.89	1.94	2.01	1.94
		2.03	2.05	[2.10]	[2.11]	2.03	2.03	2.02	2.025	2.086	[2.02]	2.01
	lb:	1.91[q]										
		1.94										
		2.03										
g value[b]	la:	1.94[c,f,r]		1.88[f]	1.86[f]							
		1.94		1.94	1.93						2.02[s]	
		2.02		2.10	2.04							
	lb:	1.93										
		1.93										
		2.02										
Temp. (K)		<100	<30	<30	<30	<120	<100	<20	<120	<30	<30	<200
Optimum T^t		10–20	10–20	10–20	10–20	10–20	10–20	5–15	10–30	10–20	5–15	10–30
Power (P) (μW) at optimum T^t		100–500	100–250	100–500	100–500	10–30	20–50	1000–10,000	100–500	100–250	100–1000	10–30
Separation T^t		50	17	10	10	50	10	5	40	10	5(30)[u]	100
Separation P^t		500	500	2000	2000	30	1000	3000	3000	1000	1000	10,000
Midpoint	1a:	−380[c,d]	−20[d]	−240[d]	−410[d]	+30[z]			+280[z,dd]	+40[z]	[< +150][s]	
potential[v,w]	1b:	−240[c,d]	−80[x,y]		[−240]			+ 80[aa,bb]				

SMP						$+120^{cc}$		
Purified complexes	1a:	$-385^{c,x,y}$	$^{x,y}-135^{x,y}$	0^{h}	-260^{h}	$+60^{ee,ff}$ (pH 7.4)	$+280^{gg,hh}$	$+35^{gg,hh}$
	1b:	$-260^{c,x,y}$						
Solubilized	1a:	$-385^{c,x,y}$	$-210^{x,y}$					
reconstitutively active	1b:	$-260^{c,x,y}$		-5^{h}	-400^{h}			
Solubilized reconstitutively inactive			$-265^{x,y}$					

[a] For the derivation of formal oxidation states see footnote *b* of Table I.

[b] The values of *g* are given in the sequence *x,y,z;* cf. footnote *c* of Table I. When the values are bracketed, they refer to peaks only of the derivative spectra.

[c] The values refer to centers 1a and 1b, respectively, as indicated.

[d] T. Ohnishi, *Biochim. Biophys. Acta* **387,** 475 (1975).

[e] N. R. Orme-Johnson, R. E. Hansen, and H. Beinert, *J. Biol. Chem.* **249,** 1922 (1974).

[f] S. P. J. Albracht, G. Dooijewaard, F. J. Leeuwerik, and B. Van Swol, *Biochim. Biophys. Acta* **459,** 300 (1977). The designation of center la and b by these authors has no relationship to that used by Ohnishi (see *d*).

[g] H. Beinert, B. A. C. Ackrell, E. B. Kearney, and T. P. Singer, *Eur. J. Biochem.* **54,** 185 (1975).

[h] T. Ohnishi, J. C. Salerno, D. B. Winter, J. Lim, C. A. Yu, L. Yu, and T. E. King, *J. Biol. Chem.* **251,** 2094 (1976).

[i] F. J. Ruzicka, H. Beinert, K. L. Schepler, W. R. Dunham, and R. H. Sands, *Proc. Natl. Acad. Sci. U.S.A.* **72,** 2886 (1975).

[j] N. R. Orme-Johnson, R. E. Hansen, and H. Beinert, *J. Biol. Chem.* **249,** 1928 (1974).

[k] The value 1.78 applies, when the preparation is reduced substantially. After reduction with ascorbate only, the value 1.81 is observed.

[m] F. J. Ruzicka and H. Beinert, *Biochem. Biophys. Res. Commun.* **66,** 622 (1975).

[n] F. J. Ruzicka and H. Beinert, *Biochem. Biophys. Res. Commun.* **58,** 556 (1974).

[p] D. Bäckström, I. Hoffström, I. Gustafsson, and A. Ehrenberg, *Biochem. Biophys. Res. Commun.* **53,** 596 (1973).

[q] According to results in *e* and *f* this value appears to be too low.

[r] Derived by computer simulation (cf. ref. *f*).

[s] T. Ohnishi, W. J. Ingledew, and S. Shiraishi, *Biochem. J.* **153,** 39 (1976).

[t] See text and Table I.

TABLE II (Continued)

[u] 5°K are useful to separate Hipip type signals from other signals of heme, Fe-S, or copper compounds, whereas separation of the signal of the soluble Hipip from that of the membrane-bound Hipip of succinate dehydrogenase can be better achieved at ~30°K.

[v] Largely determined in the presence of potential mediators and therefore subject to considerations as put forth by M. K. F. Wikström, *Biochim. Biophys. Acta* **301,** 155 (1973); P. L. Dutton and D. F. Wilson, *Biochim. Biophys. Acta* **346,** 165 (1974); W. J. Ingledew and J. B. Chappel, *Fed. Proc., Fed. Am. Soc. Exp. Biol.* **34,** 488 (1975); and R. W. Jones, T. A. Gray, and P. B. Garland, *Biochem. Soc. Trans.* **4,** 671 (1976).

[w] Unless otherwise indicated, the oxidation-reduction potentials refer to pH 7.2 and have an approximate error of ±20 mV. SMP refers to submitochondrial particles of pigeon heart (unless otherwise stated) as opposed to purified complexes of the electron-transfer system and solubilized preparations, as listed in subsequent columns. Values applying to mitochondria are in brackets.

[x] For preparations from beef heart mitochondria, measured at pH 8.0.

[y] T. Ohnishi, J. S. Leigh, C. I. Ragan, and E. Racker, *Biochem. Biophys. Res. Commun.* **56,** 775 (1974).

[z] T. Ohnishi, D. F. Wilson, T. Asakura, and B. Chance, *Biochem. Biophys. Res. Commun.* **46,** 1631 (1972).

[aa] For SMP from beef heart mitochondria, measured at pH 8.5.

[bb] W. J. Ingledew and T. Ohnishi, *FEBS Lett.* **54,** 167 (1975).

[cc] For SMP from beef heart mitochondria, measured at pH 6 to 7.

[dd] D. F. Wilson and J. S. Leigh, Jr., *Arch. Biochem. Biophys.* **150,** 154 (1972).

[ee] T. Ohnishi, J. Lim, D. B. Winter, and T. E. King, *J. Biol. Chem.* **251,** 2105 (1976).

[ff] T. Ohnishi, D. Winter, J. Lim, and T. E. King, *Biochem. Biophys. Res. Commun.* **61,** 1017 (1974).

[gg] J. S. Leigh, Jr. and M. Erecińska, *Biochim. Biophys. Acta* **387,** 95 (1975).

[hh] For preparations from pigeon breast mitochondria, measured at pH 7.4.

With dithionite as reductant the ferredoxin-type signal changes in shape and grows in intensity. It is widely accepted that these changes are due to reduction of an additional Fe-S center, which has been called center 2. The signal of center 2 can also be detected at 77° but it is more temperature-sensitive and less readily saturated than that of center 1. A quantitative estimate of this center is difficult to obtain since center 1 is also in the reduced state under all conditions when center 2 is reduced. In the author's laboratory, the ratio of the concentration of center 1 (obtained by reduction with succinate only) to that of center 1 plus 2 has never exceeded 1:1.5 for reconstitutively active and inactive as well as particulate (complex II) preparations and often has been below that.[14] There is evidence for interaction of the two centers from saturation studies,[23] but it is not certain whether this is responsible for the nonintegral intensity observed for the combined centers. Since center 2 is not reduced by succinate and since the mentioned uncertainty exists about stoichiometries, the significance of center 2 is not clear at this time.

Although both the center that appears in the 1^- state and center 1 are completely reduced by succinate in the course of minutes, only a fraction of these centers is reduced by succinate or reoxidized by *Q* analogues within the turnover time of the enzyme, indicating at least that all preparations that have been investigated are heterogeneous with respect to the reactivity of their active centers.[14,27]

ETF-ubiquinone Oxidoreductase. Evidence of various kinds has been adduced that this previously unknown Fe-S-flavoprotein functions as the link of the fatty acyl CoA dehydrogenase system to the terminal electron-transfer system of mitochondria.[28,29] It contains acid-extractable FAD and a single [4Fe-4S] center. Its EPR signal is more readily saturated than those of NADH dehydrogenase but less readily than that of center 1 of succinate dehydrogenase. In preparations of mitochondria from most sources, it appears as a shoulder or relatively minor peak just upfield of the low-field line of center 3 of NADH dehydrogenase. This signal is particularly weak in preparations from beef heart. It is more pronounced in rat and pigeon heart, and the strongest (for mitochondria) signals have been found in mitochondria from brown adipose tissue.[29]

Fe-S Center of Cytochrome b–c_1 Complex (Rieske Fe-S Center). This center is found associated with the cytochrome b–c_1 complex. After reduction, its signal is readily observed at liquid nitrogen temperature. By raising the temperature, it can thus be separated from those originating from

[27] T. P. Singer, E. B. Kearney, B. A. C. Ackrell, and H. Beinert, *Proc. FEBS Meet., 10th, 1975* Symposium No. 8, p. 173 (1975).

[28] F. J. Ruzicka and H. Beinert, *Biochim. Biophys. Res. Commun.* **66,** 622 (1975); *J. Biol. Chem.* **252,** 8440 (1977).

[29] T. Flatmark, F. J. Ruzicka, and H. Beinert, *FEBS Lett.* **63,** 51 (1976).

reduced centers 3 and 4 of NADH dehydrogenase and of ETF-ubiquinone oxidoreductase, which, at low temperature, overlap with the prominent center line ($g = 1.89$) of the Rieske center. The high-field line of the Rieske center is too broad to be useful for other than qualitative observations. It should also be noted that the high-field line shifts from $g = 1.81$ to $g = 1.78$ on extensive reduction, probably in conjunction with the reduction of one of the associated b cytochromes.[10] In purified complex III and similar preparations,[30] there are usually found some Fe-S impurities, the signals of which overlap with those of the Rieske center. Such impurities are succinate and ETF-ubiquinone oxidoreductase and possibly additional, unknown Fe-S centers.[31] (However, see Erecinska *et al.*[30])

High Potential Type Fe-S Center. In addition to that Fe-S center of succinate dehydrogenase, which occurs in the 3^+ state, there is a second Fe-S center detectable in mitochondria which naturally occurs in this oxidation state. This center is readily solubilized on sound treatment of mitochondria.[32] Its signal is not quite as sensitive to temperature as the analogous one of succinate dehydrogenase, and it is somewhat more readily saturated with microwave power. The two signals can, therefore, be differentiated to some extent by taking advantage of these differences.[33] They are, however, not very pronounced. This Fe-S center has now been shown to be part of the enzyme aconitase.[33a]

Outer Membrane Fe-S Center. A signal typical of reduced ferredoxins was found associated with the outer membrane fraction of mitochondria.[34] This signal is readily observed at liquid nitrogen temperature and is relatively easily saturated with microwave power. In integrated preparations it is elicited by addition of NADH or succinate plus phenazine methosulfate.[21] It has been shown that, unless very efficient methods of separation are used, preparations of submitochondrial particles, which are thought to be derived from the inner membrane, contain about as much of the outer membrane Fe-S center as do mitochondria.[35] Since the signal from this center is quite intense and overlaps the signals from succinate and NADH dehydrogenases, quantitative interpretations of spectra obtained from submitochondrial particles can be considerably in error.[21,35] This applies particularly to the earlier work done in the temperature range of 80–100°K,

[30] M. Erecínska, D. F. Wilson, and Y. Miyata, *Arch. Biochem. Biophys.* **177,** 133 (1976).
[31] I. Y. Lee and E. C. Slater, *Biochim. Biophys. Acta* **347,** 14 (1974).
[32] F. J. Ruzicka and H. Beinert, *Biochem. Biophys. Res. Commun.* **58,** 556 (1974).
[33] T. Ohnishi, W. J. Ingledew, and S. Shiraishi, *Biochem. J.* **153,** 39 (1976).
[33a] F. J. Ruzicka and H. Beinert, *J. Biol. Chem.* **253,** 2514 (1978).
[34] D. Bäckström, I. Hoffström, I. Gustafsson, and A. Ehrenberg, *Biochem. Biophys. Res. Commun.* **53,** 596 (1973).
[35] S. P. J. Albracht and H.-G. Heidrich, *Biochim. Biophys. Acta* **376,** 231 (1975).

where the signal of the outer-membrane protein is favored, whereas it is more likely to be strongly saturated at the powers usually applied at 10–20°K, where more recent work has been carried out. Since it has been shown that this center is part of the outer membrane, it can be used as an EPR-detectable marker of this mitochondrial fragment. The function of the center is unknown.

Stoichiometries of EPR-detectabe Components

We will consider here only stoichiometries derived by comparing results of integrations of EPR signals with quantities traditionally used for the characterization of certain component complexes or proteins of the electon-transfer system, such as, e.g., nonextractable (or histidyl-) flavin.

Cytochrome aa_3. The low-spin heme ($g \sim 3, 2, 1.5$) signal accounts for 50% of the heme of cytochrome aa_3[2,4] whereas the high-spin heme signal usually represents 5–20% of the total heme and only in rare instances up to ~40%.[3] At 9 GHz the copper signal has consistently been found to account for an amount of copper equivalent to 40% rather than the expected 50% of the total heme.

NADH Dehydrogenase. Before the heterogeneity of center 1 was recognized, it was estimated that all Fe-S centers in complex I occurred at approximately the same concentration with the stoichiometry to flavin being 2.5 to 4 Fe-S centers per acid-extractable FMN.[36] According to the more detailed investigations of Albracht *et al.*,[18] centers 1a and 1b each occur at only ~25% of the concentration of the other centers, while the ratio of FMN to total Fe-S centers falls into the range mentioned above.

Succinate Dehydrogenase. The center observed in the oxidized 3^+ state. In this spectrum the signals of the Fe-S centers in the 3^+ state and the found to occur approximately stoichiometric to the histidyl-flavin of the enzyme. In reconstitutively active preparations investigated in the author's laboratory, the oxidized center could account for maximally 35% of the flavin and usually less (~20%). Center 1 plus 2, as discussed above, always accounted for significantly less than 2 per histidyl-flavin.

ETF-ubiquinone oxidoreductase. There is one [4Fe-4S] center per acid-extractable FAD[28] and, according to a comparison of signal intensities, the ratio of ETF-ubiquinone reductase to NADH dehydrogenase in beef heart mitochondria is ~0.5 and ~2 in brown adipose tissue mitochondria from guinea pigs.[29]

Outer Membrane Fe-S Center. This center occurs at about 0.5 nmoles/mg protein in the outer membrane of rat liver mitochondria.[34]

[36] N. R. Orme-Johnson, R. E. Hansen, and H. Beinert, *J. Biol. Chem.* **249,** 1922 (1974).

Examples of EPR Spectra of Components of the Mitochondrial Electron-transfer System

In the following figures a few representative spectra are shown in which most of the EPR-detectable components discussed above can be recognized as they appear at different oxidation states of complex preparations. For details of the spectra from the purified components, the original literature will have to be consulted.

Figure 1 shows mitochondria or whole heart tissue in three different oxidation states. The preparations and oxidation states were chosen so that the most prominent Fe-S centers can be clearly recognized in at least one of the three spectra. The top spectrum shows the EPR spectrum of blowfly mitochondria which can be obtained with low endogenous substrate present. Thus, the signal of the high potential type of Fe-S center(s)[14,32] centered at $g = 2.01$ is well developed. There is some reduction of center 2 of NADH dehydrogenase ($g = 1.92$).[19,36] This center is rarely seen completely oxidized in mitochondria or submitochondrial particles. The center spectrum is that of beef heart mitochondria as isolated. In mitochondria from this source there appears to be sufficient endogenous substrate that, at the concentrations required for EPR spectroscopy, features of partial reduction clearly appear in the spectrum. In this center spectrum we can still recognize the signal of the membrane-bound Fe-S component paramagnetic in the oxidized 3^+ state. This signal is flanked on either side by a pair of lines (3309 and 3227G at 9.2 GHz) stemming from the interaction of ubisemiquinone with a paramagnetic species.[15,16] The high-field line is more pronounced. Fe-S center 2 of NADH dehydrogenase is more reduced than in the top spectrum, and the line typical for reduction of the Fe-S center of the cytochrome $b–c_1$ complex (Rieske's Fe-S center) can just be seen at $g = 1.89$.

The lowermost spectrum is that of whole pigeon heart in the reduced state. In this spectrum the signals of the Fe-S centers in the 3^+ state and the interacting species of ubisemiquinone are absent, whereas now the resonances of Fe-S centers 1–4 of NADH dehydrogenase ($g = 2.10; 2.05; 2.02; 1.94; 1.92; 1.89; 1.86$), of Rieske's Fe-S center ($g = 2.025; 1.89; 1.78$), and of ETF-ubiquinone oxidoreductase,[28] sometimes called "center 5" ($g = 2.08$),[37] stand out. Pigeon heart was chosen because the signal of ETF-ubiquinone oxidoreductase is usually very pronounced in preparations from this species.

Figure 2 shows the effect of a 100-fold change in microwave power on the signals of the Fe-S centers of NADH-ubiquinone reductase recorded at

[37] T. Ohnishi, D. F. Wilson, T. Asakura, and B. Chance, *Biochem. Biophys. Res. Commun.* **46,** 1631 (1972).

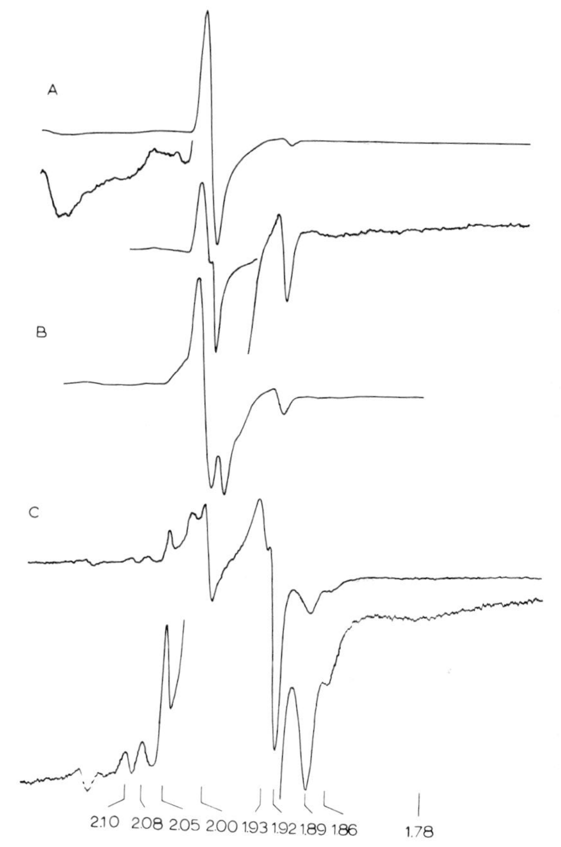

FIG. 1. EPR spectra of mitochondria or whole tissue at different states of reduction. (A) Blow-fly mitochondria, approximately 40 mg protein per ml, as prepared in 0.15 *M* KCl, 10 m*M* Tris (pH 7.3), 1m*M* EDTA and 0.5% bovine serum albumin (kindly provided by Dr. B. Sacktor). (B) Beef heart mitochondria, approximately 50 mg per ml, as prepared in 0.25 *M* sucrose, 0.02 *M* Tris, pH 7.4. (C) Whole pigeon heart frozen immediately after removal of the heart. All EPR spectra shown in this chapter represent the first derivative of the absorption (ordinate) with linearly increasing magnetic field. Unless otherwise mentioned, prominent peaks or shoulders are indicated on a g-factor scale, although this does not mean that the values given are true g-values. In many instances they will be very close to actual g-values. Conditions of EPR spectroscopy: (A) microwave power 2.7 mwatt; modulation amplitude 6.3 G; scanning rate 400 G per minute; temperature, 13.3K; enlarged wings left and right, modulation amplitude 8 G, 10× amplified; center inset, 0.09 mwatt, modulation of 6.3 G, 2.5× amplified. This inset recorded at low power shows that a radical signal is superimposed on the Hipip signal. (B) conditions as the main spectrum of (A). (C) conditions as (A) and (B), except microwave power 0.27 mwatt and scanning rate 200 G per minute. The enlarged wings are 5× amplified. Unless stated otherwise, the modulation frequency was 100 KHz and the microwave frequency 9.2 GHz for all figures of this chapter.

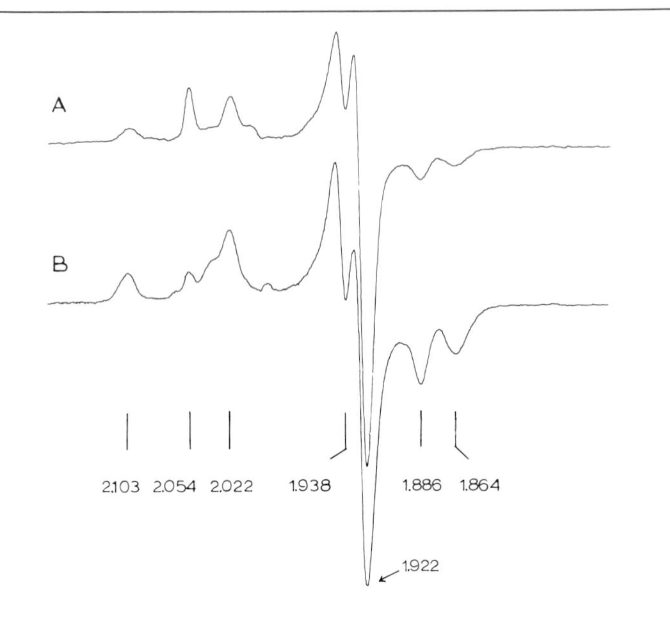

FIG. 2. EPR spectra of NADH ubiquinone reductase (18 mg of protein per ml dissolved in 0.66 *M* sucrose, 1 m*M* histidine and 0.05 *M* Tris-chloride, pH 8) after reduction with NADH (127 neq) recorded at different microwave powers to accentuate the relationships between the resonances of centers 1, 2 and 3. The conditions of EPR spectroscopy were: (A) power 3 μwatt; modulation amplitude 7.5 G; temperature 7.7°K, scanning rate 200 G/min and time constant 0.5 sec; (B) as in A but at 300 μwatt power and at 0.39 times the amplification used in (A).

7.7°K. With increasing power, the signal of center 2 disappears whereas those of centers 1, 3, and 4 stand out. In spectrum B, a line at $g \sim 2.04$ becomes visible, which, according to Albracht *et al.*,[18] is the g_z line of center 4.

Figures 3 and 4 represent a set of experiments designed to detect the Fe-S center of the outer mitochondrial membrane. The particles used for the experinent of Fig. 3 were only partly separated from outer membrane fragments, whereas those used in the experiment of Fig. 4 were separated by free-flowing electrophoresis. The condition for spectra A in both figures is addition of succinate. The signal resulting is essentially that of Fe-S center 1 of succinate dehydrogenase. The broad signal of $g = 2.05–2.10$ probably originates from a copper component of cytochrome c oxidase which was not completely reduced under the conditions of this experiment. The spectra B result following addition of phenazine-methosulfate after thawing of the samples used in A. In Fig. 3 one clearly sees a new compo-

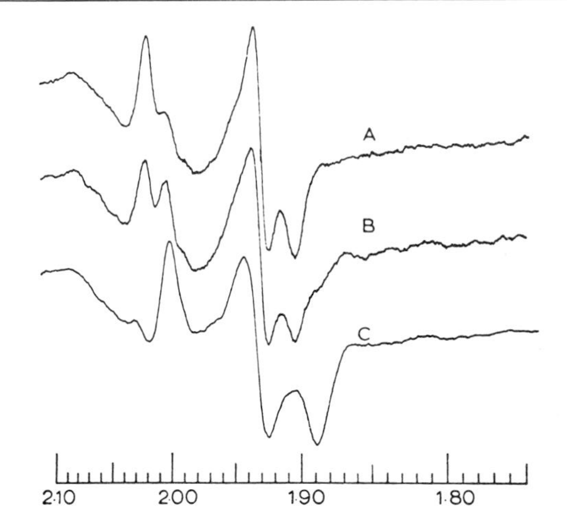

FIG. 3. Comparison of EPR spectra of mixtures of inner and outer membranes from beef heart mitochondria. The preparations were suspended in 0.25 *M* sucrose and 20 m*M* Tris-chloride (pH 7.5). (A). Succinate (25 m*M*) was added to a preparation of particles partially purified by free flowing electrophoresis. The suspension was frozen after 3 minutes at 20°C. (B). The tube used for (A) was thawed, 0.15 m*M* phenazine methosulfate was added and the suspension was frozen after 3 min at 20°. (C) A crude preparation of mitochondrial outer membranes was treated as under (B). Conditions of EPR spectroscopy: microwave power 130 mwatt; modulation amplitude 12.5G; scanning rate 250 G per minute, temperature 83K and microwave frequency 9.1 GHz. The amplification for (A) and (B) is the same. (With permission from Albracht and Heidrich, 1975.)

nent emerging at $g = 1.9$ which is missing in Fig. 4, representing the purified sample.

Effect of Alcohols on Submitochondrial Particles and Enzymes of the Respiratory Chain

Although consideration of such effects may not appear to be germane to a discussion of EPR in the study of the components of the electron-transfer system, EPR is able to provide direct evidence for them.[38] In addition, the requirement for relatively high concentration of proteins in EPR studies often makes it necessary to add reactants or inhibitors at higher concentrations than in activity assays or spectrophotometric work. Most inhibitors and some reactants (e.g., ubiquinones) are added in ethanol. While it would seem that 1–5 v/v% ethanol may be a negligible amount, this means

[38] J. C. Salerno and T. Ohnishi, *Arch. Biochem. Biophys.* **176,** 757 (1976).

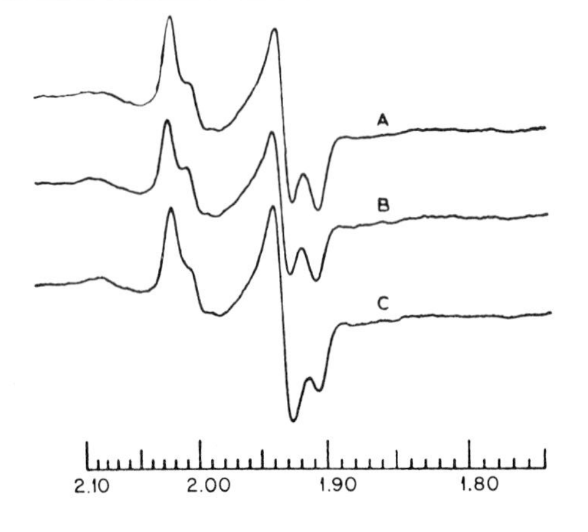

FIG. 4. EPR spectra of pure particles separated by free flowing electrophoresis. The particles were suspended as for Fig. 3. Substrate was added and the mixtures were frozen after 3 minutes at 20°C, (A) with 25 m*M* succinate; (B) with 25 m*M* succinate plus 0.15 m*M* phenazine methosulfate; (C) with 6 m*M* NADH. The conditions of EPR spectroscopy were those of Fig. 3 and the amplification was unchanged throughout. The absence of the Fe-S center of the outer membrane in the purified preparation can be clearly seen in comparison to Fig. 3. (With permission from Albracht and Heidrich, 1975.)

170–860 m*M* ethanol! Even at concentrations of 20 m*M*, several Fe-S centers show changes in signal intensity and/or shape that cannot be explained by oxidation-reduction, particularly those of NADH dehydrogenase and the center that appears in the oxidized state (3^+) of succinate dehydrogenase. Methanol as well as higher alcohols are less effective, although it should be remembered that higher alcohols may disrupt particles. Also, ethylene glycol has only about one-tenth the effect of ethanol, but its effects must be seriously considered at the concentrations used in subzero work. It remains to be shown whether the effect of ethylene glycol may be minimized by cooling the mixture to a temperature close to its freezing point. As a good alternative to ethanol, dimethyl sulfoxide is recommended as a solvent.[38] No effects on Fe-S centers have been observed with 5% dimethyl sulfoxide.

Acknowledgments

The author gratefully acknowledges the help of Drs. S. P. J. Albracht, R. C. Bray, T. Ohnishi, F. J. Ruzicka, and T. Vänngård in providing data from manuscripts in press and critical comments. The author was supported by the Institute of General Medical Sciences, USPHS, through a Research Career Award (5K6-GM-18, 442).

[12] Pulse Fourier-transform NMR Spectroscopy with Applications to Redox Proteins

By RAJ K. GUPTA and ALBERT S. MILDVAN

I. Introduction

High-resolution nuclear magnetic resonance (NMR) has proven to be a valuable spectroscopic method for the study of proteins,[1,2] hemoproteins,[3–6] and other redox systems.[7] As in other forms of spectroscopy, NMR data are generally expressed as absorbancy versus frequency (or energy). The spectral region used in NMR spectroscopy is in the radio frequency range, 60–600 MHz, for protons, corresponding to energies of 6–60 mcal/mole, requiring homogeneous magnetic fields of 14–140 kG. Because the energies involved are five orders of magnitude below those required to excite electrons in molecules, no chemical reactions are initiated. Hence NMR is a nonperturbing spectroscopic method.

High-resolution NMR spectra are usually examined by one of the following three methods:

1. The continuous-wave (CW)[1,8,9] method in which the frequency or magnetic field is continuously increased (or decreased) and the absorption of radio-frequency energy is measured.

2. The pulse Fourier-transform (pulse FT) method[10–14] in which the

[1] W. D. Phillips, Vol. 27D, p. 825; T. L. James, "Nuclear Magnetic Resonance in Biochemistry." Academic Press, New York, 1975; R. A. Dwek, "Nuclear Magnetic Resonance in Biochemistry." Oxford Univ. Press (Clarendon), London and New York, 1973.

[2] A. Allerhand and E. A. Trull, *Annu. Rev. Biochem.* **21,** 317 (1970); F. R. N. Gurd and P. Keim, Vol. 27D, p. 836; A. Allerhand, R. F. Childers, and E. Oldfield, *Ann. N.Y. Acad. Sci.* **222,** 764 (1973).

[3] K. Wuthrich, *Struct. Bonding (Berlin)* **8,** 53 (1970).

[4] C. Ho, L. W.-M. Fung, and K. J. Weichelman, this volume [13].

[5] A. G. Redfield and R. K. Gupta, *Cold Spring Harbor Symp. Quant. Biol.* **36,** 405 (1971).

[6] R. K. Gupta and A. G. Redfield, *in* "Structure and Function of Oxidation-Reduction Enzymes" (A. Åkeson and A. Ehrenberg, eds.), p. 337. Pergamon, Oxford, 1972.

[7] W. D. Phillips, M. Poe, J. F. Weiher, C. C. McDonald, and W. J. Lovenberg, *Nature (London)* **227,** 574 (1970).

[8] A. S. Mildvan and J. L. Engle, Vol. 26C, p. 654.

[9] E. D. Becker, "High Resolution NMR." Academic Press, New York, 1969.

[10] R. R. Ernst and W. A. Anderson, *Rev. Sci. Instrum.* **37,** 93 (1966).

[11] R. L. Vold, J. S. Waugh, M. P. Klein, and D. E. Phelps, *J. Chem. Phys.* **48,** 3831 (1968).

[12] A. G. Redfield, *in* "NMR: Basic Principles and Progress" (P. Diehl, E. Flück, and R. Kosfeld, eds.), Vol. 13, p. 152. Springer-Verlag, Göttingen, 1976; A. G. Redfield and R. K. Gupta, *Adv. Magn. Reson.* **5,** 81 (1971).

[13] A. S. Mildvan and R. K. Gupta, this series, Vol. 49 [15].

[14] T. C. Farrar and E. D. Becker, "Pulse and Fourier Transform NMR," Academic Press, New York, 1971.

ISBN 0-12-181954-X

transient signal or free-induction decay following a short, intense radiofrequency pulse to the sample is Fourier-transformed to yield a spectrum.

3. The rapid-scan correlation method[15,16] in which the ringing of an NMR signal following a rapid passage through resonance is used to compute the true free-induction decay which, in turn, is Fourier-transformed to yield the NMR spectrum.

Each of these methods has certain advantages and disadvantages. The continuous-wave methods require no computers and are simple and selective. However, their sensitivity is low since most CW spectra are obtained with a slow sweep rate often $\leqslant 1$ Hz/sec in order to approximate slow-passage conditions and produce lines that are undistorted by "sweep broadening" and ringing. In these experiments the magnetization vector is tilted only slightly from the equilibrium position, thus producing a relatively weak signal.

In the pulse FT method, on the other hand, a 90° pulse reorients the total magnetization of the sample, M_0, into the transverse plane; moreover, information on the entire spectrum, not just on a single line, can be obtained in a time of the order of $3T_2^*$ where T_2^* is the apparent lifetime of the transverse magnetization ($\leqslant 1$ sec), resulting in a marked improvement in sensitivity. The data thus obtained as a time response are not in an easily interpretable form, but it now is well known that simple Fourier transformation leads to the true slow-passage NMR spectrum. When long pulses are used, selected regions of the spectrum can be studied. However, the Fourier-transform method requires a computer.

In the rapid-scan correlation technique, the entire spectrum or any portion of the spectrum of the sample may be traversed under fast-passage conditions, i.e., at a sweep rate $\gg (2\pi T_2^{*2})^{-1}$ or $\geqslant 10$ Hz/sec for a line 1 Hz in width. Hence the correlation method is selective. The magnetizations corresponding to each line produce pronounced ringing, which persists for a time of the order of $3T_2^*$. In the rapid-scan experiment conditions may be chosen so as to tilt the magnetization vector all the way into the transverse plane giving a maximal signal. Hence the sensitivity is comparable to that of the pulse FT method. Data processing is most efficiently done by the use of several Fourier transformations which require a computer.

A major advantage of the pulse Fourier-transform method over the other two methods is its unique ability to add an element of time resolution to NMR spectroscopy. The ability to achieve time resolution stems from the use of pulses which divide the entire time domain into precise intervals enabling one to execute a sequence of events and to study the response of

[15] J. Dadok and R. F. Sprecher, *J. Magn. Reson.* **13,** 243 (1974).

[16] R. K. Gupta, J. A. Ferretti, and E. D. Becker, *J. Magn. Reson.* **13,** 275 (1974); **16,** 505 (1974).

the nuclear spins subjected to this sequence with high resolution with respect to time and frequency. The spin system can be prepared in a variety of useful ways by properly timed pulses of radio-frequency radiation, and the signal recovery towards magnetic equilibrium or to a steady state can be followed at later times. The act of preparation can thus be separated from the act of observation with the desired time resolution. Such time-resolved studies can be of immense value in making spectral assignments and in understanding a variety of rate processes at equilibrium in biochemical systems. Time-resolved studies also permit convenient measurements of the relaxation times of magnetic nuclei which are sensitive to their chemical environment.[8,13]

This article discusses the theoretical basis and methodological details for carrying out Fourier-transform NMR experiments in biochemical redox systems. After considering the basic Fourier-transform instrument and the preparation of samples, we discuss general experimental conditions for Fourier-transform NMR. Special methods especially useful in redox systems are then considered. For convenience a glossary of symbols used in this article is provided in Section VI.

II. Basic Description of Instrument

The essential components of a Fourier-transform NMR spectrometer are shown in Fig. 1.[10,12,14] We have omitted the field-frequency lock channel which corrects slow drifts in the magnetic field strength and is essential on

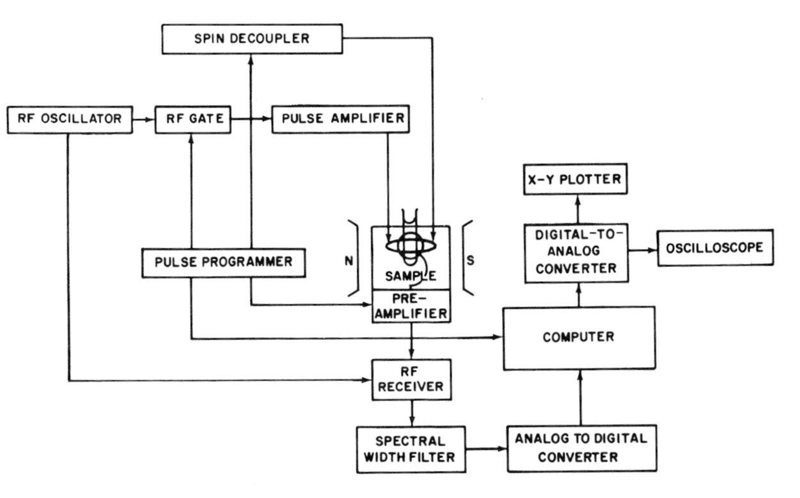

FIG. 1. Schematic diagram of a pulse Fourier-transform NMR spectrometer. For simplicity the field-frequency lock channel is not shown.

spectrometers other than those using superconducting magnets. The basic radio frequency (RF) of the spectrometer (e.g., 100 MHz for protons), synthesized from crystal-controlled oscillators, is fed by way of a high-quality RF gate into a pulse amplifier which is operated in a threshold (Class C) mode to ensure complete gating and delivers a peak power output of several hundred watts to the sample. The transient NMR signal from the sample, after passing a low-noise gated preamplifier, is converted to a lower frequency in the kilohertz range and is phase detected in the RF receiver. The audiofrequency output of the receiver is passed through a filter of adjustable spectral width to remove noise at frequencies outside the frequency range of interest. The analog-to-digital converter (ADC) digitizes the output of the spectral width filter and sends the signal to the computer for Fourier transformation and display. In a separate channel, a gated spin decoupler can be used to irradiate the sample with RF for a variety of double resonance experiments. The timing signals for all of the gating elements are provided by the pulse programmer which is synchronized to the timing circuits of the computer.

III. Samples[8,13]

With Fourier-transform spectroscopy a 0.4-ml sample volume in a 5-mm diameter sample tube can be used for obtaining spectra of $\geqslant 1$ m*M* protons of the compound of interest in ~ 5 min, and sensitivity can be further increased by a factor of ~ 3 by using 12-mm diameter sample tubes. For ^{31}P ($\geqslant 1$ m*M*) and ^{13}C ($\geqslant 10$ m*M*), larger-diameter sample tubes (10–12 mm) containing $\geqslant 1.3$ ml samples are essential for protein solutions. For locking on concentric samples, matching pairs of tubes of varying sizes are commercially available (e.g., 1 mm inside 5 mm, 5 mm inside 12 mm). In addition, effective lock can be achieved with a 1-mm shell of D_2O by placing a 10-mm tube containing the sample into a 12-mm tube containing D_2O. Probes capable of accommodating larger sample tubes (20-mm diameter containing $\geqslant 6.0$-ml volume) have been developed for ^{13}C-NMR, resulting in 4-fold greater sensitivity.[17]

Paramagnetic contamination of samples should be scrupulously avoided by using distilled deionized water, and by treating reagents with the chelating resin Chelex 100 (Biorad). Microcolumns of Chelex 100 suitable for treatment of 0.5-ml NMR samples may be made from Pasteur pipettes. After equilibration of the Chelex column with the appropriate buffer, the sample is passed through the microcolumn and the concentra-

[17] A. Allerhand, R. F. Childers, R. A. Goodman, E. Oldfield, and X. Ysern, *Am. Lab.* **4,** 19 (1972).

tion of any important component is determined. Dilution may be avoided by discarding the first 0.1 ml of the sample.

Complete deuteration of the exchangeable protons is desirable in experiments where nonexchangeable protons are being observed. This is best achieved by repeated freeze-drying of the samples (at least 3 times) for proteins which are stable to this treatment. For those proteins which denature upon freezing, repeated concentration by vacuum dialysis and dilution with D_2O or, alternatively, extensive dialysis against D_2O may be used to achieve a high level of deuteration. In observing nuclei other than protons or in observing exchangeable protons, samples are kept in predominantly H_2O solvent. The presence of 5% D_2O in the sample or a 1-mm shell of D_2O outside the sample in a concentric tube may be used for deuterium-field lock. The presence of D_2O in the sample is not essential for spectrometers operating with superconducting magnets which provide a fairly stable magnetic field without field-frequency lock. It should be pointed out that commercial D_2O often contains impurities from which it must be separated by careful vacuum sublimation to avoid contaminating the samples.

IV. General Experimental Conditions for Pulse FT NMR

A pulsed NMR experiment consists of applying a short burst of radio-frequency energy to a sample of nuclear spins in such a way as to simultaneously excite all resonances of interest and to study the time-dependent response of the spin system.[10] The manner in which the nuclear magnetic system responds contains at least as much information as that obtainable from the normal scanned spectrum in the frequency domain. The extraction of the spectral information as a function of frequency $F(\omega)$, from the free-induction decay $M(t)$, which is acquired as a function of time, requires the use of the mathematical process of Fourier transformation which interconverts the time and frequency domains according to Eq. (1).[18]

$$F(\omega) = \int_0^\infty M(t) \cos \omega t \, dt - i \int_0^\infty M(t) \sin \omega t \, dt \tag{1}$$

where i is a notation expressing the orthogonality of the two components of $F(\omega)$ in the frequency domain. In practice the experiment is restricted to a finite acquisition time AT, and the data are sampled at discrete intervals rather than continuously as expressed by Eq. (1). It may be noted that in pulse experiments the magnetization at any point in time after the pulse is a

[18] R. Bracewell, "The Fourier Transform and Its Applications," McGraw-Hill, New York, 1965.

property of the entire spin system, in contrast to continuous-wave (CW) NMR experiments, which are in the frequency domain, where the magnetization at a given point corresponds only to that particular frequency. Hence stopping a frequency spectrum halfway through results in a loss of half of the information, while stopping the pulse experiment halfway in the time domain merely degrades the quality (signal-to-noise ratio and resolution of the spectrum) but not its information content. The following considerations are required for optimization of signal-to-noise in obtaining Fourier-transform spectra.

A. *Choice of Proper RF Field Strength and Pulse Width*

The effect of a radio-frequency pulse on an ensemble of spins is most simply analyzed in terms of a coordinate system or frame of reference which rotates about the direction of the steady magnetic field at the frequency of the fluctuating RF field.[14] Each pulse of RF of amplitude H_1 and width PW nutates or flips the magnetization of the sample by an angle θ away from its equilibrium position along the steady magnetic field. The nutation occurs around the resultant field which is the vector sum of the steady field and the RF field in the rotating frame. The nutation angle θ is given by:

$$\Theta = PW \left\{ (\gamma H_1)^2 + 4\pi^2(\nu - \nu_0)^2 \right\}^{1/2} \tag{2}$$

where γ is the gyromagnetic ratio of the nucleus in question, ν is the resonance frequency, and ν_0 is the frequency of the RF pulse. Pulse experiments are often carried out using 90° pulses ($\theta = 90°, PW = PW_{90}$) to obtain a maximal signal. To simultaneously obtain maximal signals (within 2%) from all spins resonating over a given spectral width (SW), the following condition must be satisfied:

$$\gamma H_1 \geqslant 2\pi SW \tag{3}$$

or

$$PW_{90} \leqslant (4SW)^{-1} \tag{4}$$

Thus to reproduce the wide proton spectrum of cytochrome *c* extending from −35 to +25 parts per million (ppm)[19] or 6000 Hz at 100 MHz, RF fields of amplitude high enough to deliver 90° pulses of duration ≤1/4(6000) or ≤42 μsec will be required. At 220 MHz, the required 90° pulse width will be ≤19 μsec. If the above condition is not satisfied, the resulting spectrum will show phase and intensity distortions. As discussed elsewhere,[12] because of these short intense pulses, precise and effective RF gating is essential.

[19] A. Kowalsky, *Biochemistry* **4**, 2382 (1965).

B. *Proper Choice of RF Pulse Frequency*

The positioning of the RF pulse frequency (ν_0) in the pulse experiment is achieved by changing either the frequency, ν_0 (e.g., on the XL-100-FT), or the magnetic field, H_0 (e.g., on the HR-220-FT). The position of ν_0 determines the zero point of the spectrum which extends from 0 to SW Hz where SW is the desired spectral width.

As FT spectrometers using a single-phase detector cannot distinguish positive ($\nu > \nu_0$) and negative ($\nu < \nu_0$) frequencies, it is necessary to set the RF pulse frequency (ν_0) at the low- or the high-field end of the spectral width. Quadrature detection[20] which provides the capability of distinguishing between positive and negative frequencies and eliminates this need is discussed in Section IV,C,3.

C. *Optimum Data Acquisition and Analysis*

The decaying NMR signal following an RF pulse is referred to as the "free-induction decay" or FID since it arises from the free precession of the detected transverse component of the magnetization. All Fourier-transform experiments require storage of the complete free-induction decay prior to transformation, and most experiments involve the time averaging of many decays. Since transformation is often carried out via a digital computer, analog-to-digital conversion is a fundamental step in FT NMR and is carried out by a device called the analog-to-digital converter (ADC). The peculiarities, requirements, and the effects of this conversion process are discussed in this section.

1. Rate of Analog-to-Digital Conversion and Fold-Over

Let us assume that the spectrum we wish to acquire consists of a mixture of frequencies from 0 to SW Hz. A basic theorem of sampling states that to define a frequency Δ Hz, it must be sampled at a rate of at least 2Δ points/sec.[14] If analog-to-digital conversion occurs at a rate less than 2Δ, an apparent frequency shift will occur and a resonance at a frequency ν outside of the spectral width selected will be folded back to $(\Delta - \nu)$ Hz. This phenomenon is referred to as *fold-over*. To accurately digitize a free-induction signal it is therefore necessary to sample at a rate of at least twice the highest frequency present, i.e., at twice the spectral width. In fact, sampling at 2Δ points/sec is the most efficient way of acquiring data, since sampling at a slower rate will cause ambiguity arising from fold-over, while sampling faster provides no advantages and may require more computer capacity than is available. It must be noted that any noise of fre-

[20] A. G. Redfield and R. K. Gupta, *J. Chem. Phys.* **54,** 1418 (1971).

quency higher than SW Hz will be folded back as additional noise in the spectrum. Hence audio-frequency filtering of the detected signal to remove frequencies outside the spectral width before digitization is essential in order to minimize the noise fold-over problem. These filters may be selected manually or under computer control. The analog-to-digital conversion rates needed in practice for FT spectroscopy depend on the range of chemical shifts and hence on the nucleus under study. For diamagnetic proton samples at 23.5 kG with a typical chemical shift range of 10 ppm, sampling rates are on the order of 2000 points/sec, while ^{13}C at the same field requires sampling rates on the order of 10,000 points/sec. Nuclei with larger ranges of chemical shifts and instruments working at superconducting fields require even higher rates of digitization.

2. Proper Choice of Acquisition Time, Spectral Width, and Number of Data Points

The choice of the data acquisition time (AT) is governed by the resolution requirements of a given experiment. If the time response following an RF pulse is sampled for AT sec, the resulting resolution is $(1/AT)$ Hz. The number of data points (N) to be taken on a given free-induction decay depends on the acquisition time (AT) and spectral width (SW) according to the following relationship[14]:

$$N = 2\,(AT)(SW) \tag{5}$$

or

$$AT = \frac{N}{2(SW)} \tag{6}$$

In practice N is normally limited by the size of the computer core available, SW is the required spectral width, and the maximal AT, hence the resolution, is limited by Eq. (6). For example, if the computer core size is 8000 words (8K), and a spectral width (SW) of 1000 Hz is desired, then an acquisition time (AT) of 4 sec corresponding to a resolution of 1/4 Hz is the highest achievable resolution. During Fourier transformation, the N data points in the time domain are changed to $N/2$ absorption mode points and $N/2$ dispersion mode points. The resulting frequency domain spectrum has $N/2$ points with a resolution of 0.25 Hz. A large number of data points improves the definition of the resulting spectrum but requires a long acquisition time. However, since the signal decays exponentially, the longer the acquisition time, the lower the signal-to-noise ratio becomes. This is an example of the usual conflicting requirements of sensitivity and resolution.

In practice a compromise between sensitivity and resolution is often necessary. Typical values of computer core capacity are in the region of 8K memory locations. As pointed out above, the acquisition time will be set either as dictated by the spectral width and the computer core size or by the required resolution. If the parameter of choice is resolution, then, for a given spectral width, AT should be set to be several times the longest T_2^* (i.e., several times the lifetime of the longest signal), the only limitation being the available computer memory. If, however, sensitivity is the criterion, then a decision must be made as to what resolution is acceptable (or attainable) as this will then define AT. Having defined AT and SW, N is fixed by Eq. (5). The value of N fixed by these criteria may be less than the total available capacity of the computer. In this case the remaining points should be set to zero prior to Fourier transformation to enable the frequency domain spectrum to have the maximum number of points. It has been shown that doubling the data length by zero filling also complements the absorption and dispersion mode signals by transferring information from one mode to the other and vice versa. Thus if there are N actual data points, the addition of N zeroes at the end of data acquisition results in improved resolution.[21] Zero filling beyond this gives more points and thus improves the definition of the spectrum without enhancing the spectral resolution. If data are acquired for a time less than $3T_2^*$, zero filling may introduce side lobes or wiggles around each resonance.

It may be necessary to examine only a portion of the spectrum, i.e., to obtain greater resolution than is compatible with the total spectral width and the number of data points available. In this case the RF pulse frequency must be placed at one end of the spectral window desired and an audio-filter with a cutoff frequency equal to the desired spectral width must be applied before the analog-to-digital conversion to reduce the fold-over of noise and of undesired signals from outside the spectral width. If there are peaks present in the spectrum within the spectral width but on the opposite side of the RF pulse frequency, these peaks will be imaged or folded over into the final spectrum usually with a different phase, a problem which can be overcome by quadrature detection.[12,20] Noise, but not signals, from this domain can be eliminated by crystal filtering.[22] These methods are discussed in the next section.

3. Quadrature Detection[12,20] and Crystal Filtering[22]

These represent two alternative methods for eliminating the noise due to spectral imaging or fold-over from the negative half of the spectral width. As pointed out earlier, in FT NMR spectrometers using a single phase detector it is not possible to distinguish between frequencies coming from either side of the pulse frequency. Thus spins resonating at $(\nu_0 - \nu)$ Hz are

[21] E. Bartholdi and R. R. Ernst, *J. Magn. Reson.* **11,** 9 (1973).
[22] A. Allerhand, R. F. Childers, and E. Oldfield, *J. Magn. Reson.* **11,** 272 (1973).

indistinguishable from those resonating at $(\nu_0 + \nu)$ Hz. Signals at both frequencies are converted to ν Hz after subtraction of the pulse carrier frequency (ν_0). It is possible to distinguish positive resonance frequencies $(\nu_0 + \nu)$ from negative resonance frequencies $(\nu_0 - \nu)$ by means of a second phase detector in quadrature with the first, i.e., 90° out of phase with the first. The two phase detectors produce a pair of voltages which are the real and imaginary parts of $\Sigma_\nu e^{-t/T^*_{2\nu}+2\pi i\nu t}$ which is the free-induction decay signal from a set of noninteracting spins resonating over the spectral width (SW), i.e., with resonance frequencies in the range $(\nu_0 - SW)$ to $(\nu_0 + SW)$. The outputs of both detectors which are proportional to $\Sigma_\nu e^{-t/T^*_{2\nu}} \cos 2\pi\nu t$ and $\Sigma_\nu e^{-t/T^*_{2\nu}} \sin 2\pi\nu t$ respectively are audio-filtered and stored in the computer and may be Fourier-transformed as a single complex function.

It has been pointed out that the use of quadrature detection improves the signal-to-noise ratio by a factor of $\sqrt{2}$ over that obtained with a single phase detector, a 2-fold saving in running time. This gain arises from the fact that a single phase detector is incapable of distinguishing positive frequency noise $(\nu_{\text{noise}} > 0)$ from negative frequency noise $(\nu_{\text{noise}} < 0)$, and thus the latter images or folds over into the positive (or negative) frequency region normally being observed with a single phase detector. Since the noise is random, this imaging results in a factor of $\sqrt{2}$ increase in the root mean square level of the noise, which is avoided by quadrature detection. It is also possible to filter out the negative frequency noise by means of an "audio-frequency crystal filter" before data acquisition. This simpler approach retains the $\sqrt{2}$ gain in signal-to-noise ratio but, unlike quadrature detection, it does not distinguish between positive and negative resonance signals and hence does not allow positioning of the carrier in the middle of the observed spectral width of interest. The ability to achieve the latter by means of quadrature detection lowers the spectrometer RF pulse power requirement for a given spectral width by a factor of 4 or increases the maximum observable spectral width for a given RF pulse power by a factor of 2.

D. *Exponential Filtering: Sensitivity and Resolution Enhancement*[14]

By performing a few mathematical operations on the free-induction signal before Fourier transformation it is possible to improve either the resolution or the signal-to-noise ratio of the final NMR spectrum obtained. Thus multiplying the free-induction decay by the factor e^{-at} improves the signal-to-noise ratio; conversely, resolution can be enhanced at the expense of the signal-to-noise ratio by the factor e^{+at}. For optimum signal-to-noise a value for the exponential filtering constant $a = 1/T^*_2$ is used. This value, however, doubles the line width. For resolution enhancement there is no optimum value for a but a must be less than $1/T_2^*$ for the filtered signal to decrease with time, and thus be a free-induction decay.

Resolution may also be improved by a "convolution difference"[23] technique which is applicable to a spectrum containing a few narrow lines amidst intense broad lines. In this approach, the original free-induction decay (FID) is first stored in the computer. It is then filtered with a filtering constant a_1 intermediate between the large $1/T_2^*$ values of the broad lines, and the small $1/T_2^*$ values of the narrow lines. This filtering constant a_1 is chosen so as not to affect the width of the broad lines but to broaden the weak narrow lines beyond detection. A difference free-induction signal is then obtained by subtracting the filtered FID from the unfiltered FID to eliminate the broad lines in the spectrum. The difference thus obtained, in which the free-induction signals from the narrow lines predominate, may be filtered again using a filtering constant a_2 equal to $1/T_2^*$ for the narrow lines; this improves their signal-to-noise, and the resulting signal is then Fourier-transformed.

An alternative means of eliminating intense broad peaks whose free-induction signals decay rapidly can be achieved by delaying the data acquisition after the RF pulse by 3 times the T_2^* value of the broad lines.[24] Both of these methods sacrifice signal-to-noise ratio for improved resolution. Moreover, the resulting spectra are not suitable for quantitative purposes when narrow lines with variable widths are present.

E. Fourier Transformation and Phase Correction[14]

In practice, the free-induction decays are stored as a series of discrete points, and the continuous Fourier integral of Eq. (1) is replaced by a summation over a finite series of N points

$$F(\omega) = \sum_{n=0}^{N-1} M\left(\frac{n}{N} AT\right) \cos \omega \frac{n}{N} AT - i \sum_{n=0}^{N-1} M\left(\frac{n}{N} AT\right) \sin \omega \frac{n}{N} AT \tag{7}$$

where $F(\omega)$ is the Fourier-transformed point in the final spectrum at a frequency ω in the range 0 to $N\pi/AT$ rad sec^{-1}. A consequence of this digital approach is that the resulting frequency spectrum is discontinuous, defined only at frequencies $(1/AT)$ Hz apart where AT is the data acquisition time.[25]

[23] I. D. Campbell, C. M. Dobson, R. J. P. Williams, and A. V. Xavier, *J. Magn. Reson.* **11,** 172 (1973).

[24] C. H. A. Seiter, G. W. Feigenson, S. I. Chan, and M. Hsu, *J. Am. Chem. Soc.* **94,** 2535 (1972).

[25] The Fourier theorem as expressed in the discontinuous and finite form of equation (7)

The Fourier transformation is usually calculated using the fast Fourier procedure or algorithm.[25] Due to the inherent nature of this algorithm, computations are carried out most efficiently when the total number of data points is an integral power of 2. The calculation is usually performed in computer core, the original FID data being overwritten since the memory locations originally used for storing raw data are required to store the results of intermediate calculations and then the final coefficients. The "in core" nature of the calculation has the disadvantage that the original FID is not retained. Hence if the need to carry out several types of data processing is anticipated, the original FID must be saved on magnetic tape or on a disc prior to Fourier transformation. The advantage of the "in core" calculation is that N data points may be transformed in a little over N words of core, hence making the maximum use of the core available. The approximate time taken to execute an 8K transform via the fast Fourier algorithm is 10 sec with a modern computer.

In general the cosine term $C(\omega)$ and the sine term $S(\omega)$ of the Fourier-transformed signal in Eq. (7) are each a different linear combination of the absorption and dispersion mode signals. The pure absorption mode signal $A(\omega)$ is obtained by taking a linear combination of these two components, $C(\omega)$ and $S(\omega)$,

$$A(\omega) = \cos\phi \cdot C(\omega) + \sin\phi \cdot S(\omega) \tag{8}$$

where ϕ, the phase angle, is adjusted to give a pure absorption spectrum.[10] The phase angle ϕ itself may be a function of frequency due to the effect of audio filtering. In general a correction which assumes the phase to vary linearly with frequency is found to be sufficient and is achieved by having two phase controls, the initial phase which has a maximum effect at zero frequency and a minimum effect at the highest frequency, and the final phase, which has a minimum effect at zero frequency and a maximum effect at the highest frequency. An ideal solution to the nonlinear phase shifts introduced by the spectrometer system is to appropriately weight the free-induction decay according to the actual characteristics of the system. The latter may be obtained directly by feeding a δ-function impulse into the detector system and storing the response of the system.[12] Unfortunately this procedure, which requires extensive modification of the phasing programs, has not been implemented on commercial instruments.

---

requires the assumption of periodicity in the free induction decay M(t) with a time-period equal to AT, and periodicity in the NMR frequency spectrum F(ω) with a frequency interval equal to the spectral width (N/2AT); J. W. Cooley and J. W. Tukey, *Math. Comput.* **19,** 297 (1965).

F. Choice of Pulse Recycle Time

In most biochemical experiments, the signal-to-noise attainable by Fourier transforming the transient signal after a single 90° RF pulse is not sufficient to observe the resonances of the dilute sample under study. Hence, time-averaging of the transient signals from a large number of such pulses before Fourier transformation is often required to obtain an adequate signal-to-noise ratio for detecting weak resonances. The following timing sequence is normally used[14]:

$$[90^\circ \text{ RF pulse, Data acquisition time } (AT), PD]_N$$

A short 90° pulse is used to generate maximal signal which is acquired in the computer during the data acquisition time AT. A pulse delay PD is introduced after the data acquisition time to permit re-equilibration of the net magnetic vector before repeating the pulse sequence. A time interval of $5T_1$, where T_1 is the longest longitudinal relaxation time or magnetization recovery time of the nuclear spins, allows a 99% recovery of the magnetization vector between pulse cycles. Hence the pulse delay PD is normally chosen according to Eq. (9).

$$PD + AT = 5T_1 \tag{9}$$

Since the transient free-induction signal lasts only for $\sim 3T_2^*$ after the RF pulse, the AT is often set equal to $3T_2^*$. The need to wait for magnetic recovery between pulse cycles generally limits the sensitivity of the pulse FT method. Steady-state pulse FT techniques which attempt to improve signal-to-noise by eliminating the need to wait for full recovery of the magnetization between pulse cycles are discussed in the next section.

G. Steady-state Pulse FT Techniques

In situations where $T_1 \gg T_2^*$ it is possible to accelerate the accumulation of data and to obtain higher signal-to-noise per unit time by one of the following two methods which optimize either the nutation angle at a constant pulse recycle time[26] or the pulse recycle time at a constant (90°) nutation angle.[13,27,27a]

1. Optimization of Nutation Angle[10,28]

The principle of this method is simple. The following timing sequence is used:

[26] R. R. Ernst, *Adv. Magn. Reson.* **2,** 1 (1966).
[27] R. K. Gupta, *J. Magn. Reson.* **25,** 231 (1977).
[27a] J. S. Waugh, *J. Mol. Spectrusc.* **35,** 298 (1970).
[28] C. F. Midelfort, R. K. Gupta, and I. A. Rose, *Biochemistry* **15,** 2178 (1976).

$$[\theta^0 \text{ RF pulse, Data acquisition time } (AT)]_N$$

The spin system is thus subjected to rapid repetitive pulses of an appropriately chosen nutation angle θ with the constant interval AT between pulse cycles. No pulse delay is used. A steady state of magnetization is achieved after the first ~4 pulses. The value of AT is usually chosen as equal to $3T_2^*$. Following the θ° RF pulse, only partial recovery of the magnetization occurs during the time interval AT before the application of the next RF pulse. The resulting steady-state NMR signal per pulse obtained by Fourier transformation is lower in amplitude than the equilibrium value M_0 due to incomplete recovery of magnetization between pulse cycles. However, by choosing θ properly it is possible to gain significantly in signal-to-noise ratio in a given total time. Neglecting spin-echo effects, which may be eliminated by introducing a small (<100 msec) random pulse delay (PD) at the end of the acquisition time (AT), the amplitude of the steady-state signal M_θ per transient is given by[13,29]

$$M_\theta = \frac{M_0(1 - e^{-AT/T_1}) \sin\theta}{(1 - e^{-AT/T_1}\cos\theta)} \tag{10}$$

where M_0 is the fully relaxed amplitude of the signal which is equal to $M_\theta/\sin\theta$ as θ approaches 0. The optimum value of θ is obtained when $\partial M_\theta/\partial\theta = 0$ which yields:

$$\cos\theta_{opt} = e^{-AT/T_1} \tag{11}$$

Equation (11) is used to calculate the optimum value of θ to be used in the timing sequence. The signal amplitude per pulse obtained using the optimum value of θ according to Eq. (10) is:

$$M_{\theta_{opt}} = M_0\left(\frac{1 - e^{-AT/T_1}}{1 + e^{+AT/T_1}}\right)^{1/2} \tag{12}$$

The gain (G_1) in signal-to-noise over the normal pulse FT procedure [90° pulse, AT, Pulse Delay (PD)] with $PD + AT = 5T_1$ is:

$$G_1 = \left[\frac{5T_1(1 - e^{-AT/T_1})}{AT(1 + e^{+AT/T_1})}\right]^{1/2} \tag{13}$$

A plot of this gain function against AT/T_1 (Fig. 2A) indicates the maximal gain in signal-to-noise to be 1.58 corresponding to a 2.5-fold saving in the time required to attain a given signal-to-noise.

A disadvantage of the rapid pulse method is that when several lines with differing T_1 values are present in a spectrum, it is not possible to simultaneously satisfy Eq. (11) for all resonances in the spectrum. A compromise

[29] K. A. Christensen, D. M. Grant, E. M. Schulman, and G. Walling, *J. Phys. Chem.* **78,** 1971 (1974).

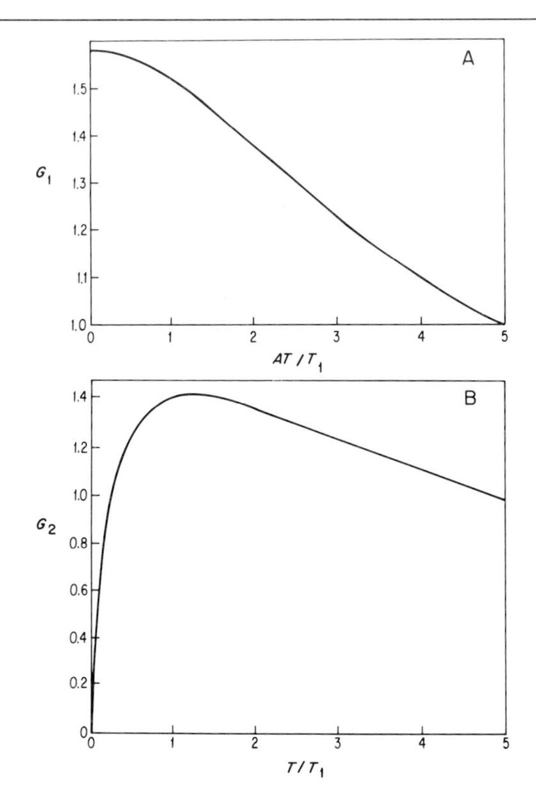

FIG. 2. The improvement in signal-to-noise using rapid-pulse, steady-state techniques. (A) Gain in signal-to-noise (G_1) as a function of AT/T_1 where AT is the data acquisition time for experiments using nutation-angle optimized pulses. (B) Gain in signal-to-noise (G_2) as a function of T/T_1 for experiments in which 90° pulses are rapidly recycled with a period $= T$.

in the choice of θ_{opt} must therefore be made in this situation. Relative intensities of different resonances with varying T_1 values will then be distorted. Hence the technique should be used with care in complex spectra. When mere observation of resonances and their chemical shifts is the object of the experiment, the technique could be profitably used even with complex spectra containing resonances with widely differing T_1 values.

2. Optimization of Pulse Recycle Time at 90° Nutation Angle[13,27]

It is possible to gain signal-to-noise per unit time simply by pulsing rapidly while using 90° pulses. Depending on the data acquisition time, no pulse delay or an appropriately chosen pulse delay is needed for optimizing the signal-to-noise by this procedure. The timing sequence is:

$$[90° \text{ RF pulse, Data acquisition time } (AT), PD]_N$$

As with the $\theta°$ pulses, a steady state of magnetization is achieved after ~4 pulses. The amplitude of the signal M_T per transient where $T = AT + PD$ is given by Eq. (10) setting $\theta = 90°$, or

$$M_T = M_0 (1 - e^{-T/T_1}) \quad (14)$$

Since signal-to-noise is proportional to $\sqrt{\text{Time}}$, the signal-to-noise acquired per unit time is given by:

$$S = \frac{M_T}{\sqrt{T}} = \frac{M_0 (1 - e^{-T/T_1})}{\sqrt{T}} \quad (15)$$

The signal-to-noise given by Eq. (15) as a function of T reaches a maximum when $\partial S/\partial T = 0$, which occurs when $T = 1.27\, T_1$. Thus in the sequence above $T = AT + PD$ should be chosen equal to $1.27\, T_1$ for optimum sensitivity. It is often desirable to set $AT = 3T_2^*$ and then $PD = (1.27\, T_1 - 3T_2^*)$. The amplitude of the steady-state signal per pulse for $T = 1.27\, T_1$ is $0.72\, M_0$. The gain in signal-to-noise over the normal pulse FT procedure using $PD = 5T_1$ is a factor of 1.43. Thus a 43% gain in signal-to-noise or a 2-fold saving in running time is possible by using this procedure.

Once again when several lines with differing T_1 values are present in a spectrum it is not possible to satisfy the condition $T = 1.27\, T_1$ simultaneously for all resonances in the spectrum. A compromise in the choice of T must therefore be made to simultaneously optimize the signal-to-noise for all resonances. A good choice in such a situation may be made by taking T as 1.27 times the average of the shortest and longest T_1. The gain in signal-to-noise for each resonance may be calculated from Eq. (16):

$$G_2 = \frac{\sqrt{5T_1}\, (1 - e^{-T/T_1})}{\sqrt{T}} \quad (16)$$

The function G_2 is plotted in Fig. 2B as a function of T/T_1. As with the previous method of optimizing the nutation angle, the relative intensities of lines in a complex spectrum may be distorted in this procedure precluding quantitative studies, but their chemical shifts will still be accurately measurable.

V. Special Methods in Fourier-Transform NMR of Redox Systems

A. *Observation of Hyperfine Shifted Resonances*[1,3-5,19,30]

Different oxidation states of redox proteins generally differ in their net magnetic moments. Thus, for example, in its oxidized state, the electron-

[30] D. R. Eaton and W. D. Phillips, *Adv. Magn. Reson.* **1**, 103 (1965).

transfer hemoprotein cytochrome *c* is paramagnetic while in its reduced state it is diamagnetic. In the paramagnetic state, a portion of the protein molecule has an unpaired electron in it, which can interact with several protons in its vicinity and shift as well as broaden their resonances. The observed hyperfine shifts arise from contact as well as dipolar interaction between unpaired electrons and protons. Contact interaction arises when the unpaired electron's wave function is nonvanishing at the site of the nucleus in question. Contact shifts can be either upfield or downfield, depending on the sign of the spin density and on the sign of the magnetic moment of the nucleus.[30] The resonance of a proton directly bonded to an aromatic ring carbon with a positive spin density is shifted upfield (due to a negative unpaired spin density on the proton). The resonance of the protons of a methyl group attached to a ring carbon with a positive spin density is shifted downfield due to a positive spin density on these protons. The difference in sign of the spin densities arises from the differing mechanisms responsible for the transmission of the spin density from the ring carbon to the protons in the two cases.[30]

Pseudocontact shifts arise from the failure of the dipolar interaction between the electron and proton to average to zero in solution due to spatial anisotropy in the magnetic moment of the unpaired electron. These shifts too can be in either direction depending on the spatial disposition of the group(s) in question.[30] The observed hyperfine shift is an algebraic sum of the contact and pseudocontact contributions.

In the diamagnetic state an important mechanism responsible for shifting the resonances of all magnetic nuclei is the interaction with ring currents which are induced to flow around the plane of aromatic rings by the applied static magnetic field. Protons located just above and below the plane of an aromatic ring are shifted upfield and those near the edge are shifted downfield by this interaction.[9] It may be noted that ring current shifts are also present in the paramagnetic forms but are overshadowed by the larger hyperfine shifts.

In addition to the classical CW methods by which hyperfine shifts were originally discovered,[19,30] two Fourier-transform approaches for observing hyperfine shifted proton resonances are especially useful—long pulse methods[12,20] and correlation spectroscopy.[15,16]

1. Long-Pulse Methods[12,20,31]

Long-pulse techniques pioneered in the laboratory of A. G. Redfield[12,20,31] have proved particularly useful in the study of hemoproteins. Unlike intense short radio-frequency pulses which have Fourier

[31] A. G. Redfield, this series, Vol. 49 [12]. A. G. Redfield, S. D. Kunz, and E. K. Ralph, *J. Magn. Reson.* **19,** 114 (1975).

components over a wide range of frequencies, weak long pulses have a narrow Fourier spectrum (spread over $1/PW$ Hz where PW is the pulse width). An advantage of long pulses which is based on their property of a narrow Fourier spectrum is their unique ability to excite only a part of a spectrum in a given experiment. By adjusting the width and strength of the pulses, it is possible to change the effective spectral window. Unlike the case of short pulses, spins outside this spectral window do not contribute appreciably to the signal. This feature is particularly useful for hemoproteins where the spectra are often spread over such a wide range that, owing to computer core limitations and resolution requirements, it is not possible to sample the entire spectrum with a single pulse. Observation of only the interesting part of a spectrum such as the weak contact-shifted proton resonances by using a smaller spectral width is often difficult in normal FT NMR since short pulses excite the entire spectrum, and the intense main aromatic and aliphatic proton bands fold over into the observed region. Long pulses avoid this problem by not perturbing the main protein bands enabling one to study any section of the spectrum at any resolution, at the same time avoiding the interferences from strong signals due to water or to other portions of the protein.

Long-pulse methods generally require specialized modifications of the spectrometer which include quadrature detection since it is desirable to set the RF pulse frequency in the center of the region of interest.[12,20,31] The computer programs also are modified to correct for the phase and intensity distortions caused by the long pulses.[12,20,31] It should be pointed out, however, that the above modifications are not essential if the pulse frequency could be at one end of the spectrum of interest and if some phase and intensity distortions could be tolerated. Constant and linear phase corrections, available on most spectrometers, may often suffice to correct for phase distortions due to the long pulse length. Only a high-quality RF attenuator is required to gain some of the practical advantages of long pulses. A simplified version of the long-pulse methods, usable with minimal modification to commercial spectrometers, involves inserting the RF attenuator in the output of the pulse amplifier. This modification is sufficient to observe hyperfine shifted proton resonances of proteins in D_2O which are well separated from the main protein bands.

A pulse width (PW) and pulse strength (γH_1) are chosen such that the desired resonances receive a $\sim\pi/2$ nutation and will therefore contribute maximally to the observed signal.

$$(\gamma H_1)(PW) = \frac{\pi}{2} \tag{17}$$

The interfering resonances, $\Delta\nu$ Hz away, receive a 2π (or 4π) nutation and therefore do not contribute to the observed signal. From Eq. (2) we may write:

$$PW\{(\gamma H_1)^2 + 4\pi^2(\Delta\nu)^2\}^{\frac{1}{2}} = 2\pi \text{ (or, } 4\pi) \tag{18}$$

Simultaneous solution of Eqs. (17) and (18) for PW yields the following:

$$PW = \left(\frac{15}{16}\right)^{\frac{1}{2}} \frac{1}{\Delta\nu} \quad \text{or} \quad \left(\frac{63}{16}\right)^{\frac{1}{2}} \frac{1}{\Delta\nu} \tag{19}$$

As an example, let us assume that we are interested in observing the two hyperfine shifted ring methyl resonances of cytochrome *c* at approximately 34.0 and 31.3 ppm downfield from tetramethylsilane (TMS) at 100 MHz. One could set the pulse frequency at 35 ppm downfield from TMS. The main protein bands extend ±5 ppm on either side of the HDO signal which occurs at ~5 ppm downfield from TMS. Hence $\Delta\nu = (35 - 5)100 = 3000$ Hz and, from Eq. (19), $PW = 323$ μsec for a 2π nutation or 661 μsec for a 4π nutation at the HDO resonance. Using the calculated PW, the pulse strength is then adjusted to nutate a resonance at the pulse frequency by $\pi/2$. Such an RF pulse would nutate the HDO line and the main protein bands by $\sim 2\pi$ (or 4π), thus effectively eliminating them from the free-induction signal, but will yield a nearly optimal signal from spins resonating about −35 ppm with a bandwidth of $\pm\Delta\nu/4$ or $\pm\Delta\nu/8$ within which the distortion of the amplitude is <2% (see Section IV). In our example this range will be ±7.5 ppm or ±4 ppm, sufficient to observe the hyperfine shifted resonances of interest without fold-over of the unwanted signals.

2. Rapid-scan Correlation Method[15,16]

This method—originally introduced by Dadok and Sprecher[15] and refined by Gupta *et al.*[16]—has been extensively used for recording NMR spectra of exchangeable and nonexchangeable proton resonances of hemoproteins in aqueous solutions by Ho and co-workers.[4] As pointed out in Section I, in this method the entire NMR spectrum of a sample, or only a part of it which excludes the water signal, is swept under fast-passage conditions (sweep rate $\gg(2\pi T_2^{*2})^{-1}$ Hz/sec). Such a fast-passage spectrum shows a ringing pattern after each resonance due to free precession of the transverse component of magnetization, which decays with a time constant T_2^*. The normal NMR spectrum of a sample can be computed from its rapid-passage response by cross-correlating its ringing pattern with that of a single sharp reference line (such as TMS) swept separately but under identical conditions or with a computer-generated reference line. The required spectrometer modification[15,16] consists of adding a sweep generator capable of sweeping either the frequency (in a field/frequency locked system) or field (in superconducting field spectrometers with no field/frequency lock) at a rate $\geqslant 10$ kHz/sec. The computer requirements are comparable to those for pulse FT NMR spectroscopy. As in the pulse FT method, the use of quadrature detection is optional and would yield a $\sqrt{2}$-fold increase in signal-to-noise.[16] The cross-correlation is computed most efficiently by the use of fast Fourier-transform procedures. The latter computations are made possible by the mathematical property that the cross-correlation of two spectra in the frequency domain is equivalent to a complex conjugate multiplication of their Fourier transforms in the time domain.[18] Thus to carry out cross-correlation, one Fourier-transforms the unknown rapid-passage spectrum of the sample under study as well as the rapid response of the reference line, multiplies the FT of the unknown response by the complex conjugate of the FT of the reference response, and then inverse Fourier-transforms the product to get the final frequency spectrum. When using the computer-generated reference line, it is best to generate its Fourier transform in the time domain directly, thus avoiding a Fourier-transformation step. In the time domain this function is simply $(\cos bt^2/2 - i \sin bt^2/2)$ where b is the sweep rate in radians/sec^2 and t is the time variable.[16]

Prior to the final inverse transformation, the data are in the form of a simple free-induction decay. Therefore, it is possible to apply digital filtering at this point as in the pulse Fourier-transformation method. Thus, for optimum signal-to-noise ratio, an exponential filter with a time constant T_2^* should be useful. In correlation spectroscopy using a theoretical function, filtering is equivalent to generating $\exp -[at + (ibt^2/2)]$ as the theoretical function, where a is positive for sensitivity and negative for resolution enhancement. In correlation spectroscopy with an experimental reference line, some line broadening is unavoidably introduced and the sensitivity is simultaneously enhanced. The line widths of the resulting spectrum in this case will be the sum of the true line widths and the line width of the reference line.[15,16]

Several other practical considerations discussed in Section IV,C for the case of pulse FT NMR also apply to rapid-scan NMR.[16] For example, the sampling rate must be chosen to be at least twice the highest frequency present in the data; here this means twice the highest ringing frequency following a peak. The total number of data points must be compatible with the frequency range to be covered and with the available computer memory. Audio-filters must be properly chosen to allow the ringing signal to pass through unattenuated. More filtering would improve signal-to-noise at the cost of broadened resonances. As in pulse FT NMR, phase shifts introduced by the filter may have to be corrected for in the computer analysis. For rapid-scan spectra cross-correlated using an experimental reference line, traversed with identical phasing, no phase corrections are needed since the phase factors cancel out in the cross-correlation process. This is a distinct advantage since use of the theoretical function requires a phasing adjustment in the final spectrum to produce a pure absorption-mode signal. Another advantage of the experimental reference line over the theoretical function approach is that in the former case, the frequency axis of the final output gets automatically shifted in the cross-correlation process so as to set the experimental reference line to zero, a desirable feature for NMR work involving chemical shift measurements. The reference line must occur (or the field must be adjusted so that the reference line occurs) either upfield or downfield from all resonances of interest in the unknown sample. If this precaution is not taken, then one runs into a problem similar to imaging in FT NMR, since resonances appearing at $(\nu - \nu_0)$ in the rapid-scan response are indistinguishable from those at $(\nu_0 - \nu)$. Such a complication does not arise when the theoretical function is used and the resonance positions remain unchanged with respect to their original positions in the rapid-scan response.

An advantage of the rapid-scan correlation method over the pulse FT method is that the dynamic range problem often encountered with pulse FT studies of samples containing an unwanted strong resonance (e.g., H_2O or HDO) can be overcome merely by avoiding scanning through the strong resonance.[4,15] In general, two separate scans would be required to cover the frequencies above and below the interfering peak. Another advantage of the rapid-scan correlation method is the simplicity of the additional instrumentation required for a laboratory already equipped with a high-resolution FT spectrometer. The method also allows recording a portion of the spectrum of a sample under high resolution without the fold-over problems encountered in short-pulse FT studies and hence is suitable for observing contact-shifted resonances in redox proteins.

Finally it should be pointed out that nuclear resonances of macromolecular systems, such as hemoproteins and other redox proteins, are fairly broad ($\geqslant$20 Hz) and therefore even fairly rapid sweeps ($\leqslant$1 kHz/sec) do not cause distortions such as ringing. In such systems, the familiar CW technique with time averaging can yield a satisfactory signal-to-noise ratio.

B. *Saturation Transfer Experiments*[5,6,32–34]

It is of interest to be able to measure the magnitudes of hyperfine shifts accurately and to determine their contact as well as pseudocontact con-

[32] R. K. Gupta and A. G. Redfield, *Science* **169,** 1204 (1970).
[33] R. K. Gupta and A. G. Redfield, *Biochem. Biophys. Res. Commun.* **41,** 273 (1970).
[33a] R. K. Gupta and S. H. Koenig, *Biochem. Biophys. Res. Commun.* **45,** 1134 (1971).
[34] R. K. Gupta, S. H. Koenig, and A. G. Redfield, *J. Magn. Reson.* **7,** 66 (1972).

tributions. To measure hyperfine shifts one needs to know the diamagnetic positions of the hyperfine shifted lines. This is a difficult task and cannot be done in a straightforward manner, particularly in cases where such lines in the diamagnetic state are not resolved from the broad overlapping region of the protein spectrum and are therefore unobservable. A saturation transfer NMR double-resonance technique has been shown to be useful in such situations.

Magnetic energy may be transferred from one chemical environment to another by a migration of the magnetic nucleus between two chemical environments (i.e., chemical exchange) or by the physical exchange of magnetic energy between two nuclei (cross-relaxation or Overhauser effects). Saturation transfer experiments make use of both of these processes to make spectral assignments and to obtain structural and kinetic information.

1. Saturation Transfer by Chemical Exchange[5,6,32–34]

This technique of spectral analysis has been extensively used to study the structure and function of the redox protein cytochrome *c*. Its usefulness lies in assigning hyperfine-shifted resonances of the paramagnetic oxidation state of a redox molecule by identifying them with the resonances in the diamagnetic redox state of the same molecule.[5,6,32–34] Such experiments are possible because the magnetic state of nuclear spins can be perturbed by a saturating RF field and the nuclear spins retain a finite memory of their state of magnetization for a time T_1. The success of such experiments requires that the rate of electron exchange between the diamagnetic and paramagnetic forms of the protein at equilibrium be comparable to $1/T_1$, the rate at which the nuclear spins lose their magnetic energy by random interactions with other spins, in the absence of the redox reaction. When this condition is fulfilled, the presence of exchange between the two redox states alternates the resonance frequencies of nuclei without giving them sufficient time to forget their perturbed magnetic state. The net result is that spins transferred to the second redox state retain the magnetic energy pumped into them by the perturbing RF field during their momentary presence in the first redox state.

The normal FT spectrometer with a gated decoupling channel is suitable for such studies.[12] Software modification to alternately add and subtract the free-induction decay signals or to accumulate an equal number of transients in the add and subtract modes is desirable. A spectrometer modification permitting control of the decoupler frequency by the computer or pulse programmer is also desirable.

The RF pulse sequence used in this technique is shown in Fig. 3. The experiments are performed on an equimolar mixture of two oxidation states

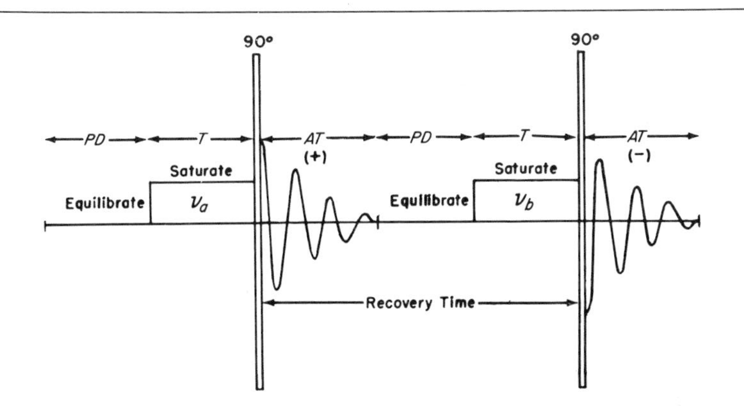

FIG. 3. The pulse sequence for saturation transfer experiments. A saturating pulse of duration T alternates between the frequencies of two resonance lines at ν_a and ν_b. The free-induction signal acquired during $AT(-)$ following saturation at ν_b is subtracted from that acquired during $AT(+)$ following saturation at ν_a. The saturated nuclei and any to which saturation has been transferred return to their full equilibrium magnetization during the recovery time between pulse cycles. The recovery time can be adjusted by varying PD, the time interval between the end of data acquisition and the start of resaturation during which the spins are allowed to equilibrate.

at equilibrium, which yields the superimposed spectra of both states. A long ($\geqslant$0.1 sec) saturating pulse of radio-frequency radiation is applied to the spins at the resonance frequency of a hyperfine shifted or otherwise easily resolved resonance of one of the two oxidation states.[5,6] Such an RF pulse tends to equalize the spin populations of the two nuclear magnetic energy levels of the irradiated resonance by short-circuiting the relaxation processes responsible for maintaining the population difference at thermal equilibrium. Consequently the saturating pulse demagnetizes the nuclear spins in resonance with it. The intensity and duration of the saturating pulse are adjusted to be no more than just sufficient to saturate the resonance at which it is applied. The entire spectrum is then scanned by a short observation pulse just after the saturating pulse to search for the saturation transfer effects. Since these effects are small, their detection is made easier by taking a difference spectrum with and without the saturating pulse or with the saturating pulse alternating in frequency between the resonance of interest and some other frequency within or outside the spectral width of interest. Such a difference technique effectively cancels all background signals not affected by the saturating pulse. This difference method is especially useful when the chemical shift difference in the two redox states is small, since it tends to cancel out the effects of direct irradiation of the observed resonances. The difference method is essential if the resonances

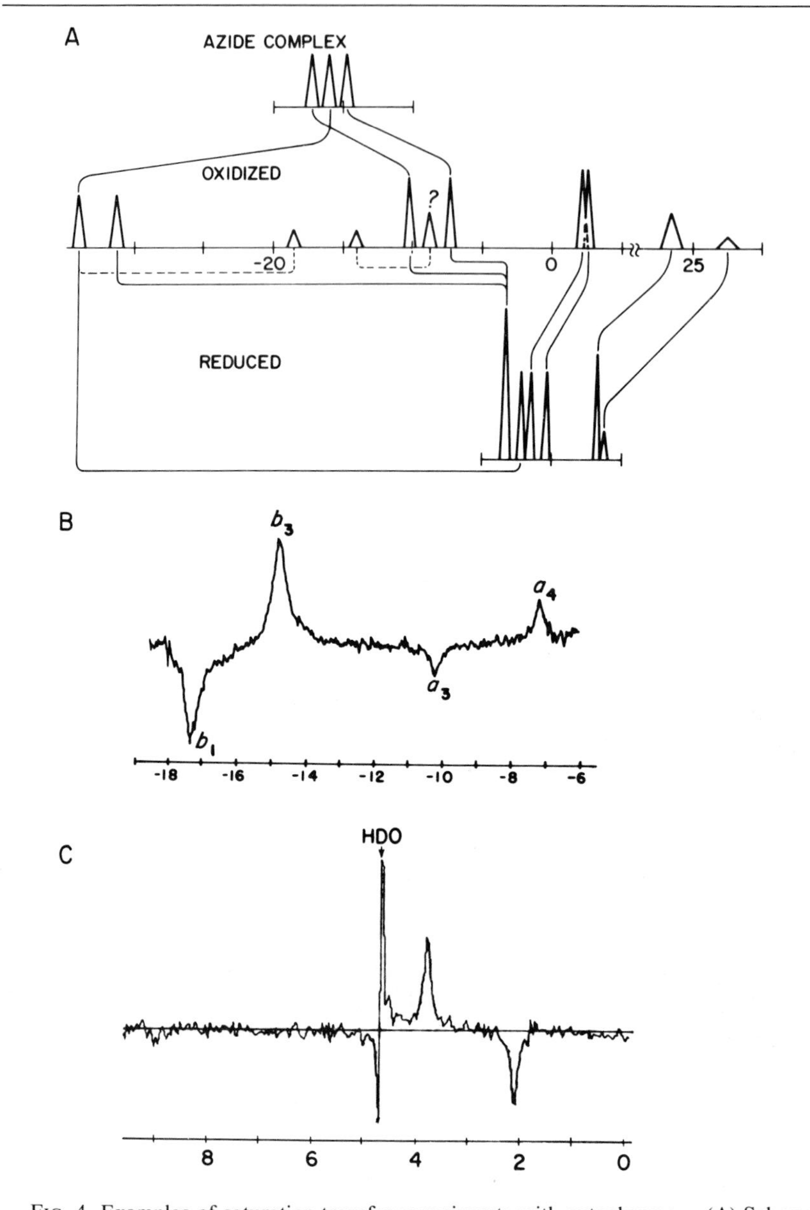

FIG. 4. Examples of saturation transfer experiments with cytochrome *c*. (A) Schematic representation of all resonances of cytochrome *c* for which saturation transfer has been observed. The resonances at −18.5, −14, and +27.4 in the oxidized and +3.8 in the reduced state have intensities characteristic of single spin; the intensity of the resonance at −8.5 ppm in the oxidized state is not known, and all the others are methyl resonances. Lines are drawn between resonances which are found experimentally to be connected by saturation transfer.

to which the saturation effects are transferred are buried under the intense aliphatic or aromatic absorption of the protein spectrum. An alternation of the saturating pulse between two resonances of interest allows the simultaneous measurement of saturation transfer effects from both resonances (Fig. 3). The difference spectrum, in such an experiment, shows positive and negative absorptions due to alternate addition and subtraction of saturation transfer effects. The difference should be taken in the time domain before Fourier transformation to save time. Alternate addition and subtraction of free-induction signals, as opposed to sequential addition and subtraction of a large number of pulses, avoids overloading the computer storage by the addition of strong signals. The observation of an NMR absorption in the difference spectrum at frequencies other than those directly saturated is indicative of the presence of cross-saturation or saturation transfer effects. Examples of such difference spectra are shown in Fig. 4. A schematic representation of all the resonances of cytochrome *c* for which saturation transfer has been observed is also included in Fig. 4.[5,33,34]

Cross-saturation occurring at resonances arising from the nuclei in a molecular state other than the one directly saturated is indicative of the presence of an exchange of protein molecules (and hence also of the saturated nuclei) between the two oxidation states present in solution with the lifetime in the state to which the saturation is transferred being of the order of the spin-lattice relaxation time in that state. Besides establishing an exchange of protein molecules between different oxidation states, the technique can be used to obtain the following information:[6]

1. It allows an accurate measurement of hyperfine shifts (or frequency shifts due to conformational change) in the spectrum. Knowing the anisotropy of the magnetic moment (i.e., *g*-tensor) from electron-spin resonance (ESR) studies, one can theoretically estimate the pseudocontact contribution to these shifts and hence by subtraction get the contact contributions.

Solid lines indicate saturation transfer occurring via chemical exchange; broken lines indicate saturation transfer occurring via cross-relaxation or Overhauser effect.[5] (B) A difference spectrum recorded with the saturating pulse alternating in frequency between the first and third of the indicated resonances of the azide complex (b_1 and b_3) in a 1:1 mixture of cytochrome *c* and azidoferricytochrome *c* (pH ~7.0, 25°C). The a_3 and a_4 resonances are those to which saturation is transferred via chemical exchange. The resonance a_4 is normally buried amidst the intense aromatic region.[33] (C) A difference spectrum showing the presence of saturation transfer via chemical exchange between the resonances of oxidized and reduced cytochrome *c* (pH ≈10, T = 25°C). The trace was obtained by Fourier-transforming the free-induction signal after time-averaging 1000 transients following 90° pulses. A saturating pulse alternated between the two methyl resonances near −35 and −32 ppm of the oxidized state. The difference spectrum was obtained by adding and subtracting the alternate free-induction signals during the averaging process. Saturation was transferred to the methyl resonances of the reduced state at −2.1 and −3.8 ppm as shown.[34]

The contact shifts may be used to map out the delocalization of the unpaired spin density over a conjugated system such as the heme ring, in the case of hemoproteins. The determination of the electron-spin wave function in this way has been used to shed light on the path of the electron transfer in cytochrome *c*.[5,35]

2. It helps to assign resonances to specific chemical groups. Three of the hyperfine shifted downfield methyls in ferricytochrome *c* transferred saturation to a resonance frequency in the diamagnetic form where one would expect ring methyls to resonate, thus virtually assigning these hyperfine shifted resonances to such methyls. One of the two upfield methyls in ferricytochrome *c* seems to shift to a frequency in the diamagnetic form where one would expect to find a porphyrin side chain methyl resonance.[6,32]

3. If some of the hyperfine shifted resonances have been identified on some other basis, it may be possible in suitable cases to derive structural information. Thus one of the two low-field methyl resonances of ferricytochrome *c*, identified as a ring methyl by its large downfield shift, was shifted to a position in the diamagnetic form 1.6 ppm upfield from the normal ring methyl resonance frequency. The observation indicated the presence of an aromatic ring close to the heme ring with its face in van der Waals contact with the methyl group in question.[5,32] This NMR prediction was later found to be consistent with the X-ray structure of ferricytochrome *c* which showed Trp-59 with its face more or less touching one of the heme methyl groups.[36,37]

4. The technique can be used to locate hyperfine shifted resonances not pulled out of the confused overlapping region of the protein spectrum. To be able to do this, one moves the double irradiation frequency across the region of interest in the spectrum. The presence of a nonvanishing difference spectrum at frequencies other than the directly saturated ones, if not arising due to cross-relaxation (see later), indicates the presence of a hyperfine shift in the resonance irradiated. If the protons double-irradiated are not affected by hyperfine interaction and there is no shift in frequency due to a possible conformational change, they would resonate at the same frequency in the two redox forms of the protein thus giving a difference signal at the directly saturated frequency only. In this way a resonance at −7.2 ppm amidst the confused aromatic region of ferricytochrome *c* was resolved and identified with a heme ring methyl group.[6]

5. The technique can be used to measure the kinetics of the redox

[35] R. K. Gupta, *Biochim. Biophys. Acta* **292,** 291 (1973).

[36] T. Takano, R. Swanson, O. B. Kallai, and R. E. Dickerson, *Cold Spring Harbor Symp. Quant. Biol.* **36,** 397 (1971).

[37] T. Takano, O. B. Kallai, R. Swanson, and R. E. Dickerson, *J. Biol. Chem.* **248,** 5234 (1973).

reaction. A study of the electron-transfer rates as a function of the concentration of exchanging species, pH, temperature, and ionic strength may throw light on the mechanism of electron transfer.[34,35] Details of the method are given in Section V,C,2.

6. In suitable cases, the experiments can be used to measure "on-off" rates (or the association and dissociation rate constants) for weakly binding ligands. The technique has been used to study the binding of azide to ferricytochrome *c*.[33] The double-resonance experiment in this case was done on the superimposed spectrum of azidoferricytochrome *c* and ferricytochrome *c* (see Section V,C,2).

2. Saturation Transfer by Cross-relaxation: The Overhauser Effect in Macromolecules[5,38–43]

Unlike the previous saturation transfer experiment, this effect will be observable in a sample of a single oxidation state. The two different mechanisms for transfer of saturation can therefore be separated by working on mixed and unmixed samples of the two oxidation states.

The basis of this cross-relaxation effect is diamagnetic in origin.[5,6,39] In a molecule the protons (or other nuclei) whose resonances are being saturated may lose a part of their magnetic energy to neighboring spins via mutual spin flips. These mutual spin flips arise from the "flip-flop" terms of the dipolar interaction Hamiltonian which dominate when the frequency of molecular motion is low.[5,6] This is generally true for proteins which tumble slowly in solution. The cross-relaxation is effective only over a short range since it varies as the inverse of the sixth power of the distance between the nuclei. The phenomenon first observed in cytochrome *c* and termed "cross-relaxation"[5] has more recently been described as a negative Overhauser effect.[40–43] Since the Overhauser or cross-relaxation effect can be used to detect magnetic nuclei in close proximity to the ones irradiated, this method has great potential for structural investigations.

Theoretical expressions have been derived which can be used to estimate the size of this effect from a knowledge of the structure of the molecule and the relevant correlation times.[39–41] These will not be reproduced here. For the simple case of two like spins, relaxing each other via dipolar interaction, the fractional decrease ($\Delta A/A$) in the intensity of a

[38] F. A. L. Anet and A. J. R. Bourn, *J. Am. Chem. Soc.* **87,** 5250 (1965).

[39] J. H. Noggle and R. E. Schirmer, "The Nuclear Overhauser Effect." Academic Press, New York (1971).

[40] P. Balaram, A. A. Bothner-By, and E. Breslow, *Biochemistry* **12,** 4695 (1973).

[41] P. Balaram, A. A. Bothner-By, and J. Dadok, *J. Am. Chem. Soc.* **94,** 4015 (1972).

[42] T. L. James and M. Cohn, *J. Biol. Chem.* **249,** 2599 (1974).

[43] T. L. James, *Biochemistry* **15,** 4724 (1976).

resonance (A) due to a saturating level of irradiation at another resonance (B) may be shown to be given by:

$$\frac{\Delta A}{A} \leqslant \frac{2T_1 - 5T_2}{2T_1 + 4T_2} \tag{20}$$

where T_1 and T_2 are the unperturbed longitudinal and transverse relaxation times of resonance A. The maximal effect is observed when nucleus A has no mechanism other than a dipolar interaction with nucleus B available for relaxation.

A gated double irradiation channel is required for carrying out these experiments and is provided by the Varian Spin Decoupler on the Varian XL-100-FT system. It is often desirable for identification purposes to turn the double irradiation off during the data acquisition time to separate cross-relaxation from decoupling effects. On the other hand, by double irradiation only during the acquisition time, decoupling is achieved but cross-saturation effects are eliminated. The detailed method for carrying out these experiments is similar to that described in the previous section on saturation transfer by chemical exchange.

The only example in a redox protein of the use of this effect is with cytochrome *c*, where it was predicted that a single proton was in van der Waals contact with one of the heme ring methyl groups in cytochrome *c*.[5,6] This NMR prediction is in agreement with the X-ray crystallographic results on this protein.[36,37] Analogous uses of the Overhauser effect in other biochemical systems have been reported.[40-43]

C. *Kinetic Measurements at Equilibrium*

When the two redox states of a given molecule are characterized by different NMR spectra, it is possible under suitable conditions to determine redox rates due to self or induced electron transfer in an equilibrium mixture of the two oxidation states from NMR measurements. Three methods have been used for such studies, NMR line broadening, saturation transfer, and spin-lattice relaxation time (T_1) measurements.

1. NMR Line-broadening Measurements[19,28,44-46a]

This technique has been applied in the two extreme situations of slow and fast exchange, defined with respect to the difference between the resonance frequencies of the two oxidation states:

[44] E. Oldfield and A. Allerhand, *Proc. Natl. Acad. Sci. U.S.A.* **70,** 3531 (1973); R. K. Gupta and T. Yonetani, *Biochim. Biophys. Acta* **292,** 502 (1973).
[45] M. S. Gutowsky and A. Saika, *J. Chem. Phys.* **21,** 1688 (1953).
[46] R. K. Gupta, T. P. Pitner, and R. Wasylischen, *J. Magn. Reson.* **13,** 383 (1974).
[46a] C. F. Midelfort, R. K. Gupta, and H. P. Meloche, *J. Biol. Chem.* **252,** 3486 (1977).

(i) Slow Exchange. When the redox rate approaches the point where the resonance lines of the same nucleus in both oxidation states broaden but do not appreciably overlap, Eq. (21) and (22) may be used to estimate the lifetime τ of a molecule in either redox state, and hence the pseudo-first-order rate constants for the redox reaction from the observed line broadenings.[28,45,46]

$$\pi \Delta \nu^{O}_{obs} = \pi \Delta \nu^{O}_{0} + \tau_O^{-1} \tag{21}$$

$$\pi \Delta \nu^{R}_{obs} = \pi \Delta \nu^{R}_{0} + \tau_R^{-1} \tag{22}$$

In Eqs. (21) and (22) $\Delta \nu^{O}_{obs}$ and $\Delta \nu^{R}_{obs}$ are the resonance line widths at half height in the oxidized and reduced state respectively. $\Delta \nu_0^{O}$ and $\Delta \nu_0^{R}$ are the line widths in the absence of the redox reaction. τ_O and τ_R are the lifetimes of a given redox molecule in the oxidized and reduced states respectively. The lifetimes are related to the redox rate constants as follows:

Self exchange: $\text{Red}_1 + \text{Ox}_2 \underset{k_{-1}}{\overset{k_1}{\rightleftharpoons}} \text{Red}_2 + \text{Ox}_1$

$$k_1 = k_{-1} = (\tau_R[\text{Ox}])^{-1} = (\tau_O[\text{Red}])^{-1} \tag{23}$$

Mediated electron transfer: $\text{Ox} + X^- \underset{k_{-1}}{\overset{k_1}{\rightleftharpoons}} \text{Red} + X^0$

Here,

$$k_1 = (\tau_O[X^-])^{-1} \tag{24}$$

and

$$k_{-1} = (\tau_R[X^0])^{-1} \tag{25}$$

where X is a different species mediating electron transfer. The line-broadening method has the advantage that it can be easily measured over a considerable range of reaction rates. Another convenient feature of Eqs. (21) and (22) is their independence of the peak separation. The disadvantage of the method is that $\Delta \nu_0^{R}$ and $\Delta \nu_0^{O}$ must be accurately known. This method has been used to estimate the self-exchange rate of cytochrome *c* as $2 \times 10^4 \ M^{-1} \ \text{sec}^{-1}$,[19] and to measure the interconversion rates of various isomeric forms of the sugar substrates of aldolases as 8–94 sec^{-1}.[28,46a]

(ii) Fast Exchange. When the redox exchange rate approaches this situation, the resonances of a given nucleus in the two oxidation states have coalesced into one resonance, whose width is given by[45,46]:

$$\Delta \nu_{obs} = \frac{\tau_O \Delta \nu_0^{O} + \tau_R \Delta \nu_0^{R}}{\tau_O + \tau_R} + \frac{4\pi(\tau_O \tau_R)^2}{(\tau_O + \tau_R)^3} (\delta \nu)^2 \tag{26}$$

where $\delta \nu$ is the separation between the resonances of the two oxidation states in the absence of the redox reaction, and τ_O and τ_R are as defined above. The two unknowns in Eq. (26), τ_R and τ_O, can be determined by measuring $\Delta\nu_{obs}$ and the equilibrium concentrations of all components using Eqs. (23) to (25).

2. Saturation Transfer Measurement of Chemical Exchange[33,34]

This method is applicable in all cases in the slow exchange limit as discussed above. In addition, it may be applicable even in situations where no significant broadening of resonances is detectable provided the redox rates are of the order of $1/T_1$, the nuclear spin-lattice relaxation rates. Such a situation is often encountered in macromolecules where $1/T_2 \gg 1/T_1$ and thus the large line widths $(1/\pi T_2)$ may be insensitive to the presence of redox transfer rates significantly slower than $1/T_2$. For example, in cytochrome *c*, a typical methyl resonance has a line width of ~20 Hz and a T_1 of ~200 msec.[5] Assuming that a 10% change in line width is accurately measurable, the lower limit of the measurable pseudo-first-order redox rate constants by the line-broadening technique will be ~6 sec^{-1}. On the other hand the ability to measure T_1 with the same accuracy (10%) would extend the lower limit to 0.5 sec^{-1}, a 10-fold improvement over the line-broadening technique. Bimolecular electron-transfer rate constants as low as 10^2 M^{-1} sec^{-1} have been measured with this technique in cytochrome *c*.[35]

The basis for the measurements is as follows. Let us define the magnetizations of a nucleus in the oxidized and reduced states respectively by M^O and M^R and their thermal equilibrium values as M_0^O and M_0^R. It may be noted that

$$\frac{\tau_O}{\tau_R} = \frac{M_0^O}{M_0^R} \tag{27}$$

When this nucleus, in the oxidized state, is irradiated with a saturating RF field, a steady state is attained such that $dM^O/dt = dM^R/dt = 0$, $M^O = 0$, and

$$M^R = \frac{\tau_R M_0^R}{\tau_R + T_1^R} \tag{28}$$

M^R and M_0^R can be determined experimentally as the amplitudes of the reduced-state resonance of the same nucleus in the presence and the absence of the saturating RF pulse.[33] Knowing T_1^R, the spin-lattice relaxation time of the nucleus in the reduced state in the absence of the redox reaction, one can measure τ_R, and hence the corresponding redox rate constant precisely. τ_O may then be calculated easily as $\tau_R (M_0^O/M_0^R)$. τ_O can also be obtained directly by saturating the reduced-state resonance and measuring the saturation transfer effect on the oxidized-state resonance, using an equation similar to Eq. (28) where the parameters characterizing the two oxidation states are interchanged.[33,33a]

Although derived for a redox reaction, the above theory is equally valid for other reactions at equilibrium such as ligand binding to and dissociation from macromolecules. In this case the parameters denoting the two oxidation states will be changed to mean liganded and unliganded states and the pseudo-first-order rate constants ($1/\tau$) will be appropriately related to the actual rate constants of the ligand-binding reaction, permitting the measurement of "on-off" rates for ligands in suitable cases. In fact, the method has been used to measure association and dissociation rate constants for azide ion binding to ferricytochrome *c*.[33,34]

3. Spin-lattice Relaxation Time (T_1) Measurement of Chemical Exchange[8,13,34,35]

This method is applicable only when a nucleus gives rise to two distinct nonoverlapping resonances in the two oxidation states and its lifetime in one of these states, e.g., the oxidized state (τ_O), is long compared to the spin-lattice relaxation time (T_1^O) in this state ($\tau_O \gg T_1^O$) while its lifetime in the other oxidation state, e.g., the reduced state, is of the order of the spin-lattice relaxation time (T_1^R) in that state ($\tau_R \sim T_1^R$). If such a situation prevails, then in mixed samples of the two oxidation states the apparent spin-lattice relaxation time (T_{1a}^R) of the nucleus in the reduced state is a function of the lifetime τ_R in that state according to the following equation:[34,35]

$$(T_{1a}^R)^{-1} = (T_1^R)^{-1} + (\tau_R)^{-1} \tag{29}$$

Hence τ_R and the redox rate constants may be obtained from a measurement of T_1^R and T_{1a}^R. The former (T_1^R) is measured as the spin-lattice relaxation time in an isolated sample of the reduced state alone while the latter (T_{1a}^R) is obtained as the spin-lattice relaxation time in a mixture of the two oxidation states. This technique has been used to measure electron-transfer rates between the two oxidation states of cytochrome *c*[34,35] as well as cytochrome c_2.[47] The T_1 measurements were made on the well-resolved methionine methyl resonance in the reduced state. Since the heme iron in the oxidized state is paramagnetic, the methionine methyl protons have a short relaxation time ($T_1^O \sim 2$ msec) in this state, in contrast to a much longer relaxation time in the diamagnetic reduced state ($T_1^R \sim 200$ msec). In mixtures containing millimolar levels of each of the two oxidation states, the lifetime of the methyl group in either state is of the order of $T_1^R (\gg T_1^O)$. Hence Eq. (29) is applicable.

[47] G. M. Smith, *Fed. Proc., Fed. Am. Soc. Exp. Biol.* **35**, 1391 (1976); G. M. Smith and M. D. Kamen, *Proc. Natl. Acad. Sci. U.S.A.* **71**, 4303 (1974).

D. Kinetic Isotope Effects[48]

The ability to do ^1H (proton), ^2H (deuterium), and ^3H (tritium) NMR (see Section V,F) may be applied to study primary and secondary kinetic isotope effects in redox systems if the relaxation rates of these nuclei are exchange-limited. As discussed in detail elsewhere, such an exchange limitation can often be demonstrated by a study of the temperature and frequency dependence of the relaxation rates.[49] An exchange-limited relaxation rate will show no frequency dependence and will increase with increasing temperature with an energy of activation $\geqslant 3$ kcal/mole. The observation of a large primary kinetic isotope effect argues in favor of the proton dissociation being the rate-limiting step for the process under examination.[48] A large primary isotope effect and a small inverse secondary isotope effect on the exchange-limited water proton relaxation rates observed upon replacing ^1H by ^2H have been used to argue in favor of rate-limiting proton dissociation rather than the dissociation of entire water molecules from the iron in methemoglobin.[48]

Pulsed[8,13] and continuous-wave methods[8,49] for measuring relaxation rates have been reviewed elsewhere.

E. Pulsed Fourier-transform NMR in H_2O Solutions: The Dynamic Range Problem

Since water is the most appropriate solvent for biochemical systems, *in vitro* biological studies, including those involving redox systems, must also be carried out in aqueous solutions. As discussed elsewhere,[31] studies in H_2O present dynamic range problems, since the water protons (111 *M*) are $\sim 10^5$-fold nore concentrated than those of the sample ($\sim$1 m*M*). Complete replacement of H_2O by D_2O cannot always be accomplished, and it is often desirable to observe exchangeable (N–H) protons in H_2O.

The strong water signal interferes with the observation of resonances of interest even when no spectral overlap occurs due to saturation of (1) the NMR receiver, (2) the analog-to-digital converter,[50] or (3) the computer

[48] R. K. Gupta and A. S. Mildvan, *J. Biol. Chem.* **250,** 246 (1975).

[49] A. S. Mildvan and M. Cohn, *Adv. Enzymol.* **33,** 1 (1970).

[50] The dynamic range of the analog to digital converter must be large enough to allow the optimal detection of a signal buried in noise, while at the same time it should accept the largest signal present without introducing any distortions. A signal may be completely recovered from noise if the peak-to-peak thermal noise in the spectrometer system exceeds the noise of the analog to digital converter.[12] For optimal retrieval of a signal buried in thermal noise, the root mean square level of the noise must represent $\sim$2 bits on the analog to digital converter.[50a] Thus a converter with a word length of 12 bits is adequate for experiments where the signal-to-noise per pulse in the time-domain is less than 1000:1. The

storage capacity.[50] While no specific solution to problem (1) has been found, an analog-to-digital converter with a word length $\geqslant 16$ bits would correct problem (2) and a computer word length $\geqslant 20$ bits would correct problem (3). Problem 3 may also be corrected by storage of each data point in two rather than one computer word (i.e., double precision) or by the technique of "block averaging" in which the signal is Fourier-transformed before the computer memory becomes saturated, and the resulting spectra from several such operations are averaged. Since none of these specific approaches, which attempt to accommodate the strong water signal, simultaneously solves problems 1, 2, and 3, none has proved satisfactory in practice. Hence the following general approaches which eliminate the water signal, and thereby solve all three problems at once, have been developed:

1. solvent saturation[12,51,52]
2. differential relaxation method[53-55]
3. long-pulse methods[12,20,31,56]
4. rapid-scan correlation spectroscopy[15,16]

1. Solvent Saturation[51,52]

This method, which has been discussed in detail elsewhere, provides the simplest solution to the problems. It consists of using the decoupler in the point decoupling mode to irradiate and thereby to equalize the populations of nuclei in the two magnetic states of water protons. This destroys the magnetization of water protons, hence its contribution to the NMR signal. The transfer of saturation to exchangeable protons limits the usefulness of this method.

ultimate dynamic range or the final signal-to-noise ratio in a spectrum after a large number of transients is limited by the word length of the computer memory. Thus if the data length for each transient coming from the analog to digital converter is 12 bits, only 16 such transients ($2^{16}/2^{12}$) will be coherently added before they would overflow in a 16 bit computer. It should be re-emphasized that overflow at any point in the time-response is not acceptable since it affects every point in the transformed spectrum.[14] Computer programs are usually written to avoid memory overflow in the handling of data either by cutting down the number of bits in use on the analog to digital converter, thus effectively reducing the converter word length and hence its dynamic range, or by using a system of weighted averaging in which the signal from the Nth transient is divided by N before being averaged. Both these procedures result in a loss in signal-to-noise ratio in experiments requiring extended averaging.

[50a] R. R. Ernst, *J. Magn. Reson.* **4,** 280 (1971).

[51] H. E. Bleich and J. A. Glasel, *J. Magn. Reson.* **18,** 401 (1975).

[52] R. K. Gupta, C. H. Fung, and A. S. Mildvan, *J. Biol. Chem.* **251,** 2421 (1976).

[53] S. L. Patt and B. D. Sykes, *J. Chem. Phys.* **56,** 3182 (1972).

[54] F. W. Benz, J. Feeney, and G. C. K. Roberts, *J. Magn. Reson.* **8,** 114 (1972).

[55] E. S. Mooberry and T. R. Krugh, *J. Magn. Reson.* **17,** 128 (1975); T. R. Krugh and W. C. Schaefer, *ibid.* **19,** 99 (1975).

[56] R. K. Gupta, *J. Magn. Reson.* **24,** 461 (1976).

2. Differential Relaxation Method[53–55]

This method, originally named "WEFT" (i.e., Water Eliminated Fourier Transform),[53] is based on the difference between the spin-lattice relaxation times of the protons of interest and those of H_2O or HDO.

The following pulse sequence is used:

[180° RF pulse, Partial recovery time τ, 90° RF pulse,
Data acquisition time (AT), Pulse delay (PD)]

A short 180° RF pulse inverts the magnetizations in the rotating frame. A partial recovery time τ allows the H_2O signal to reach its null when it is half relaxed ($\tau = T_1^{H_2O}/\ln 2$). All protons of interest are assumed to relax much more rapidly and to return to their full equilibrium intensity during the time interval τ ($\gtrsim 5T_1$ of the protons of interest) at the end of which they are sampled. When $\tau \gg 5T_1$ it is possible to save time by using a $\theta°$ pulse in place of the 180° pulse where $180° > \theta > 90°$ to invert the magnetization of all spins, followed by a short homogeneity-spoiling pulse (~10 msec duration) to destroy any transverse component of the water signal which may otherwise persist long enough to interfere with the signal from the second RF pulse. The second 90° RF pulse samples the magnetization of all spins of interest at the null point of the water signal. A major advantage of the WEFT sequence for recording spectra in H_2O or HDO solutions is that it is easily adaptable to modern commercial FT NMR spectrometers.[53]

Several disadvantages of this technique, however, are obvious[56]: (1) Studies are limited to resonances having a T_1 different from that of H_2O or HDO. For optimum performance T_1 (HDO) should be $\gtrsim 5T_1$ of the resonances of interest. When this is not the case, the relative intensities of the observed resonances may be significantly distorted by the use of WEFT. Its use therefore precludes quantitative studies based on the area under resonance absorptions. (2) Since in WEFT one is often recording partially relaxed as opposed to fully relaxed spectra, the overall sensitivity of the method is suboptimal, an undesirable feature for much biochemical work where sample quantities are limited and an optimal signal-to-noise ratio is an important consideration. (3) The intensities of exchangeable protons observed by this technique will be further distorted by the presence of cross-relaxation and chemical exchange effects. And finally, (4) the technique is not easily amenable to time-resolved studies such as measurements of spin-lattice relaxation times.

3. Long-pulse Methods[12,20,31,56]

Long observation pulses may be used to maximally excite the resonances of the nuclear spins of interest, and simultaneously to eliminate the large water signal by making the water protons execute a 2π nutation in the rotating frame. These methods, developed in the laboratory of A. G. Redfield,[31] generally require specialized modifications of the spectrometer and are discussed in Section V,A,1 and in reviews.[12,31]

A new long-pulse method has been developed which is easily adaptable to ordinary commercial spectrometers.[56] This "modified WEFT" technique is essentially free from the major disadvantages of the conventional WEFT technique, but retains the ease of adaptability to most modern NMR spectrometers and requires no instrumental modifications. The technique, unlike regular WEFT, should be useful in routine spin-lattice relaxation studies in aqueous solutions as well, with minor changes in programs.

The pulse sequence for "modified WEFT" is shown in Fig. 5.[56] All spins are allowed to equilibrate for a recovery time, *PD* ($\sim 5T_1$). A long weak nulling pulse of duration T in resonance with water spins (at ν_{H_2O}) is then applied to invert the water magnetization in the rotating frame by $\sim 180°$, followed by a waiting time τ_{null} to allow the longitudinal component of water magnetization to half relax, i.e., to reach its null. At the null point of H_2O, a short intense 90° pulse samples the magnetization of all spins over the spectral region of interest. The free-induction signal acquired during the acquisition time (*AT*) is Fourier-transformed in the usual manner. A homogeneity-spoiling (*HS*) pulse lasting for <10 msec within the interval τ_{null} may be needed if $\tau_{null} <$ ($5T_2^*$ of the H_2O resonance) to destroy any residual transverse component of the water magnetization. It is not necessary to make the long nulling pulse exactly 180°. In fact, by adjusting the amplitude and duration of the nulling pulse to correspond to a nutation angle θ in the range 90–180°, followed by an *HS* pulse, it is possible to arbitrarily shorten τ_{null}, the only limitation being the recovery of the magnetic field homogeneity after the *HS* pulse which in our system (a Varian XL-100-15 FT) requires ~ 10 msec. For any desired τ_{null}, the nutation angle

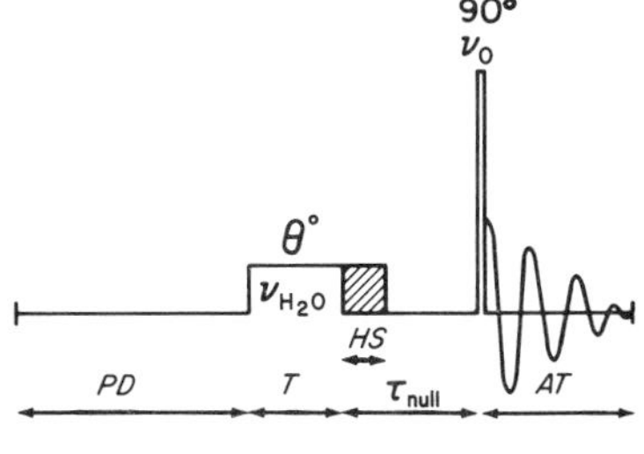

FIG. 5. Pulse sequence for elimination of water signal by a long-pulse or "modified WEFT" technique.[56]

in the rotating frame induced by the nulling pulse may be calculated using the following equation:

$$\cos \theta = 1 - e^{\tau_{null}/T_1} \quad (30)$$

When θ is known, the required length of the long pulse may then be calculated as $(\theta/90°) \times$ (the 90° pulse length). It may be noted that in the above pulse sequence, the nulling pulse leaves the magnetization of all spins unaffected except those very close to H_2O (within $\sim 1/T$ Hz where T is the long-pulse duration). Hence, unlike the regular WEFT, the modified WEFT method does not distort the relative intensities of resonances under observation. Like WEFT, however, modified WEFT permits the observation of resonances on either side of the H_2O signal in a single run. The long nulling pulse may be obtained by using the gated spin decoupler on the Varian XL-100-15 FT system. τ_{null} may be determined empirically by initially setting the long-pulse width to $\sim$0.1 sec, adjusting its RF amplitude to invert the water magnetization, and measuring the time required for the H_2O signal to reach its null. Using this method, the residual H_2O signal was reduced from 111 M to 0.3 M and the noise in the spectrum was entirely random. Hence the level of nulling achieved was sufficient to study $\sim$1 mM samples in H_2O solutions.

Although named "modified WEFT," the basis of the H_2O resonance elimination in this technique is quite different from that in WEFT. For example, the modified WEFT technique is not dependent on the differential relaxation behavior of H_2O and the protons of interest and hence is applicable to systems with any T_1 value for water, particularly in working with paramagnetic biochemical systems where the T_1 of water protons may be considerably shortened by electron–nuclear interactions. Further, since the choice of the nulling pulse width is limited only by the separation $\Delta\nu$ of the resonances of interest from water, the only requirement being that the nulling pulse not affect the spins of interest, the pulse width could be made as short as $1/\Delta\nu$ to save time. Thus, for observing tryptophan NH,[57] and imidazole NH protons[20,58] in proteins, which are $>$500 Hz from the H_2O resonance (at 100 MHz), the width of the nulling pulse need only be $\geqslant$2 msec. Since τ_{null} can be shortened to any value ($>$20 msec) by choosing the nulling pulse nutation angle appropriately, it is possible to eliminate intensity distortions due to cross-relaxation and saturation transfer effects which occur on a time scale $\lesssim$100 msec. Interestingly, in favorable cases it may be possible to determine proton exchange rates with H_2O or cross-

[57] J. D. Glickson, C. C. McDonald, and W. D. Phillips, *Biochem. Biophys. Res. Commun.* **35,** 492 (1969).

[58] R. K. Gupta and J. M. Pesando, *J. Biol. Chem.* **250,** 2630 (1975).

relaxation with H_2O by studying the intensities of exchangeable proton resonances as a function of τ_{null} adjusting the nulling nutation angle in the range 90–180° according to Eq. (30). Modified WEFT is in principle also adaptable to spin-lattice relaxation time measurements with minor software modifications.[13]

4. Rapid-scan Correlation Spectroscopy[15,16]

This method of eliminating water signals merely scans other regions of the spectrum. It has been discussed in detail in Section V,A,2.

F. Variable-frequency NMR of a Wide Variety of Nuclei[13,59–61]

It is often desirable in structural studies of biochemical systems, including redox proteins, to carry out NMR studies of a variety of nuclei at several frequencies.[62–66] Several approaches have been used to achieve this on modern commercial spectrometers with minimal instrumental modifications.[59–61] We have developed a scheme of field-frequency locking on different nuclei to resonate the observed nucleus at the appropriate frequencies. The minor spectrometer modification needed to achieve this has been described in detail elsewhere.[13] This scheme which has been successfully tested on the Varian XL-100-FT spectrometer makes use of the spin decoupler as the source of the variable lock-frequency in the range 14–17 MHz, over which the lock channel of the spectrometer is tunable. With the commercial availability of variable-frequency decouplers, observation frequencies continuously variable in the range 9–33 MHz in addition to several fixed frequencies are available for observing a wide variety of nuclei at full field (23.5 kG) while maintaining field-frequency lock on deuterium. A combination of the variability in the observation frequency with our scheme of locking on different nuclei makes it possible to carry out Fourier-transform NMR studies of many nuclei at several frequencies.

No spectrometer modifications are required. The usual tuning procedures used with various nuclei are sufficient. The lock channel operates at

[59] R. K. Gupta, *J. Magn. Reson.* **16,** 185 (1974).
[60] C. Peters, H. Codrington, H. Walsh, and P. Ellis, *J. Magn. Reson.* **11,** 431 (1973).
[61] P. Ellis, H. Walsh, and C. Peters, *J. Magn. Reson.* **11,** 426 (1973).
[62] C. M. Grisham, R. K. Gupta, R. E. Barnett, and A. S. Mildvan, *J. Biol. Chem.* **249,** 6738 (1974).
[63] C. F. Springgate, A. S. Mildvan, R. Abramson, J. L. Engle, and L. A. Loeb, *J. Biol. Chem.* **248,** 5987 (1973).
[64] E. Melamud and A. S. Mildvan, *J. Biol. Chem.* **250,** 8193 (1975).
[65] F. Kayne and J. Reuben, *J. Am. Chem. Soc.* **92,** 770 (1970).
[66] C. H. Fung, R. K. Gupta, and A. S. Mildvan, *Biochemistry* **15,** 85 (1976).

the standard fixed frequency (15.351 MHz), while the observation frequency is provided by the variable-frequency source of RF.

The static magnetic field is lowered to bring an appropriately selected lock nucleus (^{1}H, ^{2}H, ^{11}B, ^{7}Li, ^{31}P, ^{13}C, ^{203}Tl, ^{205}Tl, ^{19}F, or ^{23}Na) into resonance at the standard lock-channel frequency (15.351 MHz), and the corresponding frequency for the observed nucleus, calculated according to the following equation, is obtained from the decoupler.

$$\nu_{\text{obs(at lower field)}} = \frac{15.351 \cdot \nu_{\text{resonance for observed nucleus at full field}}}{\nu_{\text{resonance for selected lock nucleus at full field}}} \tag{31}$$

By varying the choice of the lock nucleus it is possible to carry out a frequency-dependence study of any given observed nucleus over the frequency range covered by the decoupler and tunable elements of the spectrometer. Possible combinations of lock and observed nuclei on a 100-MHz system are given in Table I. When the compound used for locking cannot be mixed with the sample, concentric sample tubes may be used as discussed in Section III.

G. Distance Measurements in Hemoproteins and Other Redox Systems

High-resolution NMR studies in solution, like X-ray diffraction in the crystalline state, can be used to measure distances from the individual atoms of a molecule in solution to a nearby paramagnetic reference point.[8,13,49] The distances are most accurately obtained by the measurement of paramagnetic effects on the longitudinal relaxation rates of the nuclei. Using this method distances as large as 24 Å from a paramagnetic metal to a nucleus of a ligand in the same complex have been calculated with precisions of better than ±10% in hemoproteins.[67]

The distance-dependence of the paramagnetic component of the longitudinal relaxation rate ($1/T_{1p}$), which forms the basis of distance calculations, is given by the following Solomon-Bloembergen equation:

$$\frac{1}{T_{1p}} = \frac{2fqS(S+1)g^2\beta^2\gamma_I^2}{15r^6}\left(\frac{3\tau_{c_1}}{1+\omega_I^2\tau_{c_1}^2} + \frac{7\tau_{c_2}}{1+\omega_S^2\tau_{c_2}^2}\right) \tag{32}$$

where f is the ratio of concentrations of the paramagnet and the relaxing ligand, q is the relative stoichiometry of the bound ligand and bound paramagnet (i.e., the coordination number), S is the electron spin, γ_I is the nuclear gyromagnetic ratio, g is the electronic g factor, β is the Bohr magneton, r is the metal–nucleus distance, ω_I and ω_S are the nuclear and electron precession frequencies, and τ_{c_1} and τ_{c_2} are the correlation times for

[67] R. K. Gupta, *J. Biol. Chem.* **251**, 6815 (1976).

TABLE I

COMBINATIONS OF THE OBSERVED AND LOCK NUCLEI ON THE VARIAN XL-100-FT SYSTEM EQUIPPED WITH GYROCODE OBSERVE ACCESSORY[a]

Observed nucleus	Lock nuclei (ν_{obs})
^{1}H	$^{2}H(100)$, $^{205}Tl(26.601)$
^{2}H	$^{1}H(15.351)$
^{3}H	$^{205}Tl(28.373)$, $^{31}P(40.448)$
^{7}Li	$^{205}Tl(10.338)$, $^{11}B(18.594)$, $^{23}Na(22.553)$
^{11}B	$^{2}H(32.084)$, $^{23}Na(18.619)$
^{13}C	$^{2}H(25.144)$, $^{7}Li(9.932)$
^{19}F	$^{205}Tl(25.025)$, $^{2}H(94.077)$
^{23}Na	$^{31}P(10.031)$, $^{7}Li(10.449)$, $^{2}H(26.452)$
^{27}Al	$^{31}P(9.881)$, $^{7}Li(10.293)$, $^{2}H(26.057)$
^{29}Si	$^{2}H(19.865)$, $^{11}B(9.505)$, $^{23}Na(11.528)$
^{31}P	$^{2}H(40.481)$, $^{205}Tl(10.768)$, $^{11}B(19.369)$, $^{23}Na(23.493)$
^{51}V	$^{31}P(9.966)$, $^{7}Li(10.381)$, $^{2}H(26.280)$
^{55}Mn	$^{7}Li(9.743)$, $^{2}H(24.664)$
^{59}Co	$^{11}B(11.298)$, $^{2}H(23.614)$
^{63}Cu	$^{31}P(10.051)$, $^{7}Li(10.470)$, $^{2}H(26.506)$
^{65}Cu	$^{31}P(10.767)$, $^{7}Li(11.216)$, $^{2}H(28.394)$
^{75}As	$^{23}Na(9.939)$, $^{1}H(17.127)$
^{77}Se	$^{23}Na(11.067)$, $^{2}H(19.07)$
^{79}Br	$^{7}Li(9.897)$, $^{2}H(25.054)$
^{81}Br	$^{31}P(10.241)$, $^{7}Li(10.668)$, $^{2}H(27.006)$
^{87}Rb	$^{23}Na(18.989)$, $^{2}H(32.720)$
^{93}Nb	$^{7}Li(9.655)$, $^{2}H(24.443)$
^{111}Cd	$^{2}H(21.204)$, $^{11}B(10.145)$
^{113}Cd	$^{2}H(22.181)$, $^{11}B(10.613)$
^{117}Sn	$^{205}Tl(9.477)$, $^{23}Na(20.675)$
^{119}Sn	$^{205}Tl(9.915)$, $^{11}B(17.833)$, $^{23}Na(21.630)$
^{121}Sb	$^{11}B(11.450)$, $^{2}H(23.931)$
^{127}I	$^{11}B(9.573)$, $^{23}Na(11.611)$, $^{2}H(20.007)$
^{195}Pt	$^{11}B(10.287)$, $^{2}H(21.50)$
^{199}Hg	$^{23}Na(10.346)$, $^{2}H(17.827)$
^{203}Tl	$^{19}F(9.325)$, $^{31}P(21.672)$, $^{7}Li(22.575)$, $^{11}B(27.344)$, $^{23}Na(33.166)$
^{205}Tl	$^{19}F(9.417)$, $^{31}P(21.884)$, $^{7}Li(22.796)$, $^{11}B(27.612)$, $^{23}Na(33.491)$
^{207}Pb	$^{11}B(10.010)$, $^{2}H(20.922)$

[a] All of the lock nuclei listed have been successfully used. The compounds used have been H_2O, D_2O, boron trifluoride etherate, trimethylborate, lithium chloride, phosphoric acid, thallium acetate, hexafluoro benzene, and sodium chloride.

the dipolar interaction.[8,13,49,68,69] The use of the above equation involves the following assumptions which must be justified in individual cases.[13]

1. The $1/T_{1p}$ is occurring in a single complex.
2. The outer sphere contribution to $1/T_{1p}$ is negligible.
3. The $1/T_{1p}$ is not exchange-limited.
4. The hyperfine contact contribution to $1/T_{1p}$ is negligible.
5. The effects of electron delocalization and of the spatial anisotropy of the unpaired electron's magnetic moment are small. These assumptions can be justified by appropriate experiments as discussed elsewhere.[13]

The present generation of commercial NMR spectrometers such as the Varian XL-100-FT, the Bruker WHX-90, and JEOL PFT-100 are suitable for $1/T_1$ measurements. The inversion recovery, demagnetization recovery, or one of the steady-state methods discussed elsewhere may be used for T_1 measurements.[13] The $1/T_{1p}$ is obtained by subtracting the relaxation rate measured in the presence of a diamagnetic system, such as the enzyme–reduced spin-label–substrate complex, from that measured in the presence of the corresponding paramagnetic system, the enzyme–spin-label–substrate complex. The correlation time for the dipolar interaction in Eq. (32) is most appropriately determined by the frequency dependence of T_{1p} of the ligand or from the frequency dependence of T_{1p} of water in the same complex when the correlation process is dominated by electron-spin relaxation. Further details of the methodology are discussed elsewhere.[8,13] Studies of paramagnetic effects on the longitudinal relaxation rates have been used to detect the presence or absence of fast-exchanging water ligands of the metal in hemoproteins,[48] to map out the binding site of allosteric effectors in hemoglobin,[67] to establish the origin of contact-shifted resonances in cytochrome *c*,[5] to detect the movement of the distal histidine during the R-T transition in methemoglobin,[48] to measure intersubstrate distances and substrate conformations on dehydrogenases using spin-label,[70–72] as well as in a wide variety of other problems.[73]

VI. Glossary of Symbols

$A(\omega)$	Absorption mode NMR signal as a function of frequency.
AT	Data acquisition time (sec).
β	Bohr magneton.

[68] I. Solomon, *Phys. Rev.* **99,** 559 (1955).

[69] N. Bloembergen, *J. Chem. Phys.* **27,** 572 (1957).

[70] A. S. Mildvan and H. Weiner, *J. Biol. Chem.* **244,** 2465 (1969).

[71] A. S. Mildvan, L. Waber, J. J. Villafranca, and H. Weiner, *in* "Structure and Function of Oxidation-Reduction Enzymes" (A. Åkeson and A. Ehrenberg, eds.), p. 745. Pergamon, Oxford, 1972.

[72] D. L. Sloan, and A. S. Mildvan, *Biochemistry* **13,** 1711 (1974).

[73] A. S. Mildvan, *Annu. Rev. Biochem.* **43,** 357 (1974).

$C(\omega)$	Fourier cosine transform of $M(t)$.
$\delta\nu$	Frequency separation between oxidized- and reduced-state resonances.
$\Delta\nu = \nu - \nu_0$	Difference between the frequency of resonance (ν) and that of the RF pulse (ν_0).
$\Delta\nu_0^O$, $\Delta\nu_0^R$	Line widths of the oxidized- and reduced-state resonances in the absence of the exchange reaction, respectively.
$\Delta\nu^O_{obs}$, $\Delta\nu^R_{obs}$	Observed line widths in the oxidized and reduced state, respectively.
f	Ratio of concentrations of the paramagnet and the relaxing ligand.
$F(\omega)$	NMR spectrum as a function of frequency.
g	Electronic g factor.
γ or γ_I	Gyromagnetic ratio of a nucleus (sec^{-1} $gauss^{-1}$).
H_1	Amplitude of the radio-frequency field.
H_0	Static magnetic field in gauss.
HS	Homogeneity-spoiling pulse.
M_0	Total equilibrium magnetization of the sample.
M_0^O, M_0^R	Equilibrium magnetization of the oxidized and reduced states, respectively.
M^O, M^R	Magnetization of the oxidized and reduced states, respectively, after perturbation of the oxidized state by RF irradiation.
$M(t)$	Magnetization as a function of time or free-induction decay.
M_T	Steady-state magnetization of a sample subjected to rapid 90° pulses with a recycle time T.
M_θ	Transverse component of the magnetization vector of a sample subjected to rapidly recycled $\theta°$ RF pulses.
N	Number of data points.
ν	Resonance frequency of a nucleus in MHz or Hz.
ν_{H_2O}	Resonance frequency of water protons.
ν_{noise}	A frequency in the noise spectrum.
ν_0	Frequency of the RF pulse in MHz or Hz.
ω	Angular frequency in radians/sec.
ω_I	Nuclear precession frequency.
ω_s	Electron precession frequency.
PD	Pulse delay in sec.
ϕ	Phase angle.
PW	Pulse width (μsec or msec).
PW_{90}	Pulse width corresponding to a 2π flip or nutation.
q	Relative stoichiometry of the bound ligand and bound paramagnet.
r	Metal–nucleus distance.
k_1, k_{-1}	Forward and reverse redox rate constants.
S	Electron-spin quantum number.
$S(\omega)$	Fourier sine transform of $M(t)$.
$\sum_\nu$	Summation over all resonance frequencies ν.
SW	Spectral width in Hz.
T_1	Longitudinal (spin-lattice) relaxation time.
T_2	Transverse (spin-spin) relaxation time.
T_2^*	Apparent lifetime of the transverse magnetization in sec.
T_1^O, T_1^R	Spin-lattice relaxation times of oxidized and reduced states, respectively.
T_{1a}^R	Apparent spin-lattice relaxation time of the reduced state in the presence of the redox reaction.
T_{1p}^{-1}	Paramagnetic contribution to the longitudinal relaxation rate.
τ	Partial recovery time.

τ_c	Correlation time for dipolar interaction.
τ_{null}	The partial recovery time resulting in a null in NMR signal.
τ_O, τ_R	Lifetimes in oxidized and reduced states, respectively.
θ	Nutation or flip angle of the magnetization vector away from its equilibrium position in the rotating frame.
θ_{opt}	Nutation or flip angle corresponding to a maximal steady-state signal.

Acknowledgments

This work was supported by National Institutes of Health Grants AM-19454 and AM-13351, by National Science Foundation Grant BMS-74-03739, by grants to this Institute from the National Institutes of Health (CA-06927 and RR-05539), and by an appropriation from the Commonwealth of Pennsylvania.

R. K. Gupta is the recipient of a Research Career Development Award, AM (NIH)-00231, from the U.S. Public Health Service.

[13] NMR of Hemoproteins and Iron-Sulfur Proteins*

By CHIEN HO, LESLIE W.-M. FUNG,† and KAREN J. WIECHELMAN‡

Introduction

During the past decade, nuclear magnetic resonance (NMR) spectroscopy has been developed into a very powerful tool not only for the detailed elucidation of the structures and interactions of small molecules but also for the study of biological macromolecules in solution. The most general and potentially the most useful NMR approach to correlate the structure–function relationship in a protein is the direct examination of the resonances of the hydrogen atoms (protons) of the molecule. NMR spectroscopy is the only technique currently available that provides the possibility of monitoring the conformations and dynamics of individual atoms of a complex macromolecule such as a protein in solution. However, in practice one usually finds that there is a severe overlapping of the individual proton resonances in the NMR spectrum of a complex molecule, making it very difficult to resolve all of the individual proton resonances in the spectrum of a native protein molecule. The proton resonances which arise from hydrogen atoms of the amino acid residues in a protein occur over a relatively

* The writing of this chapter was supported by research grants from the National Institutes of Health (HL-10383) and the National Science Foundation (PCM 76-21469).

† A recipient of a National Research Service Award of the National Institutes of Health, U.S. Public Health Service (GM-05164). Present address: Department of Chemistry, Wayne State University, Detroit, Michigan 48202.

‡ Present address: Department of Chemistry, University of Southwestern Louisiana, Lafayette, Louisiana 70504.

ISBN 0-12-181954-X

narrow spectral region occupying approximately 10 parts per million (ppm). Since there are a large number of protons contributing to the ^{1}H NMR spectrum in this region, the resonances usually overlap severely. The spectrum is further complicated by the unusually broad line width of the resonances which are believed to arise from anisotropic dipole–dipole interactions between protons in the slowly rotating large protein molecules. These dipole–dipole interactions average to zero in rapidly rotating small molecules so their proton resonances are generally narrow in comparison to those of the larger and more rigid macromolecules. For recent reviews on the applications of NMR spectroscopy to proteins, refer to Roberts and Jardetzky,[1] McDonald and Phillips,[2] Dwek,[3] James,[4] and Wüthrich.[5]

A general approach to improve the resolution in ^{1}H NMR spectroscopy of proteins is to obtain their spectra at higher magnetic fields since the resolution of diamagnetic proton resonances is a linear function of the applied magnetic field strength. In those proteins which contain paramagnetic metal ions, the resonances in the vicinity of the paramagnetic centers are shifted away from the majority of the diamagnetic proton resonances. Thus, some spectral regions of such proteins are better resolved. The profound effects of paramagnetic centers on the characteristics of nuclear resonances (e.g., chemical shifts and nuclear relaxation times) are well established.[6,7] Such effects are due to the presence of unpaired electrons in the paramagnetic species. The magnetic moment of an electron is 658 times larger than that of a proton; consequently the electron–nucleus spin interactions (the so-called hyperfine interactions) can cause extensive variations of the resonance positions (chemical shifts) and nuclear relaxation times (the nuclear spin-lattice relaxation time, T_1, and the nuclear spin-spin relaxation time, T_2) in paramagnetic molecules. On the basis of the electronic spin-lattice relaxation times (T_{1e}) of paramagnetic metal ions, there are two general effects of the paramagnetic center on the NMR spectrum of a protein. The presence of paramagnetic ions with relatively long electronic spin-lattice relaxation times, i.e., $T_{1e} \geqslant 10^{-10}$ sec, will result

[1] G. C. K. Roberts and O. Jardetzky, *Adv. Protein Chem.* **24,** 447 (1970).

[2] C. C. McDonald and W. D. Phillips, *in* "Fine Structure of Proteins and Nucleic Acids" (G. D. Fasman and S. N. Timasheff, eds.), p. 1. Dekker, New York, 1970.

[3] R. A. Dwek, "Nuclear Magnetic Resonance in Biochemistry: Applications to Enzyme Systems." Oxford Univ. Press (Clarendon), London and New York, 1973.

[4] T. L. James, "Nuclear Magnetic Resonance in Biochemistry: Principles and Applications." Academic Press, New York, 1975.

[5] K. Wüthrich, "NMR in Biological Research: Peptides and Proteins." North-Holland Publ., Amsterdam, 1976.

[6] D. R. Eaton and W. D. Phillips, *Adv. Magn. Reson.* **1,** 103 (1965).

[7] G. N. La Mar, W. D. Horrocks, Jr., and R. H. Holm, eds., "NMR of Paramagnetic Molecules: Principles and Applications." Academic Press, New York, 1973.

in line broadening in the nuclear resonances, making these paramagnetic species useful as relaxation probes. The presence of paramagnetic ions with short electronic spin-lattice relaxation times, i.e., $T_{1e} \leqslant 10^{-10}$ sec, results in a large shift in the positions of the resonances, making these paramagnetic species useful as shift probes. These resonances with unusually large chemical shifts are known as hyperfine shifted (hfs) resonances. For recent discussions of the effects of relaxation in paramagnetic systems of biological interest, the reader should consult Dwek[3] and Wüthrich.[5]

Chemical Shift

Proton chemical shifts in compounds containing paramagnetic metal ions with short electronic spin relaxation times can be unusually large. For example, Fig. 1 shows the ^{1}H NMR spectra of hemoglobin (Hb) samples in which the iron atoms are in either the ferrous or ferric forms in both the high- and low-spin states. The observed chemical shifts cover a range from +25 to −90 ppm from the proton resonance of water. The observed chemical shifts (δ_{obs}) in Fig. 1 can be expressed as the sum of three terms:

$$\delta_{obs} = \delta_d + \delta_c + \delta_p \tag{1}$$

where δ_d is the diamagnetic chemical shift, i.e., those chemical shifts in the absence of unpaired electrons. δ_c is the contact shift (or Fermi shift)[8-10] and δ_p is the pseudocontact shift.[11] For example, the chemical shift range in diamagnetic moleucles, such as oxyhemoglobin (HbO_2) or carbonmonoxyhemoglobin (HbCO), is no more than 18 ppm (including the low-field exchangeable and the up-field ring-current shifted proton resonances). The observed chemical shifts in deoxyhemoglobin (deoxy-Hb) (high-spin ferrous) and methemoglobin (met-Hb) (high-spin ferric) span a range of about 100 ppm. In order to account for the observed shifts in deoxy-Hb and met-Hb we need to use all three terms given in Eq. (1). The last two terms (δ_c and δ_p) are the hyperfine shifted contributions arising from electron–nucleus spin interactions.

The contact shifts (δ_c) result from protons that have direct contact with the iron atoms and interact with the unpaired electrons either through chemical bonds or by hyperconjugation. The magnitude of δ_c is proportional to the unpaired electron density at the nucleus. The shift for nucleus i is given by the following equation[9,10].

[8] E. Fermi, *Z. Phys.* **60,** 320 (1930).

[9] N. Bloembergen, *J. Chem. Phys.* **27,** 595 (1957).

[10] H. M. McConnell and D. B. Chesnut, *J. Chem. Phys.* **28,** 107 (1958).

[11] H. M. McConnell and R. E. Robertson, *J. Chem. Phys.* **29,** 1361 (1958).

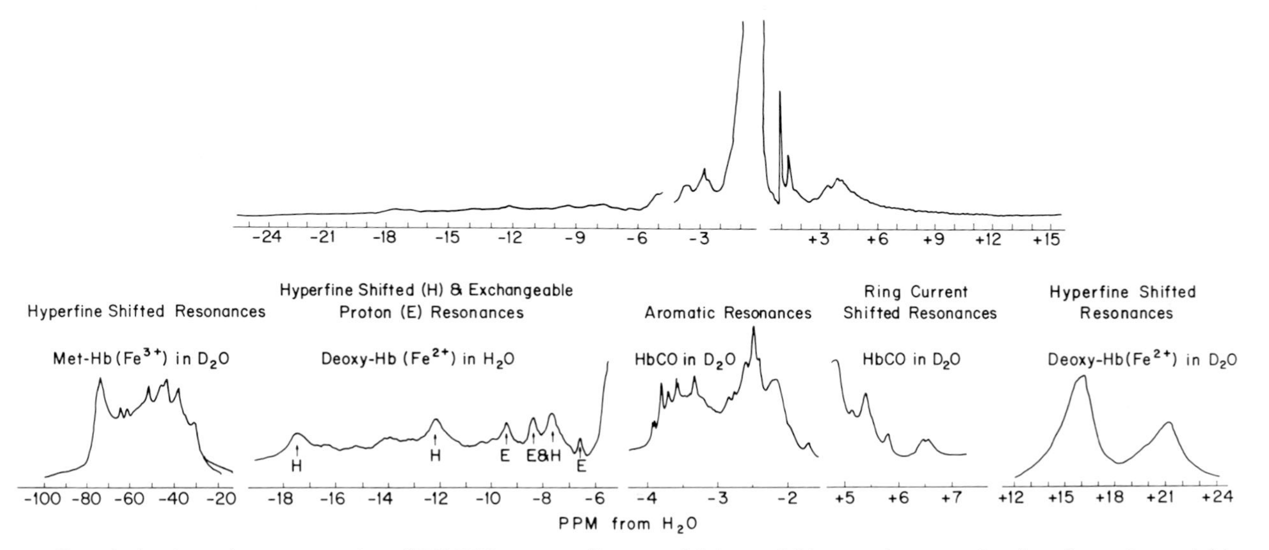

FIG. 1. A schematic representation of ^1H NMR spectra of human adult hemoglobin over the spectral regions for methemoglobin (high-spin ferric, $S = 5/2$), deoxyhemoglobin (high-spin ferrous, $S = 2$), and carbonmonoxyhemoglobin (low-spin ferrous, $S = 0$) in D_2O and H_2O. The vertical scales at various spectral regions are not comparable.

$$(\delta_c)_i = -A_i \frac{\gamma_e}{\gamma_I} \frac{g\beta S\ (S+1)}{3kT} \tag{2}$$

where A_i is the hyperfine coupling constant (in frequency units), γ_e and γ_I are the respective electronic and nuclear gyromagnetic ratios, g is the electronic g-factor, β is the Bohr magneton, S is the total electronic spin, k is the Boltzmann constant, and T is the absolute temperature. Equation (2) is valid for isotropic g-factors and for systems in which only the electronic ground state is populated. For systems with several thermally populated excited electronic states, the temperature dependence has been found to differ from T^{-1}. Modified expressions for Eq. (2) that take into account anisotropic g-values and additional temperature dependence have been reported.[5,12,13]

Pseudocontact shifts (δ_p) arise from protons which have no direct contact with the iron atoms but have dipolar interactions with the unpaired electrons of iron atoms through space. The interaction is anisotropic. As a consequence, this electron–nucleus dipole–dipole interaction does not average to zero under the rotational motions of molecular tumbling in solution, i.e., $\tau_R \gg T_{1e}$ (where τ_R is the rotational tumbling time of the molecule). The expression for pseudocontact shifts in a paramagnetic system is given by:[11]

$$(\delta_p)_i = -\frac{3\ cos^2\ \theta_i - 1}{r_i^3} \cdot \frac{g_\parallel - g_\perp}{3} \cdot \frac{g_\parallel - g_\perp}{3} \cdot \frac{\beta^2 S(S+1)}{3kT} \tag{3}$$

where r_i is the distance from the paramagnetic metal ion to the nucleus i, $g_\parallel$ and $g_\perp$ are the g-tensor components parallel and perpendicular respectively to the molecular symmetry axis, and θ_i is the azimuthal angle. In deriving the above equation, it is assumed that the g-tensor is axially symmetric and that the possible effects of spin delocalization on the pseudocontact interactions are negligible. Equations have also been derived for cases where the g-tensor is not axially symmetric.[13,14] Hence, the pseudocontact shift is determined by the molecular geometry and the principal values of the electronic g-tensor.[11] In Eq. (3), the angular dependence of the pseudocontact shift means that different nuclei in the same paramagnetic molecule can have shifts of different magnitude and sign since $(3\cos^2\theta - 1)$ has positive values for $\theta < 54°44'$, and negative values for $\theta > 54°44'$. Due to the r^{-3} dependence pseudocontact shifts are important only for nuclei near the paramagnetic center. In addition, Eq. (3) shows that δ_p can be calculated for all nuclei in a paramagnetic molecule if the molecular

[12] J. P. Jesson, *J. Chem. Phys.* **47,** 579 (1967).
[13] R. J. Kurland and B. R. McGarvey, *J. Magn. Reson.* **2,** 286 (1970).
[14] J. P. Jesson, *J. Chem. Phys.* **47,** 582 (1967).

geometry and the components of the *g*-tensor are known. Conversely, molecular structural information can be determined from the pseudocontact shifts.

Three principal criteria can be used to distinguish between diamagnetic and hyperfine shifted resonances. The chemical shifts of the hfs resonances are, in general, much larger than those of diamagnetic resonances. In the absence of conformational changes, the chemical shifts of diamagnetic resonances are only slightly temperature sensitive while both contact and pseudocontact shifts are generally proportional to the reciprocal of temperature as shown by Eqs. (2) and (3). However, in certain iron-sulfur proteins (such as the oxidized form of the eight-iron ferredoxin from *Clostridium pasteurianum*[15]), the observed hfs proton resonances actually increase with temperature in contrast to the relations given in Eqs. (2) and (3). We will discuss this behavior in a later section. The relaxation times (T_1 and T_2) of a paramagnetic system are usually much shorter than those of the corresponding diamagnetic analogue. In other words, the resonance line widths of those nuclei which are sufficiently near a paramagnetic center to experience electron–nucleus interactions are appreciably broadened.

It is often desirable to separate the contributions of the contact (δ_c) and pseudocontact (δ_p) shifts since the individual terms taken separately can yield information about the electron-spin distribution and the molecular geometry. However, it is usually difficult to distinguish between pseudocontact and contact effects in an NMR spectrum. In some instances the mechanism responsible for the hyperfine shifts may be ascertained by (1) comparing the NMR spectra of similar molecules[2,4] or (2) considering the molecular geometry in terms of the transmission of the effects "through bonds" (contact shift) or "through space" (pseudocontact shift). It may also be possible to determine if the shift is due to contact interactions on the basis of the different dependence of T_1 and T_2 on the hyperfine coupling constant.[16]

Line Width

The line width of a resonance is related to the spin-spin relaxation time by

$$T_2 = \frac{1}{\pi \Delta \nu} \tag{4}$$

where $\Delta\nu$ (in Hz) is the full line width at half height of a given resonance. The presence of a paramagnetic center can greatly influence the nuclear

[15] M. Poe, W. D. Phillips, C. C. McDonald, and W. Lovenberg, *Proc. Natl. Acad. Sci. U.S.A.* **65**, 797 (1970).

[16] K. Wüthrich, *Struct. Bonding (Berlin)* **8**, 53 (1970).

relaxation times T_1 and T_2 through dipole–dipole interaction and scalar coupling. For certain paramagnetic systems with short T_{1e}, such as in hemoproteins with iron atoms in the high-spin ferrous state, it has recently been found that there is an additional mechanism for providing line broadening through electron–nucleus interaction.[17-20] Modulation of this dipolar field due to the spin polarization (known as the "Curie spin") by rotational diffusion can introduce an extra field-dependent term into the expression for the spin-spin relaxation,[20] giving

$$\frac{1}{T_2} = \frac{4}{5}\frac{\gamma_I^2 g^4\beta^4}{r^6}\frac{S^2(S+1)^2H_0^2}{9k^2T^2}\cdot\tau_R + \frac{7}{15}\frac{\gamma_I^2 g^2\beta^2}{r^6}\cdot S(S+1)T_{1e} + \frac{1}{T_2^{\text{diamag}}} \tag{5}$$

where τ_R is the rotational correlation time of the molecule. The first term in Eq. (5) is the Curie spin contribution, the second term is the standard dipolar term, and the third term is the diamagnetic contribution to the spin-spin relaxation time (T_2). From this expression, we would expect that the Curie spin line-width contribution is proportional to the square of the applied magnetic field, H_0^2, and has a temperature dependence of τ_R/T^2. In the standard dipolar term, T_{1e} would be expected to show much smaller temperature and magnetic field dependence than τ_R, thus making this term essentially magnetic field and temperature independent in comparison to the first term of Eq. (5).

This article contains a brief review of several aspects of high-resolution NMR studies of the hyperfine shifted proton resonances of hemoproteins and iron-sulfur proteins. These two classes of metalloproteins all contain iron atoms which can exist in multiple oxidation and spin states. Four different electronic configurations of the iron atom can occur in these proteins. The high-spin Fe(III) ($S = 5/2$), low-spin Fe(III) ($S = 1/2$), and high-spin Fe(II) ($S = 2$) forms are all paramagnetic, whereas the low-spin Fe(II) ($S = 0$) form is diamagnetic. Since the T_{1e} values for both the low-spin ferric and high-spin ferrous forms are less than 10^{-11} sec, the presence of these two forms of the iron atom in these protein molecules can produce large hfs in proton resonances affected by interactions with the iron atom.[3,5,16,20] The T_{1e} values of the high-spin ferric compounds approximate 10^{-11} sec so one would expect to find both sizeable line broadening and large hyperfine shifts of the resonances affected by the iron atom.

[17] M. Gueron, *J. Magn. Reson.* **19,** 58 (1975).
[18] K. Wüthrich, J. Hochmann, R. M. Keller, G. Wagner, M. Brunori, and C. Giacometti, *J. Magn. Reson.* **19,** 111 (1975).
[19] A. J. Vega and D. Fiat, *Mol. Phys.* **31,** 347 (1966).
[20] M. E. Johnson, L. W.-M. Fung, and C. Ho, *J. Am. Chem. Soc.* **99,** 1245 (1977).

The main emphasis of this article is on the kinds of structural and functional information which can be elucidated from the hyperfine shifted proton resonances of hemoproteins and iron-sulfur proteins. A number of recent reviews on hemoproteins[5,21–23] and iron-sulfur proteins[24,25] have appeared recently.

Experimental Section

Sample Preparation

Assuming that a purified protein sample of interest can be obtained, sample preparation for ^{1}H NMR studies is relatively straightforward. Due to the intense proton resonance of H_2O, it is common to replace most of the water in a protein sample with D_2O. In many cases, this can be conveniently done by repeated dilution with D_2O and subsequent concentration through an ultrafiltration device.

Anaerobic samples can be prepared in a nitrogen-filled glove bag and transferred into a NMR sample tube fitted with a rubber stopper. For some hemoproteins and iron-sulfur proteins, the addition of a chemical reducing agent (such as dithionite, ascorbate, etc.) may be needed to convert these proteins from the ferric to the ferrous state. However, when dithionite is used, oxygen should be excluded from the system before the dithionite is added. In the case of hemoglobin (Hb) it is often desirable to observe the hfs resonances at different oxygen saturations; however, the addition of oxygen to many mutant hemoglobins can result in the formation of significant amounts of methemoglobin (met-Hb). The methemoglobin reductase system of Hayashi *et al.*[26] can reduce the iron atom and consequently convert methemoglobin to the reduced or ferrous form. The reductase system works efficiently in D_2O if NADPH is substituted for NADP.[27] Studies of human adult hemoglobin (Hb A) in the presence or absence of the reductase system have shown that neither the ^{1}H NMR spectrum nor the oxygenation properties of the hemoglobin molecule are affected by the reductase system.[26,27] Ferredoxins and other iron-sulfur proteins are known to be very

[21] C. Ho, T. R. Lindstrom, J. J. Baldassare, and J. J. Breen, *Ann. N. Y. Acad. Sci.* **222,** 21 (1973).

[22] R. G. Shulman, J. J. Hopfield, and S. Ogawa, *Q. Rev. Biophys.* **8,** 325 (1975).

[23] J. S. Morrow and F. R. N. Gurd, *Crit. Rev. Biochem.* **3,** 221 (1975).

[24] W. D. Phillips and M. Poe, this series, Vol. 24 [26].

[25] W. D. Phillips and M. Poe, *in* "Iron-Sulfur Proteins" (W. Lovenberg, ed.), Vol. 2, p. 255. Academic Press, New York, 1973.

[26] A. Hayashi, T. Suzuki, and M. Shen, *Biochim. Biophys. Acta* **310,** 309 (1973).

[27] K. J. Wiechelman, S. Charache, and C. Ho, *Biochemistry* **13,** 4772 (1974).

sensitive to oxygen, the interaction with oxygen being related to the inorganic sulfide present in these proteins.[28] These proteins are relatively stable under anaerobic conditions.

With a modern NMR spectrometer one can detect hfs proton resonances of deoxyhemoglobin down to approximately 10^{-4} *M* concentration range using a 5-mm NMR tube (the usual sample volume is ~0.3 ml). With bigger sample tubes, lower concentration limits (say to approximately 10^{-5} *M*) can be achieved, especially with smaller proteins.

NMR Techniques

During the past several years, significant advances have been achieved in both NMR instrumentation and techniques. For a recent discussion on these two topics, the reader should refer to James.[4] Briefly, there are two techniques that are commonly used for high-resolution NMR studies of proteins, namely Fourier transform (FT) NMR[29] and NMR correlation spectroscopy.[30] Both techniques require spectral accumulation by means of a computer of average transients so as to improve the signal-to-noise ratios of dilute protein samples. NMR correlation spectroscopy is especially well suited for ^{1}H NMR studies of biological macromolecules that contain a sizeable amount of solvent peak, such as H_2O. With the proper selection of a pulse sequence, FT NMR can also be used to obtain ^{1}H NMR spectra of proteins in water.[31] The positions of resonances (chemical shifts) are usually expressed as parts per million with respect to the resonance of a standard. For ^{1}H NMR studies of proteins in water, the usual standard is the sharp resonance of the methyl protons of the sodium salt of 2,2-dimethyl-2-silapentane-5-sulfonate (DSS), which is 4.75 ppm up-field from the proton resonance of water at 30°C. In many ^{1}H NMR studies, it is often quite convenient not only to use the residual water proton signal in the sample as a proton lock but also as an internal reference for proton chemical shift calibration. It should be noted that the chemical shift of water is quite sensitive to temperature. In this article, we have used the following convention for the sign of chemical shifts. Chemical shifts down-field from the resonance signal (such as DSS or H_2O) are assigned negative values and those up-field from the reference are assigned positive values. Due to the low intensity and broadness of hfs resonances, the accuracy of chemical shift measurements is about ± 0.1 ppm.

[28] R. Malkin, *in* "Iron-Sulfur Proteins" (W. Lovenberg, ed.), Vol. 2, p. 1. Academic Press, New York, 1973.

[29] T. C. Farrar and E. D. Becker, "Pulse and Fourier Transform NMR: Introduction to Theory and Methods." Academic Press, New York, 1971.

[30] J. Dadok and R. T. Sprecher, *J. Magn. Reson.* **13,** 243 (1974).

[31] A. G. Redfield, S. D. Kunz, and E. K. Ralph, *J. Magn. Reson.* **19,** 114 (1975).

Determination of the number of protons in a resonance can be obtained through intensity calibration against a known resonance of a reference sample. This is often made by the use of precision coaxial NMR tubes. The inner and outer tubes contain the reference and protein solutions respectively, both with a known volume and concentration. The reference solution should provide a well-resolved resonance with a known intensity and chemical shift in the region of the hfs resonances of the protein sample under investigation. For example, organic compounds in solution doped with a NMR shift reagent have been used to calibrate the intensity of the hfs resonances of deoxyhemoglobin (deoxy-Hb).[32]

In general, both sensitivity and resolution for ^1H resonances are improved by obtaining the NMR spectrum using a higher frequency spectrometer. Commercial instruments operating at high frequencies, such as the Bruker 360 MHz NMR spectrometer, are now commercially available. In addition, even higher frequency spectrometers in the range 500–600 MHz as the observation frequency for proton resonances are now being developed in several laboratories. However, as mentioned earlier, this is not necessarily the case for certain metalloproteins with short T_{1e} values, such as high-spin ferrous hemoproteins.[18,20] In such cases, the line width of the hfs resonances are proportional to the square of the applied magnetic field, H_0^2, as given in Eq. (5). In other words, the resolution would be increased linearly with increasing magnetic field, but the line width of these resonances would be increased by H_0^2. Thus, in such cases, both resolution and sensitivity would be worsened by going to very high frequency. Figure 2 shows the line width of the hfs proton resonances of deoxy-Hb A as a function of frequency to illustrate this point.[20] It should be noted that the variation of the line width of hfs proton resonances of certain iron-containing proteins as a function of temperature would be more complex than the relations given in Eqs. (2) and (3). Figure 3 shows the variation of the line width of hfs proton resonances of deoxy-Hb A as a function of temperature. The rate of line width increase is slightly faster than the η/T^3 predicted by Eq. (5) where η is the solvent viscosity. For a recent discussion of the variation of the line width of hfs proton resonances of deoxymyoglobin and deoxyhemoglobin, refer to Johnson *et al.*[20] The results given in Figs. 2 and 3 clearly show that for certain paramagnetic metalloproteins, there is another relaxation mechanism which can produce line broadening of the hfs proton resonances upon increasing the magnetic field or decreasing the temperature. Hence, it is essential to measure ^1H NMR parameters of systems containing paramagnetic centers at more than

[32] L. W.-M. Fung, A. P. Minton, T. R. Lindstrom, A. V. Pisciotta, and C. Ho, *Biochemistry* **16,** 1452 (1977).

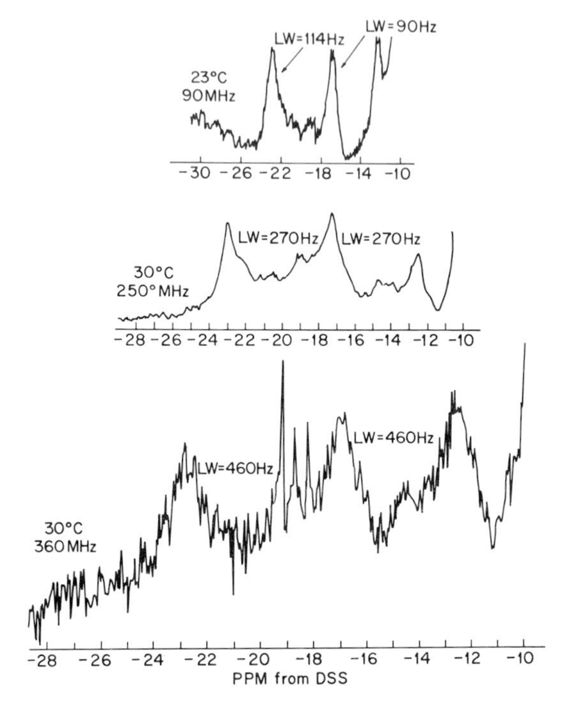

FIG. 2. Hyperfine shifted proton resonances of deoxyhemoglobin at 90, 250, and 360 MHz. Hemoglobin samples are 14% at pD 7.0 in 0.1 *M* bis-Tris with 10 m*M* inositol hexaphosphate. From previous work, the resonances at −23 and −17 ppm from DSS have been respectively assigned to heme methyls of the β and α chains.[32-35] Line width (LW) values shown are corrected for the broadening added in the exponential multiplication of the free-induction decay before Fourier transformation (90 and 360 MHz). The spikes near the center of the 360 MHz spectrum are due to receiver nonlinearities which permit off-frequency "ringing" from residual water in the Fourier transform mode. [Taken from Fig. 7 of Johnson *et al.*,[20] with permission.]

one frequency and temperature before interpreting the meaning of the line width of a given signal.

Selected Examples

Iron-Sulfur Proteins

Iron-sulfur proteins occur in anaerobic bacteria, in the chloroplasts of higher plants, and in mitochondrial membranes of animal tissues. They are

[33] D. G. Davis, T. R. Lindstrom, N. H. Mock, J. J. Baldassare, S. Charache, R. T. Jones, and C. Ho, *J. Mol. Biol.* **60,** 101 (1971).

[34] T. R. Lindstrom, C. Ho, and A. V. Pisciotta, *Nature (London), New Biol.* **237,** 263 (1972).

[35] L. W.-M. Fung, A. P. Minton, and C. Ho, *Proc. Natl. Acad. Sci. U.S.A.* **73,** 1581 (1976).

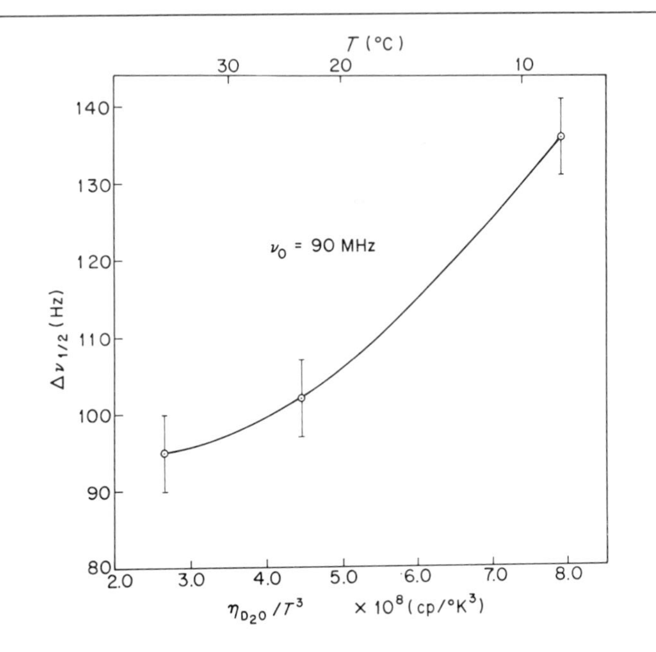

FIG. 3. Temperature dependence of the deoxyhemoglobin line width at 90 MHz. The line widths of the α and β resonances have been observed to be approximately equal under all experimental conditions in deoxyhemoglobin; thus they have been averaged to reduce statistical uncertainty. The averaged values are plotted here. η_{D_2O} is the viscosity of D_2O. [Taken from Fig. 8 of Johnson *et al.*,[20] with permission.]

low-molecular-weight proteins which appear to function as electron carriers by undergoing reversible Fe(II)–Fe(III) transitions. In the iron-sulfur proteins studied to date there appear to be three general classes of iron-sulfur complexes. In rubredoxin, which contains one iron atom and no inorganic sulfur, the iron atom is held in the protein through coordinate linkages with four cysteine residues in the protein molecule. The plant-type ferredoxins contain two iron atoms and two inorganic sulfur atoms per protein molecule. Each iron atom bonds to two cysteine residues on the protein and to the two inorganic sulfur atoms, forming a binuclear cluster. The bacterial-type ferredoxins and the high-potential-iron protein from *Chromatium* contain one or more tetranuclear iron clusters composed of four iron atoms, four inorganic sulfur atoms, and the sulfide group from four cysteine residues. The iron and inorganic sulfur atoms are located in alternating corners of a cube with each iron atom bonded to a cysteine residue in the polypeptide chain. For a recent review on iron-sulfur proteins, refer to Orme-Johnson.[36]

[36] W. H. Orme-Johnson, *Annu. Rev. Biochem.* **42,** 159 (1973).

At least one of the oxidation states of all of the iron-sulfur proteins studied to date has been found to be paramagnetic. The existence of paramagnetic centers in these proteins should make them ideal candidates for study by ^1H NMR spectroscopy. Since these proteins vary from strong oxidizing agents to strong reducing agents, we might be able to relate their structure, as sensed by NMR spectroscopy, to their functional properties. Excellent reviews of the NMR studies of the iron-sulfur proteins[24,25] were published in 1972 and 1973. Although only a few papers on iron-sulfur proteins have been published since that time, much work remains to be done on this class of proteins.

As noted above, in all of the iron-sulfur proteins, the iron atom is bonded to cysteine residues in the polypeptide chain. The eight-iron ferredoxin from *Clostridium pasteurianum* contains two tetranuclear iron clusters with eight cysteine residues involved in binding the iron atoms to the protein. The β-CH_2 protons of the cysteine residues appear to be the protons most likely to be affected by hyperfine interactions with the iron atoms.[15] Sixteen hfs resonances each with an intensity of one proton per molecule are observed in the NMR spectrum of oxidized *C. pasteurianum* ferredoxin,[15] suggesting that the two β-CH_2 protons on each cysteine are nonequivalent. Figure 4 shows the ^1H NMR spectrum of oxidized ferredoxin from *C. pasteurianum* as a function of temperature to illustrate the NMR spectrum for this type of protein. This observation can be most easily explained by postulating that rotation about the C_β—S bond is hindered by the bonds between the C_β atom and the protein molecule and between the sulfur atom and the iron atom. The two protons on the β carbon would then be expected to make different angles with the paramagnetic center. Eq. (3) shows that the pseudocontact shift depends on both r and θ, and Eq. (2) shows that the contact or Fermi shift induced by a paramagnetic center depends on A_i, the hyperfine coupling constant. The angular dependence of A_i for this type of situation has been discussed.[15,37] Thus, both the pseudocontact and contact shifts should differ for nonequivalent protons attached to the β carbon of the cysteine residues. Subsequent studies of the two-iron ferredoxins from spinach, parsley, alfalfa, and soybeans,[38,39] the four-iron high-potential-iron protein from Chromatium,[40] the four-iron ferredoxin from *Bacillus polymyxa*, and the eight-

[37] E. W. Stone and A. H. Maki, *J. Chem. Phys.* **37,** 1326 (1967).

[38] M. Poe, W. D. Phillips, J. D. Glickson, C. C. McDonald, and A. San Pietro, *Proc. Natl. Acad. Sci. U.S.A.* **68,** 68 (1971).

[39] J. D. Glickson, W. D. Phillips, C. C. McDonald, and M. Poe, *Biochem. Biophys. Res. Commun.* **42,** 271 (1971).

[40] W. D. Phillips, M. Poe, C. C. McDonald, and R. G. Bartsch, *Proc. Natl. Acad. Sci. U.S.A.* **67,** 682 (1970).

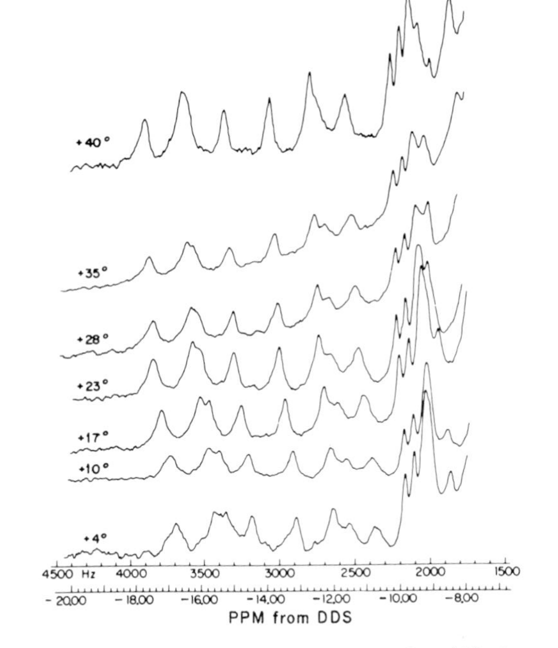

FIG. 4. Temperature dependence of the low-field hyperfine shifted proton resonances of oxidized ferredoxin form *Clostridium pasteurianum* at 220 MHz. These ^{1}H NMR spectra were obtained from a single D_2O solution of 15.4 m*M* oxidized ferredoxin. The vertical scales at the various temperatures are not comparable. The sign of the chemical shifts has been changed from the original data so as to follow the sign convention used in this chapter. [Taken from Fig. 5 of Phillips and Poe,[25] p. 267, with permission.]

iron ferredoxin from *C. acidi-urici*[41] have shown that the protons attached to the β carbon of the cysteine residues in all of these protons are also nonequivalent.

The relationships in Eqs. (2) and (3) show that both δ_c and δ_p should follow the Curie law. As the temperature of the system is increased, the hfs resonances are supposed to shift up-field. The resonances of the oxidized form of ferredoxin from *C. pasteurianum* were found to shift down-field as the temperature was increased[15] (e.g., see Fig. 4). This temperature dependence of the hfs resonances does not follow the Curie law, which suggests that extensive antiferromagnetic exchange coupling occurs between component iron atoms.[15,42] (For a recent discussion of antiferromagnetic ex-

[41] C. C. McDonald, W. D. Phillips, W. Lovenberg, and R. H. Holm, *Ann. N. Y. Acad. Sci.* **222,** 789 (1973).

[42] G. Palmer, *in* "Iron-Sulfur Proteins" (W. Lovenberg, ed.), Vol. 2, p. 285. Academic Press, New York, 1973.

change interaction, consult Palmer[42]). The hfs resonances of the oxidized forms of all of the two-, four-, and eight-iron ferredoxins studied to date and the reduced form of the high-potential-iron protein from *Chromatium* demonstrate this deviation from the Curie law,[15,38-40,43,44] indicating that there is antiferromagnetic exchange coupling between the iron atoms of each of these proteins.

The two-iron ferredoxins have provided an excellent opportunity not only for some interesting NMR studies but also for speculation on the nature of the active site in these proteins. The redox center is believed to be a binuclear cluster containing both iron atoms, each of which is coordinated by four sulfur atoms. Two of these sulfur atoms, the inorganic sulfur atoms, are bonded to both iron atoms, and the other four sulfur atoms, which are external to the binuclear center, come from cysteine residues in the polypeptide chain.[45,46] In the oxidized form, both iron atoms are high-spin with a total of ten unpaired electrons. However, the magnetic susceptibility measurements have indicated that oxidized ferredoxins are essentially diamagnetic as a result of the antiferromagnetic exchange interaction between the two component iron atoms.[47] In the reduced form, one iron atom remains high-spin ferric and the other is converted to the high-spin ferrous state.[46,48]

In the ^1H NMR spectra of the oxidized form of spinach, parsley, alfalfa, soybean, and algal ferredoxins, two hfs proton resonances, one at ~ -15 ppm from DSS with an intensity of one proton[38,39,49,50] and the other at ~ -35 ppm with approximately eight protons,[49,50] have been observed. The exact assignment of these resonances has not been made, but it has been suggested that the broad resonance at ~ -35 ppm is due to the eight protons of the β-CH_2 of the four cysteine residues in the redox center.[49] Both of these resonances are found to have a positive temperature dependence. The presence of these two hfs proton resonances in the spectrum indicates that there must be some residual paramagnetism associated with

[43] M. Poe, W. D. Phillips, C. C. McDonald, and W. H. Orme-Johnson, *Biochem. Biophys. Res. Commun.* **47,** 705 (1971).

[44] W. D. Phillips, C. C. McDonald, N. A. Strombaugh, and W. H. Orme-Johnson, *Proc. Natl. Acad. Sci. U.S.A.* **71,** 140 (1974).

[45] H. Brintzinger, G. Palmer, and R. H. Sands, *Proc. Natl. Acad. Sci. U.S.A.* **55,** 397 (1967).

[46] W. R. Dunham, G. Palmer, R. H. Sands, and A. J. Bearden, *Biochim. Biophys. Acta* **253,** 373 (1971).

[47] T. H. Moss, D. Petering, and G. Palmer, *J. Biol. Chem.* **244,** 2275 (1969).

[48] J. F. Gibson, D. O. Hall, J. H. M. Thornley, and F. R. Whatley, *Proc. Natl. Acad. Sci. U.S.A.* **56,** 987 (1966).

[49] I. Salmeen and G. Palmer, *Arch. Biochem. Biophys.* **150,** 767 (1972).

[50] R. E. Anderson, W. R. Dunham, R. H. Sands, A. J. Bearden, and H. L. Crespi, *Biochim. Biophys. Acta* **408,** 306 (1975).

the oxidized form of these ferredoxins. These findings are in good agreement with the results of Gibson *et al.*[48] indicating that the two iron atoms in spinach ferredoxin strongly interact with one another through the bridging ligands. An antiferromagnetic exchange interaction between the ferric iron atoms could couple the spins to give a total spin of $\frac{1}{2}$ in the ground state.[48] A similar antiferromagnetic coupling has also been observed for the reduced form of the high-potential-iron protein[40] and the oxidized form of the eight-iron ferredoxins of *C. pasteurianum*[15] and *C. acidi-urici*.[43] In the ^1H NMR spectra of the reduced form of ferredoxins, hfs resonances over the range from -10 to -45 ppm have been observed.[38,39,49,50] The lowest field resonance is the one observed at ~ -44 ppm from DSS with two protons for spinach and algal ferredoxins.[49,50] The nature of this resonance is not known. In the reduced form of parsley ferredoxin, eight hfs proton resonances (each with unit intensity) over the region from -10 to -22 ppm from DSS have been observed.[38] They have been attributed to the eight β-CH_2 protons of four cysteine residues.[38] The β-CH_2 protons of the reduced ferredoxin appear to be distributed into two classes on the basis of the temperature dependence of their hyperfine interactions. With increasing temperature, four of the resonances between -10 and -17 ppm from DSS move slowly up-field (i.e., decrease contact shift) while the remaining four resonances move rapidly down-field (i.e., increase contact shift).[38,39] In the reduced forms of spinach and alfalfa ferredoxins, only six hfs resonances have been observed over the spectral region from -10 to -22 ppm from DSS.[38,39] These resonances also show two types of temperature dependence similar to those observed for parsley ferredoxin. Neither set of resonances follows the Curie law dependence observed at low temperatures for magnetic susceptibility.[38,51] It is believed that in the reduced state one iron atom remains high-spin ferric and the other is converted to high-spin ferrous.[48,51] A consequence of this model is that protons which "sense" the spin density on the ferric iron will show a positive temperature dependence while those which "sense" the ferrous iron will have a negative temperature dependence.[46] The present NMR results provide strong support to the popular model for the structure of the active center of two-iron sulfur proteins, namely, that each iron atom bonds to two cysteine residues on the protein and to the two inorganic sulfur atoms, forming a binuclear cluster. The final proof of the proposed redox center of two-iron sulfur proteins will have to await the structural determination from single-crystal X-ray diffraction studies.

The four-iron ferredoxins from the facultative nitrogen-fixing bacterium

[51] G. Palmer, W. R. Dunham, J. Fee, R. H. Sands, T. Iizuka, and T. Yonetani, *Biochim. Biophys. Acta* **245,** 201 (1971).

Bacillus polymyxa have a molecular weight of about 9000 daltons and contain four iron atoms per molecule.[52] The iron and labile sulfur atoms in the four-iron ferredoxin are arranged into a single tetrameric cluster similar to the two tetrameric clusters found in the eight-iron ferredoxins.[53] In oxidized *B. polymyxa* ferredoxin there are seven readily identifiable hfs resonances, each with an intensity of one proton per molecule.[44] These resonances appear in the same spectral region and have widths and temperature dependences similar to the oxidized forms of the eight-iron ferredoxins from *C. pasteurianum* and *C. acidi-urici*. Resonances which correspond to 11 protons can be identified in the hfs spectrum of the reduced ferredoxin. Assignment of these resonances in the reduced protein is somewhat complicated since there are only eight β-CH_2 cysteine protons involved in binding the Fe_4S_4 cluster. The positions of these β-CH_2 protons are dominated by hyperfine interactions. The α-CH protons are also subject to a similar, but smaller, perturbation. It has not been possible to assign any of the resonances to either the α-CH or β-CH_2 protons.

Recent studies indicate that the protein portion of the iron-sulfur proteins is important for cluster formation and activity in the bacterial-type ferredoxins. It has been suggested that the existence of possible hydrogen bond donors in ferredoxin may contribute to the stabilization of a more negatively charged cluster.[54] To investigate this problem, deuterated ferredoxin was purified from blue-green alga *Synechococcus lividus* which were grown in D_2O. This deuterated ferredoxin was equilibrated with H_2O, and ^1H NMR spectra were obtained.[55] Only slowly exchangeable amide protons involved in the hydrogen bonds were observed. Comparing the spectra of oxidized, partially reduced, and reduced proteins, two signals from the slowest exchanging protons and one from the intermediate rate shifted considerably in the reduced protein. The large shifts suggest that the resonances may be hfs resonances from protons near the iron-sulfur centers. If this is confirmed, it would indicate that the NH—S stability is of the same order as that of NH—O bonds, as suggested by hydrogen isotope exchange studies[56] and by X-ray studies.[57]

[52] N. A. Strombaugh, R. H. Burris, and W. H. Orme-Johnson, *J. Biol. Chem.* **248,** 7951 (1973).

[53] L. C. Sieker, E. Adman, and L. H. Jensen, *Nature (London)* **235,** 40 (1972).

[54] C. W. Carter, Jr., J. Kraut, S. T. Freer, and R. A. Alden, *J. Biol. Chem.* **249,** 6339 (1974).

[55] H. L. Crespi, A. G. Kostke, and U. H. Smith, *Biochem. Biophys. Res. Commun.* **61,** 1407 (1974).

[56] J. S. Hong and J. C. Rabinowitz, *J. Biol. Chem.* **245,** 4995 (1970).

[57] E. Adman, K. D. Watenpaugh, and L. H. Jensen, *Proc. Natl. Acad. Sci. U.S.A.* **72,** 4854 (1975).

Hemoproteins

The hemoproteins consist of a variety of proteins and enzymes which play vital roles in all forms of living organisms. The cytochromes are electron carriers in the respiratory chain in all organisms. In vertebrates the hemoglobin molecule is responsible for carrying oxygen from the lungs to the tissues, where the oxygen molecule is transferred to myoglobin, which holds it until it is needed for metabolic oxidation. All of these proteins contain heme as their prosthetic group. Heme consists of four pyrrole units connected by methenyl bridges (porphyrin) with an iron atom in the center held in position by bonds to four of the pyrrole nitrogens. Various hemoproteins may contain heme groups which differ in the nature and arrangement of the side chains on the porphyrin ring. The heme iron in these proteins can undergo reversible changes in oxidation state [Fe(II) to Fe(III)] and/or spin state (high-spin to low-spin). At least one oxidation- and spin-state combination in these hemoproteins is paramagnetic, making it possible to use NMR spectroscopy as a unique tool to investigate the geometrical and functional properties of the active sites (i.e., the heme groups) of hemoproteins. In 1965, Kowalsky[58] first reported the hfs resonances of horse heart cytochrone *c* (Cyt *c*). Since then well over 100 articles on ^{1}H NMR studies of hemoproteins have appeared in the scientific literature. Due to space limitations, two hemoproteins, cytochrome *c* and hemoglobin, have been selected as examples to illustrate the unique information that can be obtained from the hfs proton resonances of these proteins. For further details on the application of NMR spectroscopy to hemoproteins, the reader should refer to several review articles,[2,4,5,16,21-23,59] and the references therein.

CYTOCHROME *c*

Cytochromes of the *c*-type are a group of related proteins (such as c, c_1, c_2, etc.) with molecular weights in the range 12,000 to 40,000 daltons. Each of these proteins has a heme group (iron-protoporphyrin IX) which is the same prosthetic group as in myoglobin and hemoglobin. In cytochromes *c*, the heme groups are covalently attached to the protein by thioether linkages. These linkages are formed by the addition of sulfhydryl groups of two cysteine residues to the vinyl groups of the heme. (This is in contrast to myoglobin and hemoglobin, whose heme groups are not attached by covalent bonds to the protein molecules, but by a coordinate bond between the

[58] A. Kowalsky, *Biochemistry* **4,** 2382 (1965).

[59] C. Ho, L. W.-M. Fung, K. J. Wiechelman, G. Pifat, and M. E. Johnson, *Prog. Clin. Biol. Res.* **1,** 43 (1975).

heme iron and one of the axial ligands, the proximal histidine residue of the protein moiety). This group of proteins all undergoes reversible Fe(II)–Fe(III) valency changes during the electron-transport process. The ferric form of cytochromes *c* is in low-spin state with one unpaired electron per iron atom ($S = \frac{1}{2}$) and that of the ferrous form is low-spin diamagnetic ($S = 0$). This section will emphasize two aspects of ^{1}H NMR studies of Cyt *c*, namely structural and functional information derived from hfs resonances.

Kowalsky[58] first observed the hfs proton resonances of horse heart ferricytochrome *c* (ferric-Cyt *c*), which is a small protein consisting of a single polypeptide chain with a molecular weight of about 12,400 daltons. Further NMR studies of this protein have been concerned with (1) assigning a resonance to the sixth axial ligand and identifying that ligand, (2) investigating the mechanism of electron transfer, and (3) elucidating the conformation and conformational changes of Cyt *c* under different conditions (e.g., see Refs. 60–70).

Figure 5 gives two typical ^{1}H NMR spectra of ferro- and ferricytochrome *c*. The ^{1}H NMR spectrum (Fig. 5B) of the diamagnetic ferrocytochrome *c* (ferro-Cyt *c*) illustrates the severe overlapping of proton resonances in the spectral region from 0 to −10 ppm from DSS, which is a characteristic feature of the spectra of diamagnetic proteins.[1,2] In paramagnetic ferric-Cyt *c*, a number of well-resolved hfs resonances are observed in the spectral regions from −10 to −35 ppm and 0 to +25 ppm from DSS as shown in Fig. 5A. The resonances at −34.0, −31.4, −10.3, +2.1, +2.5, and +23.2 ppm correspond in intensity to three protons each while the resonances at −18.1, −13.8, −1.8, −11.5, +4.0, and +6.0 ppm all have an intensity of one proton.[60,61] A number of resonances in the NMR spectrum have been assigned by comparing the spectra of cytochromes *c* from eukaryotic and prokaryotic species with known amino acid sequences,[60,61,69,71-73] or chemically modified Cyt *c*,[61,74] by using the

[60] C. C. McDonald, W. D. Phillips, and S. N. Vinogradov, *Biochem. Biophys. Res. Commun.* **36,** 442 (1969).

[61] K. Wüthrich, *Proc. Natl. Acad. Sci. U.S.A.* **63,** 1071 (1969).

[62] R. K. Gupta and A. G. Redfield, *Science* **109,** 1204 (1970).

[63] R. K. Gupta and A. G. Redfield, *Biochem. Biophys. Res. Commun.* **41,** 273 (1970).

[64] A. G. Redfield and R. K. Gupta, *Cold Spring Harbor Symp. Quant. Biol.* **36,** 405 (1971).

[65] K. Wüthrich, I. Aviram, and A. Schejter, *Biochim. Biophys. Acta* **253,** 98 (1971).

[66] R. K. Gupta, S. H. Koenig, and A. G. Redfield, *J. Magn. Reson.* **7,** 66 (1972).

[67] R. K. Gupta, *Biochim. Biophys. Acta* **292,** 291 (1973).

[68] E. Stellwagen and R. G. Shulman, *J. Mol. Biol.* **80,** 559 (1973).

[69] C. C. McDonald and W. D. Phillips, *Biochemistry* **12,** 3170 (1973).

[70] J. S. Cohen, W. R. Fisher, and A. N. Schechter, *J. Biol. Chem.* **249,** 1113 (1973).

[71] G. E. Krejcarek, L. Turna, and K. Dus, *Biochem. Biophys. Res. Commun.* **42,** 983 (1971).

[72] R. M. Keller, G. M. Pettigrew, and K. Wüthrich, *FEBS Lett.* **36,** 151 (1973).

[73] G. M. Smith and M. D. Kamen, *Proc. Natl. Acad. Sci. U.S.A.* **71,** 4303 (1974).

[74] K. Wüthrich, I. Aviram, and A. Schejter, *Biochim. Biophys. Acta* **253,** 98 (1971).

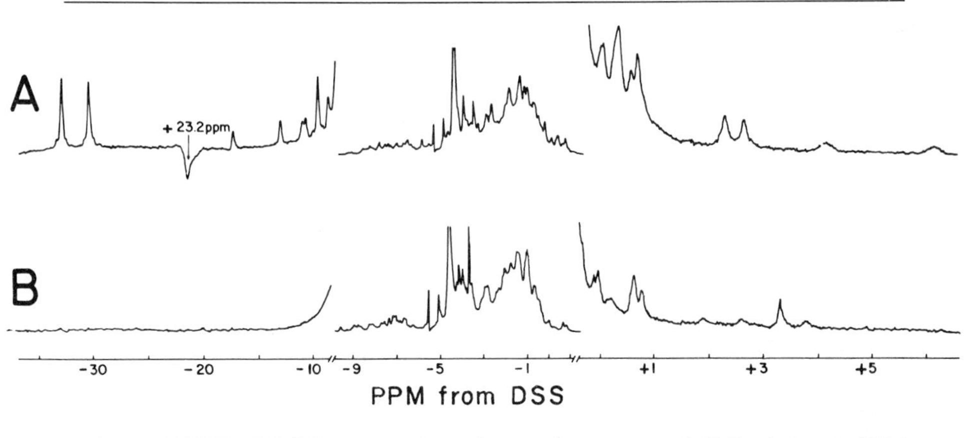

FIG. 5. 220 MHz ^{1}H NMR spectra of cytochrome *c* from guanaco in D_2O solution at pD 7.0 and 35°C. Spectrum A is a sample of ferricytochrome *c* and spectrum B is a sample of ferrocytochrome *c*. The sharp resonances between −4 and −6 ppm from DSS correspond to the HDO resonance and its first and second spinning side bands. The vertical and horizontal scales are different for the three parts of the spectrum. The high-field line at +23.2 ppm is observed as an inversed resonance of the center band of the spectrum. The sign of the chemical shifts has been changed from the original data so as to follow the convention used in this article. [Taken from Fig. VI-19 of Wüthrich,[5] p. 257, with permission.]

double-resonance technique[62-64] and by correlating these NMR data with the structure of horse heart Cyt *c* determined by X-ray diffraction.[75] Assignment of the hfs resonances to protons of specific amino acids are of special interest to this article. (For details, see Wüthrich[5] and McDonald and Phillips[69].) The resonance at +23.2 ppm from DSS (pD 7.0 and 30°C) of ferric-Cyt *c* has been assigned to the methyl group of methionine at position 80 (Met-80) in the polypeptide chain, which was found to be the sixth axial ligand of the heme group.[61,64] The corresponding resonance is found at +3.3 ppm from DSS in the spectrum of ferro-Cyt *c* under the same conditions. The resonances at −31.4 and −34.0 ppm have been assigned to two of the methyl groups on the porphyrin ring. The two hfs resonances at +2.1 and +2.6 ppm from DSS (pD 7.0 and 35°C) of ferric-Cyt *c* have been assigned to the methyl groups of Cys-14 and Cys-17 respectively, which are covalently attached to the porphyrin ring via thioether linkages.

The information content of a NMR spectrum of protein will be greatly increased if individual resonances can be assigned to specific amino acid residues in the polypeptide chain. According to Eqs. (2) and (3), hfs resonances can be calculated if the molecular geometry and the *g*-tensor are known. Based on X-ray and electron paramagnetic resonance (EPR)

[75] R. E. Dickerson, T. Takano, D. Eisenberg, O. B. Kallai, L. Samson, A. Cooper, and E. Margoliash, *J. Biol. Chem.* **246**, 1511 (1971).

data, the pseudocontact shifts of ferric-Cyt *c* have been estimated.[72] This calculation suggests that the pseudocontact shifts should occur in the spectral region between +2 to −10 ppm from DSS. The hfs proton resonances outside this range are believed to be due to contact or Fermi interactions. However, the *g*-values obtained from EPR measurements are those of crystals of ferric-Cyt *c* at low temperature. The *g*-values are not known for native ferric-Cyt *c* at room temperature. Thus, it is difficult to evaluate the exact influence of pseudocontact interactions on the ^{1}H NMR spectrum of ferric-Cyt *c*. On the basis of available information on the anisotropy of the *g*-values of Cyt *c* and other low-spin ferric heme compounds,[16,62,64,76] it has been pointed out that the resonances of protons near the face of the porphyrin ring should be shifted to low field and those protons around the edge of the heme shifted to high field.[69]

More recently, by using resolution enhancement and spin decoupling techniques, it has been possible to assign a number of hyperfine shifted aromatic proton resonances of Cyt *c*.[77,78]

Several cytochromes *c* obtained from bacteria, yeast, and other species have been found to be similar in both their primary and tertiary structures to those of horse heart Cyt *c*,[5,69,73] and these similarities are reflected in their ^{1}H NMR spectra. In Cyt *c*-557 from *Crithidia oncopelti*, Cys-14 is replaced by alanine but the remainder of the primary structure is similar to other cytochromes *c*.[72] The heme group is believed to be covalently attached to the remaining cysteine residue while the vinyl group normally involved in a bond to Cys-14 is free. The hfs proton spectrum of ferric-Cyt *c*-557 was found to be similar to the spectra of cytochromes *c* from vertebrates, indicating that the loss of one thioether bond in this protein does not significantly alter the electronic environment of the heme group. The cytochromes c_3 from *Desulfovibrio vulgarius* and *Desulfovibrio gigas* are similar in size to vertebrate cytochromes *c*. However, each cytochrome contains three or four hemes, and their primary structures show little homology with vertebrate cytochromes *c*. The ^{1}H NMR spectra of these two Cyt c_3 molecules are similar to each other but show little resemblance to the spectra of other cytochromes *c*.[79,80]

The ^{1}H NMR spectra of ferric- and ferro-Cyt *c* differ significantly as shown in Fig. 5, and the spectrum of a partially reduced sample consists of superimposed resonances of the two forms.[62,64] Redfield and Gupta have

[76] R. G. Shulman, S. H. Glarum, and M. Karplus, *J. Mol. Biol.* **57,** 93 (1971).

[77] C. M. Dobson, G. R. Moore, and R. J. P. Williams, *FEBS Lett.* **51,** 60 (1975).

[78] G. R. Moore and R. J. P. Williams, *FEBS Lett.* **53,** 334 (1975).

[79] C. C. McDonald, W. D. Phillips, and J. LeGall, *Biochemistry* **13,** 1952 (1974).

[80] C. M. Dobson, N. J. Hoyle, C. F. Geraldes, P. E. Wright, R. J. P. Williams, M. Bruschi, and J. LeGall, *Nature (London)* **249,** 425 (1974).

developed an elegant technique using double-resonance NMR which not only allowed them to make spectral assignments, but also to investigate the mechanism of electron transfer in the respiratory chain.[62–64,66,67] A partially reduced sample was obtained by adding a calculated amount of ascorbic acid to a 10% solution of ferric-Cyt *c*. The presence of superimposed resonances of oxidized and reduced Cyt *c* in the ^{1}H NMR spectrum of a partially reduced sample indicates that the rate of electron exchange between the oxidation states is similar to the spin-lattice relaxation time for protons. The double-resonance experiment consists of applying a radio-frequency pulse for 0.1 sec to the resonance frequency of a hfs resonance of ferric-Cyt *c*. The magnitude of the pulse is set so that it is just sufficient to saturate the hfs resonance. When the saturating radio-frequency pulse is applied to a given resonance, the spin populations of the two nuclear Zeeman energy levels responsible for the NMR signal become equal since relaxation of the spins cannot occur. Since the interconversion rate between the two oxidation forms of the Cyt *c* is of the proper order of magnitude, the saturation of the resonance in the ferric form is transferred to the resonance of the same proton in ferro-Cyt *c*. As a result of this transfer of saturation, the resonance corresponding to the irradiated hfs resonance does not appear in the spectrum of the reduced protein when the spectrum is scanned by an observation pulse immediately after the saturating pulse. Since the resonance of interest is often obscured by the presence of other resonances in the spectrum, changes in the spectrum are monitored by means of a difference spectrum. The difference spectrum is obtained either with and without the saturating pulse or with the saturating pulse jumping between the resonance of interest and some other point in the spectrum. When the observed resonance is close to the saturated resonance, the effect of direct irradiation of the saturating pulse (due to a finite frequency spread) can be minimized by alternating the position of the saturating pulse when the difference spectrum is obtained.

This double-resonance technique makes it possible to map the position of a resonance in the oxidized state onto its position in the reduced state. Using this technique it was found that the resonance at +23.2 ppm in the spectrum of the oxidized protein comes from the same group as the resonance at +3.3 ppm in reduced Cyt *c*. The X-ray structure of ferric-Cyt *c* shows that this resonance must come from the Met-80 methyl group since it is the only group close enough to the heme to be so broadened by the electron spin. Using this technique it has been possible to correlate the positions of a number of resonances in ferro-Cyt *c* with resonances in ferric-Cyt *c*.[64] (As an illustration of this technique, refer to Fig. 4 of this volume [12], "Pulse Fourier-transform NMR Spectroscopy of Redox Proteins" by Gupta and Mildvan.) The same technique has been applied to mixtures of ferri-Cyt *c* and azidoferricytochrome *c* to identify a number of resonances in the spectrum of azidoferricytochrome *c*.[63]

Based on kinetic theory, the rate constant for the transfer of electrons from reduced to oxidized molecules of Cyt *c* has been estimated by measuring the fractional decrease in the resonance at +3.3 ppm from DSS and its T_1 value.[63,64,66,67] Due to hyperfine interactions, the proton spins of the methionine methyl group in ferric-Cyt *c* have a shortened T_1 (~2 msec), compared to a significantly longer time in the reduced state (~200 msec).[66] The electron exchange between Cyt *c* molecules is found to involve binary collisions, and the second-order rate constant for this process has a value between 1×10^3 to 1×10^5 $mole^{-1}$ sec^{-1} for horse heart Cyt *c*, depending on pH, ionic strength, and temperature.[66] Thus, NMR spectroscopy, in suitable cases, offers the possibility of investigating certain dynamic as well as functional aspects of the electron-transfer processes in hemoproteins.

Hemoglobin

Hemoglobin is the oxygen-transporting pigment present in the red blood cell. It is a heme-containing protein with a molecular weight of approximately 65,000 daltons and is composed of four subunits (normally two α chains and two β chains). Each of the subunits contains a protein part and a prosthetic group, heme, which can combine reversibly with oxygen when the iron atom is in the ferrous state. The oxygenation of Hb is a cooperative process with a Hill coefficient (n) of approximately 3 for normal adult hemoglobin. For recent reviews on the structure–function relationship in hemoglobin, consult Antonini and Brunori[81] and Baldwin.[82]

When the Hb molecule is in the deoxy (unliganded) form, the ferrous iron atom is paramagnetic with four unpaired electrons per iron atom ($S = 2$), giving rise to a number of hyperfine shifted proton resonances. When ligands (such as O_2 or CO) bind to the Hb molecule, the heme iron becomes diamagnetic and the resonances of the protons which give rise to the hyperfine shifted resonances return to their positions buried in the diamagnetic proton resonances. The Hb molecule can be oxidized to the ferric state (known as methemoglobin) with five unpaired electrons per iron atom ($S = 5/2$). The binding of certain ligands (such as cyanide or azide) causes the heme iron of met-Hb to change from the high-spin ferric to the low-spin ferric state with one unpaired electron per heme ($S = \frac{1}{2}$). Figure 1 shows the ^1H NMR spectra of hemoglobin under different conditions. The hfs proton resonances of deoxyhemoglobin occur over the spectral regions

[81] E. Antonini and M. Brunori, "Hemoglobin and Myoglobin in their Reactions with Ligands." North-Holland Publ., Amsterdam, 1971.

[82] J. M. Baldwin, *Prog. Biophys. Mol. Biol.* **29**, 225 (1975).

between −7 to −18 ppm from HDO (or −12 to −23 ppm from DSS) and between +12 to +24 ppm from HDO. The hfs resonances of met-Hb occur over the spectral region between −30 to −80 ppm from HDO. Since these hfs proton resonances are due to proton groups on the porphyrin ring and/or proton groups from the amino acid residues in the vicinity of the heme group, they are excellent spectroscopic probes for investigating the structure–function relationship in hemoglobin. During the past 10 years, a very large number of papers on ^{1}H NMR studies of hemoglobin have been published in scientific journals. For detailed information, the reader should refer to recent reviews,[21-23,59] and the references therein.

Figure 6 shows the low-field region of the hfs proton resonances of deoxy Hb A in D_2O which contains three prominent hfs proton resonances

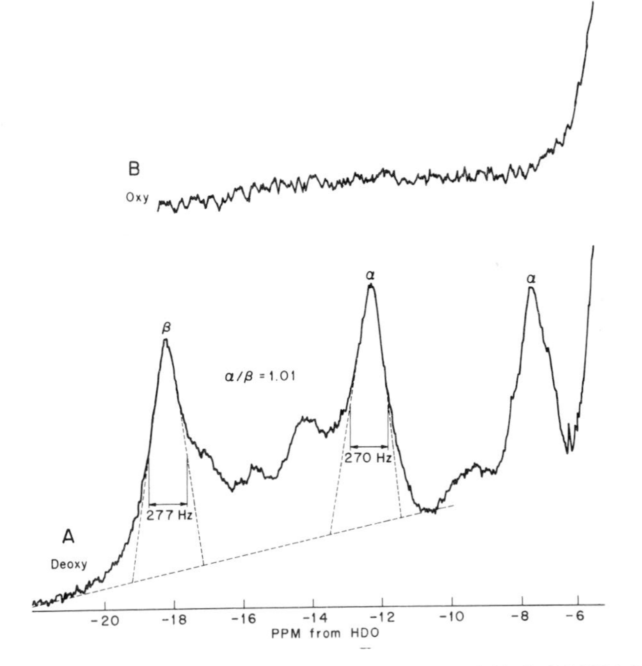

FIG. 6. 250 MHz ^{1}H NMR spectra of 16% human adult hemoglobin in 0.1 *M* bis-Tris plus 30 m*M* 2,3-diphosphoglycerate at pD 7.0 and 26°C over the spectral region between −6 and −20 ppm from HDO. Spectrum A represents a 100% deoxy sample and spectrum B a fully oxy sample. The dotted line in A illustrates the method for drawing the base line and determining the area ratio of the two resonances at −18 and −12 ppm. The α/β ratio is the area ratio of the α resonance at −12 ppm and the β resonance at −18 ppm. Full line widths at half-height of the peaks at −18 and −12 ppm are also shown in B. [Taken in part from Fig. 1 of Johnson and Ho,[84] with permission.]

at ~ -18, ~ -12, and ~ -8 ppm from HDO. The exact chemical shifts of these resonances depend on temperature, pH, and the presence of organic phosphate. The relative intensities of these resonances indicate that the two lowest field lines are probably heme methyls and the -8 ppm resonance probably comes from a methylene group.[32,83,84] By using a number of mutant hemoglobins, it has been possible to ascertain that the major resonance at -18 ppm comes from β-chain protons and that the resonances at -12 and -8 ppm are from α-chain protons.[32,85,86] The finding that the α and β heme resonances of Hb A occur at different positions in the ^{1}H NMR spectrum suggests that the α and β hemes are magnetically or structurally nonequivalent.

Since the α and β hemes are structurally nonequivalent, it is possible that they are also functionally nonequivalent. The hfs proton resonances should be ideally suited for such a study since they are present only in the unliganded hemoglobin molecule (Fig. 6). Studies of the binding of oxygen or carbon monoxide to hemoglobin in the absence and presence of organic phosphates, such as 2,3-diphosphoglycerate (P_2-glycerate) and inositol hexaphosphate (Ins-P_6), indicate that under some conditions the α and β hemes may, in fact, be functionally nonequivalent.[84,85] If the binding of oxygen to the α and β hemes is random (i.e., no preference), one would expect the α and β heme resonances to lose intensity at the same rate. If, on the other hand, one of the chains has a greater affinity for ligand, the resonance arising from its heme group should lose intensity more rapidly than the resonance of the low-affinity heme.

To gain the maximal amount of information from these saturation studies, it is necessary to be able to monitor the extent of ligand saturation accurately. Two different optical techniques have been devised to measure the degree of ligand saturation. Since light scattering from the round NMR tube can cause artifacts in the optical measurements, care must be taken to minimize the light scattering. Measurements can be made directly through the 5-mm NMR sample tube which is held in the light path by a specially designed cuvette holder in the near-infrared region of the optical spectrum using the deoxy absorption band at 757 nm.[84] In an alternative method, a movable lucite insert is placed in each NMR tube.[86] The lower part of the insert is thoroughly polished to eliminate light scattering and is about 0.4 mm smaller in diameter than the inside of the 5-mm NMR tube. The middle section, which fits snugly inside the tube, has four grooves to allow

[83] D. G. Davis, N. H. Mock, T. R. Lindstrom, S. Charache, and C. Ho, *Biochem. Biophys. Res. Commun.* **40,** 343 (1970).

[84] M. E. Johnson and C. Ho, *Biochemistry* **13,** 3653 (1974).

[85] T. R. Lindstrom and C. Ho, *Proc. Natl. Acad. Sci. U.S.A.* **60,** 1707 (1972).

[86] T.-H. Huang and A. G. Redfield, *J. Biol. Chem.* **251,** 7114 (1976).

the solution to pass through freely when the insert is moved. When the lucite spacer is in place, the absorption of the sample is reduced enough so that optical measurements can be made in the visible region of the spectrum. Using both of these techniques, if the optical measurements are made over a large enough spectral region to include isosbestic points, one can be reasonably confident that no artifacts are introduced by either light scattering or methemoglobin formation.

Early studies of Hb A using hfs proton resonances have shown that the functional properties of the α and β chains within an intact tetrameric hemoglobin depend on the nature of the oxidation state of the heme iron, ligand, pH, and phosphate.[84,85,87–89] These NMR results suggest that there is not much difference in the affinity for CO between the α and β chains of Hb A at neutral pD, both in the presence and absence of organic phosphates.[84] On the other hand, when O_2 is the ligand, it exhibits a very slight preferential binding to the α hemes in 0.1 *M* bis-Tris buffer at neutral pD, a marked preferential binding to the α hemes in ~30 m*M* P_2-glycerate, and nearly exclusive binding to the α hemes in the presence of ~13 m*M* Ins-P_6 at low saturation.[84] The hemoglobin concentration in these studies is about 10%. As an illustration of this technique to study the hemoglobin–ligand reaction, Fig. 7 shows the hfs protons of deoxy-Hb A as a function of CO and O_2 in 13 m*M* Ins-P_6. A major difficulty in obtaining accurate oxygen-saturation data of Hb A by the NMR method in the earlier studies was due to the uncertainty in ascertaining the baselines of the NMR spectra as shown in Fig. 7. Based on these results, it is difficult to eliminate the possibility that there are kinetic effects which cause line broadening rather than a loss of intensity due to preferential ligation in the NMR spectra of Hb A as a function of oxygenation. With improved techniques in the treatment of data, it has been possible to obtain more reproducible baselines and to obtain more accurate measurements of the peak intensity using difference spectroscopy. Figure 8 shows the difference spectra of deoxy-Hb A in Ins-P_6 as a function of O_2 saturation.[90] These recent results have confirmed the early conclusion of Johnson and Ho,[84] namely that the α chains of Hb A have a higher affinity for O_2 than the β chains in the presence of organic phosphates.

The functional properties of a number of human mutant hemoglobins have also been investigated by the NMR technique. For Hb Chesapeake (α92FG4 Arg→Leu), CO binds essentially at random to the α and β hemes

[87] T. R. Lindstrom, J. S. Olson, N. H. Mock, Q. H. Gibson, and C. Ho, *Biochem. Biophys. Res. Commun.* **45,** 22 (1971).

[88] J. J. Breen, D. Bertoli, J. Dadok, and C. Ho, *Biophys. Chem.* **2,** 49 (1974).

[89] D. G. Davis, S. Charache, and C. Ho, *Proc. Natl. Acad. Sci. U.S.A.* **63,** 1403 (1969).

[90] G. Viggiano and C. Ho, *Biophys. J.* **17,** 235a (1977).

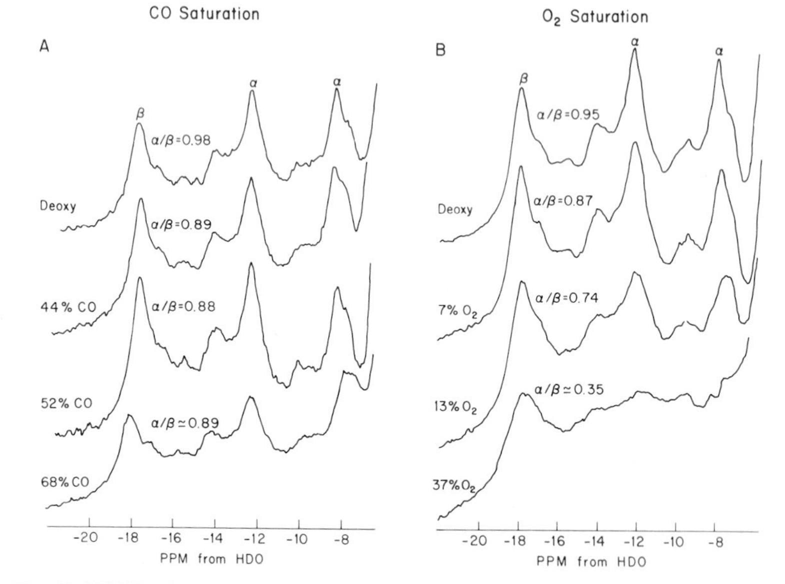

FIG. 7. 250 MHz hyperfine shifted proton resonances of 11% human adult hemoglobin in 0.09 *M* bis-Tris plus 13 m*M* inositol hexaphosphate at pD 7.0 and 27°C as a function of carbon monoxide (A) and oxygen (B) saturation. α/β ratios are area ratios of the α resonance at -12 ppm and the β resonance at -18 ppm. Due to slight variations in scale expansion during computer printout of the spectra, the chemical shift scale shown above is exact for the bottom spectrum only. To an accuracy of $\sim$0.2 ppm no chemical shift variations were observed during any saturation series. [Taken from Fig. 3 of Johnson and Ho,[84] with permission.]

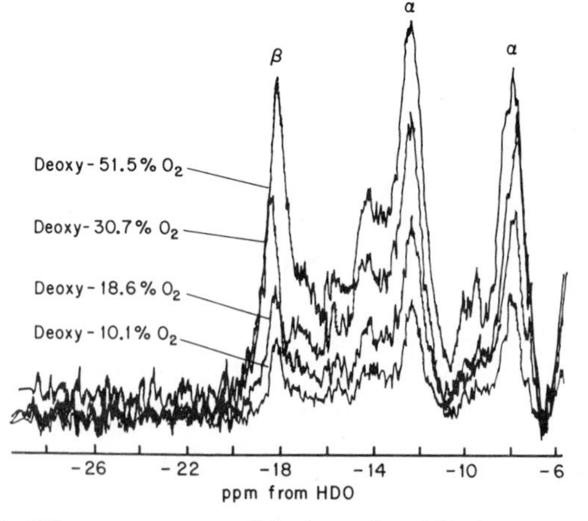

FIG. 8. 250 MHz difference spectrum of the hyperfine shifted proton resonances of 11.5% human adult hemoglobin in 0.1 *M* bis-Tris plus 10 m*M* inositol hexaphosphate at pD 6.8 and 27°C as a function of oxygenation. [Taken from G. Viggiano and C. Ho, unpublished results.]

in 0.1 *M* bis-Tris at pD ~7, in the presence of P_2-glycerate, and in the presence of Ins-P_6. Oxygen binds to the α and β hemes of Hb Chesapeake essentially at random in 0.1 *M* bis-Tris and in the presence of P_2-glycerate; however, oxygen shows a preference for the α hemes in the presence of Ins-P_6.[27,59] Studies of Hb Malmö (β97FG4 His→Gln) show that the affinity of the α and β hemes for CO are nearly equal in the absence and presence of organic phosphates. The hemes of Hb Malmö in 0.1 *M* bis-Tris buffer or in the presence of P_2-glycerate have very similar affinities for oxygen; however, in the presence of Ins-P_6 the α heme has a slightly higher affinity for oxygen.[91] In Hb Yakima (β99G1 Asp→His), X-ray crystallographic studies have shown that the β heme is more exposed than the β heme of Hb A as a result of the amino acid substitution.[92] It has been shown that in both Hb Yakima and Hb Kempsey (β99G1 Asp→Asn), the β hemes have a greater affinity for both oxygen and carbon monoxide in the absence of organic phosphate. In the presence of Ins-P_6, both of these mutant hemoglobins bind CO and O_2 to the α and β hemes randomly.[21,59,93]

A number of studies have confirmed that the positions of the hfs proton resonances are sensitive to the heme environment and to hemoglobin interactions.[21,33,59,94-97] As a result of this finding, it was possible to investigate the quaternary structures of valency hybrids under different conditions. These studies involved the deoxy cyanomet valency hybrids Hb $(\alpha^+CN\beta)_2$ and Hb $(\alpha\beta^+CN)_2$.[94,98] The hfs proton resonances of the hemes in either the cyanomet or deoxy states were used to monitor structural changes around the hemes. The deoxy-β heme spectrum of the hybrid Hb $(\alpha^+CN\beta)_2$ in the absence of phosphate was different from that of deoxy-Hb A. When 1 mole of P_2-glycerate per Hb tetramer was added, there were significant changes in the resonances of α^+CN and the resonance of the deoxy-β subunits became similar to the β heme resonance in deoxy-Hb A. These results were believed to be due to a change from the oxy to the deoxy quaternary structure resulting from the preferential binding of P_2-glycerate to the deoxy quaternary structure.[94,98] A similar change was observed in the spectrum of the Hb $(\alpha\beta^+CN)_2$ hybrid upon the addition of Ins-P_6, a stronger allosteric effector. NMR studies of the mutant Hb M Iwate (α87F8 His→Tyr), in which the iron atoms of the α chains are

[91] K. J. Wiechelman, V. F. Fairbanks, and C. Ho, *Biochemistry* **15,** 1414 (1976).
[92] P. D. Pulsinelli, *J. Mol. Biol.* **74,** 57 (1973).
[93] T. R. Lindstrom, J. J. Baldassare, H. F. Bunn, and C. Ho, *Biochemistry* **12,** 4212 (1973).
[94] S. Ogawa and R. G. Shulman, *J. Mol. Biol.* **70,** 315 (1972).
[95] M. F. Perutz, J. E. Ladner, S. R. Simon, and C. Ho, *Biochemistry* **13,** 2163 (1974).
[96] S. Ogawa, D. J. Patel, and S. R. Simon, *Biochemistry* **13,** 2001 (1974).
[97] T. Asakura, K. Adachi, J. S. Wiley, L. W.-M. Fung, C. Ho, J. V. Kilmartin, and M. F. Perutz, *J. Mol. Biol.* **104,** 185 (1976).
[98] S. Ogawa and R.G. Shulman, *Biochem. Biophys. Res. Commun.* **42,** 9 (1971).

permanently in the ferric state, showed that the $\beta^{+}CN$ heme resonances of Hb M Iwate are very similar to the $\beta^{+}CN$ resonances of Hb $(\alpha\beta^{+}CN)_2$ in the presence of Ins-P_6. Since X-ray crystallographic studies have shown that both liganded and unliganded Hb M Iwate are in the deoxy quaternary structure,[99] this is additional proof that the changes observed in the spectrum of Hb $(\alpha\beta^{+}CN)_2$ upon the addition of Ins-P_6 are due to a change in the quaternary structure brought about by the binding of Ins-P_6.[100]

Studies of mutant or chemically modified hemoglobins which remain in the oxy quaternary structure even when unliganded have shown that the hfs proton resonances are altered drastically in these hemoglobins.[93,95,96] From these studies a pattern emerges which can be used to identify those hemoglobins which are in the oxy quaternary structure even when they are unliganded. The hfs proton spectra of these hemoglobins generally have their β heme resonance shifted to -15.5 ppm and their α heme resonances occurring at -11.5 and -7.5 ppm from HDO, compared to chemical shifts of ~ -18, ~ -12, and ~ -8 ppm for hemoglobins in the deoxy quaternary structure. The superimposed spectra of isolated α and β chains, if added, would produce peaks at about -14.3, -10.6, and -8.1 ppm.[95] Thus the paramagnetic shifted resonances of an unliganded hemoglobin in the oxy quaternary structure are similar to those of the free deoxy chains, although there are small differences in the magnitudes of the chemical shifts. This similarity in the hfs resonances is in agreement with the finding that the electronic spectra of the sum of free deoxy α and β chains correspond to the spectrum of a hemoglobin in the oxy quaternary structure.[95] The addition of Ins-P_6 to hemoglobins in the oxy quaternary structure usually results in drastic changes in their hyperfine shifted proton spectra, which become similar to those of deoxy-Hb A, suggesting that the hemoglobins have been switched to the deoxy quaternary structure. This switch to the deoxy quaternary structure which is induced by Ins-P_6 is also reflected in the functional properties of these hemoglobins which have lowered oxygen affinities and increased values of the Hill coefficient.

Although much effort has been devoted to the hemoglobin problem, the detailed molecular mechanism for the cooperative oxygenation of hemoglobin is not yet fully understood. The main experimental difficulty is that partially oxygenated intermediate species are only present as a small fraction of the total Hb content due to the cooperative nature of the oxygenation process. It is necessary to understand the nature and role of these intermediates before one can fully understand the function of hemoglobin as an oxygen carrier. For recent discussions on this topic, the reader should

[99] J. Greer, *J. Mol. Biol.* **59,** 107 (1971).

[100] R. G. Shulman, S. Ogawa, A. Mayer, and C. L. Castillo, *Ann. N. Y. Acad. Sci.* **222,** 9 (1973).

consult Shulman *et al.*,[22] Fung *et al.*,[32,35] and Baldwin.[82] A recent ^{1}H NMR investigation of Hb M Milwaukee (β67E11 Val→Glu), which is a naturally occurring valency hybrid with the β hemes permanently oxidized, has provided significant new insight into this problem.[32,35] In this mutant, the two abnormal β chains cannot combine with oxygen, whereas the two α chains are normal and combine with oxygen cooperatively with a Hill coefficient of approximately 1.3. As shown in Fig. 1, the hfs proton resonances of high-spin met-Hb occur at spectral regions different from those due to high-spin deoxy-Hb. Thus, it is possible to monitor the hfs proton resonances of the abnormal ferric-β chains of Hb M Milwaukee over the spectral region from -30 to -60 ppm from HDO as a function of oxygenation of the two normal α chains as shown in Fig. 9. When the α chains are

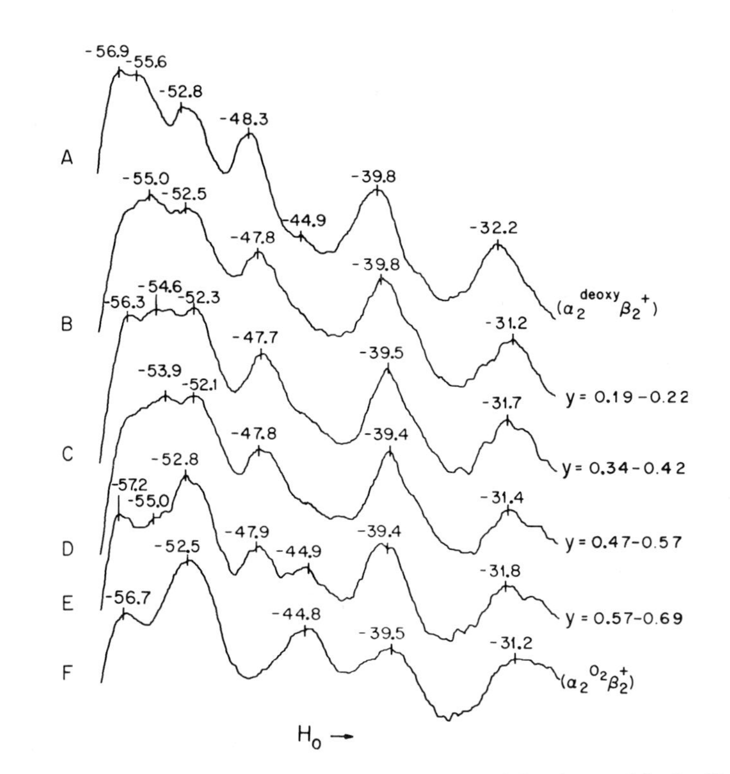

FIG. 9. 250 MHz hyperfine shifted proton resonances of the abnormal ferric-β hemes of 15% Hb M Milwaukee (β67E11 Val→Glu) in 0.1 *M* bis-Tris at pD 7.0 and 30°C as a function of the oxygenation of the normal α chains. The units are in ppm from HDO. The range of fractional saturation is based on p_{50} = 5 torr or p_{50} = 10 torr. H_0 represents the applied magnetic field, which increases from left to right as shown in the figure. [Taken from Fig. 5 of Fung *et al.*,[32] with permission.]

deoxygenated, distinct resonances are observed at about −48, −53, and −55 to −57 ppm from HDO. When the α chains are fully oxygenated, three major resonances at about −45, −52, and −57 ppm are present. Upon the step-by-step addition of O_2 to the α chains of Hb M Milwaukee, the NMR spectrum changes gradually from the deoxy spectrum (Fig. 9A) to that of the fully oxy one (Fig. 9F). When the α chains are about 60% oxygenated, the −48 and −45 ppm resonances coexist with approximately equal intensities and the spectral feature in the region between −52 to −57 is already converted to that characteristic of the fully oxygenated species (Fig. 9F). Thus, the two spectral changes observed in the two regions, −45 to −48 ppm and −52 to −57 ppm, are asynchronous. At intermediate values of O_2 saturation, the hfs proton resonances of the ferric-β chains in Hb M Milwaukee show some features which are not found in either the deoxy or oxy spectrum. In other words, the ^1H NMR spectra of partially oxygenated Hb M Milwaukee cannot be described as an appropriately weighted average of the spectra of Fig. 9A and F. These results are not consistent with a two-structure model for the oxygenation of this mutant hemoglobin. In view of the similarities between Hb A and Hb M Milwaukee, it is suggested that a two-state concerted allosteric model does not provide an adequate description of the structure–function relationships in normal adult hemoglobin.[32,35]

Conclusion

High-resolution proton nuclear magnetic resonance spectroscopy has made important as well as unique contributions to our current knowledge of iron-sulfur proteins and hemoproteins. In addition to the hyperfine shifted proton resonances discussed in this article, there are other resonances, such as aromatic, aliphatic, ring-current shifted, and exchangeable proton resonances, that can provide valuable information about the conformations and conformational changes of these proteins. In fact, it is difficult, at the present time, to find another technique that can provide so many well-characterized probes to investigate the structure–function relationship of a protein in solution. Obviously, much work still remains to be done on iron-sulfur proteins and hemoproteins, and NMR spectroscopy can be expected to yield further insights into these proteins. However, a word of caution is needed in the interpretation of the results obtained from NMR or other forms of spectroscopy. Optical spectroscopic changes provide information on the free-energy differences between the ground and excited states of a molecule. NMR spectroscopic changes are related to the energy differences between nuclear spin states. On the other hand, the free-energy changes associated with biochemical reactions reflect the changes of the free energies of the ground state of the system under consideration. Although the energy differences in NMR spectroscopy are less than the thermal energy at room temperature, there is no simple, direct relationship between the observed chemical shift changes and the free-energy changes associated with biochemical reactions. The relevancy between the observed NMR spectral changes and the structure–function relationships of a given protein can only be inferred from the experimental data. However, carefully designed experiments to correlate the NMR spectral data with the structural and functional properties of a given protein can bridge this gap.

Acknowledgments

We wish to extend our gratitude to Dr. E. A. Pratt, Dr. M. P. N. Gent, and Dr. G. Palmer for helpful discussions during the preparation of this manuscript.

[14] The Hydride Transfer Stereospecificity of Nicotinamide Adenine Dinucleotide Linked Enzymes: A Proton Magnetic Resonance Technique

By LYLE J. ARNOLD, JR., and KWAN-SA YOU

The pyridine nucleotide linked enzymes are an extensive class of enzymes which transfer hydride between substrate and the pyridine 4 position of NAD^+ or $NADP^+$. These enzymes can distinguish the diastereotopic methylene protons at the dihydropyridine 4 position of NADH or NADPH, transferring the hydride to the substrate stereospecifically (for reviews, see Popják[1] and Bentley[2]). As a result of the stereospecificity, the enzymes belonging to this class have been classified into two groups: The A-stereospecific enzymes which transfer hydride between substrate and the A(*pro*-R) side of the dihydropyridine ring and the B-stereospecific enzymes which transfer hydride to the B(*pro*-S) side.

Pioneering work on the stereospecificity of oxidoreductases was carried out by mass spectroscopy, employing deuterium-labeled coenzymes.[3] Subsequently, Jarabak and Talalay[4] introduced a technique utilizing

[1] G. Popják, *in* "The Enzymes" (P. D. Boyer, ed.), 3rd ed., Vol. 2, p. 115. Academic Press, New York, 1970.

[2] R. Bentley, "Molecular Asymmetry in Biology," Vol. 2, p. 1. Academic Press, New York, 1970.

[3] H. R. Levy and B. Vennesland, *J. Biol. Chem.* **228,** 85 (1957).

[4] J. Jarabak and P. Talalay, *J. Biol. Chem.* **235,** 2147 (1960).

ISBN 0-12-181954-X

tritium-labeled coenzymes, and this method has been used predominantly in recent years. However, these conventional techniques are generally laborious, time-consuming, and require multiple purification steps.

The proton magnetic resonance (^{1}H NMR) technique[5] described here is accurate, safe, and much easier than conventional techniques. Furthermore, this ^{1}H NMR technique is superior since the exact oxidoreduction site is directly monitored, rather than the entire coenzyme molecule, and crude extracts containing a given enzyme can often be employed.

Principle

Specifically deuterated pyridine nucleotides are reacted with the enzyme whose stereospecificity is to be determined in the presence of the appropriate substrate. The examination of the ^{1}H NMR spectra of the product coenzyme readily reveals which hydride is transferred to (or from) the coenzyme.

Shown below are the three methods which can be employed for these studies. The method used will depend on the availability of substrates and on the equilibrium of a particular enzymic reaction.

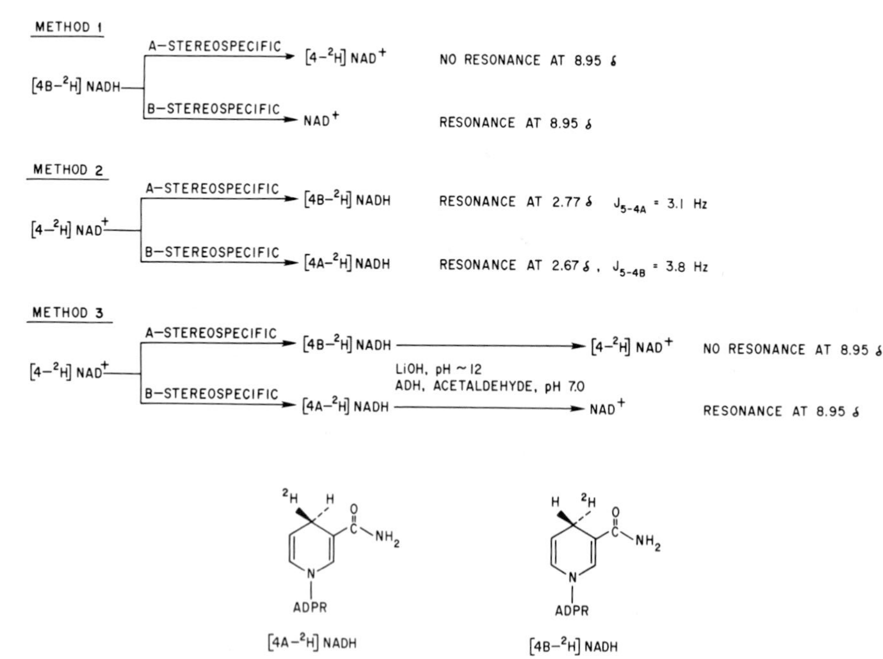

[5] L. J. Arnold, Jr., K. You, W. S. Allison, and N. O. Kaplan, *Biochemistry* **15,** 4844 (1976).

Method 1

In method 1, which is the simplest method of the three, [4B-^{2}H]NADH is oxidized with the enzyme under investigation. The ^1H NMR spectrum of the product NAD^+ is then analyzed for the retention of deuterium at the pyridine 4 position. If an enzyme is B-stereospecific, it will remove deuteride and the ^1H NMR spectrum of the resulting NAD^+ will show the pyridine 4 proton resonance which appears at 8.95δ (pD 3, 23°C) as a doublet; an A-stereospecific enzyme, on the other hand, removes hydride and the spectrum will not show this resonance (see Fig. 1).

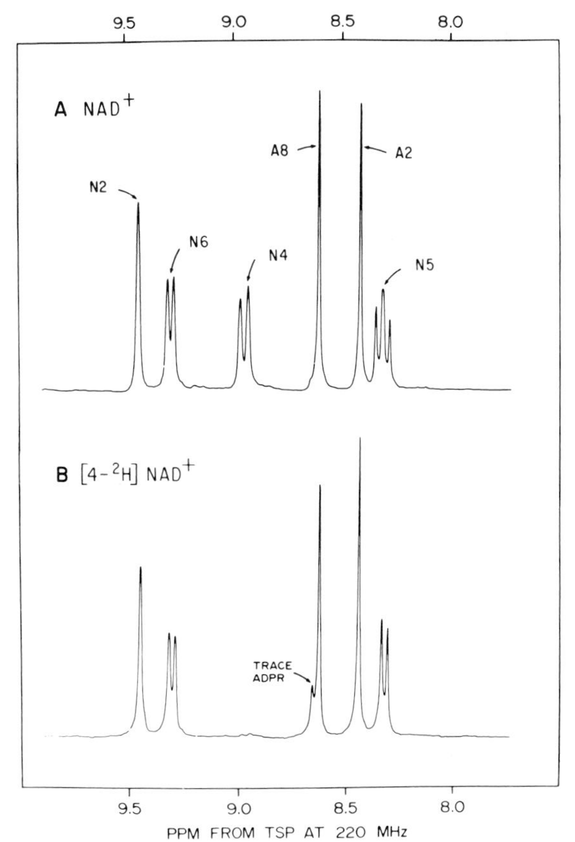

FIG. 1. The ^1H NMR aromatic region of (A) NAD^+ and (B) [4-^{2}H]NAD^+ (20 mM, pD 3, 23°C, 100 scans). The deuterium content at the pyridine 4 position was found to be 97 ± 1% by integration. The resonances designated by N2, N4, N5, and N6 are from the protons at carbon 2, 4, 5, and 6 of the pyridine ring, and A2 and A8 are those at carbon 2 and 8 of the adenine ring. The small resonance on the left side of the A8 resonance in [4-^{2}H]NAD^+ is caused by contaminating ADP ribose. Taken from Arnold *et al.*,[5] reprinted with permission from *Biochemistry* **15,** 1976, copyright, The American Chemical Society.

Method 2

Method 2, which utilizes the oxidized coenzyme, can be applied to enzymes that catalyze reactions predominantly in the direction of coenzyme reduction and in cases where only the reduced form of the substrate is available. In this method, the position of the transferred hydride in the product NADH can be identified since the 4A-proton has a chemical shift of 2.77δ and an *unresolved* coupling constant (J_{5-4A}) of 3.1 Hz, whereas the 4B-proton has a chemical shift of 2.67δ and a *resolved* coupling constant (J_{5-4B}) of 3.8 Hz[5,6] (see Fig. 3).

Method 3

Method 3, a combination of methods 1 and 2, is devised to alleviate the handling of NADH, which is somewhat unstable and requires a careful determination of chemical shifts and coupling constants. In this method, [4-^{2}H]NAD$^+$ is first reduced to NADH with the enzyme under study and then it is reoxidized with a well-known A-stereospecific enzyme, yeast alcohol dehydrogenase. The interpretation of the ^{1}H NMR spectrum is the same as that of method 1.

Procedure for the Stereospecificity Determination

Preparation of Deuterated Pyridine Nucleotides

[4B-^{2}H]NADH is prepared by reducing 300 μmole of NAD$^+$ with 10 μmole of lipoamide and 10 units of lipoamide dehydrogenase (a B-stereospecific enzyme) in the presence of 600 μmole of dithiothreitol and 0.05 *M* $(NH_4)_2CO_3$ in 50 ml of ^2H_2O; the pD is maintained at 8.5 by additions of [U-^{2}H] ammonium hydroxide throughout the reaction. When A_{260}/A_{340} is less than 3.0, the reaction mixture is diluted to 200 ml and applied to a DEAE-ll column (carbonate form, 2.5 × 25 cm). The [4B-^{2}H]NADH is then purified by eluting the column with a linear 1-liter gradient of 0–0.5 *M* $(NH_4)_2CO_3$. The [4B-^{2}H]NADH elutes at approximately 0.15 *M* $(NH_4)_2CO_3$, and those fractions with A_{260}/A_{340} of less than 2.4 are pooled and lyophilized.

[4-^{2}H]NAD$^+$ is obtained by oxidizing the [4B-^{2}H]NADH thus prepared with yeast alcohol dehydrogenase (an A-stereospecific enzyme) in the presence of excess acetaldehyde and 0.05 *M* $(NH_4)_2CO_3$. The reaction

[6] One should be aware that the chemical shifts and coupling constants of NADH are somewhat dependent upon both temperature and concentration. Thus the values given are accurate only at 20 mM and 23°C.

mixture is then applied to a Dowex-1 column (formate form, 2.5 × 25 cm), and the [4-^{2}H]NAD^+ is purified by eluting the column with a 1-liter linear gradient of 0–1.5 *M* formic acid. The [4-^{2}H]NAD^+, which elutes at approximately 0.4 *N* formic acid, is assayed by reducing it with yeast alcohol dehydrogenase in the presence of ethanol. Those fractions giving A_{260}/A_{340} of less than 2.6 upon assay are pooled and lyophilized. The [4-^{2}H]NAD^+ thus prepared has a deuterium content of 97 ± 1% at the pyridine 4 position as determined by ^{1}H NMR spectroscopy (see Fig. 1).

Enzymic Reactions

Method 1. Typical reaction mixtures contain 5–20 units of enzyme, 10–15 μmole of [4B-^{2}H]NADH, 50–100 μmole of substrate, and 150 μmole of potassium phosphate in a final volume of 30 ml at pH 7.0–8.0. In this and other methods, any other substances[7] which are required for a particular enzyme reaction are also added. It should be noted, however, that these reaction conditions may be varied depending on the properties of the enzyme under investigation. When there is no further decrease in A_{340}, the reaction is quenched by lowering the pH of the medium to 2 with concentrated HNO_3 and the mixture is lyophilized. Usually it is not necessary to purify the product NAD^+ from the reaction mixture, but when crude extracts are used as enzyme sources or when paramagnetic ions are present in the reaction mixture, NAD^+ is purified after lyophilization by acid–acetone precipitation.[8] The samples are then treated as described in the section below on ^{1}H NMR Instrumentation and Measurements.

Method 2. Typical reaction mixtures contain 5–20 units of enzyme, 10–15 μmole of [4-^{2}H]NAD^+, 50–100 μmole of substrate, and 150 μmole of Tris-HCl in a final volume of 30 ml at pH 8.0–9.0. When there is no further increase in A_{340}, 1 *N* LiOH is added to raise the pH to approximately 12 and the solutions are lyophilized. After lyophilization samples are dissolved in 0.5 ml of H_2O, and the NADH is purified by barium thiocyanate–acetone precipitation and converted to its sodium salt.[9] Samples are then treated according to the procedure in ^{1}H NMR Instrumentation and Measurements.

Method 3. The initial reaction mixtures have the same compositions as that for method 2. When the reactions are completed, 1 *N* LiOH is added to

[7] If paramagnetic ions such as Mn^{2+} are required for enzyme activity, it will be necessary to remove them by acid-acetone precipitation of NAD^+[8], since the ^{1}H NMR resonances are markedly broadened by paramagnetic impurities.

[8] A. Kornberg, Vol. 3, p. 876.

[9] N. J. Oppenheimer, L. J. Arnold, Jr., and N. O. Kaplan, *Proc. Natl. Acad. Sci. U.S.A.* **68**, 3200 (1971).

raise the pH to approximately 12 and the mixtures are placed in boiling water for 90 sec. This procedure denatures the protein and rapidly destroys any residual NAD^+, which would complicate the subsequent analysis of NADH, but leaves NADH intact. The mixtures are then neutralized (pH 7–8) by adding 0.1 *N* HCl with stirring, and the formed NADH is oxidized with excess acetaldehyde by yeast alcohol dehydrogenase. When there is no further decrease in A_{340}, the pH is lowered to 2 with concentrated HNO_3 and the samples are lyophilized for the preparation of ^1H NMR samples.

^1H NMR Instrumentation and Measurements

These measurements can be made with most commonly available NMR instruments. However, the instrument employed may affect the methods used as well as the quantity of material required. For methods 1 and 3, any instruments having a magnetic field of 40 MHz or more can be used, but for method 2 a 100 MHz or higher field instrument is recommended in order to achieve accurate resolution of chemical shifts.

Furthermore, the minimum quantity of material required for analysis by any of the methods depends upon the strength of the magnetic field and upon the availability of signal-averaging capabilities. From our experience when using a 5-mm sample tube, the following formula expresses the minimum concentration of coenzyme needed for the reliable determination of signal intensities for all instruments:

$$\text{Minimum concentration} = \frac{2\ M}{S/N \times (\text{transients})^{1/2}}$$

where S/N represents the signal-to-noise ratio of the instrument. Thus, if the S/N is 50 and the number of transients is 400, the sample concentration required is 2 m*M*. If vortex plugs are employed in order to reduce the sample volume to 0.3 ml, only 0.6 μmole of the coenzyme is needed for a concentration of 2 m*M*. Where possible, however, higher concentrations should be employed in order to eliminate time-consuming signal accumulations.

In our studies the samples prepared according to methods 1, 2, and 3 are lyophilized twice from 99.8% ^2H_2O to remove water and then dissolved in 100% ^2H_2O containing 1 m*M* tetramethylammonium chloride (TMAC) and 1 m*M* EDTA. The sample volumes are maintained at 0.3 ml with Wilmad Teflon vortex plugs. Spectra are taken at 23°C (the ambient temperature of the probe) with a Varian HR-220 nuclear magnetic resonance spectrometer interfaced to a Transform Technology 220 Fourier transform system. To improve the signal-to-noise ratio, as few as 16 transients are accumulated, but when sample concentrations are low, as many as 200 transients are acquired.

Spectra are generally taken at a sweep width of 2500 Hz with a recycle time of 1.6 sec and processed with an exponential multiplier of 0.2–0.4 Hz. Chemical shifts are obtained by determining chemical shifts relative to internal TMAC, and adding 3.2 ppm, the chemical shift difference between TMAC and trimethylsilyl sodium [U-^{2}H]propionate (TSP).

The aromatic ^{1}H NMR spectral regions for NAD$^+$ and [4-^{2}H]NAD$^+$ are shown in Fig. 1. Results obtained applying method 1 for the stereospecificity determination of *Pseudomonas* melilotate hydroxylase (E.C. 1.14.13.4) and *Rhodopseudomonas spheroides* D-β-hydroxybutyrate dehydrogenase (E.C. 1.1.1.30) are shown in Fig. 2. Figure 3 shows the ^{1}H NMR spectra of stereospecifically deuterated NADH as well as the spectrum obtained with glucose 6-phosphate dehydrogenase according to method 2. Finally, Fig. 4 presents the results with scallop octopine dehydrogenase (E.C. 1.5.1.11) and *Pseudomonas* 4-amino-butanal dehydrogenase following method 3.

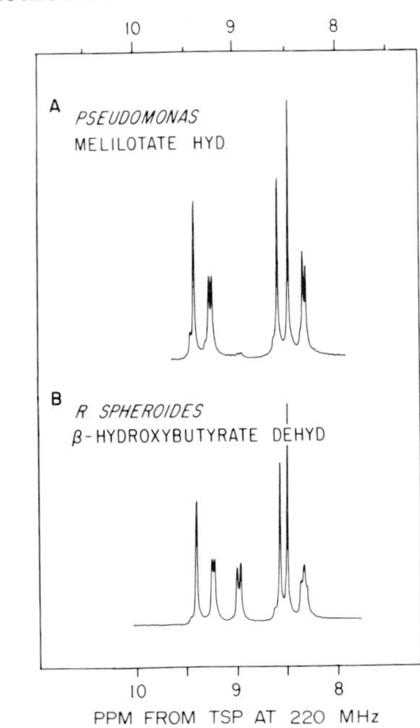

FIG. 2. The ^{1}H NMR aromatic region of NAD$^+$ formed from [4B-^{2}H]NADH according to method 1 (pD 3, 23°C). The enzymes employed are: (A) *Pseudomonas* melilotate hydroxylase, 100 scans; and (B) *Rhodopseudomonas spheroides* D-β-hydroxybutyrate dehydrogenase, 64 scans.

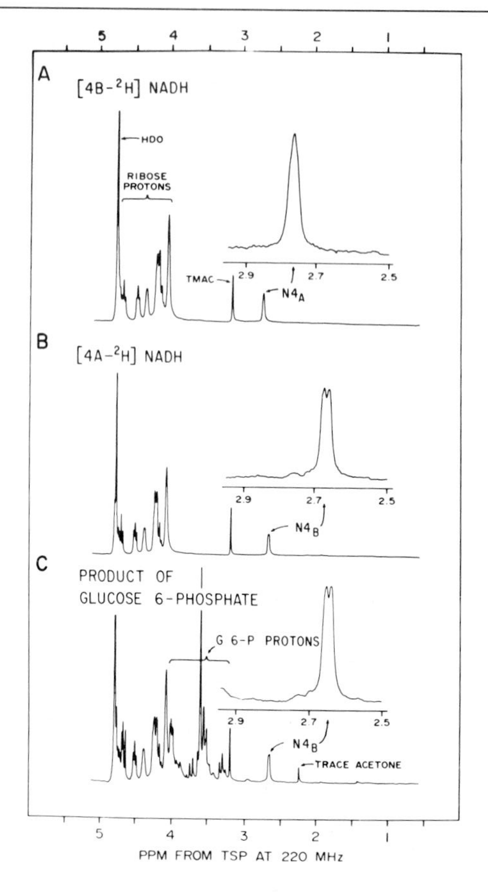

FIG. 3. The up-field portion of the ^{1}H NMR spectra of (A) [4B-^{2}H]NADH, (B) [4A-^{2}H]NADH, and (C) NADH formed from [4-^{2}H]NAD$^+$ by *Leuconostoc mesenteroides* glucose 6-phosphate dehydrogenase according to method 2. The resonances seen between 3 and 4 ppm in spectrum (C) are from glucose 6-phosphate (G 6-P) protons. The inserts show the expansions of the dihydropyridine 4 proton region in order to clearly reveal the differences in chemical shift and coupling constants of the stereospecific labels (20 m*M*, pD 8.0, 23°C, 100 scans). Taken from Arnold *et al.*[5]

The retention of a proton at the pyridine 4 position in methods 1 and 3 can be quantitatively determined by comparing the integration of its resonance to that of the resonance for the proton at the pyridine 6 position. In this regard, one should not use the pyridine 2 proton resonance as an integration standard. The pyridine 2 proton undergoes partial deuterium exchange with ^2H_2O when NAD$^+$ is subjected to alkaline conditions[10,11]

[10] A. San Pietro, *J. Biol. Chem.* **217,** 589 (1955).

[11] H. E. Dubb, M. Saunders, and J. H. Wang, *J. Am. Chem. Soc.* **80,** 1767 (1958).

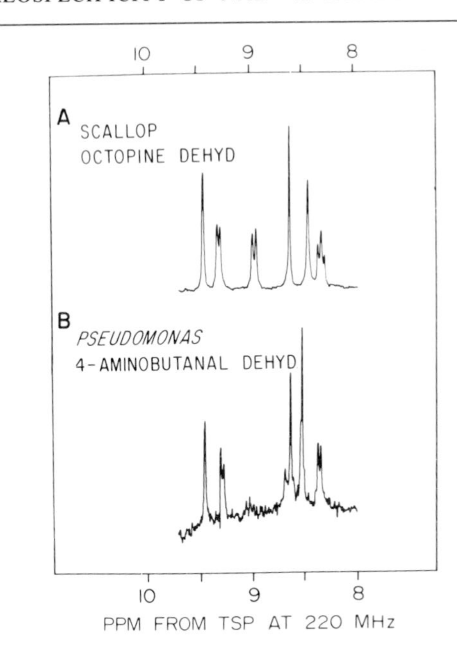

FIG. 4. The ^1H NMR aromatic region of NAD^+ formed from $[4\text{-}^2H]NAD^+$ according to method 3 (pD 3, 23°C). The enzymes employed are: (A) scallop octopine dehydrogenase, 16 scans (a crude extract of abalone muscle was used as the source of this enzyme); and (B) *Pseudomonas* 4-aminobutanal dehydrogenase, 200 scans.

while it is being reduced to $[4B\text{-}^2H]NADH$ and consequently shows decreased proton integration.

Discussion

The enzymic reactions described here are purposely carried out at low to moderate concentrations of coenzyme in order to minimize the nonstereospecific intermolecular exchange of the pyridine 4 proton between NAD^+ and NADH, which is known to occur at elevated concentrations of coenzyme. Ludowieg and Levy,[12] for example, found that a nonstereospecific exchange occurred between 30 m*M* $[4\text{-}^3H]NAD^+$ and 30 m*M* NADH to the extent of 39% in 8 hr at pH 8.0 and 30°C. However, when the coenzyme concentrations were lowered to 0.3 m*M*, the exchange was only 7% under the same conditions. The combined concentrations of the oxidized and reduced coenzymes are lower than 0.5 m*M* under the reaction conditions described here, and the reactions are generally completed within 30 min.

[12] J. Ludowieg and A. Levy, *Biochemistry* **3,** 373 (1964).

Although all these ^1H NMR methods can be applied easily, methods 1 and 3 are preferred to method 2 because: (1) NAD^+ is more stable than NADH, since NADH is susceptible to nonstereospecific air oxidation; (2) the determination of the presence or absence of the pyridine 4 proton resonance (8.95δ) is much simpler than determining the chemical shifts and coupling constants of the closely situated pyridine 4A- and 4B-proton resonances of NADH; (3) purification of NAD^+ from the reaction mixture is not usually required (especially if the reaction components are known) since impurities having absorptions in the aromatic proton region are limited.

Moreover, if one desires to use [4A-^2H]NADH instead of [4B-^2H]NADH in method 1, it can be prepared from [4-^2H]NAD^+ by reducing it with lipoamide dehydrogenase in H_2O.

When compared with conventional mass spectroscopic and radioisotopic techniques, the described ^1H NMR technique has numerous advantages. These include: (1) The accurate and rapid monitoring of the *exact* oxidoreduction site rather than the entire coenzyme molecule (the accuracy of this ^1H NMR technique is limited by the accuracy of integration which is approximately ±2% in spectra with good signal-to-noise ratios); (2) the elimination of extensive and tedious purification steps; (3) the use of crude extracts (especially if a control experiment, i.e., an experiment without substrate, is carried out as shown in the case of octopine dehydrogenase, Fig. 4); (4) the alleviation of the hazards associated with the use of radioisotopes; and (5) the use of micromole quantities of coenzyme when instrumentation with Fourier-transform analysis is available.

Employing the technique presented here, it is possible to determine the stereospecificity of $NADP^+$-linked enzymes as well. It was previously shown that at pD 7.0 the resonance for the pyridine 4 proton of $NADP^+$ appears at 8.78δ,[13] and at pD 8.5 the pyridine 4A- and 4B-protons of NADPH appear at 2.81δ (J_{5-4A} = 3.1 Hz) and 2.72δ (J_{5-4B} = 3.9 Hz), respectively.[5,14] [4B-^2H]NADPH may be readily prepared with the use of $NADP^+$-linked glutathione reductase in the presence of glutathione and dithiothreitol in ^2H_2O.

[13] R. H. Sarma and N. O. Kaplan, *Biochem. Biophys. Res. Commun.* **36,** 780 (1969).

[14] N. J. Oppenheimer, L. J. Arnold, Jr., and N. O. Kaplan, *Biochemistry* **17** (in press).

[15] Resonance Raman Spectra of Hemoproteins

By THOMAS G. SPIRO

I. Introduction

The forte of vibrational spectroscopy in molecular structure analysis is its high resolution, together with its sensitivity to alterations in molecular bonding and geometry. Neither resolution nor sensitivity are as high as in nuclear magnetic resonance (NMR) spectroscopy, but the vibrational spectrum is not subject to the many interaction and relaxation effects that complicate and frequently obscure the NMR spectrum. The frequencies are characteristic of molecular groupings in specific conformational states, and are relatively insensitive to environmental influences. Vibrational spectra can be obtained on solids and dispersions as well as solutions.

The analysis of molecular vibrations is a classical discipline, with firm theoretical and experimental underpinnings.[1-3] Two forms of spectroscopy, infrared (IR) and Raman, provide the needed data. IR spectroscopy involves direct absorption of light at frequencies appropriate for molecular vibrations. Since water is a strong absorber of IR radiation over much of the spectral range of interest, the use of IR spectroscopy in biological studies is limited. Raman spectroscopy involves a frequency scan of light scattered from a sample which is illuminated with a monochromatic source. Occasionally an incident photon lifts a molecule to a vibrationally excited state, emerging with its own energy lowered by a vibrational quantum. The Raman spectrum, illustrated schematically in Fig. 1, contains a series of peaks whose *shifts* from the incident frequency are the same as the vibrational frequencies of the sample. Raman scattering is inherently inefficient, but the high power densities afforded by lasers makes acquisition of high-quality spectra straightforward. Water is a poor Raman scatterer and provides little interference.

When the sample is transparent to the incident light the Raman spectrum contains all the vibrational modes of the sample that produce a change in the polarizability. For biological macromolecules, these are exceedingly numerous. Repeating structural units, e.g., peptide and nucleotide links, give rise to recognizable packets of vibrational modes, and Raman spec-

[1] G. Herzberg, "Infrared and Raman Spectra of Polyatomic Molecules." Van Nostrand-Reinhold, Princeton, New Jersey, 1945.

[2] E. B. Wilson, J. C. Decius, and P. C. Cross, "Molecular Vibrations." McGraw-Hill, New York, 1955.

[3] N. B. Colthup, L. H. Daly, and S. E. Wiberly, "Introduction to Infrared and Raman Spectroscopy," 2nd ed. Academic Press, New York, 1975.

ISBN 0-12-181954-X

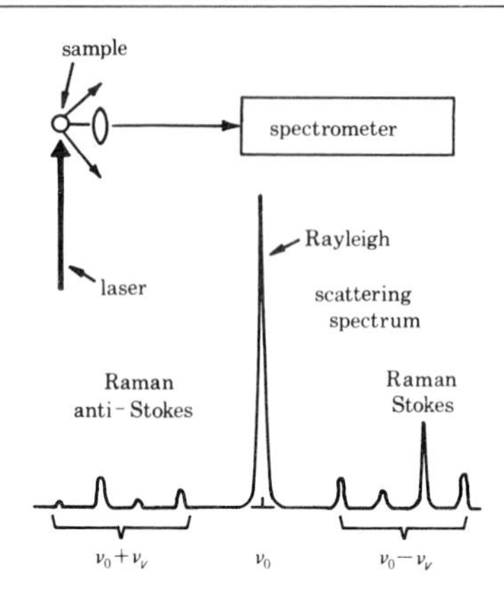

FIG. 1. Schematic illustration of laser Raman scattering. Most photons are scattered without shift from their initial frequency, ν_0, producing the Rayleigh peak (which is several orders of magnitude larger than the Raman peak in an actual spectrum). Photons which raise molecules to vibrationally excited states emerge with their frequencies decreased by the vibrational frequencies, ν_v, producing the Stokes peaks. Alternatively, molecules which are already in excited states can add energy to incident photons, producing the anti-Stokes peaks. The latter are weaker than the Stokes peaks because the population of molecules in excited states decreases exponentially with increasing vibrational frequency (and decreasing temperature). From Spiro.[8a]

troscopy can be applied to the analysis of protein or nucleic acid conformations.[4-7] However, information about specific sites in a biological sample is generally lost in a sea of vibrational bands. A way out of the difficulty can be provided by *resonance enhancement* of the Raman spectrum, if the site of interest gives rise to an electronic absorption band. Tuning the wavelength of the laser to the absorption band can produce a large increase in the scattering cross-section for certain vibrational modes of the chromophoric group. These can then be monitored for evidence of structural change associated with biochemical activity, free from interference from the vibrational bands of the matrix.

Hemoproteins have proven to be fertile subjects for resonance Raman (RR) studies, results of which have recently been reviewed elsewhere.[8,8a]

[4] T. G. Spiro, *in* "Chemical and Biochemical Applications of Lasers" (C. B. Moore, ed.), Academic Press, New York, 1974.

[5] B. G. Frushour and J. L. Koenig, *in* "Advances in Infrared and Raman Spectroscopy" (R. J. H. Clark and R. E. Hester, eds.), Vol. 1, p. 35. Heyden, London, 1975.

[6] T. G. Spiro and B. P. Gaber, *Annu. Rev. Biochem.* **46,** 553 (1977).

[7] S. C. Erfurth, P. J. Boyd, and W. L. Peticolas, *Biopolymers* **14,** 1245 (1975).

[8] T. G. Spiro, *Biochim. Biophys. Acta* **416,** 169 (1975).

[8a] T. G. Spiro, *Proc. R. Soc. London, Ser.* **A 345,** 89 (1975).

The heme group has strong absorption bands in the visible and near-ultraviolet region, accessible to the argon ion laser (4579–5286 Å) which is the most common Raman light source. RR spectra, containing several vibrational modes of the heme group, are readily obtainable at concentrations of 10^{-3}–10^{-5} M, well below the concentrations at which normal polypeptide Raman bands can be observed.

II. Resonance Enhancement

A. Scattering Mechanisms

In order to use RR spectroscopy with greatest effect, it is important to understand, at least qualitatively, what sorts of enhancement patterns to expect. The intensity of a Raman band is proportional to the square of the change in molecular polarizability associated with the molecular vibration. The polarizability derivative is a tensor quantity, whose elements are given by the Kramers-Heisenberg-Dirac dispersion equation[9]

$$(\alpha_{ij})_{mn} = \frac{1}{h} \sum_e \frac{(M_j)_{me}\,(M_i)_{en}}{\nu_e - \nu_0 + i\Gamma_e} + \frac{(M_i)_{me}\,(M_j)_{en}}{\nu_e + \nu_s + i\Gamma_e}$$

Here m and n are the initial and final (vibrational) states of the molecule, while e is an excited electronic state, and the summation is over all excited states. The quantities $(M_j)_{me}$ and $(M_i)_{en}$ are electric dipole transition moments, along the molecular axes j and i, while ν_e is the transition frequency, $i\Gamma_e$ is a damping term which is a measure of the electronic bandwidth, and ν_0 and ν_s are the frequencies of the incident and scattered photons. When ν_0 approaches ν_e, the left-hand term in the summation can be very large for an allowed electronic transition. This is the origin of resonance enhancement. Its magnitude is predicted to vary directly with the intensity of the resonant absorption band and inversely with its breadth.

The vibrational modes that undergo enhancement are those that significantly modulate the electronic transition. The transition moment is a product of a purely electronic part, and a vibrational part, involving the (Frank-Condon) overlap of the excited and ground-state vibrational wavefunctions. These overlaps are largest for those vibrational modes that most directly lead to the molecular distortions experienced by the excited states. These are the vibrations that show maximal Raman enhancement[10] for resonance with a single excited state.[11] For example, π–π^* transitions,

[9] J. H. Van Vleck, *Proc. Natl. Acad. Sci. U.S.A.* **A15,** 759 (1929).

[10] A. Y. Hirakawa and M. Tsuboi, *Science* **188,** 359 (1975).

[11] J. Tang and A. C. Albrecht, *in* "Raman Spectroscopy" (H. A. Szymanski, ed.), Vol. 2, Chapter 2. Plenum, New York, 1970.

which produce the dominant absorption bands of polyene and aromatic chromophores, show greatest enhancement for stretching vibrations of the π bonds, which are weakened in the excited state. Likewise, ligand $\rightarrow$ metal charge-transfer transitions, often exhibited by metalloproteins, provide enhancement for stretching vibrations of the metal–ligand bonds, which are weakened in the excited state, or of internal ligand vibrations, whose force constants are affected by the loss of the transferred electrons.

Only totally symmetric vibrations can be enhanced by resonance with a single excited state (*A* term scattering).[11,12] Nontotally symmetric vibrations are subject to enhancement when two nearby excited states are vibronically mixed (*B* term scattering).[11,12] The mixing vibrations, which lend intensity to both the absorption and RR spectra, may have any symmetry contained in the direct product of the two excited-state representations.

B. Heme Proteins

The electronic spectra of heme proteins are dominated by the aromatic system of the porphyrin ring. The visible and near-ultraviolet region contain two π–π^* transitions.[13] Both are of E_u symmetry under the D_{4h} point group, which applies approximately to metalloporphyrins. Being nearly degenerate, they undergo considerable interaction, with the transition dipoles adding for the higher energy transition and nearly cancelling for the lower one. The result is a very intense band near 400 nm, called the Soret, γ, or *B* band, and a much weaker one near 550 nm, called the α, or Q_0, band. The lower energy transition, however, gains back some 10% of the Soret intensity through vibronic mixing, with the formation of a vibronic sideband, called the β, or Q_v, band some 1300 cm^{-1} above the center of the α band.

The appearance of the RR spectrum depends markedly on the laser wavelength, as illustrated for oxy- and deoxyhemoglobin in Fig. 2.[14] For wavelengths below about 500 nm, the intense Soret transition dominates the scattering mechanism, via an *A* term.[15] The enhanced modes are totally symmetric. For wavelengths within the α and β absorption envelopes, *B* term scattering is predominant,[16] due to the strong vibronic mixing. Allowed vibrational symmetries are $E_u \times E_u = A_{1g} + A_{2g} + B_{1g} + B_{2g}$, but

[12] W. L. Peticolas, L. Nafie, P. Stein, and B. Fanconi, *J. Chem. Phys.* **52,** 1576–1584 (1970).

[13] M. Gouterman, *J. Chem. Phys.* **30,** 1139 (1959); *J. Mol. Spectrosc.* **6,** 138 (1961).

[14] T. G. Spiro and T. C. Strekas, *J. Am. Chem. Soc.* **96,** 338 (1974).

[15] T. C. Strekas and T. G. Spiro, *J. Raman Spectrosc.* **1,** 197–206 (1973); L. A. Nafie, M. Pezolet, and W. L. Peticolas, *Chem. Phys. Lett.* **20,** 563–568 (1973).

[16] T. G. Spiro and T. C. Strekas, *Proc. Natl. Acad. Sci. U.S.A.* **69,** 2622–2626 (1972).

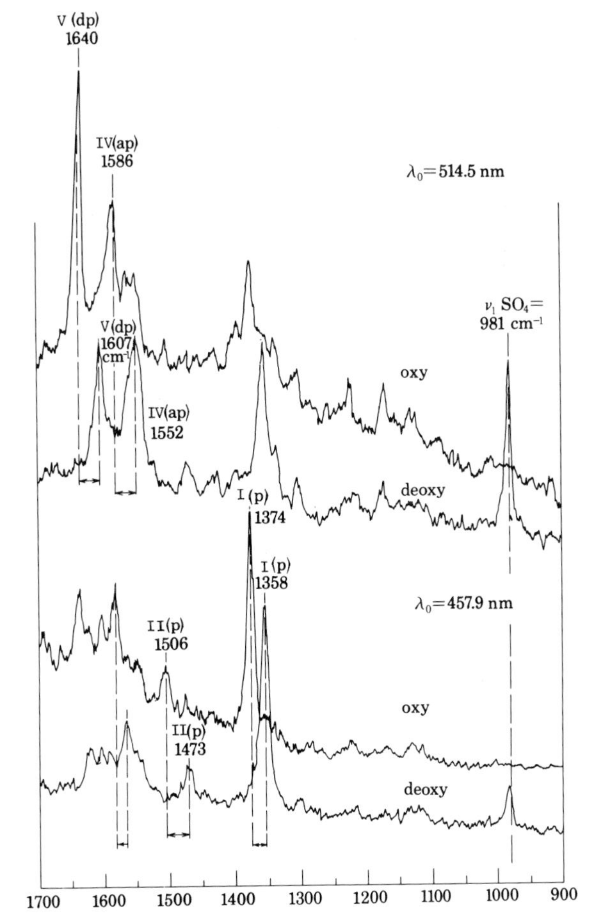

FIG. 2. Resonance Raman spectra of oxy- and deoxyhemoglobin in the α–β ($\lambda_0 = 514.5$ nm) and Soret ($\lambda_0 = 451.9$ nm) scattering regions. The solutions were 0.68 m*M* and 0.34 m*M* in heme for oxy- and deoxyhemoglobin, respectively, and the latter contained 0.4 *M* $(NH_4)_2SO_4$, the $\nu_1(SO^{2-}_4)$ band of which is indicated. Bands are labeled as in Fig. 4, and their frequency shifts are marked by the arrows. Adapted from Spiro and Strekas.[14]

A_{1g} (totally symmetric) modes are ineffective in mixing the two electronic transitions.[17]

III. Polarizations

The vibrational symmetries can be determined by the polarizations of the Raman bands. For the geometry shown in Fig. 1, the laser polarization

[17] M. H. Perrin, M. Gouterman, and C. L. Perrin, *J. Chem. Phys.* **50**, 4137 (1969).

is oriented perpendicular to the scattering direction. The scattered light can be analyzed into two components, with polarization parallel or perpendicular to the incident polarization. For a sample of randomly oriented molecules, the intensities of these components, $I_{\parallel}$ and $I_{\perp}$, can be expressed in terms of rotational invariants of the Raman tensor:[16]

$$I_{\parallel} = K(45\bar{\alpha}^2 + 4\gamma_S^2)$$
$$I_{\perp} = K(3\gamma_s^2 + 5\gamma_{as}^2)$$

where K is a constant of the experiment. $\bar{\alpha}^2$ is the square of the trace, or mean value, of the tensor

$$\bar{\alpha} = \tfrac{1}{3}(\alpha_{xx} + \alpha_{yy} + \alpha_{zz})$$

and γ_s^2 and γ_{as}^2 are the symmetric and antisymmetric anisotropies:

$$\gamma_s^2 = \tfrac{1}{2}[(\alpha_{xx} - \alpha_{yy})^2 + (\alpha_{xx} - \alpha_{zz})^2 + (\alpha_{yy} - \alpha_{zz})^2] + \tfrac{3}{4}[(\alpha_{xy} + \alpha_{yx})^2 + (\alpha_{xz} + \alpha_{zx})^2 + (\alpha_{yz} + \alpha_{zy})^2]$$

and

$$\gamma_{as}^2 = \tfrac{3}{4}[(\alpha_{xy} - \alpha_{yx})^2 + (\alpha_{xz} - \alpha_{zx})^2 + (\alpha_{yz} - \alpha_{zy})^2]$$

Normally the Raman tensor is symmetric, and $\gamma_{as}^2 = 0$. In this case the depolarization ratio, $\rho = I_{\perp}/I_{\parallel}$, is $\frac{3}{4}$ for nontotally symmetric modes, which always have a zero trace. Totally symmetric modes have a finite (usually large) value for $\bar{\alpha}$, and $\rho < \frac{3}{4}$. It is possible for ρ to exceed $\frac{3}{4}$, a condition called anomalous polarization,[16] if $\gamma_{as}^2 \neq 0$. Nonzero values of γ_{as}^2 are possible if vibronic mixing occurs via a mode that has an asymmetric Raman tensor, i.e., $\alpha_{ij} \neq \alpha_{ji}$. (Tensor patterns for various molecular symmetries have been tabulated by McClain.[18]) Such modes have the symmetry of a rotation about the principal molecular axis. They are relatively uncommon, and situations in which they mix nearby electronic states are expected to be quite rare. One does arise in metalloporphyrins, however: the A_{2g} modes have $\alpha_{xy} = -\alpha_{yx}$, and they are active in vibronic mixing. They are inactive in infrared and nonresonance Raman spectroscopy (off resonance, γ_{as}^2 goes to zero because of cancellation of contributions from adjacent vibronic levels[19,20]), and their appearance in the RR spectra provides vibrational data that are otherwise unavailable. It has recently been observed that spin-orbit coupling effects can also give rise to anomalous polarization in heavy-metal complexes.[21,22]

If γ_{as}^2 is nonzero, then two intensity measurements are insufficient to determine the three tensor invariants. (It is possible to have any value of ρ if $\bar{\alpha}^2$, γ_s^2, and γ_{as}^2 all contribute.) The required additional data can be provided

[18] W. M. McClain, *J. Chem. Phys.* **55**, 2789 (1971).
[19] D. W. Collins, D. B. Fitchen, and A. Lewis, *J. Chem. Phys.* **59**, 5714 (1973).
[20] L. D. Barron, *Mol. Phys.* **31**, 129 (1976).
[21] H. Hamaguchi, I. Harada, and T. Shimanouchi, *Chem. Phys. Lett.* **32**, 103 (1975).
[22] H. Hamaguchi and T. Shimanouchi, *Chem. Phys. Lett.* **38**, 370 (1976).

by circular polarization measurements.[18,23,24] If the incident light is circularly polarized, then the scattered light can be analyzed into co- and contrarotating components (a back-scattering geometry must be used to maintain the circular polarization).[23,24] Their intensities depend on a different combination of the tensor invariants:

$$\begin{aligned} I_{\text{co}} &= K(6\gamma_s^2) \\ I_{\text{contra}} &= K(45\bar{\alpha}^2 + \gamma_s^2 + 5\gamma_{as}^2) \end{aligned}$$

Any three of the intensity measurements, $I_{\parallel}$, $I_{\perp}$, I_{co}, and I_{contra} can provide the relative magnitudes of the tensor invariants, while the fourth provides a check. The combination of $I_{\perp}$ and I_{co} is particularly useful in evaluating antisymmetric scattering, since

$$\gamma_{as}^2 = \frac{I_{\perp} - \frac{1}{2}I_{\text{co}}}{K}$$

In the case of cytochrome *c* the detection of nonzero γ_{as}^2 contributions to some Raman bands, which could not be assigned to A_{2g} modes, provided a direct indication of reduction in heme symmetry.[24]

IV. Mode Assignments

A. *In-Plane Modes*

The resonant π–π^* transitions are polarized in the heme plane, and RR enhancement is expected for those vibrations which alter the in-plane polarizability. Since the π bonds are weakened in the excited states, the largest enhancement is expected for modes involving the stretching of π bonds (C—C and C—N; C—H deformation modes are of similar energy and also mix) in the porphyrin ring, which are expected in the region 1100–1700 cm^{-1}. This is indeed where the dominant heme RR bands are found. The observed spectra are complex, but overlapping bands can be resolved and cataloged via their differing polarization properties and excitation wavelength dependencies. For cytochrome *c*,[25] all the Raman bands observed in this region can be accounted for by the expected ring modes, without reference to the peripheral pyrrole substituents, all of which are saturated and therefore not part of the aromatic chromophore. The frequencies are calculated with reasonable accuracy with a simple approximate force field.[26] Only the B_{2g} modes fail to appear in the observed spectrum.[16,26]

[23] M. Pezolet, L. A. Nafie, and W. L. Peticolas, *J. Raman Spectrosc.* **1,** 455 (1973).
[24] J. R. Nestor and T. G. Spiro, *J. Raman Spectrosc.* **1,** 339 (1973).
[25] T. G. Spiro and T. C. Strekas, *J. Am. Chem. Soc.* **97,** 2309 (1975).
[26] P. Stein, M. J. Burke, and T. G. Spiro, *J. Am. Chem. Soc.* **97,** 2304 (1975).

Peripheral substituents are expected to contribute to the RR spectrum if they extend the ring conjugation. Proteins containing protoheme (protoporphyrin IX) show extra bands[25,27] that are attributable to the influence of the two vinyl substituents. Vinyl C=C stretching and C—H bending modes might be enhanced. In addition, the asymmetric disposition of the vinyl group destroys the symmetry center of the chromophore and may induce Raman activity into infrared-active (E_u) porphyrin ring modes. A preliminary report of heme a RR spectra in cytochrome oxidase[28] includes the suggested assignment of the formyl C=O stretching mode to a band at 1660 cm^{-1}, observed in the reduced form. The 9-keto C=O stretch has been observed in chlorophyll RR spectra.[29]

Enhancement is weaker for heme bands below 1100 cm^{-1}, presumably because involvement of the porphyrin π bonds is slight for modes in this region. Nevertheless a number of bands of weak to moderate intensity can be observed, particularly with excitation below 500 nm, where totally symmetric modes are enhanced in resonance with the Soret transition. The low-frequency region should include modes involving Fe–N stretching. Some assignments of these modes have been suggested.[30]

B. *Out-of-Plane Modes*

Out-of-plane porphyrin modes that are nontotally symmetric cannot be resonance enhanced since they cannot mix the (in-plane) $\pi-\pi^*$ transitions. Leaving aside for a moment the iron axial ligands, planar porphyrins have no out-of-plane modes that are totally symmetric. If the porphyrin is distorted from planarity (e.g., via doming, see below) then out-of-plane deformation modes can be totally symmetric. Their enhancement will depend on the extent to which they alter the in-plane polarizability (α_{xx} and α_{yy}), which is expected to be slight.

The most interesting out-of-plane modes are those that involve the axial ligands, since these ligands largely determine the biological activity of heme proteins. In this respect it is unfortunate that the heme electronic spectrum is so heavily dominated by in-plane transitions. Significant enhancement of axial ligand modes may nevertheless be produced by resonance with out-of-plane charge-transfer transitions, which may be buried in

[27] F. Adar and M. Erecinska, *Arch. Biochem. Biophys.* **165,** 570 (1974).

[28] I. Salmeen, L. Rimai, D. Gill, T. Yamamoto, G. Palmer, C. R. Hartzell, and H. Beinert, *Biochem. Biophys. Res. Commun.* **52,** 1100 (1973).

[29] M. Lutz and J. Breton, *Biochem. Biophys. Res. Commun.* **53,** 413 (1973); M. Lutz, *J. Raman Spectrosc.* **2,** 417 (1974); M. Lutz, J. Kleo, and F. Reiss-Husson, *Biochem. Biophys. Res. Commun.* **69,** 711 (1976).

[30] H. Brunner and H. Sussner, *Biochim. Biophys. Acta* **310,** 20 (1973).

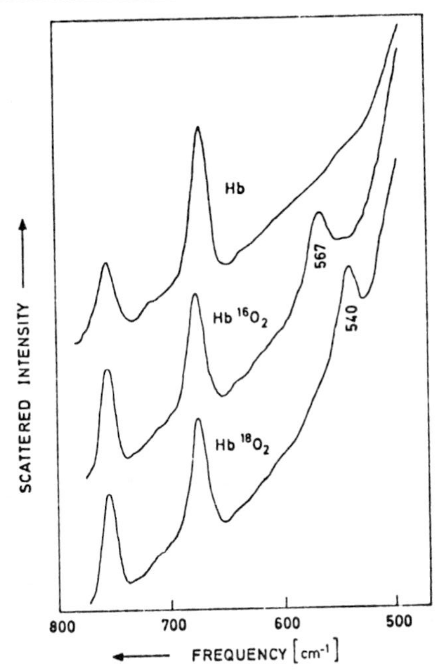

FIG. 3. Resonance Raman spectra of 0.03 m*M* hemoglobin (Hb), $Hb^{16}O_2$, and $Hb^{18}O_2$, respectively, at 15°C, pH 7.4, using 488.0 nm excitation. From Brunner.[32]

the π–π^* bands. A well-documented case is that of bis-pyridine Fe(II) heme,[31] for which modes of the bound pyridine are strongly enhanced by resonance with an Fe → pyridine charge-transfer transition, which can be observed as a shoulder on the Soret band. An important out-of-plane mode, detected by Brunner,[32] is the Fe–O stretch in oxyhemoglobin. As shown in Fig. 3, it shifts from 567 to 540 cm^{-1} on substitution of $^{18}O_2$ for $^{16}O_2$. Its enhancement may arise from resonance with Fe → O_2 charge-transfer transitions, which have been detected in oriented crystal absorption spectra.[33] The relatively high Fe–O frequency (compare 500 cm^{-1} for ν_{Fe-O} in oxyhemerythrin,[34] which contains bound peroxide) suggests substantial multiple bond character in the Fe–O bond, in line with recent X-ray structural data for a protein-free heme–O_2 complex.[35] Other axial Fe–*X* modes have been observed for the fluoride adduct of octaethylporphyrin

[31] T. G. Spiro and M. J. Burke, *J. Am. Chem. Soc.* **98,** 5482 (1976).
[32] H. Brunner, *Naturwissenschaften* **61,** 129 (1974).
[33] M. W. Makinen and W. A. Eaton, *Ann. N.Y. Acad. Sci.* **206,** 210 (1973).
[34] J. B. R. Dunn, D. F. Shriver, and I. M. Klotz, *Proc. Natl. Acad. Sci. U.S.A.* **70,** 2582 (1973).
[35] J. P. Collman, R. R. Gagne, C. A. Reed, W. T. Robinson, and G. A. Rodley, *Proc. Natl. Acad. Sci. U.S.A.* **71,** 1326 (1974).

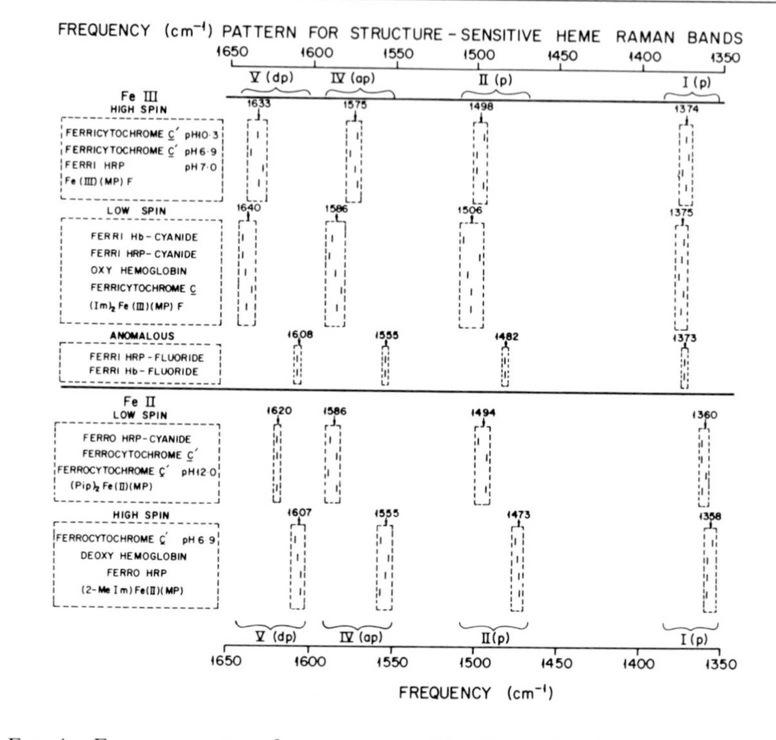

FIG. 4. Frequency pattern for structure-sensitive Raman bands of various heme proteins, and iron-mesoporphyrin IX analogs, p = polarized, dp = depolarized, and ap = anomalously polarized. From Spiro and Burke.[31]

Fe(III) ($\nu_{Fe-F} = 581\ cm^{-1}$)[36] and the μ-oxo dimer of tetraphenylporphyrin Fe(III) ($\nu_{Fe-O(sym)} = 363\ cm^{-1}$)[37] using the Fe^{54}–Fe^{56} isotope shift. Axial modes have also been reported[38] for etioporphyrin Mn(III)(X^-) (where X = Cl, Br, I).

V. Structural Interpretation

Even though the high-frequency porphyrin modes that dominate the RR spectra do not involve significant movement of the central iron atom, they are sensitive to the iron chemistry. In particular the frequencies of four of the bands correlate reliably with the iron oxidation or spin-state, or both, as diagrammed in Figure 4.[31]

[36] J. Kincaid and K. Nakamoto, *Spectrosc. Lett.* **9,** 19 (1976).

[37] J. M. Burke, J. R. Kincaid, and T. G. Spiro, *J. Amer. Chem. Soc.* (in press).

[38] S. Asher and K. Sauer, *J. Chem. Phys.* **64,** 4120 (1976).

A. Oxidation State

Oxidation from Fe(II) to Fe(III) increases the frequencies of the bands labeled I and V by 10–30 cm^{-1}. Band I stands out in particular, since it is the strongest band in the Soret-enhanced spectrum, and its frequency seems to be independent of spin-state. An extra 7 cm^{-1} increase in this band has been observed for Compound II of horseradish peroxidase,[39] which is believed to contain Fe(IV). Band I was first proposed as an oxidation-state marker by Yamamoto *et al.*,[40] who also concluded that oxyhemoglobin should be formulated as a superoxide (O_2^-) complex of Fe^{3+}, in line with Weiss's suggestion,[40a] since it gives a band I frequency, 1377 cm^{-1}, typical of Fe(III) hemes. Similar frequency increases are observed for bands I and V when π-acid ligands, such as phosphines and CO, are bound to low-spin Fe(II) hemes, and the extent of the shifts varies with the π-acidity.[31] A reasonable interpretation is that these frequencies are sensitive to back donation from the filled Fe(II) π orbitals to the empty porphyrin π^* orbitals, which is relieved by the competition for these same Fe(II) electrons by the axial π acceptor ligands, or by actual removal of an electron to produce Fe(III) heme. O_2 shows the largest shifts, and appears to be the best π acceptor; this inference is supported by the high Fe–O frequency, referred to earlier, which is evidence of Fe–O multiple bonding. Whether on this account the complex should be formulated $Fe^{3+}–O_2^-$ remains problematical; the actual electron distribution depends on both the σ and π bonding interactions.

B. Spin State

Conversion of Fe(II) hemes from low- to high-spin state is accompanied by decreases in the frequencies of bands II, IV, and V by 10–40 cm^{-1}. For Fe(III) hemes the same bands shift between low- and high-spin forms, but the frequency decrements are appreciably smaller than for Fe(II). The exceptions are the high-spin forms of methemoglobin and myoglobin, whose frequency shifts from the low-spin values are as large as for Fe(II).[31]

These shifts are believed to reflect the stereochemical changes that are known to accompany heme spin-state changes.[41] Low-spin hemes are planar, but in high-spin hemes the iron atom lies out of the heme plane toward one of the axial ligands, while the other axial ligand is bound weakly, or is absent. The pyrrole rings tilt slightly as the nitrogen atoms

[39] G. Rakshit, T. G. Spiro, and M. Uyeda, *Biochem. Biophys. Res. Commun.* **71,** 803 (1976).

[40] T. Yamamoto, G. Palmer, D. Gill, I. T. Salmeen, and L. Rimai, *J. Biol. Chem.* **248,** 5211 (1973).

[40a] J. Weiss, *Nature (London)* **203,** 83 (1964).

[41] J. L. Hoard, *in* "Porphyrins and Metalloporphyrins" (K. M. Smith, ed.), p. 317. Am. Elsevier, New York, 1975.

follow the iron atom out of the plane, producing an overall doming of the porphyrin ring. This doming is expected to disrupt the π-conjugation, and might thereby account for the observed frequency shifts. Preliminary normal coordinate calculations on domed heme show that slight disruption of π-conjugation can indeed produce the observed shifts.[26] The smaller shifts for Fe(III) than for Fe(II) hemes suggest less doming for the high-spin forms of the former,[31] for which there is some X-ray crystallographic support.[41]

Raman frequencies identical to those of deoxyhemoglobin or myoglobin are given by protein-free high-spin Fe(II) heme, with 2-methylimidazole bound as the fifth ligand.[31] If it is accepted that these frequencies are sensitive to porphyrin stereochemistry, then this observation implies that the globin pocket accepts high-spin Fe(II) heme without distortion, and makes it unlikely that molecular tension in deoxyhemoglobin (*T* state) is accommodated by extra doming of the porphyrin ring.[42] Presumably the tension is expressed elsewhere in the molecule, possibly in many small distortions of the polypeptide chain.[43] Consistent with this conclusion, no Raman spectral changes are observed when deoxyhemoglobin is switched from the *T* to the *R* quaternary structure by chemical modification[44] or by pH change in the case of protein from carp.[45] On the other hand when nitrosyl hemoglobin is switched from *R* to *T* by the addition of inositol hexaphosphate, band V, at 1633 cm^{-1}, splits into two components, one at 1633, the other at 1643 cm^{-1}.[46] This has been interpreted[47] as reflecting the formation of 5-coordinate NO-heme via breaking of the iron–imidazole bonds in the α chains.

The extra frequency shifts observed for methemoglobin and myoglobin do demonstrate a protein influence on porphyrin conformation, and the above considerations suggest that doming is enhanced relative to that found normally for high-spin Fe(III).[31] This distortion is plausible if it is assumed that the globin pocket stabilizes the high-spin Fe(II) porphyrin conformation and resists relaxation of the doming upon oxidation to Fe(III). This would destabilize the Fe(III) form and provide a structural basis for resistance to oxidation, which is an essential aspect of hemoglobin and myoglobin in their roles as reversible oxygen carriers.

Spaulding *et al.*[48a] have advanced a different explanation for the spin-

[42] J. L. Hoard and W. R. Scheidt, *Proc. Natl. Acad. Sci. U.S.A.* **70,** 3919 (1973); **71,** 1578 (1974).

[43] J. J. Hopfield, *J. Mol. Biol.* **77,** 207 (1973).

[44] H. Sussner, A. Mayer, H. Brunner, and H. Fasold, *Eur. J. Biochem.* **41,** 465 (1974).

[45] D. M. Scholler, B. M. Hoffman, and D. F. Shriver, *J. Amer. Chem. Soc.* **26,** 7866 (1976).

[46] L. D. Barron and A. Szabo, *J. Am. Chem. Soc.* **97,** 660 (1975).

[47] M. F. Perutz, J. V. Kilmartin, K. Nagai, A Szabo, and S. R. Simon, *Biochemistry* **15,** 378 (1976).

[48a] L. D. Spaulding, C. C. Chang, N. T. Yu, and R. H. Felton, *J. Am. Chem. Soc.* **97,** 2517 (1975).

state frequency shifts, noting that the band IV frequency is also lowered in planar porphyrin complexes of heavy metals, such as Sn(IV) or Ag(II), for which X-ray crystal structures show expansion of the porphyrin core, presumably forced by the large size of the ions. They suggest that the frequency decrements for high-spin hemes likewise reflect expansion of the porphyrin core. The anomalously low band IV frequency of methemoglobin is interpreted by Spaulding *et al.* as reflecting a position for the iron atom closer to the heme plane than normal, given the large implied porphyrin expansion. They suggest that this expansion may be forced by the distal ligand (H_2O) pulling the iron atom toward the plane. Recent X-ray crystallographic results do suggest that the iron atoms are closer to the heme plane in aquo-methemoglobin than in five-coordinate Fe(III) porphyrins.[48b]

It is likely that both doming and porphyrin expansion would produce similar frequency shifts, since both weaken the π bonds (expansion would do this by stretching them).[48c] Which distortion actually occurs in methemoglobin and metmyoglobin remains an open question. With respect to the iron atom moving into the heme plane, this is known to produce spin-pairing instead of porphyrin expansion for Fe(II), as evidenced by the crystal structure of Fe(II)TPP.[41] For Fe(III) however, expansion appears to be a viable distortion, since a recent crystal structure of high-spin bis-aquo $[Fe(III)TPP]^+$ shows the iron atom to be in the plane of an expanded porphinato core.[48d] However no close analog to the methemoglobin complex, i.e., a Fe(III) porphyrin with imidazole and water axial ligands, has yet been isolated or characterized.

VI. Prospects

For well-characterized heme proteins, it is possible to use the RR spectra to evaluate protein influences on heme structure, if any. Short of this, the Raman signatures can be used to identify oxidation with spin state. It is feasible to observe both high- and low-spin forms in a spin-state mixture.[49] Also different hemes can be distinguished by selective excitation. Thus Adar and Erecinska[27] were able to determine separately the RR spectra of the *b* and *c* cytochromes in succinate-cytochrome *c* reductase by taking advantage of the ~10-nm difference in the α and β band maxima, and tuning to one or the other. Similarly Lutz and Breton[29] were able to excite chlorophyll a and b selectively in chloroplasts. And Salmeen *et al.*[28] produced cytochrome *a* RR spectra in electron-transport particles, without

[48b] R. C. Ladner, E. J. Heidner and M. F. Perutz, *J. Mol. Biol.* **114,** 385 (1977).
[48c] A. Warshel, *Ann. Rev. Bioeng. and Biophys.* **6,** 273 (1977).
[48d] M. E. Kastner, W. R. Scheidt, T. Mashiko and C. A. Reed, *J. Am. Chem. Soc.* **100,** 666 (1978).
[49] T. C. Strekas and T. G. Spiro, *Biochim. Biophys. Acta* **351,** 237 (1974).

interference from other cytochromes, by exciting directly in the heme *a* Soret band.

The recent observation of some axial ligand modes gives hope that the axial linkages in heme proteins can be studied by RR spectroscopy. What is needed is a systematic search for the laser wavelengths best suited to bring out such modes, since the absorption spectrum, dominated as it is by porphyrin π–π^* transitions, provides little guidance. Since a number of charge-transfer transitions are likely to be found in the ultraviolet region, it is likely that ultraviolet excitation will prove fruitful in this regard.

VII. Experimental Considerations

A. *Laser Sources*

RR spectra are obtainable with standard laser Raman spectrometers, consisting of a laser source, a sample compartment with fore-optics designed to focus the scattered light on the spectrometer entrance slit, a scanning double monachromator to disperse the light with high stray-light rejection, and a high-quality photomultiplier with either DC detection of the amplified signal or photon-counting circuitry.[50] The most common light source is the continuous-wave (cw) argon ion laser, which gives the highest average power levels of commercially available lasers, at reasonable cost, and is stable and reliable. Its strongest lines are at 488.0 and 514.5 nm, but weaker lines are available from 457.9–528.6 nm, which are useful for RR studies. A pair of near-UV lines, at 351.1 and 363.8 nm, can also be extracted with UV mirrors. Many laboratories use a krypton ion laser, often in addition to the Ar^+ laser, which has strong lines in the yellow (568.2 nm) and red (647.1 nm) regions. The Ar^+ laser can also be used to pump dye lasers which are tunable in the yellow and red regions. The most efficient dye is rhodamine 6G, which has a practical tuning range of 560–630 nm, but other dyes can extend this range, although at lower power levels, from about 530–700 nm. With sufficient power in the near-UV lines it is also possible to pump dyes which are tunable in the blue and violet region.

Greater continuous-tuning ranges are available with pulsed laser systems. Flash-lamp dye lasers are tunable from 435–730 nm, while the pulsed N_2 laser, at 337.1 nm, can be used to pump dyes from 350–750 nm. In both cases the high peak power of the pulses makes possible efficient frequency doubling of the dye laser output, giving tunable ultraviolet pulses. The maximum average power of pulsed lasers is well below that of available cw

[50] M. C. Tobin, "Laser Raman Spectroscopy." Wiley (Interscience), New York, 1970; T. R. Gilson and P. J. Hendra, "Laser Raman Spectroscopy." Wiley (Interscience), New York, 1970.

lasers, and the bunching of the photons requires the use of integrating detectors rather than photon counting. The greater tunability, especially into the ultraviolet, make these lasers attractive for RR studies, however.

B. Problems Associated with Absorbing Samples

Although RR spectra can be obtained with standard Raman instrumentation, special precautions are required to deal with the side effects of illuminating an absorbing sample. The absorption itself is the first problem, since it attenuates both incident and scattered photons. It is therefore essential to minimize both incident and scattered light paths through the sample. For the 90° scattering geometry shown in Fig. 1 the laser beam must be positioned as close as possible to the front surface of the sample. Alternatively the beam can be scattered directly off a flat front surface (back-scattering) either at 180°, using a small mirror to direct the laser and block the reflected beam but not the scattered light, or at a small angle that directs the reflected beam away from the spectrometer. It may also be important to adjust the concentration of the sample since scattering increases linearly with concentration, but absorption increases exponentially. For 90° scattering there is an optimum concentration, corresponding to an absorbance (optical density) of roughly 13 per centimeter of path length, beyond which the signal decreases.[51] For samples with molar extinction coefficients of 10^3–10^4, this places the optimum concentration in the millimolar range. The situation is somewhat different for back-scattering, because the light penetration depth also varies with concentration. The result is that the signal strength approaches an asymptotic limit with increasing concentration.[51a] Highly absorbing samples should therefore be run in back-scattering.

Absorption also produces heat. This can be carried away from the absorption path by solvent molecules (especially water) with reasonable efficiency, but only the most robust absorbing molecules can withstand prolonged laser irradiation in a stationary sample. (For solutions, this applies primarily to microsamples in capillary tubes. Larger liquid samples apparently undergo convection in the laser beam,[52] and are not really stationary.) Local heating can be greatly alleviated by spinning[53] or circulating[54] the sample through the laser beam; in the latter case the sample can also be circulated through a bath for better temperature control.

[51] T. C. Strekas, D. H. Adams, A. Packer, and T. G. Spiro, *Appl. Spectrosc.* **28,** 324 (1974).
[51a] D. F. Shriver and J. B. R. Dunn, *Appl. Spectrosc.* **28,** 319 (1974).
[52] L. Rimai, I. T. Salmeen, and D. H. Petering, *Biochemistry* **14,** 378 (1975).
[53] W. Kiefer and H. J. Bernstein, *Appl. Spectrosc.* **25,** 500 (1971).
[54] W. H. Woodruff and T. G. Spiro, *Appl. Spectrosc.* **28,** 74 (1974).

Other problems are associated with the subsequent fate of absorbed photons, since the absorption probability is far greater than the scattering probability. If the ground state is efficiently regenerated through radiationless processes (internal conversion and collisional deactivation), then the Raman signal is unaffected, unless the photon flux is large enough to depopulate the ground state significantly. If absorption leads to photochemistry, then the Raman spectrum will reflect the presence of a usually ill-defined mixture of photoproducts. (The definition can be improved by the use of a second laser, at a different wavelength, to establish a photostationary state,[55] or, in principle, by the use of timed laser pulses, if the photokinetic parameters are known.) A general method to circumvent photoreactivity is to flow the sample with sufficient velocity to minimize the fraction of sample molecules that are photolyzed in the beam at any instant.[56,57]

Fluorescence is the most aggravating interference, since even low levels of photoemission can obscure the inherently weaker Raman scattering. Frequently luminescence arises from impurities, which can be removed by purification procedures or occasionally can be burned out by prolonged laser irradiation. (This procedure is more applicable to nominally transparent samples with fluorescent impurities, than to absorbing samples, which are unlikely to withstand prolonged irradiation.) In RR studies intrinsic fluorescence from the irradiated absorption band is to be expected, with its intensity depending on the efficiency of radiationless deactivation processes. Addition of external quenching agents (e.g., sodium iodide) can lower the fluorescence background appreciably, although care must be taken to avoid chemical interactions with the quencher. Since the fluorescence spectrum is usually shifted to the red from the absorption band, it frequently helps to tune the laser to the blue side of the band. Even though resonance enhancement may be reduced, the Raman signal-to-fluorescence background is usually improved.

It should be possible to separate fluorescence from scattering by time resolution, since the latter process is much faster. Very short laser pulses and very fast detection may be required, however, since fluorescent lifetimes are frequently in the subnanosecond range. The needed instrumentation is still under development.[58,59]

A promising new technique for obtaining fluorescence-free Raman

[55] A. R. Oseroff and R. H. Callendar, *Biochemistry* **13,** 4243 (1974).

[56] R. Mathies, A. L. Oseroff, and L. Stryer, *Proc. Natl. Acad. Sci. U.S.A.* **73,** 1 (1976).

[57] R. H. Callendar, A. Doukas, R. Crouch, and K. Nakanishi, *Biochemistry* **15,** 1621 (1976).

[58] R. B. Van Duyne, D. L. Jeanmarie, and D. F. Shriver, *Anal. Chem.* **46,** 213 (1974).

[59] J. M. Harris, R. W. Chrisman, F. E. Lytle, and R. S. Tobias, *Anal. Chem.* **48,** 1937 (1976).

spectra is coherent anti-Stokes Raman scattering (CARS) in which the Raman signal is generated as a coherent beam of light easily separated from the nondirectional fluorescence.[60] Resonance enhancement can still be exploited, and resonance CARS spectra of dilute aqueous solutions of cytochrome *c* and vitamin B_{12} have recently been obtained.[61]

[60] R. F. Begley, A. B. Harvey, R. L. Byer, and B. S. Hudson, *J. Chem. Phys.* **61**, 2466 (1974).
[61] J. R. Nestor and T. G. Spiro, *Proc. Natl. Acad. Sci. U.S.A.* (in press).

[16] Circular Dichroism Spectroscopy of Hemoproteins

By YASH P. MYER

I. Introduction

A comprehensive understanding of the functioning of proteins, and biological macromolecules in general, rests in the concurrent elucidation of structure, conformation, and function and of the interrelationships between various permutations of the three fundamental properties. X-ray diffraction provides information on interrelationships of protein structure and conformation, while functional investigations in conjunction with structural perturbation clarify interrelationships of structure and function. During the past several decades, various optical, magnetic, and electrical methods have been developed which permit the conformational probing of macromolecules in solution. Optical activity spectroscopy—optical rotatory dispersion (ORD) and circular dichroism (CD)—is a well-accepted technique for probing the conformation of macromolecules in solution, since it is inextricably bound to the interaction between groups, rather than the groups themselves, as is the situation in ordinary absorptive spectroscopy. In the case of proteins, the exceedingly high conformational sensitivity of the optical activity spectrum stems from the fact that although the protein groups (e.g., the amide bonds, the absorbing amino acid side chains, etc.) are inherently optically inactive, the conformational organization of the molecule itself confers activity upon these groups. The presence of the heme group in hemoproteins provides an additional locale for the conformational probing of these proteins, and indeed, this fact has been exploited during the last 10–12 years. The subject of ORD and/or CD spectral studies of proteins and polypeptides, including details of instrumentation, measurement, procedures, etc., has been extensively cov-

ISBN 0-12-181954-X

ered through a variety of review articles and books.[1-9] The basic principles and theory of optical activity may be found in Refs. 10–16 with emphasis on the optical activity of polypeptides. Some recent reviews also include the conjugated proteins in general,[17-20] but the information regarding hemoproteins, except for recently published reviews,[21,22] is scattered throughout the literature. This article is an attempt to provide the necessary background for interpreting the CD spectra of hemoproteins, with emphasis on the CD spectrum of heme, some aspects of methodology, and several specific applications. No claim is made to comprehensive coverage of the CD studies of hemoproteins, or of the theoretical bases for the interpretation of CD spectra. The reader is thus advised to pursue these aspects in references cited in appropriate sections of the article.

[1] Beychok, S., *Science* **154,** 1288–1299 (1966).

[2] Beychok, S., *Annu. Rev. Biochem.* **37,** 437–462 (1968).

[3] Beychok, S., *in* "Poly-α-Amino Acids" (G. D. Fasman, ed.), pp. 293–337. Dekker, New York, 1967.

[4] Timasheff, S. N., Susi, H., Townend, R., Stevens, L., Gorbunoff, M. J., and Kumosinski, T. G., *in* "Conformation of Biopolymers" (G. N. Ramachandran, ed.), Vol. 1, pp. 173–196. Academic Press, New York, 1967.

[5] Adler, A. J., Greenfield, N. J., and Fasman, G. D., Vol. 27, pp. 675–735.

[6] Tinoco, I., and Cantor, C. R., *Methods Biochem. Anal.* **18,** 81 (1970).

[7] Fasman, G. D., *PAABS Rev.* **2,** 587–650 (1973).

[8] Velluz, L., and Legrand, M., "Optical Circular Dichroism; Principles, Measurements and Applications." Verlag-Chemie, Weinheim, 1965.

[9] Crabb, P., "Optical Rotatory Dispersion and Circular Dichroism in Organic Chemistry." Holden-Day, San Francisco, California, 1965.

[10] Moscowitz, A., *in* "Optical Rotatory Dispersion" (C. Djerassi, ed.), pp. 150–177. McGraw-Hill, New York, 1960.

[11] Moscowitz, A., *in* "Optical Rotatory Dispersion and Circular Dichroism in Organic Chemistry" (G. Snatzke, ed.), pp. 41–70. Sadler Res. Labs., Philadelphia, Pennsylvania, 1967.

[12] Schellman, J. A., *Acc. Chem. Res.* **1,** 144–151 (1969).

[13] Caldwell, D. J., and Eyring, H., "The Theory of Optical Activity." Wiley, New York, 1971.

[14] Bayley, P. M., *Prog. Biophys. Mol. Biol.* **27,** 1–76 (1973).

[15] Condon, E. V., Alter, W., and Eyring, H., *J. Chem. Phys.* **5,** 753 (1937).

[16] Kirkwood, J. W., *J. Chem. Phys.* **5,** 479 (1937).

[16a] Kuhn, W., *Trans. Faraday Soc.* **26,** 293 (1930).

[17] Ulmer, D. D., and Vallee, B. L., *Adv. Enzymol.* **27,** 37 (1967).

[18] Gratzer, W. B., and Cowburn, D. A., *Nature* (*London*) **222,** 426 (1969).

[19] Urry, D. W., ed., "Spectroscopic Approaches to Biomolecular Conformation." Am. Med. Assoc., Chicago, Illinois, 1970.

[20] Vallee, B. L., and Wacker, W. E. C., *Proteins, 2nd Ed.* **5,** 94 (1970).

[21] Blauer, G., *Struct. Bonding* (*Berlin*) **18,** 69–129 (1974).

[22] Myer, Y. P., and Pande, A. J., *in* "The Porphyrins" (D. Dolphin, ed.). Academic Press, New York, 1978 (in press).

II. Instrumentation and Measurement of CD Spectra

A. *Instrumentation*

The first commercial instrument capable of measuring CD spectra below 220 nm became available about 1965, and since then significant advances in the technology have occurred. Present-day commercial instruments cover the region from 185 to approximately 1000 nm and have a limiting sensitivity of about 0.2 mdeg and a precision of better than 10–15% at highest sensitivity. Models made in or later than 1975—the JASCO-J-40, Cary 61, Roussel-Jouan (the three primary instruments)—are all either computerized or have incorporated capabilities for easy coupling. This feature allows on-line normalization of data, correction of base-line contributions, the determination, electronically, of difference spectra (attachments are available for direct measurements as well), and signal averaging, a definite asset when measuring samples with poor absorptivity-to-activity characteristics.

A generalized instrumental set-up is shown in Fig. 1. Detailed considerations are to be found in Velluz and Legrand,[8] Crabb,[9] and Woldbye.[23] The

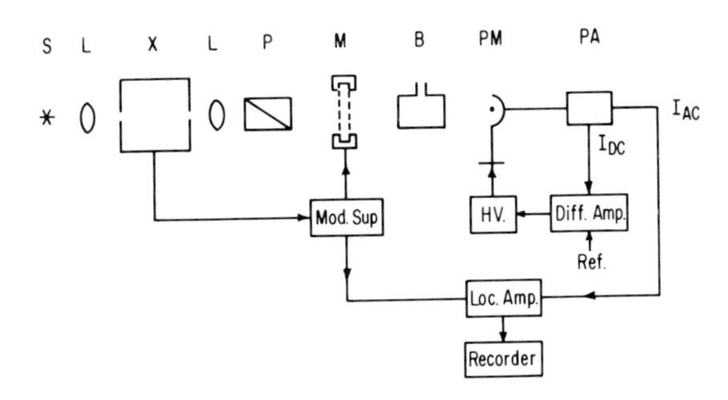

FIG. 1. Block diagram of a circular dichroism spectrophotometer.

set-up comprises a light source (S), a monochromator (X), a polarizer (P), a modulator (M), a sample holder (B), and a photomultiplier (PM). The plane-polarized light is generated by the polarizer and is transformed by the modulator into alternating right- and left-handed polarized light, which, after passing through the sample, is monitored through a photomultiplier amplification system. The CD signal is then extracted from the PM output as the AC component of the outcoming signal, which is normalized either through division by the amplitude of the signal, i.e., the DC component, or by maintenance of a constant DC component of the output. The informa-

[23] Woldbye, F., *in* "Optical Rotatory Dispersion and Circular Dichroism in Organic Chemistry" (G. Snatzke, ed.), pp. 85–99. Sadler Res. Labs, Philadelphia, Pennsylvania, 1967.

tion, in the form of the difference in absorptivity between left- and right-handed circularly polarized light, is recorded either directly as ΔOD or changed to the appropriate dimensions in degrees. A substance frequently used for the calibration of the instrument is d-camphor-10-sulfonic acid in aqueous solution, using a molar ellipticity of +4260 deg at maximum wavelength, λ_{max}, of 290.5 nm. Since no standardized values of ellipticity are available for any suitable materials, detailed calibration of the instrument is essential, as has been pointed out recently by Cassim and Yang.[24]

B. Measurements

The procedures and precautions necessary for obtaining reliable CD spectra of proteins in solution have been covered in detail by Adler *et al.*;[5] these considerations are applicable to CD measurements of hemoproteins as well. The proper selection of solvent for a measurement in a given region of the spectrum, a concentration of protein providing an adequate signal for accurate measurement, and the proper selection of cell path length to yield optimal optical density and clarity of the solution are some considerations relevant to CD spectroscopic measurements. The proper manipulation of the scanning speed of the instrument, the time constant for response, and the required signal-to-noise ratio are interrelated features which must be understood through use of the instrument. As a rule of thumb, the time constant of the instrument and the scanning speed should be set in such a way that the interruption of the scanning of the instrument results in a change of less than 5% in the signal.

C. Normalization of Raw Data

The CD spectrum of a hemoprotein, as of other proteins, in the region below 250 nm is expressed in terms of the contribution from the amide bonds of the molecule as residue ellipticity, and has the dimensions of deg-cm²/decimole of amide bonds. The observed quantity, θ_{obs}, is transferred to residue ellipticity $[\theta]_\lambda$ using the relation:

$$[\theta]_\lambda = \frac{\theta_{obs}}{l \cdot C'} \tag{1}$$

where θ_{obs} is in degrees, the cell path length l is in centimeters, and the concentration C' is in decimoles of residues/milliliter [$C' = C/(MRW \times 100)$ where C is concentration in gm/liter and MRW is the mean residue molecular weight]. Above 250 nm, the ellipticity is expressed as molar ellipticity, which is calculated by Eq. 1, except the concentration C' is taken as decimoles of protein per milliliter.

[24] Cassim, J. Y., and Yang, J. T., *Biochemistry* **8,** 1947–1951 (1969).

III. Characterizing Parameters for CD Spectral Data

Like any other absorptive phenomenon, the circular dichroism band of a given transition, the kth transition, is defined by the position of the maximum ellipticity, λ_k^0, the magnitude of the corresponding ellipticity, $[\theta]_k^0$, and the band half-width, Δ_k, i.e., the band width at the eth height. Corresponding to the dipole strength, D_k, of an electronic transition, the integrated band area of the dichroic band yields the rotatory strength, R_k, and is obtained by the expression:

$$R_k = \frac{3hC}{8\pi^3 N} \int_0^\infty \frac{[\theta]_k^\lambda}{\lambda} \, d\lambda \tag{2}$$

where $[\theta]_k$ is the ellipticity of the kth transition as a function of wavelength λ and N is the number of optically active molecules per milliliter of solution. Using the approximation that CD bands are nearly Gaussian, an approximation true to the first order, the above expression reduces to:

$$R_k \simeq 1.23 \times 10^{-42} \frac{[\theta]_k^0 \cdot \Delta k}{\lambda_k^0} \tag{3}$$

where the three quantities, $[\theta]_k^0$, Δk, and λ_k^0 are as defined above. The normal dimensions of rotatory strength are cgs units, but the frequently used unit for R_k is the Debye magneton (DM) unit (DM $= 0.927 \times 10^{-38}$ cgs units).

Another quantity which has been found useful for identification and thus for assignment of optically active transitions of a molecule is the anisotropy factor, g_k, which for the kth transition is defined as $g_k = 4R_k/D_k$.

IV. Classification of Protein Functional Groups

A chromophore is defined primarily as a structurally and spectroscopically identifiable moiety of the system; e.g., the helically arranged polypeptide backbone, hexahelicine, the aromatic side chains, the heme group, etc. are all categorized as chromophores, although they differ in molecular complexity. A spectroscopically and structurally identifiable group under given conditions may in and of itself be categorized as a chromophore, (e.g., the phenyl group of phenylalanine in a protein), or it may conform to the role of an intrinsic constituent of a gross chromophore (e.g., phenyl groups of hexahelicine), provided the combination renders the entire molecular moiety an intrinsically dissymmetric molecule.

In considering the interrelationships of optical activity and molecular conformation, further classification of chromophores into the following two broad groups has been found useful:[10] (1) inherently symmetric chromophores but asymmetrically perturbed by the environment, e.g., the carbonyl chromophore in steroids, the amino acid side chains in proteins, the heme group in hemoproteins, etc.; and (2) chromophores which are inherently dissymmetric, e.g., hexahelicine. For ease of consideration, protein chromophores are further differentiated as intrinsic and extrinsic. The amide bonds constituting the polypeptide backbone and exhibiting absorption and CD bands in the region below 250 nm are referred to as intrinsic chromophores, and consequently, the spectral region below 250 nm is referred to as the intrinsic spectral region. All other groups, the amino acid side chains, the prosthetic groups, etc. are referred to as extrinsic chromophores. The spectral region 250–300 nm is dominated by contributions from aromatic side chains, and it is justifiably labeled the aromatic absorption region, although this region also contains contributions from the heme group.[22] In hemoproteins the region 300–380 nm is referred to as the Δ- or the *N*-band region (porphyrin $\pi-\pi^*$ transitions; $\epsilon = 10^3-10^4$); 390–430 nm, the Soret region (doubly degenerate porphyrin $\pi-\pi^*$ transitions; $\epsilon = 10^4-10^5$); 510–580 nm, the α-, β-band region (metal-ligand charge-transfer region); 590–630 nm, the *D*-band region; and above 630 nm, the near-IR region (porphyrin-metal charge-transfer bands and metal *d–d* transition bands.[22,25] The characterization of the various heme transitions and their assignments have been considered in some detail in a recent review of spectra of heme systems.[25]

V. Analysis of CD Spectra of Hemoproteins

The analysis of CD spectra of proteins in terms of the conformation of the chromophores is accomplished through two broad approaches, the theoretical approach and the phenomenological approach. Both the theoretical and phenomenological approaches as applied to proteins have been the subject of a number of treatments.[1,7,13–16a] Only an outline will be attempted here.

A. *Theories of Optical Activity*

Theoretical developments during the last two decades have led to three main approaches dealing with the existence of optical activity in symmetric chromophores. These theories are (1) the one-electron theory or the Condon, Alter, and Eyring theory,[15] (2) the dipole–dipole coupling mechanism (Kuhn-Kirkwood mechanism),[16,16a] and (3) the $\mu-m$ mechanism.[12]

(1) The one-electron theory considers the localization of the interacting electric and the magnetic transition moments in one and the same

[25] Williams, R. J. P., *Struct. Bonding (Berlin)* **7**, 1–45 (1970).

chromophore. It adequately predicts the optical activity of the symmetric amide chromophores in helical and other organized structures. One apparent consequence of mixing is the development of rotatory strength in both interacting states, of identical magnitude, but opposite in sign. The resulting expression for the rotatory strength from these considerations turns out to be adequately represented as simply the dot product of the electric and the magnetic transition moments of the chromophore, i.e., $R_k = \overrightarrow{\mu_e} \cdot \overrightarrow{\mu_m} = |\mu_e| \times |\mu_m| \cos \theta$. This relationship provides some insight into the symmetry requirements for the existence of optical activity. The vanishing of activity occurs when either of the two transition moments is zero ($|\mu_e| = 0$ or $|\mu_m| = 0$) or when the directions of the electric and the magnetic transition moments are perpendicular to one another ($\theta = 90°$).

(2) The dipole–dipole coupling mechanism requires coupling of electric and magnetic transition moments of two identical or nonidentical chromophores, and this applies in general to symmetric chromophores which are asymmetrically perturbed by environment. The basic premise of this approach is that if two chromophores, identical or nonidentical, are sufficiently close so as to result in coupling of the electronic transition moments of the two chromophores, then the optical activity generated in both the coupled states is of equal magnitude but opposite in sign. The opposition in sign is usually regarded as the reciprocal relationship; the equality of magnitude, the conservation of rotatory strength. This is consistent with the well-documented sum rule,[16a] which states that the sum overall of the rotatory strengths in a molecular system must be equal to zero.

Differentiation between the two commonly occurring situations, coupling among identical chromophores and nonidentical chromophores, can be made, since the resulting effects in the two situations are clearly distinct. The apparent effect of transition dipole–dipole coupling of two nonidentical chromophores is simply the generation of optical activity in both the coupled states with conservation of rotatory strength, without any significant perturbation of transition energies. Only if the interaction is strong enough to produce significant potential-energy perturbation will the energy of the observed states become different from that of the isolated chromophores.

The rotatory strength of isolated nonidentical chromophores through dipole–dipole coupling is given by the expression:

$$R = \pm\frac{2}{hC} \cdot \frac{V_{12}\,\epsilon_1\epsilon_2}{(\epsilon_2^2 - \epsilon_1^2)}\,[R_{12} \cdot \vec{\mu}_1 \times \vec{\mu}_2] \qquad (4)$$

where V_{12} is the perturbation energy; ϵ_1 and ϵ_2, the energy of the coupled electronic states; and $\vec{R}_{12}$, the vector distance between the transition dipole moments, $\vec{\mu}_1$ and $\vec{\mu}_2$, respectively. Since the rotatory strength is inversely proportional to the difference of the squares of the energy of the coupling transitions, i.e., $R_k \propto 1/(\epsilon_2^2 - \epsilon_1^2)$, the resulting activity of coupled transitions which are widely separated on the energy scale will be relatively small.

The rotatory strength of identical chromophores through the dipole–dipole coupling mechanism can be expressed by:

$$R = \pm\frac{\pi}{2\lambda}[\vec{R}_{12} \cdot \vec{\mu}_1 \times \vec{\mu}_2] \tag{5}$$

where all the quantities are the same as for Eq. 4, and λ refers to the transition of the noninteracting chromophores. The result expected from such a case is almost the same as for nonidentical chromophores, but the interaction energy of the excited states may produce significant splitting of the transitions, which will be seen as a couplet. This consequence of transition dipole interaction is the so-called "exciton splitting." Regarding the assignment of sign to the two bands of the couplet, a positive Cotton effect is assigned to the high-energy band for a case of positive coupling energy and to the low-energy band for a negative coupling energy system. It has also been shown that if the arrangement of the transition moments has the sense of a screw—i.e., right- or left-handedness—the alteration of the relative orientation of the transition moments could result in a positive-to-negative interaction energy.[12] The effect of the relative orientation of two identical chromophores or the nature of the stacking is thus intimately bound to the magnitude of splitting between the two resulting transitions as well as to the magnitude of the rotatory strength and the sign of the Cotton effect. A general, but typical, case involving dipole–dipole coupling of identical chromophoric transitions is the α-helical array of the protein backbone.[11-14] This model has been successfully used to elucidate conformation of nucleotides and conformational aspects of dipeptides, and recently it has been used to answer the question of the presence or absence of heme–heme interaction in hemoproteins with multiple hemes,[26,27] as well as to resolve the nature of stacking in some heme model systems.[28]

(3) The μ–m mechanism involves the coupling of the electric transition

[26] Drucker, H., Campbell, L. L., and Woody, R. W., *Biochemistry* **9,** 1519–1527 (1970).

[27] Urry, D. W., and VanGelder, B. F., *in* "Structure and Function of Cytochromes" (K. Okunuki, M. D. Kamen, and I. Sekuzu, eds.), pp. 210–214. Univ. Park Press, Baltimore, Maryland, 1968.

[28] Urry, D. W., *J. Am. Chem. Soc.* **89,** 4190–4196 (1967).

$$R = \pm \frac{V_{12}}{\epsilon_2 - \epsilon_1} (\vec{\mu}_e \cdot \vec{\mu}_m) \tag{6}$$

moment of one chromophore with the magnetic transition moment of the second. The resulting expression for the rotatory strength is:

The dependence of rotatory strength on the dot product of the electric and the magnetic transition moments of the two coupling chromophoric transitions invokes requirements similar to those of the one-electron approach; i.e., the existence of optical activity requires finite values of both μ_e and μ_m as well as a relative orientation of the two transition moment vectors other than 90°. Although the possibility of such a mechanism being operational in proteins and polypeptides was acknowledged at quite an early date, only recently has any appreciable attention been given to this model.[12] Direct coupling between the electric and magnetic dipoles is usually not thought to occur to any appreciable degree; thus the relative contributions from such a mechanism, if operational, are generally considered to be rather minor, especially in the case of proteins and large synthetic polypeptides.

B. Phenomenological Approach for Analysis of Intrinsic CD Spectra

The phenomenological approach involves the analysis of the observed properties of unknown systems with respect to those exhibited by well-characterized systems, the characterization of which is usually carried out through probes other than those under consideration. In proteins the phenomenological approach has been utilized for the analysis of the intrinsic dichroic spectrum, and, to a lesser degree, the aromatic region. Through the efforts of both experimentalists and theoreticians, today there are detailed explications of as many as five to six different model polypeptide systems which conform to one or another of the commonly observed conformations in proteins. Reproduced in Fig. 2 are the well-accepted CD spectra of the three commonly occurring organized polypeptide structures: α helix, antiparallel β sheet, and the so-called disordered form. The realization that there are more than two common conformational forms of the polypeptide in proteins, the α-helical and disordered forms, has lead to approaches utilizing "multiple-component analysis,"[7] which have superceded the earlier approaches, such as analysis based on ellipticity at a single wavelength, e.g., the 222-nm minimum of the α-helical form.[1-3] "Multiple-component analysis" assumes a linear relationship between the fraction of residues in a given form and the contribution it makes to the total spectrum of the protein. Of the various multiple-component analytical procedures utilizing information on synthetic polypeptides as models, two have been frequently used, the "isodichroic method," as described by

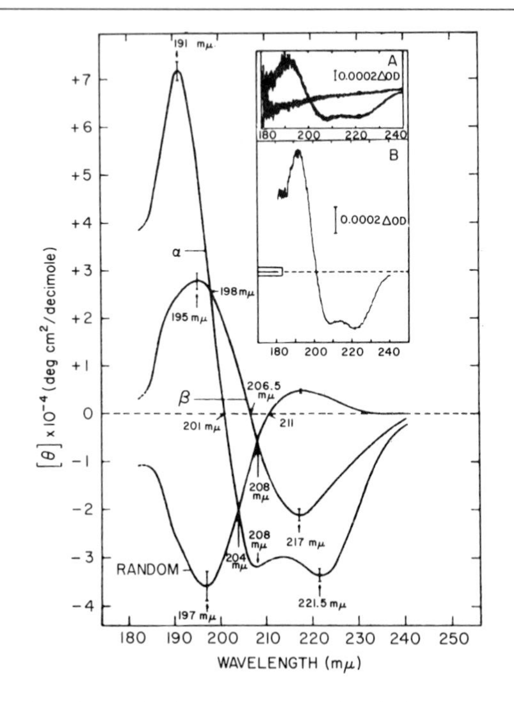

FIG. 2. CD spectra of a synthetic polypeptide, poly-L-lysine, corresponding to the three characterized structures. Random, random coil form at pH 7.4; α, α-helix at pH 11.4; β, β structure at pH 11.4 heated to 51° and cooled to 22°. Inset: Comparison of instrumental performance with and without time-averaging capabilities. (A) Direct trace of instrumental output from JASCO-J-5; (B) output after 16 repeated scans. Spectrum is of poly-L-lysine in water, pH 11.4, temperature 22° (From Myer.[29])

the author[29] and later improved by Rosenkranz and Scholtan,[30] and the second outlined by Greenfield and Fasman[31] and extended by Straus *et al.*[32] using computer fitting of data to models. Recently, a new approach has been developed by Saxena and Wetlaufer.[33] This approach utilizes experimental CD data from a number of proteins and the estimated conformational composition from X-ray diffraction in order to generate a CD spectrum of a set of proteins. Solution of three simultaneous equations at a series of wavelengths yielded the expected CD spectra of the

[29] Myer, Y. P., *Res. Commun. Chem. Pathol. Pharmacol.* **1,** 607–616 (1970).

[30] Rosenkranz, H., and Scholtan, W., *Hoppe-Seyler's Z. Physiol. Chem.* **352,** 896–904 (1971).

[31] Greenfield, N., and Fasman, G. D., *Biochemistry* **8,** 4108–4116 (1969).

[32] Straus, J. H., Gordon, A. S., and Wallach, D. F. H., *J. Biochem.* (*Eur.*) **11,** 201–212 (1969).

[33] Saxena, I. P., and Wetlaufer, D. B., *Proc. Natl. Acad. Sci. U.S.A.* **68,** 969–972 (1972).

three implicated conformations, α helix, β structure, and the so-called "random" form, which could then be used for the analysis of CD spectra of proteins of unknown conformational composition. A similar procedure, but involving several proteins in direct permutations, was also applied by Chen and Yang.[34] It has recently been further extended to include the contributions from structures other than those accounted for, e.g., the "tailing" effects of organized structures.[35]

It must be emphasized that none of the phenomenological approaches thus far developed has produced results which can be regarded as a true picture of the conformational characterization of proteins. In general all these procedures are found to work fairly well for those systems which have a relatively high proportion of organized structure, but they fail to varying degrees in systems with a lower proportion of ordered structure. There are multiple reasons for the inadequacy of all these procedures. They have been discussed in detail by the authors of each procedure, and also summarized by Fasman[7] and Bayley[14] in their recent review articles. Some of the apparent reasons are (1) the uncertainty of the CD parameters of the three conformations of synthetic polypeptides; (2) the applicability of information from long homopolymers to proteins, which are heteropolymers and exhibit relatively small segments of organized structure; (3) the presence of other ordered structures, such as the 3_{10} helix and triple helix, which are not accounted for in any of the procedures; and (4), the lack of accountability for contributions from chromophores other than the amide groups of the protein. In hemoproteins factor (4) is further enhanced, since there are contributions associated with the heme group in this region of the spectrum as well, which is dependent on the valence state of the metal atom and its coordination configuration. Evidence also exists indicating coupling perturbation in this region of the spectrum.[22] Thus, for most practical purposes, the analyses of the intrinsic CD spectra of hemoproteins, and proteins in general, should be considered only as parameters equivalent to those of the model used, which may or may not approximate the actual system.

C. *Heme Activity and Analysis*

The CD spectrum of hemoproteins above 300 nm is the product of the heme group, although the group in and of itself is optically inactive. The inherent D_{4h} symmetry of the primary unit of heme, the porphyrin dianion, and the lowering of symmetry to C_{2v} or D_{2h}, in the case of iron protoporphyrin IX, are still significant enough to prohibit the generation of any

[34] Chen, Y., and Yang, J. T., *Biochem. Biophys. Res. Commun.* **44,** 1285–1291 (1971).

[35] Chen, Y., Yang, J. T., and Martinez, H. M., *Biochemistry* **11,** 4120–4131 (1972).

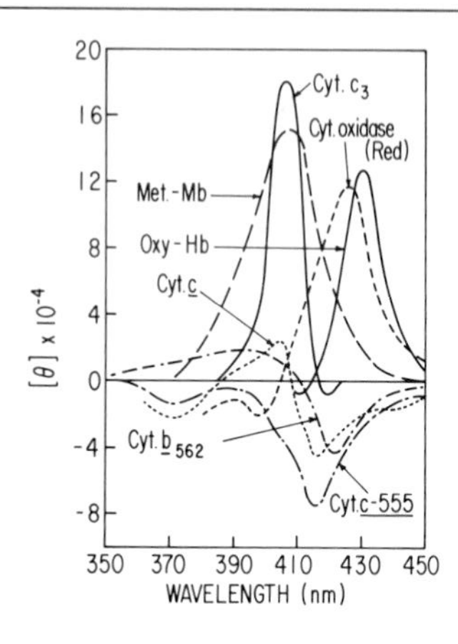

FIG. 3. Comparison of Soret CD spectra of a variety of hemoproteins. Data reproduced from the following references: . hemoglobin (Sugita *et al.*[83]); myoglobin (Straus *et al.*[32]; R. A. Frankel, personal communication); ferric cytochrome *c* (personal data; Myer[51]); ferric cytochrome *c* oxidase, personal data (Myer[45]); ferric cytochrome c_3 (Drucker *et al.*[26]); ferric cytochrome *c*-555 (G. C. Hill and Y. P. Myer, unpublished data); cytochrome b_{562} from *E. coli* (unpublished data).

appreciable rotatory power in this chromophore.[12] The fact is, however, that the spectra of the hemoproteins are very rich in detail over the entire spectral range above 300 nm. In Fig. 3 the Soret CD spectra of a variety of representative hemoproteins are compared. The diversity of these Soret CD spectra, and the similar degree of variation in other parts of the spectrum, necessitate a detailed comprehension of the factors determining optical activity for adequate analysis. The following section outlines the contemporary bases underlying heme activity, with the focus on Soret activity. CD spectral investigations of some well-understood model systems will be examined, along with some specific applications.

The first suggestion that the origin of activity in the heme transition is a result of its arrangement in the polypeptide helices stems from the studies of CD spectra of complexes of protoporphyrin IX with poly-L-lysine.[36,37] The heme–polylysine complex at pH 11, in which the polypeptide is α-helical,[3,4,29,31] exhibits a large negative Cotton effect in the Soret region, but this effect is absent at lower pH, where poly-L-lysine attains a random form. Similar Cotton effects have also been observed for heme–poly-L-ornithine[38] as well as heme–poly-L-histidine complexes.[3] The heme–

[36] Blauer, G., *Nature* (*London*) **189,** 396 (1961).

[37] Blauer, G., *Biochim. Biophys. Acta* **79,** 547 (1967).

[38] Blauer, G., Ehrenberg, A., and Zvilichovsky, B., *in* "Structure and Function of Oxidation-Reduction Enzymes" (A. Åkenson and A. Ehrenberg, eds.), p. 205. Pergamon, Oxford, 1972.

helices arrangement could result in heme transition dipole interaction with the polypeptide backbone,[3] although other types of operational mechanisms, such as induction of dissymmetry in the heme group, mixing of the porphyrin transitions with those of the metal atom, etc. may be involved.

A better understanding of heme optical activity in hemoproteins, particularly for myoglobin and hemoglobin and corresponding types, comes from the studies reported by Hsu and Woody.[39] Using the available coordinates for both of these proteins and employing a theoretical approach, they computed the Soret CD spectra. Of the various possible contributing mechanisms, dipole–dipole coupling involving heme transitions and allowed $\pi-\pi^*$ transitions of nearby aromatic side chains was found to be the dominant factor in determining the Soret activity. The near degeneracy of the Soret transitions adds further sensitivity to the spectrum, since this results in generation of a spectrum with a higher order of complexity, i.e., with negative and positive Cotton effects, but of unequal magnitude. They also showed that the complexity of the Soret Cotton effect is dependent upon the direction of polarization of the porphyrin transitions. A polarization direction of about 45° from the pyrrol nitrogens generated the single-banded spectrum of ferric myoglobin, and the complex Soret spectrum of oxyhemoglobin (a large positive Cotton effect at longer wavelengths followed by a small negative Cotton effect) required a directional orientation of about 75°. In addition, these investigations showed that (1) the tilting of the proximal histidine with respect to the heme plane or its planar orientation has little effect; (2) the dichroic contributions originating from aromatic chromophores located as far away as 12 Å, which extends the effective region to neighboring chains, is significant; and (3) the relative orientation of the aromatic groups in the vicinity of the distal histidine can alter the Soret rotatory strength by as much as 100%. Although there is significant disparity between the calculated and the observed rotatory strengths for both of these systems, 0.3 and 0.1 DM vs. 0.5 and 0.2 DM for myoglobin and hemoglobin respectively, these studies provide important information regarding the interrelationships of the optical activity and various structural and conformational aspects of the protein. One obvious inference from these studies is that the interpretation of the dichroic changes of the heme transition upon perturbation of the heme group—such as binding of extrinsic ligands, alteration of the spin state of the metal atom and its oxidation state, and the differences observed among preparations of the same protein from different species as well as the extrapolation of results from one hemoprotein to another—should be performed cautiously, since simple alterations of the polarization direction or orientation of a distal aromatic

[39] Hsu, M., and Woody, R. W., *J. Am. Chem. Soc.* **93,** 3515–3535 (1971).

chromophore could result in significant alteration of both the complexity and magnitude, without implying any major conformational change in either the protein or the heme group.

The analysis of heme activity in hemoproteins with covalently bonded heme groups, cytochromes of type *c*, or of multiple heme proteins, such as cytochrome *c* oxidase, is based primarily on the phenomenological approach, i.e., extrapolation of results from simple heme models, although some use of theory has been made for the interpretation of complex Soret CD spectra of these systems. The CD spectra of heme *c* fragments from cytochrome *c* as a model for these proteins have been investigated by various groups,[28,40–43] but recent studies with modified models seem to provide some indication of the primary reason for activity in these proteins. Myer and Harbury[44] first showed that in the case of heme *c* octapeptide (heme-containing fragment from residue 14–21 of horse heart cytochrome *c*), the replacement of His-18 from heme iron coordination results in an almost complete elimination of Soret Cotton effects. Recent CD studies with modified heme *c* systems varying in protein complexity (i.e., from systems with only eight amino acids to 104 amino acids, as in native protein) and with and without aromatic chromophores generally confirmed the earlier findings.[22,41,42] The CD spectra of one of the systems, both with and without His-18, are reproduced in Fig. 4 for reference. The Soret Cotton effect of the heme *c* systems with identical ligands at both positions 5 and 6 of heme iron, irrespective of the spin state of the metal atom, is found to be a small negative Cotton effect, R_k about 0.02–0.07 DM. Heme *c* systems with His-18 at one of the coordinating positions, on the other hand, exhibit strong positive Cotton effects, R_k of 0.24–0.55, depending on the nature of the sixth liganding group, the largest strength for the weakest ligand field group, H_2O, and the lowest, for the strongest ligand field group, imidazole.[22] Since the simplest heme *c* systems contain no aromatic side chains, e.g., heme octa- and undecapeptides, and since the length of the polypeptide chain, the spin state of the metal atom, or its oxidation state are found to generate a small negative band of almost comparable rotatory strength, as long as the two axial positions are occupied by identical liganding groups, the observed small negative ellipticities have been attributed to sources other than transition-dipole coupling of identical (hemes) or of nonidentical chromophores. The decreasing rotatory strength of the Soret Cotton effect with increasing ligand field strength of the liganded

[40] Urry, D. W., and Pettegrew, J. W., *J. Am. Chem. Soc.* **89,** 5276–5287 (1967).

[41] MacDonald, L. H., Ph.D. Thesis, State University of New York at Albany (1974).

[42] Pande, A. J., MacDonald, L. H., and Myer, Y. P., *Biophys. J.* **15,** 286a (1975).

[43] Zand, R., and Vinogradov, S. N., *Biochem. Biophys. Res. Commun.* **26,** 121–127 (1967).

[44] Myer, Y. P., and Harbury, H. A., *J. Biol. Chem.* **241,** 4299–4303 (1966).

group at position six of heme *c* iron in systems containing His-18 at position five (Myer and Pande,[22] Tables V and VI) has been regarded as an indication that the optical activity originates, or is determined by, lowering of the symmetry of the heme group because of differences in the liganding field strength of the two axial ligands, rather than the transition-dipole coupling mechanism operational in hemoglobin and myoglobin.[22,41]

The Soret CD spectra of type *c* cytochromes, however, deviate appreciably from those of heme *c* models, especially in the complexity of the spectrum, which clearly demonstrates sensitivity to chromophore conformation and environment. The Soret spectrum of the simple heme *c* models, a single-banded positive peak (Fig. 4), is relatively uncomplicated when compared to a variety of cytochromes and their complexes (Fig. 3). There are several explanations for the doubly-inflected CD spectrum of a chromophore in a strongly absorbing region, $\epsilon \simeq 10^4–10^5$, the most apparent being exciton splitting,[12,16] a consequence of transition-dipole coupling of identical chromophores. This could be the explanation for multiple heme systems such as cytochrome c_3,[26] the CO complex of cytochrome oxidase,[27]

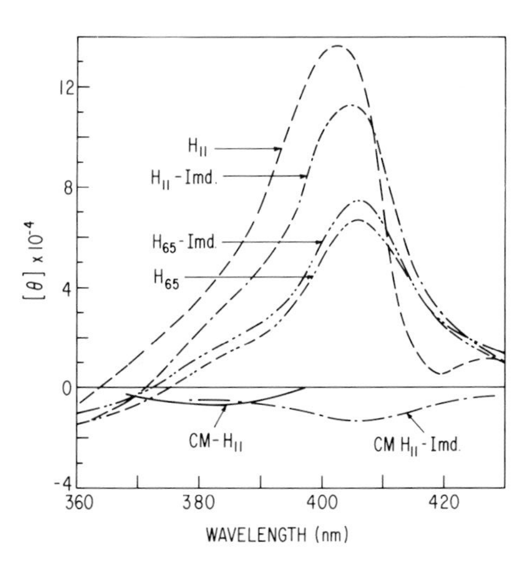

FIG. 4. Soret CD spectra of heme *c*–peptide systems with identical and nonidentical axial ligands;—-—, carboxymethylated heme *c* undecapeptide in 0.05 *M* imidazole (coordination configuration of Imd-Fe-Imd); ——, carboxymethylated heme *c* undecapeptide in 8 *M* urea (configuration, water-Fe-water); — — - — —, heme *c* undecapeptide in 0.05 *M* imidazole (configuration, Imd-Fe-His); — — —, heme *c* undecapeptide in 8 *M* urea (configuration, water-Fe-His; — — - - — —, H_{65} fragment of cytochrome *c* (configuration, ? [possibly lysine]-Fe-His);----, H^{65} fragment in 0.05 *M* imidazole (configuration, Imd-Fe-His). All measurements for the ferric form in 0.1 *M* phosphate buffer, pH 7.0, temperature 22°, unless otherwise stated (taken from MacDonald[41]).

and cytochrome *c*-552[44a] since each of them contain two or more heme groups per protein molecule, but not for the cytochromes *c* or for the complex of HR peroxidase as these proteins contain a single heme group per protein molecule and are known to be monomeric under the conditions of these investigations. The application of the sum rule, which dictates the development of equal and opposite rotatory strengths of the two transitions, rules out a simple explanation for cytochrome c_3, as well as for ferrous cytochrome *c* oxidase.[45]

Complexity of the type seen in cytochrome *c* or ferric oxidase can be explained on the basis of inherently lower symmetry of the heme group, possibly because of coordination configuration, environment, etc. resulting in splitting of the Soret transition, following which each transition then independently gains or loses optical activity through a dipole–dipole coupling mechanism involving the transition dipole of other chromophores in the molecule.[12] In such a case, one would expect a maximum of two Cotton effects for a heme group, and the sign and magnitude of each Cotton effect to be independently governed by the protein conformation surrounding the group. Therefore Soret spectra of any complexity can be the result. Splitting of the Soret transition of cytochrome *c* has indeed been documented by Eaton and Hochstrasser,[46,47] and the thermal and urea denaturation studies of this protein have produced evidence for selective governance of the ellipticity of the negative Cotton effect at 418 nm through porphyrin–polypeptide dipole interaction.[48] The latter renders the Soret CD spectra of these proteins extremely sensitive to the conformational integrity of the molecule, especially in the immediate vicinity of the heme group. Whether a similar explanation can be invoked for the CD spectra of cytochrome *c* oxidase, its CO complex, etc. is a subject of investigation, although there are some indications that for a particular type of preparation (i.e., a preparation exhibiting an asymmetric Soret Cotton effect in the ferric form and a single-peaked spectrum in the ferrous form),[22,45] the two hemes of the molecule are conformationally independent and distinct, and the asymmetry of the ferric form is because of significant splitting of the heme *a* group.[49,50]

[44a] Bartsch, R. G., Meyer, T. E., and Robinson, A. B., *in* "Structure and Function of Cytochromes" (K. Okunuki, M. D. Kamen, and I. Sekuzu, eds.), pp. 443–451. Univ. Park Press, Baltimore, Maryland, 1968.

[45] Myer, Y. P., *J. Biol. Chem.* **246,** 1241–1248 (1971).

[46] Eaton, W. E., and Hochstrasser, R. M., *J. Chem. Phys.* **46,** 2533 (1967).

[47] Eaton, W. E., and Hochstrasser, R. M., *J. Chem. Phys.* **49,** 985 (1968).

[48] Myer, Y. P., *Biochemistry* **7,** 765–776 (1968).

[49] Myer, Y. P., and King, T. E., *Biochem. Biophys. Res. Commun.* **34,** 170–175 (1969).

[50] Myer, Y. P., *Proc. Am. Chem. Soc.* p. 250a (1968); *Fed. Proc., Fed. Am. Soc. Exp. Biol.* **27,** 314a (1968).

In addition to the two preceding possibilities, the complexity of the Soret CD spectrum could be a reflection of significant vibrational contributions and/or mixing of the metal–porphyrin transitions. The latter can be readily determined, since it should impart spin-state and oxidation-state sensitivity to the Soret spectrum. The following are the idealized forms of the Soret CD spectra of heme systems corresponding to various possibilities. (1) If the origin of activity is due to transition-dipole coupling between two conformationally and configurationally identical heme groups, then (a) in the event the Soret degeneracy is maintained, one should expect a maximum of two Cotton effects of equal and opposite sign, or (b) if the degeneracy is revoked, possibly because of configurational perturbation of one or both heme groups, the spectrum to be expected in this region may be four CD bands, two positive and two negative for a two-heme system; (2) if the resulting activity is due to the interaction of a heme transition with nonheme transitions, then (a) if the degeneracy is maintained, i.e., a higher order of symmetry than C_{2v}, one should expect in the main a single-banded Soret CD spectrum with the sign determined by the sign of the coupling energy, or (b) if degeneracy is inherently absent, a two-banded spectrum is expected, with the rotatory strength determined almost independently for each of the transitions, depending upon the nature of the interacting groups, the distance, etc. Both theoretical and phenomenological considerations of CD spectra of hemoproteins provide some understanding of the interrelationships between optical activity, conformation, and configuration of the heme group. Detailed analysis of the spectrum, not only in one region, but over the entire range, is essential for precise interpretation of results. Unfortunately, this has been done in only a few cases, which will be described in the following section.

VI. CD Spectra of Hemoproteins

Almost all hemoproteins which have been purified to an acceptable level have been subjected to CD spectral measurements, but coverage in this article is limited to only those proteins which have been extensively investigated through this technique and are well characterized through other techniques, and to the CD spectrum in the region below 250 nm, the intrinsic absorption region, and the region from 380–450 nm, the Soret region. The coverage of the CD spectrum is limited because these two regions are the best understood in terms of the origin of optical activity and thus most susceptible to interpretation. Not that measurements in other parts of the spectrum are unimportant—significant contributions have been made through investigations in other regions, e.g., the conforma-

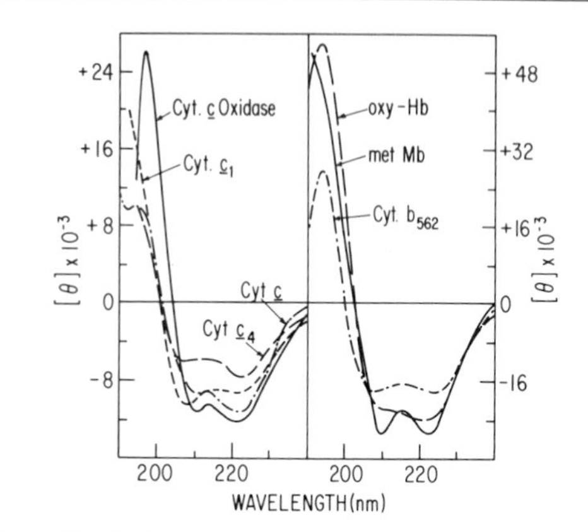

FIG. 5. Comparison of intrinsic CD spectra of a variety of hemoproteins. Information taken from references as indicated: oxyhemoglobin (Straus *et al.*[32]); metmyoglobin (Straus *et al.*[32]); ferric cytochrome *c* (personal data; Myer[60]); ferric cytochrome *c* oxidase (personal data; Myer[45]); ferric cytochrome c_1 (Kaminsky *et al.*[91]); ferric cytochrome c_4 (VanGelder *et al.*[97]); cytochrome b_{562} (personal unpublished data).

tion of aromatic side chains in cytochrome *c*[51] and the coordination configuration of myoglobin complexes with extrinsic ligands[52,53]—but in general the interpretation of the CD spectrum in regions other than the Soret has been far from satisfactory. The exceedingly great complexity of the CD spectrum in the region 250–300 nm, owing to contributions from multiple chromophores, the aromatic side chains, and the heme group,[4,22] render this region difficult to interpret unambiguously. The lack of understanding of the interdependence of circular dichroic properties on the conformation and configuration of the heme group in the Δ-absorption region, in the α- and β-absorption region, as well as in the near-IR region also makes quantitative interpretations difficult. For a detailed consideration of the CD spectrum of hemoproteins in regions other than the Soret region, the reader is directed to consult two recent reviews[21,22] or the original papers on the system of particular interest.

A. *Intrinsic CD Spectra*

The CD spectra of a variety of hemoproteins show that all the hemoproteins in general exhibit characteristics typical of α-helical organization of the polypeptide backbone, i.e., minima at 220–222 nm and 208–211 nm and a maximum in the region below 200 nm (Fig. 5). The larger differences in the

[51] Myer, Y. P., *J. Biol. Chem.* **243,** 2115–2122 (1968).
[52] Bolard, J., and Garnier, A., *Biochim. Biophys. Acta* **263,** 535–549 (1972).
[53] Garnier, A., Bolard, J., and Danon, J., *Chem. Phys. Lett.* **15,** 141–143 (1972).

ellipticities of the negative extrema and the positional differences of the positive peak from those observed for α-helical models (Fig. 2) are reflections of the varying degrees of organization among these proteins and the presence of structures other than the α-helical and the so-called random coil form. The intrinsic CD spectrum has been subjected to analysis through one or more of the phenomenological approaches; the results from these studies and from X-ray diffraction, where available, are listed in Table I.[29-32,34,35,45,54-68] It is apparent that the estimates of organized structure in proteins which are predominantly helical, hemoglobin and myoglobin, are in excellent agreement with those from X-ray diffraction studies, but the same is not true in those cases in which the proportion of organized structure is low. Although the proportion of α-helicity for cytochrome *c* estimated through analysis of the CD spectrum is in excellent agreement with X-ray estimates, the procedure yields a relatively high proportion of β structure. Similarly, high estimates for β structure seem to be the case in other hemoproteins which are low in helical structure. Whether this reflects an inherent difference between the conformation of the protein in solution vs. the crystalline state, or contributions from transitions other than the amide bond, or the inadequacy of the models used, etc., could not be discerned. This aspect of the analysis has been the subject of a variety of articles.[7,14,22,29,60] While these estimations may not provide the precise conformational composition of the secondary structure of the polypeptide, they do constitute an exceedingly sensitive probe for comparative studies of these systems.

The near identity of the intrinsic CD spectra of cytochromes *c* from horse, beef, pigeon, tuna, chicken, and turtle, in which the sequential

[54] Melki, G., *Biochim. Biophys. Acta* **263,** 226–243 (1972); *J. Phys.* (*Paris*) **33,** 37–47 (1972).
[55] Kendrew, J. C., *Science* **139,** 1259 (1962).
[56] Beychok, S., Tyuma, I., Benesch, R. E., and Benesch, R., *J. Biol. Chem.* **242,** 2460–2462 (1967).
[57] Waks, M., Yip, Y. K., and Beychok, S., *J. Biol. Chem.* **248,** 6462–6470 (1973).
[58] Anders, S. F., and Atassi, M. Z., *Biochemistry* **9,** 2268–2275 (1970).
[59] Samejima, T., and Kita, M., *J. Biochem.* (*Tokyo*) **65,** 759–766 (1969).
[60] Myer, Y. P., *Biochim. Biophys. Acta* **154,** 84–90 (1968).
[61] Myer, Y. P., *Biochim. Biophys. Acta* **221,** 94–106 (1970).
[62] Dickerson, R. E., Takano, T., Eisenberg, D., Kallai, O. B., Samson, L., Cooper, A., and Margoliash, E., *J. Biol. Chem.* **246,** 1511 (1971).
[63] Takano, T., Kallai, O. B., Swanson, R., and Dickerson, R. E., *J. Biol. Chem.* **248,** 5234 (1973).
[64] Kaminsky, L. S., Chiang, Y., and King, T. E., *J. Biol. Chem.* **250,** 7280–7287 (1975).
[65] Sturtevant, J. M., and Tsong, T. Y., *J. Biol. Chem.* **244,** 4942–4950 (1969).
[66] Huntley, T. E., and Strittmatter, P., *J. Biol. Chem.* **247,** 4641 (1972).
[67] Strickland, E. H., Kay, E., and Shannon, L. M., *J. Biol. Chem.* **245,** 1233–1238 (1970).
[68] Samejima, T., and Kita, M., *Biochim. Biophys. Acta* **175,** 24–30 (1969).

TABLE I

CONFORMATIONAL COMPOSITION OF VARIOUS HEMOPROTEINS AND DERIVATIVES

Analytical method	$[\theta]_{222}$	Linear combination of synthetic polypeptide organized structures: Isodichroic method[29,31]			Linear combination of synthetic polypeptide organized structures: Computer fitting[31]			Proteins as reference forms[34,35]			X-ray diffraction		
Percent composition	α-Helix	α-	β-	R-	α-	β-	R-	α-	β-	R-	α-	β-	R-
System													
Hemoglobin													
Oxyhemoglobin	55–80[32,54]	—	—	—	—	—	—	—	—	—	72	—	28[55]
Deoxyhemoglobin	80[54]	—	—	—	—	—	—	—	—	—	—	—	—
α-Chain	72[56]	—	—	—	—	—	—	—	—	—	—	—	—
β-Chain	67[56]	—	—	—	—	—	—	—	—	—	—	—	—
Apohemoglobin	52[57]	—	—	—	—	—	—	51	—	—	—	—	—
α-Globin	27[57]	—	—	—	—	—	—	19	—	—	—	—	—
β-Globin	52[57]	—	—	—	—	—	—	—	—	—	—	—	—
Myoglobin													
Metmyoglobin	61–66[32,58]	66	0	40[30]	68	6	27[31]	77	—	23	65–72	0	32–23[55]
Apomyoglobin	54[59]	—	—	—	48[28]	—	—	—	—	—	—	—	—
Ferromyoglobin	70[59]	—	—	—	—	—	—	—	—	—	—	—	—
Cytochromes c													
HH ferri-cyt. *c*	11–32[60,61]	9	58	30[29]	—	—	—	27	6	27	11–39	—	—[62]
		11	21	66[30]									
HH ferro-cyt. *c*	14–36[60,61]	14	45	32[29]	—	—	—	—	—	—	11–39	—	—[63]
		17	20	63[30]									
Ferri- & ferro-cyt. c_1	—	—	—	—	20–25	25	—[64]	—	—	—	—	—	—

Cytochrome c oxidase[45]													
Ferri-oxidase	39[45]	13	60	—[45]	—	—	—	—	—	—	—	—	—
		15[a]	42	—									
Ferro-oxidase	44[45]	18	60	—[45]	—	—	—	—	—	—	—	—	—
		21[a]	40	—									
Cytochrome b_2[65]													
Ferric & ferrous	35–40	—	—	—	—	—	—	—	—	—	—	—	—
Cytochrome b_5[66]													
Native	—	—	—	—	20–30	10–15	45	—	—	—	—	—	—
Apoprotein	—	—	—	—	19[b]	36	45	—	—	—	—	—	—
Peroxidase													
HRP-A1 & -C[67]	—	—	—	—	30–40[b]	—	—	—	—	—	—	—	—
HRP-Apoenzyme[67]	—	—	—	—	25–30[b]	—	—	—	—	—	—	—	—
Catalase[68]	54	—	—	—	47[b]	—	—	—	—	—	—	—	—

[a] Using poly-L-serine as reference model for disordered form.

[b] Using ellipticities at 208 nm, the position of the isodichroic point for random and β-form poly-L-lysine.[29] This constitutes the basis of the isodichroic method as well.

variation is as much as 40–45%, has been the basis for acceptance of protein conformational homology of these systems.[61,69] The insensitivity of the intrinsic CD spectrum to structural alteration among myoglobin preparations from sperm whale, goat, lamb, beef, and camel[70] reflects a nonconformational role for the variable structures involved. The comparative studies of mammalian hemoglobins also generally confirm the polypeptide conformational insensitivity accompanying structural variations in mammalian systems. On the other hand, the intrinsic CD spectra of deoxy-Hb C and deoxy-Hb E show them to be 10–11% more helical than deoxy-Hb A; Hb S and Hb F are found to be less helical than Hb A;[54] lamprey oxy- and deoxy-Hb's were found to contain only about 70% of the α-helical structure of the corresponding forms of Hb A;[71] and the preparation from *L. terrestris* was found to contain only about one-half the organization.[72] Making use of the intrinsic dichroic spectrum, Myer and co-workers investigated the conformational effects of chemical modification of a variety of protein functional groups of cytochrome *c*. They concluded that the nitration and iodination of two of the four tyrosyl side chains, and formylation of the invariant tryptophanyl side chain, all result in significant, almost complete, derangement of the polypeptide conformation,[73,74] whereas the modification of the tryptophan residue with *N*-bromosuccinimide[73] or of the methionyl side chains[41] is accompanied by imperceptible perturbation of the protein secondary structure. Based on these observations, and conclusions from studies in other parts of the spectrum, they were able to discern which preparations should be considered important for the elucidation of structural–functional relationships of the protein and which control the function through a conformational role.

The interrelationships of polypeptide conformation and the coordination configuration of the heme group and/or the spin state of the metal atom are other aspects studied through CD spectroscopic measurements in the intrinsic absorption region. Samejima and Kita[59] observed the insensitivity of the intrinsic CD spectrum of myoglobin to the change of valence state of the metal atom as well as to the formation of complexes with cyanide and azide of ferric myoglobin and the oxy- and CO complexes of ferro-myoglobin. Thus they concluded an absence of interdependence of

[69] Zand, R., and Vinogradov, S. N., *Arch. Biochem. Biophys.* **125,** 94–97 (1968).
[70] Atassi, M. Z., *Biochim. Biophys. Acta* **221,** 612–622 (1970).
[71] Sugita, Y., Dohi, Y., and Yoneyama, Y., *Biochem. Biophys. Res. Commun.* **31,** 447–452 (1968).
[72] Harrington, J. P., Pandolfelli, E. R., and Herskovits, T. T., *Biochim. Biophys. Acta* **328,** 61–73 (1973).
[73] O'Hern, D. J., Pal, P. K., and Myer, Y. P., *Biochemistry* **14,** 382–391 (1975).
[74] Pal, P. K., Verma, B., and Myer, Y. P., *Biochemistry* **14,** 4325–4334 (1975).

the polypeptide conformation of myoglobin on both the nature of the coordination configuration of the heme group and the spin state of the metal atom. Similar studies by these authors[68] on catalase, however, led to conclusions to the contrary. The oxygenation of hemoglobin is found to accompany a small, but definite, decrease of the intrinsic negative extremum, which, when taken at face value, is regarded as evidence of the alteration of the polypeptide conformation.[75] Myer has shown that the formation of the cyanide and imidazole complexes of ferric cytochrome *c* enhances the ellipticity of the 222-nm band,[48] and similarly, the cyanide complex of ferric cytochrome *c* oxidase, the CO complex of ferrous oxidase, and the reduced ferric cyanide complex all exhibit enhanced ellipticities in this region of the spectrum.[45] The change of valence state of the metal atom for both HRP and JRP[76,77] and cytochrome b_2[78] did not affect the intrinsic CD spectra, while the reduction of cytochrome *c*[43,51,60] as well as of cytochrome *c* oxidase[45] is found to be accompanied by increases of the intrinsic extrema. In all these instances the magnitude of alteration of ellipticities of the intrinsic dichroic inflections was relatively small, i.e., within 5–10% of the magnitudes, which in the absence of any complexities could be, and in most cases has been, taken to reflect the perturbation of the polypeptide conformation accompanying alterations of heme configuration and/or spin state of the metal atom. As stated earlier, this region contains contributions from groups other than the polypeptide backbone, especially heme contributions, and there is evidence for dipole–dipole coupling between the heme and polypeptide chromophores, all of which could alter the ellipticities of the intrinsic extrema.[60,69,75] Therefore interpretation of small variations in the intrinsic CD spectrum upon perturbation of the heme group should be conducted with caution. In this regard the intrinsic dichroic changes associated with reduction of cytochrome *c* have been interpreted not as a reflection of the alteration of polypeptide conformation, but as an alteration of heme contributions.[60,69]

Considerable interest has been directed toward the intrinsic CD spectrum in order to ascertain the role of the heme group in determining the protein conformation. The removal of heme from both sperm whale and horse myoglobins generates intrinsic CD spectra reflecting a reduction of helicity by about 10–20%.[58,59] Similarly, the removal of heme from isolated α and β chains of hemoglobin is shown to produce a CD spectrum reflecting

[75] Urry, D. W., *J. Biol. Chem.* **242,** 4441–4448 (1967).
[76] Strickland, E. H., Kay, E., Shannon, L., and Horwitz, J., *J. Biol. Chem.* **243,** 3560–3565 (1968).
[77] Hamaguchi, K., Ikeda, K., Yoshida, C., and Morita, Y., *J. Biochem. (Tokyo)* **66,** 191–201 (1969).
[78] Sturtevant, J. M., and Tsong, T. Y., *J. Biol. Chem.* **244,** 4942–4950 (1969).

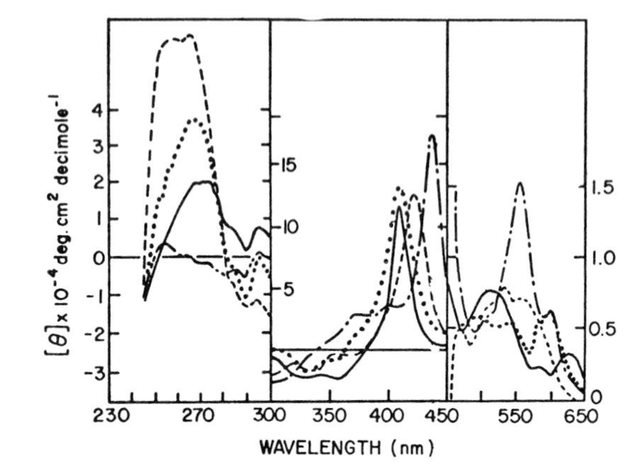

Fig. 6. CD spectra of sperm whale myoglobin, in phosphate buffer, pH 7.05. ——, ferric; · · · · ·, fluoroferric;—·—·—, ferrous;-----, cyanoferric. (Adapted from Nicola *et al.*,[91] where conditions are given.)

a decrease in organized structure, the most dramatic effects being for the α chain.[57,79] Intrinsic CD studies have also provided evidence indicating that the refolding of α-globin is induced not only by substitution of heme, but also upon binding with β-subunits.[57] The effect of removal of heme from cytochrome *c* is found to be an almost complete transformation of the native intrinsic CD spectrum to that of a protein without any organized structure;[80] HRP exhibits a reduction of helicity to about 18% from 35–40% upon removal of heme;[76] and JRP is transformed to a β-sheeted protein structure from a native helical structure with removal of heme.[77] Studies of the intrinsic CD spectrum of cytochrome b_2 upon removal of FMN and/or the nucleotide component, however, showed the insensitivity of the polypeptide conformation to the presence or absence of these prosthetic groups.[78,81]

B. Visible CD Spectra

1. Myoglobins and Hemoglobins. The Soret dichroic spectrum of the myoglobins and their complexes is typified by a positive Cotton effect, with varying degrees of complexity toward the lower wavelengths (Fig. 6). The distinctness of the CD spectrum of each complex is a reflection of the conformational sensitivity of the heme group. Willick *et al.*[82] have analyzed

[79] Yip, Y. K., Waks, M., and Beychok, S., *J. Biol. Chem.* **247,** 7237–7244 (1972).
[80] Fisher, W. R., Taninchi, H., and Anfinson, C. B., *J. Biol. Chem.* **248,** 3188–3195 (1973).
[81] Tsong, T. Y., and Sturtevant, J. M., *J. Biol. Chem.* **244,** 2397–2402 (1969).
[82] Willick, G. E., Schonbaum, G. R., and Kay, C. M., *Biochemistry* **8,** 3729–3734 (1969).

the Soret CD spectra of various complexes through Gaussian fitting and comparison with similar features from absorption spectra, and they concluded that the heme conformation remains relatively invariant through alteration of oxidation state of the metal atom. The small variations of the Soret CD spectrum as well as those seen in the aromatic region and the visible region[52,53] from one molecular form to another are indeed reflections of the sensitivity of the coordination configuration of the heme group and its CD spectrum. The dependence of the Soret CD spectrum on the immediate environment and the coordination configuration of the heme group is best illustrated by Hsu and Woody[39] through theoretical calculation of the CD spectrum (see Section V,C). These studies confirm in general the conformational sensitivity of the heme group to its environment and also indicate that the observed alterations of the Soret CD spectra upon complexation with extrinsic ligands could simply be a reflection of changes in the direction of polarization of the porphyrin $\pi-\pi^*$ transition components, which can easily be brought about by the relative orientation of the liganding group, etc. Garnier and co-workers[52,53] have analyzed the visible CD spectra of sperm whale myoglobin and its complexes and correlated them to the magnetic CD spectra and the absorption properties. They concluded that the observed splitting of the α and β bands in the azide and nitro complexes of met-Mb and in the carbonyl and oxygen complexes of ferromyoglobin can be interpreted as an indication of the lowering of the symmetry of the heme group below C_{4v} because of the formation of a nonlinear Fe–ligand bond in each case, while the lack of splitting of α and β Cotton effects of MbNO and $MbCN^-$ conforms to the idea of the formation of linear complexes.

The CD spectra of hemoglobins, like those of all other heme systems, are again rich in detail (Fig. 7). The Soret CD spectra of mammalian hemoglobins and of their derivatives with extrinsic ligands in both the ferric and ferrous states of the metal atom are generally doubly inflected, with a large positive peak at higher wavelengths and a small negative peak in the lower end of the region.[54,56,83-86] The Soret CD peaks of most forms of hemoglobin A, except CO-Hb A, were found to coincide with or were relatively close to absorption wavelengths.[83] Likewise, the dichroic peaks in the visible region, 500–600 nm, though complex, correspond rather well with the positions of the absorption maxima.[87] The CO-Hb A spectrum in the visible region is distinct because of splitting of the α and β Cotton

[83] Sugita, Y., Nagai, M., and Yoneyama, Y., *J. Biol. Chem.* **246,** 383–388 (1971).
[84] Ueda, Y., Shiga, T., and Tyuma, I., *Biochim. Biophys. Acta* **207,** 18–29 (1970).
[85] Ueda, Y., Shiga, T., and Tyuma, I., *Biochem. Biophys. Res. Commun.* **35,** 1–5 (1969).
[86] Nagai, M., Sugita, Y., and Yoneyama, Y., *J. Biol. Chem.* **244,** 1651–1658 (1969).
[87] Nagai, M., Sugita, Y., and Yoneyama, Y., *J. Biol. Chem.* **247,** 285–290 (1972).

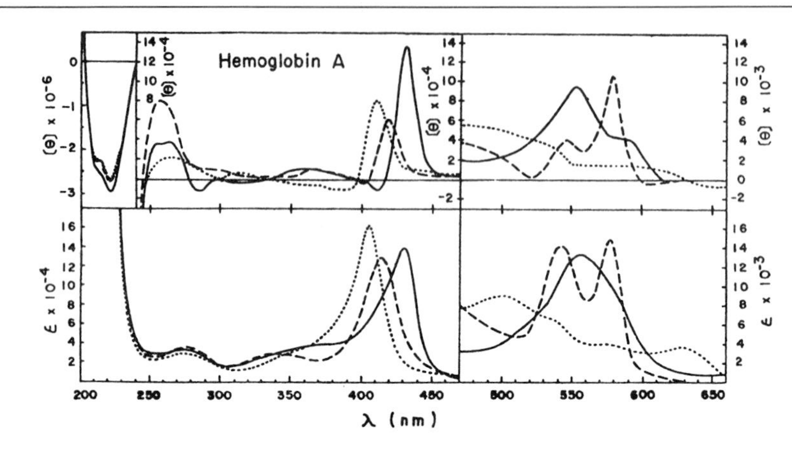

FIG. 7. CD and absorption spectra of native human hemoglobin. ——, deoxygenated hemoglobin; -----, oxygenated hemoglobin; · · · · ·, methemoglobin. (Adapted from Sugita *et al.*[83] in which conditions are given.)

effects. The Soret CD spectrum of CO-Hb A is also distinguished by its complexity, i.e., two negative peaks at 415 and 445 nm and a positive peak at 435 nm, none of which corresponds to the absorption maximum.[84,85]

The lowered symmetry of the heme group, possibly because of heme–heme interaction, was a commonly used reason for the complexity of the Soret Cotton effect in a variety of hemoglobin systems.[40,82] Theoretical consideration of the Soret optical activity of oxyhemoglobin, however, showed that while maintaining the near degeneracy of the porphyrin transitions, i.e., a relatively high degree of symmetry of the heme group, the observed complexity of the Soret Cotton effects of various derivatives of hemoglobin (the exception being CO-hemoglobin) could be generated by simply altering the polarization direction of the porphyrin transitions, which is dependent upon the nature of the coordination configuration of the heme iron, the direction of the coupling dipole, etc.[39] Thus the apparent differences in Soret CD spectra among the various forms of hemoglobin may not be a reflection of significant conformational perturbation of the heme group *per se*, but rather a consequence of the alteration of coordination configuration, of the protein environments, or both. The splitting of the Soret Cotton effects as well as of the visible CD bands upon formation of the CO complex, on the other hand, is an indication of significant lowering of the heme symmetry because of the formation of a nonlinear coordination configuration of the CO-Fe linkage.

Nagai *et al.*[87] and Sasazuki *et al.*[88] have reported comparative dichroic studies of normal and abnormal hemoglobins. On the basis of the identity of the dichroic spectrum of oxy-Rainier hemoglobin with that of oxy-Hb A, conformational identity was ascertained, while the observed dichroic differences between the deoxy forms were attributed to the single structural change in the β chains of the protein.[87] The dichroic spectrum of Berts hemoglobin is similar to the isolated β chains of Hb A, which agrees with the finding that this molecule is a composite of four identical β-like chains and lacks the cooperativity of normal Hb A.[88] Harrington *et al.*[72] reported comparative CD investigations of a multiple-chain hemoglobin from the earthworm, *Lumbricus terrestris,* of a four-chained hemoglobin from *Glycera dibranchiata,* and a corresponding single-chain myoglobin-like protein. Dichroic differences, such as the negative Soret Cotton effects and the complex visible spectra of the lower animal hemoglobin complexes, are considered to be reflections of heme environmental differences which may also include modified heme geometry, the coordination configuration of heme iron, etc. The negative Soret Cotton effect is also present in methemoglobin from *Chironomus thummi thummi,*[89] oxy- and deoxy-lamprey hemoglobins,[71,90] and all the forms of ferric and ferrous soybean leghemoglobins.[91] Generally, the negative Soret Cotton effect has been attributed to explanations of the sort outlined above, but Sugita *et al.*[71] ruled out the possibility of differences of coordination configuration of heme iron in leghemoglobin derivatives as an explanation of the observed variation of the Soret Cotton effect on the basis of structural characterization of this protein. Instead, they subscribe to the idea that the tertiary structural differences of these hemoglobins are primarily reflected in the CD spectra of the heme chromophore.

The CD studies of hemoglobin subunits and of apoproteins have proven to be very useful for the elucidation of not only the conformational aspects of subunit interaction of the protein, but also the interdependence of heme conformation in this system. The CD spectrum of each of the isolated chains differs from that of the native protein, both in the Soret[86] and the visible regions of the spectrum[83–85,92] in all oxidation and liganded states. The dichroic differences between the spectra of the α and β chains reflect

[88] Sasazuki, T., Isomoto, A., and Nakajima, H., *J. Mol. Biol.* **65,** 365–369 (1972).

[89] Wollmer, A., Buse, G., Sick, H., and Gersonde, K., *J. Biochem. (Eur)* **24,** 547–552 (1972).

[90] Lampe, J., Rein, H., and Scheler, W., *FEBS Lett.* **23,** 282–284 (1972).

[91] Nicola, N. A., Minasian, E., Appleby, C. A., and Leach, S. J., *Biochemistry* **14,** 5141–5149 (1975).

[92] Goodall, P. T., and Shooter, E. J., *J. Mol. Biol.* **39,** 675–678 (1969).

the conformational differences of the heme group of the two chains. The spectral differences from native Hb A, especially the positions of the bands, are indicative of conformational differences of the heme groups. The finding that a simple arithmetic summation of contributions of the α and β chains fails to reproduce the CD spectrum of native Hb A, especially the large disparity found for the deoxy forms of the protein and subunits,[83,85,86,92] is a reflection that the reconstitution of the protein involves conformational alteration of certain subunit structures and/or of heme. In this regard, the inference from theoretical treatments that aromatic residues as far away as 12 Å for a given heme group in hemoglobin contribute significantly to the generation of the Soret CD spectrum,[39] which clearly extends the effective domain of heme to neighboring subunits, provides a feasible molecular explanation of the interchain dependence of heme optical activity.

2. *Cytochromes c*. The CD spectra of mammalian-type cytochromes *c* in the Soret absorption region and in the ferric state of the metal atom are dominated by a positive peak at 403–406 nm and a larger negative peak at 416–418 nm, none of which corresponds to the absorption maximum of the protein, 408 nm.[48,51,60,93] The change of valence state of the metal atom generates a spectrum which more or less reflects an image of the ferric spectrum, a positive peak at 424–432 nm and a negative inflection at about 390 nm, with a well-defined shoulder at 408–411 nm (Fig. 8).[51,60,93] Except for small differences in the relative magnitudes of the positive and negative Soret CD spectra, the dichroic curves of both the ferric and ferrous forms of cytochrome c_2 from *Rhodospirillum molischianum*,[94] ferrous *Pseudomonas* cytochrome *c*,[93] and ferric cytochrome *c* from *Candida krusei*[95] are all indistinguishable from those seen for the corresponding forms of the mammalian-type cytochromes *c*. The CD spectrum of ferric *Pseudomonas* cytochrome *c*, however, is different from mammalian cytochromes above 404 nm; i.e., the negative peak is replaced by a large positive peak.[93] The Soret spectrum of the ferrous form of cytochrome *c* from *C. krusei* is an almost symmetrical S-shaped curve with a maximum at 432 nm and a minimum at about 411 nm.[95] An almost symmetrical S-shaped Soret spectrum, though inverted, is also characteristic of cytochrome *c*-552 from *Chromatium D*.[44a] A doubly-inflected spectrum, but with significant differences in relative magnitudes, is found to be the case for cytochromes

[93] Vinogradov, S. N., and Zand, R., *Arch. Biochem. Biophys.* **125,** 902–910 (1968).

[94] Flatmark, T., Dus, K., DeKlerk, H., and Kamen, M. D., *Biochemistry* **9,** 1991–1996 (1970).

[95] Hamaguchi, K., Ikeda, K., and Narita, K., *in* "Structure and Function of Cytochromes" (K. Okunuki, M. D. Kamen, and I. Sekuzu, eds.), pp. 328–334. Univ. Park Press, Baltimore, Maryland, 1968.

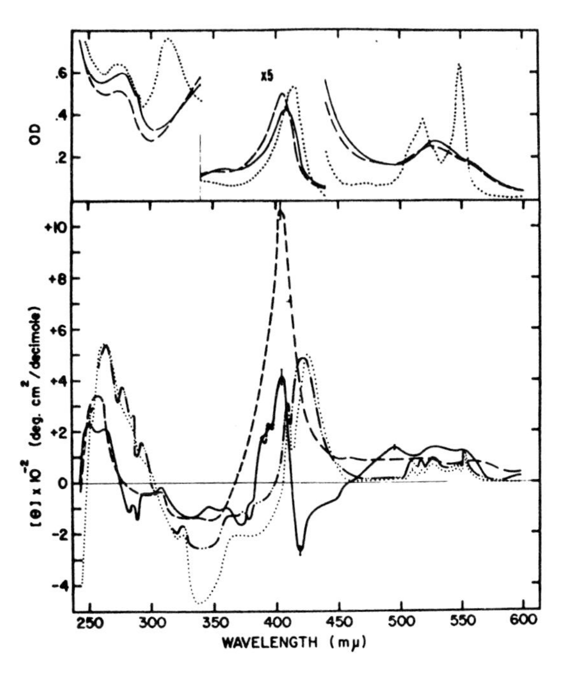

FIG. 8. CD and absorption spectra of horse heart cytochrome *c*. ——, ferricytochrome *c*; ·····, ferrocytochrome *c*;---, ferricytochrome *c* in 8 *M* urea;–··–··, ferrocytochrome *c* in 8 *M* urea. (Adopted from Myer[51] where conditions for measurements are to be found.)

c_3 from *Desulfovibrio desulfuricans* and *D. salexigenes*,[26] while a positive single-banded spectrum for both the ferrous and the ferric forms of cytochrome c_2 from *Rhodospirillum rubrum*,[96] mammalian cytochrome c_1,[64] and cytochrome c_4 from *Azotobacter vinelandii*[97] is the case. The apparent effect of the change of valence state of the metal atom of cytochromes c_1, c_2, c_3, c_4, and *Chlorobium* *c*-553 is simply a red shift of the spectrum maintaining consistency between the absorption maximum and the Soret CD maximum in each case. The CD differences of the Soret and visible spectra among cytochrome *c* preparations from different species[61,69] and from one

[96] Flatmark, T., and Robinson, A. R., *in* "Structure and Function of Cytochromes" (K. Okunuki, M. D. Kamen, and I. Sekuzu, eds.), pp. 318–327. Univ. Park Press, Baltimore, Maryland, 1968.

[97] VanGelder, B. F., Urry, D. W., and Beinert, H. *in* "Structure and Function of Cytochromes" (K. Ikunuki, M. D. Kamen, and I. Sekuzu, eds), pp. 328–334. Univ. Park Press, Baltimore, Maryland, 1968.

type of cytochrome *c* to another are reflections of conformational differences of the heme moiety/moieties from one protein to another; and conversely, the identity and/or similarity of dichroic curves are indicative of conformational homology for these systems.[61] The Soret dichroic differences between ferric mammalian cytochromes *c* and the preparation from *Pseudomonas aeruginosa* are regarded as a reflection of different coordination configurations of heme iron.[93] Myer has established that the negative limb of the Soret Cotton effects, the 418-nm negative peak, of mammalian cytochromes is a direct reflection of heme–polypeptide interactions, and thus a measure of the integrity of the heme crevice.[48] The S-shaped Soret spectrum of cytochrome *c*-552 from *Chromatium D* is considered to be a reflection of heme–heme interaction; and the single-peaked Soret CS spectrum of cytochrome c_1, cytochrome c_2 from *R. rubrum*, cytochrome c_4 from *A. vinelandii*, and the negative single-peaked spectrum of cytochrome *c*-555[22] are indications of a relatively higher symmetry of heme group in these proteins, possibly as in myoglobin. The existence of small, but definite, differences between the CD spectra of ferric mammalian cytochromes, on the one hand, and among the ferrous forms, on the other, has lead to the conclusion that, although there are small differences in the heme environments among various preparations, a gross conformational homology exists.[61] Similarly, the dichroic differences between ferric mammalian cytochrome *c* and ferric cytochrome c_2 from *R. rubrum* were attributed to subtle differences in the heme environments, rather than to differences in the polypeptide conformation.[96]

The oxidoreduction-linked alteration of the Soret CD spectra of cytochrome c_1, cytochrome c_2, cytochrome c_3, and cytochrome c_4—a simple displacement of the dichroic spectrum to the red while maintaining the identity of the dichroic and absorption maxima for both the ferric and ferrous forms—is interpreted as evidence of the absence of oxidation-reduction-linked conformational perturbation of the heme group in these proteins. The inversion of the Soret CD spectrum of mammalian cytochromes *c* upon change of valence state of the metal atom,[51] as well as the transformation of the asymmetric Soret CD spectrum of ferric cytochrome *c* from *C. krusei* to a symmetrical S-shaped spectrum upon reduction,[95] on the other hand, are cases in which oxidoreduction-linked conformational alterations of the heme group have been invoked. Myer[51] has proposed that the nature of the oxidoreduction-linked conformational alteration of the heme group in mammalian cytochromes *c* could be the flipping of iron from one side of the heme plane to the other or the distortion of the porphyrin structure or both; i.e., the apparent effect is the alteration of the sign-determining parameter for the optical activity.

The spectral region above 600 nm is the least investigated region of these proteins. Eaton and Charney[98] have analyzed the CD spectrum of horse heart ferrocytochrome *c* in this region of the spectrum. Using spectral criteria, such as polarization, absorption characteristics, and anisotropicity, they assigned the bands in this region to the lowest crystal fields, single-state transitions, some of which are shown to be magnetic-dipole-allowed metal *d–d* transitions. They also concluded that iron in the ferrous form may be further displaced from the porphyrin ring.

Like the CD spectra of hemoglobin and myoglobin, the CD spectrum of cytochrome *c* has been used to monitor conformational changes accompanying perturbations such as ligand binding; pH variation; denaturation with urea, temperature,[48,51] and alcohols;[99] chemical modification of protein functional groups;[41,73,74] etc. Using urea and temperature denaturation studies and employing the analysis of changes over the entire spectral range of 200–600 nm, Myer[48] has shown the existence of a two-step denaturation process for horse heart ferric cytochrome *c*: the first, primarily the uncoupling of heme polypeptide interaction, and the second, the unfolding of the protein accompanied by alteration of coordination configuration. Using the dichroic characteristics of the intermediate form, it was determined that the conformational alteration resulting from replacement of intrinsic ligand by extrinsic ligands, imidazole, cyanide, and azide, was merely a loosening of the heme crevice, without alteration of the polypeptide conformation.

Recently, a similar comparison of the CD spectra of ferric and ferrous formyl- and iodo-cytochromes *c* with those of the native and the partially perturbed form has lead to a conformational interpretation for the functional alteration of these preparations, rather than the usual structural interpretation.[73,74] From similar types of investigations, Myer *et al.*[73] have shown that the modification of the invariant tryptophanyl side chain with *N*-bromosuccinimide results in only a slight loosening of the heme crevice, without any alteration of either the polypeptide conformation or that of the heme group. The splitting of the visible CD spectrum of horse heart ferric cytochrome *c* upon formation of a complex with NO has been considered to be due to the formation of a nonlinear ligand–Fe linkage, while the absence of splitting of the α and β Cotton effects of ferrous NO-cytochrome *c* is interpreted as due to the presence of a linear complex configuration.[52]

3. Cytochrome c Oxidase. The CD studies of cytochrome *c* oxidase have been limited to a single species and a few preparations (i.e., preparations from different laboratories) and primarily to the Soret absorption

[98] Eaton, W. E., and Charney, E., *J. Chem. Phys.* **51,** 4502 (1969).

[99] Kaminsky, L. S., Yang, F. C., and King, T. E., *J. Biol. Chem.* **247,** 1354–1359 (1972).

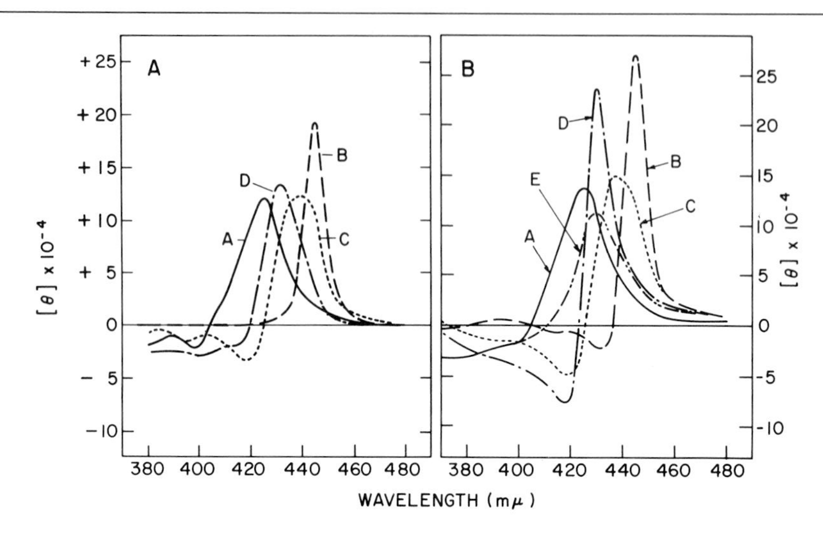

FIG. 9. CD spectra of cytochrome *c* oxidase and its complexes. Preparations from two different laboratories. (A) Preparation from Dr. King's laboratory (Myer[45]). (B) Preparation from Dr. Beinert's group (Myer[102]). Curve A, oxidized oxidase; curve B, reduced oxidase; curve C, CO complex of reduced oxidase; curve D, ferricyanide–oxidized CO complex; curve E, produced upon oxygenation of the ferricyanide–oxidized CO complex. Data for (A) taken from Myer;[45] (B) based on unpublished data, in part from Myer.[102] Conditions are to be found in appropriate references.

region,[27,45,49,50,100–103] although a few reports have appeared dealing with the visible spectrum.[104,105] The asymmetric Soret CD spectrum of ferric oxidase (Fig. 9; maximum at about 426 nm, shoulder at about 410 nm, and negative band at about 400 nm) has been attributed to a multiplicity of contributing transitions from the two heme groups. The absence of features such as concurrent negative and positive bands of comparable magnitude rules out the possibility of heme–heme interaction between the two ferric hemes of the protein.[27,45,50,100] Myer and King[101] also reported a lack of sensitivity of the ferric Soret spectrum to both aging and the addition of ionic detergents, thus establishing the conformational insensitivity of ferric hemes to both processes, which are known to result in alteration of the

[100] Urry, D. W., Wainio, W. W., and Grebner, D., *Biochem. Biophys. Res. Commun.* **27,** 625–631 (1967).

[101] Myer, Y. P., and King, T. E., *Biochem. Biophys. Res. Commun.* **33,** 43–48 (1968).

[102] Myer, Y. P., *Biochem. Biophys. Res. Commun.* **49,** 1194–1200 (1972).

[103] Tiesjema, R. H., Hardy, G. P. M. A., and VanGelder, B. F., *Biochim. Biophys. Acta* **375,** 24–33 (1974).

[104] Yong, F. C., and King, T. E., *Biochem. Biophys. Res. Commun.* **40,** 1445–1451 (1970).

[105] Love, B., and Auer, H. E., *Biochem. Biophys. Res. Commun.* **41,** 1437–1442 (1970).

biological function of the protein. The Soret CD changes resulting from formation of the cyanide complex, i.e., alteration of only selected transitions,[45] are found to be in conformity with the accepted view that only one of the two heme *a*'s, heme *a*, binds cyanide.

The reported Soret CD spectrum of reduced cytochrome *c* oxidase differs from one laboratory to another. The preparation investigated earlier by this laboratory exhibited a single-banded symmetrical spectrum with a maximum at 445 nm[50,101] (Fig. 9A). Parallel reports from other laboratories using a preparation from a different group were of a complex spectrum with a large positive peak at about 446 nm and a negative minimum at about 432 nm[25,100] (Fig. 9B). Later studies from this laboratory[102] with a preparation from the Wisconsin group (Fig. 9B) and the report of Tiesjema *et al.*[103] also confirmed the differentiation as outlined above. The possible cause for the differentiation has been resolved through investigations reported by this laboratory[47,101] showing that the complex Soret CD spectrum of reduced cytochrome *c* oxidase, like that shown in Fig. 9B, is induced upon addition of ionic detergent, deoxycholate, which is present or used during isolation of preparations exhibiting complex spectra.[27,47,100] The transformation of the single-banded Soret CD spectrum of ferrous cytochrome *c* oxidase in nonionic detergents to a complex spectrum upon addition of deoxycholate is indeed a direct indication of the conformational sensitivity of heme to the nature of the detergent. Whether it is a consequence of the alteration of one of the two heme *a* moieties or of both is still unresolved.

The conclusions regarding the conformational implications associated with change of valence state of metal atoms and other reducible components (two copper atoms) depend upon the preparation under consideration. Based on studies of systems with deoxycholate, Urry and coworkers[27,102] have interpreted the transformation of the ferric Soret spectrum to the complex ferrous Soret spectrum (Fig. 9B) as an indication of significant conformational alteration of the heme group, i.e., from a situation with no interaction to one with heme–heme interaction, whereas the oxidation-reduction-linked transformation for the preparation free of ionic detergents (Fig. 9A) reflects a change from a lower to a higher symmetry situation of heme groups.[47,101] Similar opposing conclusions regarding the presence or absence of heme–heme interaction and the interdependence of heme conformation on valence state of the metal atoms and the coordination configuration are also the case for the two types of systems outlined above. Myer and King,[49] using the ionic detergent-free preparation, observed identity between the difference Soret CD spectra for the change of valence state of the heme *a* component, on the basis of which they concluded that the two heme groups of oxidase must be not only conformationally independent, but also distinct. Similar studies, but using a preparation with a complex Soret CD spectrum (Fig. 9B), failed to produce the reported identity of the difference CD spectra.[22] A similar observation has also been made by Yong and King.[104] Thus it is apparent that the conformation of the heme chromophores of reduced cytochrome oxidase is dependent upon the nature of the detergent used during isolation and/or dispersion in solution. Which form of oxidase (the form exhibiting a single symmetrical Soret CD spectrum as in Fig. 9A or the form exhibiting a complex Soret spectrum as in Fig. 9B) is the physiologically relevant form is yet to be established. These CD studies do indicate that careful characterization of the preparation is a must for a noncontroversial interpretation of the results from any investigation of this hemoprotein.

Yong and King[104] have reported the CD spectra in the 600-nm region for oxidized, reduced, and corresponding complexes with extrinsic ligands. Like the spectra in the Soret region, CD spectra of this region also exhibit oxidation–reduction dependence as well as dependence on the coordination configuration of the metal atom. They tentatively assign the origin of these Cotton effects to interaction between the protein and the chromophores.

The formation of the CO complex of oxidase generates a Soret spectrum which is characterized by concurrent positive and negative peaks (Fig. 9A and B). A similar situation is observed upon addition of ferricyanide to the reduced CO complex, where the change is the oxidation of both the coppers and the heme *a* components. The observed nature of the Soret complexity and the presence of two heme *a* groups in the molecule led to an inference of the existence of heme–heme interaction,[27,100] which however may not be the case, since similar complexity has been observed for other hemoproteins which are monomeric and contain a single heme group per protein molecule (see section on cytochromes *c*). Love and Auer[105] have shown from detailed investigations of CD spectra of the carbon monoxide complex that the state of aggregation of oxidase determines its spectroscopic properties, and thus the observed complexity of the CD spectrum.

The Soret CD spectrum of oxidase has also been the basis for characterization of the so-called "oxygen" complex of cytochrome *c* oxidase. Through comparison of the CD spectrum of the oxygenated complex generated upon oxygenation of the ferricyanide-oxidized form of the reduced CO complex of the protein, with the Soret CD spectrum of the product obtained directly upon oxygenation of reduced cytochrome *c* oxidase, Myer[102] has concluded that the "oxygenated" complex of oxidase is a composite of oxidized heme *a* and oxidized copper atoms, with heme a_3 as either a_3^{2+}—O_2 or a_3^{3+}—O_2^{1-}; this has recently been confirmed through more direct studies.

4. *Cytochromes b*. The CD spectra of cytochromes b_2 from yeast

(cytochrome b_2 "S") and from *Hansenula* (cytochrome b_2 "H"), and of their derivatives, DNA-free enzyme, b_2 core, and apoenzyme, have been reported primarily by two groups, the Yale group[65,81] and the French group.[106] Based on dichroic changes in the region 200–600 nm upon removal of the FMN moiety and the change of valence state of the metal atom, the Yale group concluded that the FMN and heme moieties are conformationally linked; i.e., they are in close proximity and exhibit spin-state dependence. Heme–FMN conformational dependence was confirmed in general by the French group, but in addition they used fluorescence quenching of tryptophan and tyrosine and concluded that FMN–heme conformational dependence is not because of direct FMN–heme interaction, but because of indirect effects transmitted through the protein moiety of the molecule.

The CD spectra of larval cytochrome *b*-555 and *b*-563 over the region 200–600 nm has been reported by Okada and Okunuki,[107,108] and like other hemoproteins, they exhibit spectral details over the entire region. For both the oxidized and reduced forms, the Soret CD spectra as well as the visible CD spectra were found to be negative, and the positions of the CD bands were consistent with those of the absorption bands. We have recently investigated cytochrome *b*-562 from *E. coli,* and the Soret spectral details are very similar to those reported above,[109,110] except that the position of the reduced CD band was about 5 nm to the red for the reduced ferrous protein. The identity of the CD extrema with the absorption peaks and the single-banded nature of the Soret spectrum in general establish that the heme moieties in these proteins are located in a significantly less asymmetrical conformational state, while the inverted nature of both the Soret and visible spectra indicates a significant departure of the heme environment from those of hemoglobins, myoglobins, and cytochromes in general. Okada and Okunuki, on the basis of available information, have proposed a generalized set of rules for CD spectra of cytochromes of type *b*,[107,108] which thus far appear to hold for the systems investigated.

5. *Other Hemoproteins*. The conformational implications accompanying alteration of the oxidation state of the metal atom and of the coordination configuration for both ferric and ferrous horse radish peroxidase[67,76,82,111] and Japanese radish peroxidase,[77] of native and liganded forms of bovine liver catalase, and also of the effects of denaturation

[106] Iwatsuho, M., and Risler, J. L., *J. Biochem.* (*Eur.*) **9,** 280–285 (1969).
[107] Okada, Y., and Okunuki, K., *Biochem. J.* **67,** 603–605 (1970).
[108] Okada, Y., and Okunuki, K., *Biochem. J.* **67,** 487–496 (1970).
[109] Bullock, P., and Myer, Y. P., *Biochemistry* **17,** (1978) (in press).
[110] Myer, Y. P., and Bullock, P., *Biochemistry* **17,** (1978) (in press).
[111] Strickland, E. H., *Biochim. Biophys. Acta* **151,** 70–75 (1968).

and pH changes on the CD spectra, have been reported. Localized conformational alteration of the heme binding site upon change of valence state of the metal atom is concluded for HRP, on the basis of analysis and comparison of the Soret CD spectra.[82] The transformation of the two-peaked Soret spectrum, peaks at 405 and 413 nm, of ferric JRP-*a* upon reduction to a doubly-peaked spectrum with peaks at 404 and 424 nm, with an additional negative inflection at about 441 nm,[77] could also be indication of oxidation-reduction-linked conformational alteration of the heme group. The nature of the complexity of the Soret CD spectrum for both valence states of the metal atom, on the other hand, has led to the suggestion of the coexistence of multiple spin-state forms of heme, which seems to be the case, since JRP is found to be in a mixed-spin state at neutral pH. Strickland[109] has reported CD studies of HRP–substrate complexes, and on the basis of dichroic differences in the 280-nm region, he has suggested the possibility of the occurrence of alteration of the orientation of the aromatic side chains upon formation of the enzyme–substrate complex. Comparative CD studies of two peroxidase isoenzymes from HR root and their apoenzymes by Strickland *et al.*[67] have shown that isoenzymes have essentially similar active sites.

Acknowledgments

This work was supported by a grant (PCM 7301090 and 77-07441) from the National Science Foundation. My thanks to my wife (Ruth D. Myer) for editing, typing, and organization of material in this manuscript.

[17] Spin States of Heme Proteins by Magnetic Circular Dichroism

By LARRY E. VICKERY

Introduction

The optical spectra of heme proteins have been known for some time to be sensitive to the spin state of the heme iron, but no quantitative analysis of either absorption-band intensity or wavelength position has proved generally applicable for their assignment. Recent findings on the magneto-optical activity of heme proteins, however, suggest that this technique can provide a relatively straightforward spectroscopic method for determining spin states. The sensitivity of the method to both ferrous and ferric redox states and the ability to obtain measurements at physiological as well as cryogenic temperatures complement the more commonly used electron paramagnetic resonance technique. In addition measurements can be ob-

ISBN 0-12-181954-X

tained on dilute solutions (micromolar range) in the presence of other paramagnetic species, offering advantages over direct susceptibility measurements, the Mössbauer effect, and the Evans nuclear magnetic resonance shift method.

The phenomenon of magneto-optical activity was discovered over 100 years ago by Michael Faraday who found that the presence of a magnetic field applied parallel to the direction of the measuring light beam could induce optical rotation. In contrast to natural optical activity this property is exhibited by all matter and does not require either an inherent molecular asymmetry or the presence of an asymmetric environment. Considerable progress toward understanding the theoretical basis of the origin of the Faraday effect has been achieved recently, and several reviews on the subject are available.[1-3] The magnetically induced optical activity can be expressed and determined either as magnetic optical rotatory dispersion (MORD) or as magnetic circular dichroism (MCD). MORD measures the difference in refractive index between left and right circularly polarized light (L and RCPL), while MCD measures the difference in absorption. Just as in natural optical activity the two phenomena can be related by the Kramers-Kronig dispersion relations; only MCD will be discussed here because it leads to less complex band shape and overlap, and the transitions of interest occur in an optically accessible region. General discussions of biochemical applications of both MORD[4] and MCD[5] have appeared in this series.

Three types of magneto-optical effects can be distinguished according to their origin, and each gives rise to a characteristic spectroscopic band shape (see Fig. 1). Faraday *A* terms arise from transitions to orbitally degenerate excited states when the applied magnetic field splits the energy levels into components of opposite angular momentum (Zeeman effect); these then selectively absorb either L or RCPL because of the opposite angular momentum associated with the polarized photons. This effect can only occur in chromophores possessing sufficient symmetry to have molecular orbitals of similar energy, and since most metalloporphyrins have approximate D_{4h} symmetry, *A* terms are expected for the heme $\pi-\pi^*$ transitions. The shape of *A* terms in an MCD spectrum resembles the derivative of the absorption band, and their intensity is a sensitive function of the excited-state orbital angular momentum and the transition bandwidth and oscillator strength.

[1] A. D. Buckingham and P. J. Stephens, *Annu. Rev. Phys. Chem.* **17,** 399–432 (1966).

[2] P. N. Schatz and A. J. McCaffrey, *Q. Rev., Chem. Soc.* **23,** 552–584 (1969).

[3] P. J. Stephens, *Annu. Rev. Phys. Chem.* **25,** 201–232 (1974).

[4] V. E. Shashoua, this series, Vol. 27 [31].

[5] B. Holmquist and B. L. Vallee, , this series, Vol. 49 [6].

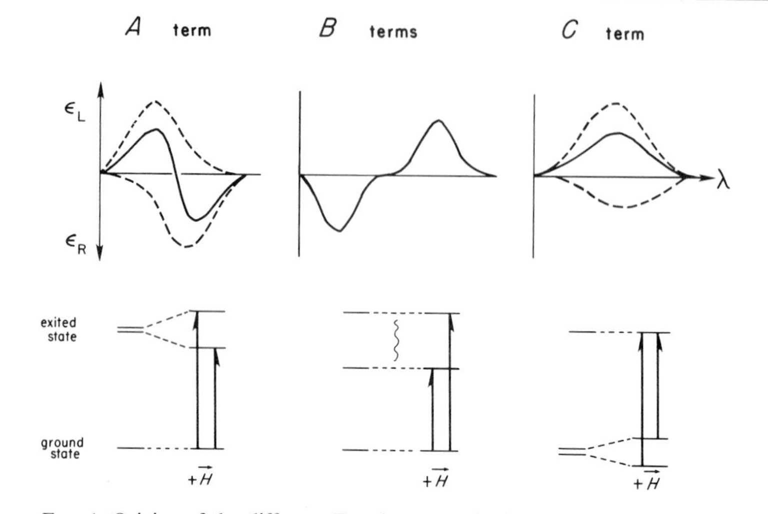

FIG. 1. Origins of the different Faraday terms in magnetic optical activity.

Faraday B terms arise from a magnetic-field-induced mixing of transitions. These occur in all compounds and are most intense for close-lying, orthogonally polarized transitions. They are in general weak for symmetric *a*-, *b*-, and *c*-type hemes but dominate the spectra of *d* heme and other porphyrin complexes of D_{2h} symmetry in which the x and y directions are not equivalent.Except for the change in sign these B terms resemble the absorption bands in shape.

Finally, Faraday C terms occur in paramagnetic materials which have spin-degenerate ground states. As with A terms a Zeeman splitting of the energy levels leads to a shift in the absorption of L and RCPL, but in this case the two initial levels will be differently populated according to a Boltzmann distribution. This gives rise to different absorption intensities for L and RCPL which are temperature dependent. For bands whose width is greater than the splitting induced (generally a few wave numbers), the *net* difference in absorption between L and RCPL yields an MCD shape resembling the absorption curve. The intensity of the MCD is directly proportional to the population difference of the two ground states and therefore to the reciprocal of the absolute temperature. All heme complexes with the exception of diamagnetic low-spin ferrous forms have unpaired electrons and hence can be expected to exhibit C terms. This paramagnetic C-type MCD is not restricted to optical transitions directly involving the iron atom (porphyrin-to-iron change transfer or iron d–d transitions) but is also observed for the predominantly π–π^* porphyrin bands in the visible and near-UV Soret regions due to (iron) spin–(porphyrin) orbit

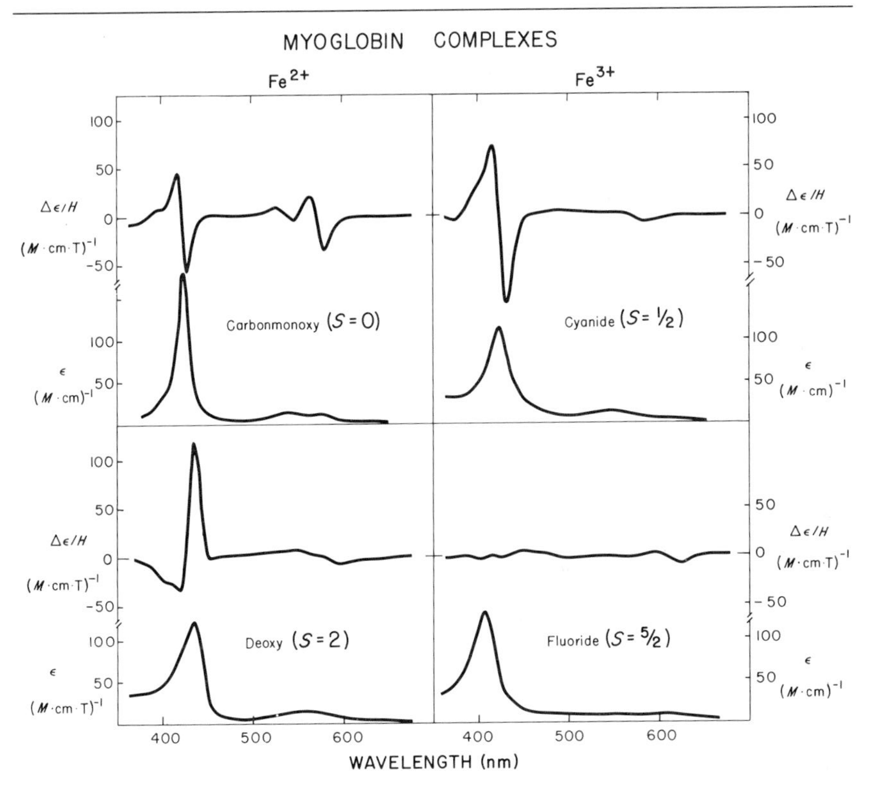

FIG. 2. MCD and absorption spectra of myoglobin complexes in various redox and spin states (data replotted from Vickery *et al.*[9]).

coupling.[6-8] It is this sensitivity of the MCD spectrum to the spin of the iron which makes the technique a useful probe of the heme protein's active center. This article will describe methods of measuring the temperature dependence of the MCD which are necessary to establish the paramagnetic origin of C terms and investigate spin states in heme proteins.

Examples of the types of MCD curves observed with heme proteins in different redox and spin states are illustrated by the spectra of the myoglobin complexes shown in Fig. 2.[9] These curves are all taken from the room

[6] P. J. Stephens, J. C. Sutherland, J. C. Cheng, and W. A. Eaton, *in* "The Excited States of Biological Molecules" (J. B. Birks, ed.), pp. 434–444. Wiley, New York, 1975.

[7] J. Treu and J. J. Hopfield, *J. Chem. Phys.* **63,** 613–623 (1975).

[8] M. A. Livshitz, A. M. Arutyunyan, and Y. A. Sharanov, *J. Chem. Phys.* **64,** 1276–1280 (1976).

[9] L. Vickery, T. Nozawa, and K. Sauer, *J. Am. Chem. Soc.* **98,** 343–350 (1976).

temperature results and hence by themselves do not distinguish C terms, but dramatic changes in the spectra with changes in the paramagnetism of the iron are evident. In the ferrous case the diamagnetic carbonmonoxide complex exhibits only simple A terms typical of many hem*o*chromes,[10] while the MCD of the deoxy form is extremely temperature dependent allowing one to distinguish between the low ($S = 0$) and high ($S = 2$) spin states. Both ferric forms are paramagnetic, but the much more intense C terms of the low-spin ($S = 1/2$) when compared with the high-spin ($S = 5/2$) complex allow one to determine the amount of hem*i*chrome present. A more detailed discussion of these observations will follow the sections on experimental procedures.

Instrumentation

Commercially available CD spectrophotometers can be adapted for MCD measurements simply by placing a magnet in the sample compartment so that the magnetic field is parallel to the measuring light beam. In the case of electromagnets this means that holes must be located through the pole pieces, and for superconducting magnets the bore must go through the solenoid. Superconducting magnets which attain fields of 50,000 G or greater provide high sensitivity since MCD is directly proportional to field strength; this is advantageous when signals are weak in an absolute sense due either to a small magnetic anisotropy ratio ($\Delta\epsilon_{MCD}/\epsilon$) or to low concentrations and when the magnetic anisotrophy is small relative to the natural anisotropy ($\Delta\epsilon_{CD}/\epsilon$). Electromagnets capable of generating near 15,000 G, however, offer an important advantage in providing a more open and accessible sample region for large or complexly shaped cells, temperature-control accessories, and side illumination. Sensitivity problems can be met by time-averaging; in addition, they are less expensive to purchase and operate. The relatively intense signals of hemes together with the need for sample manipulations weigh in favor of the electromagnet for experiments on heme proteins—all of the spectra reported here were obtained at fields from 9000–15,000 G—and the few modified procedures for working with superconducting magnets will not be discussed.

At least three types of improvements over commercial spectrometers can be made in laboratory-designed instruments. The first of these involves the use of a piezo-optical (stress-plate) modulator rather than the conventional electro-optical modulator (Pockels cell) as a programmable quarter wave plate. The quartz photoelastic device usually operates at frequencies near 50 kHz (making kinetic experiments possible) and can be driven at

[10] L. Vickery, T. Nozawa, and K. Sauer, *J. Am. Chem. Soc.* **98,** 351–357 (1976).

amplitudes sufficient to work in the near IR. Stephens and co-workers have assembled an MCD instrument which operates to 2 μm utilizing this and a solid-state detector;[11] the spectrometer described by Sutherland *et al.* performs well to about 1.2 μm using a photomultiplier tube having an S-1 type photocathode.[12] A second type of improvement has been achieved by incorporating monochrometers with better dispersion in the visible and near-IR regions.[11,12] While the double-prism monochrometers in most commercial CD instruments function well in the UV, prism-grating combinations make it easier to resolve the often sharp heme bands which occur in the visible region and provide increased intensity and wavelength accuracy. Finally, interfacing a computer to any spectrometer system can greatly facilitate the large amount of data processing necessary to extract the MCD curves by making the calculations necessary to correct for the presence of natural CD and base lines as described below. In addition the computer can be used for signal-averaging or smoothing, to convert raw data to the units desired, and to accurately calculate even very small difference spectra.

Methods of Measurement

Units

In recording a spectrum what is actually measured is the difference in absorption of L and RCPL:

$$\Delta A = A_L - A_R \quad (1)$$

This value can be expressed in terms of molar absorptivity (or extinction) difference and placed on a unit magnetic field basis using the expression

$$\frac{\Delta\epsilon}{H} = \frac{\Delta A}{M \cdot l \cdot H} \quad \text{with the units } (M \cdot \text{cm} \cdot \text{T})^{-1} \quad (2)$$

where M = molar concentration, l = sample cell pathlength in cm, and H = magnetic field strength in Tesla where one T = 10,000 G. The sign convention is such that $\Delta\epsilon/H$ is positive when the field is parallel to the direction of propagation of the measuring beam and negative when antiparallel. These units are readily comparable to absorption $\epsilon(M \cdot \text{cm})^{-1}$ units and generally give $\Delta\epsilon/H$ values for heme proteins from 1–100 $(M \cdot \text{cm} \cdot \text{T})^{-1}$. As an extension of early optical rotatory studies some results are reported in units of degrees ellipticity which are related to the absorption difference by

[11] G. A. Osborne, J. C. Cheng, and P. J. Stephens, *Rev. Sci. Instrum.* **44,** 10–16 (1973).
[12] J. C. Sutherland, L. E. Vickery, and M. P. Klein, *Rev. Sci. Instrum.* **45,** 1089–1094 (1974).

$$\theta \text{ (degree)} = 33 \, \Delta A \tag{3}$$

The molar ellipticity is then often defined on a unit magnetic field (G^{-1}) basis:

$$[\theta]_m = \frac{100 \cdot \theta}{M \cdot l \cdot H} \quad (\text{deg} \cdot \text{cm}^2 \cdot \text{decimole}^{-1} \cdot \text{G}^{-1}) \tag{4}$$

The two MCD expressions are thus related by:

$$\frac{[\theta]_m}{\text{Gauss}} = \frac{0.33 \, \Delta\epsilon}{\text{Tesla}} \tag{5}$$

Calibration

The sensitivity of the spectrometer can be checked using D-camphor sulfonic acid, preferably recrystallized several times from benzene or ethyl acetate and well dried. Reported values of the CD of this compound vary from $\Delta\delta_{290} = 2.20$ to $2.49 \ (M \cdot \text{cm})^{-1}$ (see Chen and Yang[13]), but most work has been reported on the basis of the lower value so that at a concentration of 1 mg/ml, $\Delta A_{290} = 9.3 \times 10^{-3} \ (\text{cm})^{-1}$.

Magnetic field strength can be measured directly with a Hall probe or Gaussmeter or can be determined with chemical standards. Since field inhomogeneity can result from having holes in the tapered pole pieces, calibration with solutions has the advantage of averaging the magnetic flux over the volume actually measured, providing similar path lengths and slit or beam widths are used. Potassium ferricyanide exhibits no natural CD but has an easily measurable MCD in the visible with $\Delta\epsilon_{422}/H = 3.0 \ (M \cdot \text{cm} \cdot \text{T})^{-1}$ at room temperature, so that a $10^{-3}M$ solution in a 1-cm cell will exhibit a ΔA of $3 = 10^{-3}/10{,}000$ field.[12] Fresh samples should be used and can be prepared gravimetrically or assuming $\epsilon_{420} = 1040 \ (M \cdot \text{cm})^{-1}$, and since the MCD signal arises from a C term the sample should be within a few degrees of 20°C. Cobaltous sulfate, $\Delta\epsilon_{510}/H = 1.86 \times 10^{-2} \ (M \cdot \text{cm} \cdot \text{T})^{1}$, has also been recommended for this purpose.[14]

Calculations

When analog results are read directly from charts, the MCD is most simply determined as the difference between the field-on ($\Delta A_{+} = \Delta A_{\text{MCD}} + A_{\text{CD}}$) and field-off ($\Delta A_{\text{CD}}$ only) spectra, a solvent base line being necessary only if CD results are desired. Alternatively, the two spectra can both be

[13] G. C. Chen and J. T. Yang, *Anal. Lett.* **10,** 1195–1207 (1977).

[14] G. Barth, J. H. Dawson, P. M. Dolinger, R. E. Linder, E. Bunnenberg, and C. Djerassi, *Anal. Biochem.* **65,** 100–108 (1975).

recorded with the field on but reversed in direction. This changes the sign of the MCD without affecting the CD. The MCD is then calculated as one-half the difference between the two curves:

$$\Delta A_{\mathrm{MCD}} = \frac{\Delta A_{+} - \Delta A_{-}}{2} \tag{6}$$

When digital data storage and a computer are available it is advantageous to use the latter method since the field is on in both scans and the MCD signal-to-noise ratio is improved by a factor of $\sqrt{2}$; in addition, the CD, also signal averaged, can be calculated from the same experiment:

$$\Delta A_{\mathrm{CD}} = \frac{\Delta A_{+} + \Delta A_{-}}{2} \tag{7}$$

With high-quality digital data, weak CD (or MCD) spectra can be obtained by this method even in the presence of an intense MCD (or CD). Figure 3 shows the raw data and computer-generated MCD and CD plots of such an example. In this case base-line spectra were not recorded, but precaution must be taken to insure a flat instrumental base line and to provide proper shielding of the photomultiplier (mu metal is convenient) to prevent field lines of force from affecting the dynode chain amplification.

Recording Spectra

Certain aspects of MCD spectroscopy make it necessary to pay more careful attention to operating parameters than is usually necessary in absorption spectroscopy. One problem, common to CD as well, is that because very small differences in absorption of L and RCPL (generally one part in 10^3 to 10^5) are measured, large numbers of photons are needed to obtain good statistics. Consequently, a high-intensity light source and wide slit widths are generally used to improve signal-to-noise levels. However, care must be taken to maintain an instrumental bandpass less than 1/5 of the natural spectral bandwidth to resolve narrow transitions; the slit width required can be determined from the dispersion curve provided with the instrument. The best procedure is to determine the slit width, response time constant, and scan rate parameters for a given band empirically by decreasing each until no further sharpening or increase in intensity is observed. It is also recommended that multiple scans always be recorded (even when signal averaging is not used) to insure reproducibility and as a check for sample stability during the measurement.

Sample Handling

Room-temperature MCD measurements can be obtained with the normal rectangular or cylindrical cuvettes used for absorption spectroscopy.

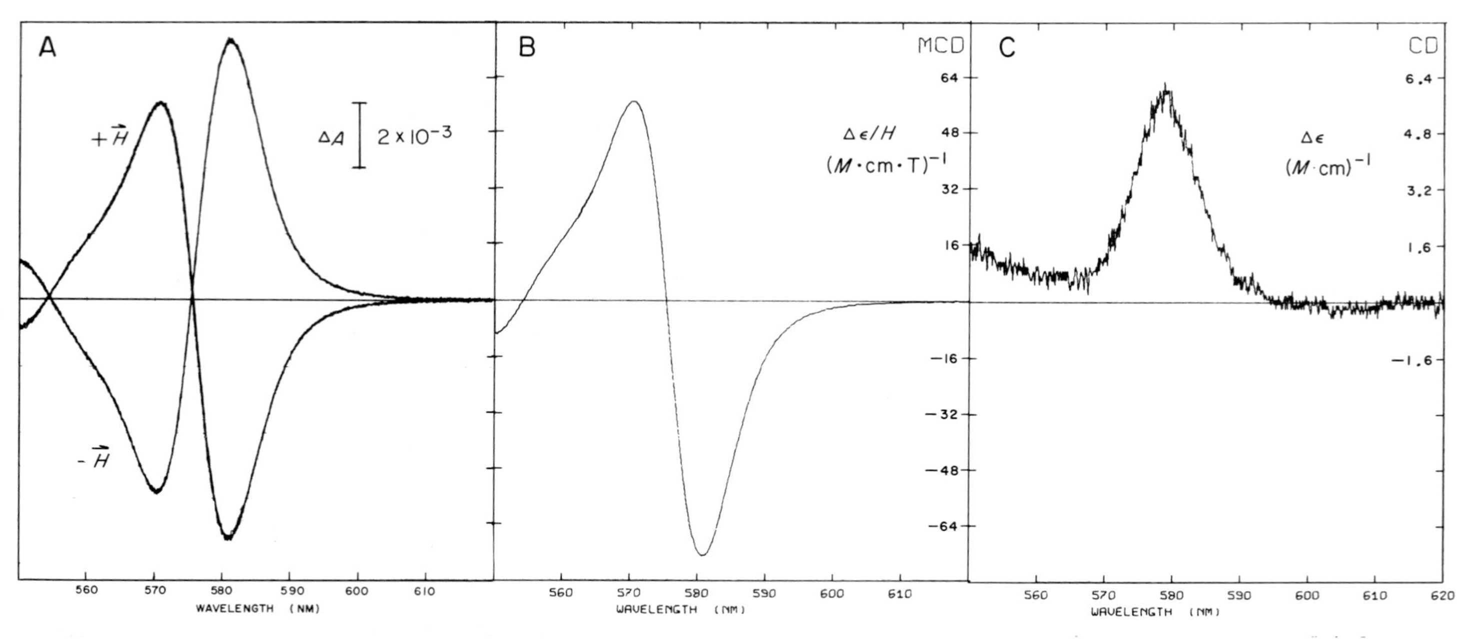

FIG. 3. MCD and CD spectra of oxyhemoglobin: (A) Chart showing original traces; four passes were recorded for each field direction. (B) Signal-averaged MCD, Eq. (6). (C) Signal-averaged CD, Eq. (7), assuming no base-line correction. The sample contained 72 μM hemoglobin A_0 in 0.1 M HEPES, 1 mM EDTA, and 1 mM inositol hexaphosphate; pH 7.0, 4°C, path = 1 cm, field = 1.42 T.

Since MCD arises only from absorption and not refractive differences, strain or imperfections in the optical windows are not important and homemade cells of Pyrex or Lucite are quite satisfactory in the visible spectral region. Sample concentrations should be adjusted so that at least 5% of the light is transmitted (OD < 1.3) at all wavelengths; optimal signal-to-noise ratio is obtained at an OD near 0.8–0.9, and at OD values above 2.0 ($<1\%$ transmittance) not only is the noise level unnecessarily high, but there is a danger of nonlinear dynode gain. For a typical heme protein this means that a total heme concentration of about 10 μM is desirable for measurements in the Soret region (350–470 nm), 50–100 μM for the visible bands (470–670 nm), and approximately 1 mM for the near-IR transitions (0.7–1.7 μm) if a 1-cm cell is to be used. Short-path cells for samples of high absorbance can be accommodated in most magnets, but the actual pole piece gap may restrict the length of longer-path cells.

For measurements in the 0–37° range or higher, jacketed cells or cell holders can easily be placed in the electromagnet gap with a stream of dry nitrogen flushed over the window to prevent condensation in the lower range. If the temperature needs to be maintained only slightly below the freezing point of water, 0 to $-10°$, to stabilize some reactive compounds, freezing-point-depression solvents of the type recommended by Douzou[15] can be used in connection with refrigerated circulators. For determination of C terms in the MCD spectra, however, a much wider range of temperature variation is preferred, and this requires both special glass-forming solvent mixtures and cryostats.

The solvent system of choice for C-term determination depends on the lower temperature limit desired and compatability with the sample to be studied. Solutions of glycerol in water or buffer (75% glycerol) have proved convenient for a number of heme proteins and usually allow measurements to temperatures as low as -100 to $-150°$C before cracking and scattering the incident beam. Solutions of ethylene glycol also remain sufficiently fluid below 0°C for low-temperature mixing experiments,[15] and a 50% aqueous mixture will form a clear glass down to about $-140°$C. Potassium glycerophosphate, glycerol, and buffer in equal-volume amounts yield a glass stable to liquid nitrogen temperature ($-196°$C), but not all proteins are stable in this high-ionic-strength mixture. Care must also be taken to neutralize the alkaline glycerophosphate, and it is recommended that all glycol solvents be added slowly with sufficient mixing to prevent high local concentrations. The lowest-temperature measurements, down to 4°K, have been obtained using *saturated* solutions of sucrose. All of the solvent systems seem to form stable glasses best in short-path-length cells (1–5 mm).

[15] G. Hui Bon Hoa and P. Douzou, *J. Biol. Chem.* **248,** 4649–4654 (1973).

This may serve to minimize the strain induced by small volume changes; sample cells with thin, flexible Lucite windows can be easily made and used.

Vacuum dewars of almost any geometry can be constructed from glass, quartz, or metal with windows for use down to liquid-nitrogen temperature. The main restriction is a short total optical path to fit between the magnet pole pieces, a gap which may be just greater than 1 cm. We have found it convenient to have interchangeable pole pieces to provide wider gaps of 1 to 2 inches at the expense of some field strength in order to accommodate cryostats and other long-path cells. Temperature control can be achieved by regulating the flow rate of dry nitrogen through coils immersed in liquid nitrogen before flowing into the dewar and over the sample. Figure 4 shows a sketch of one way to set this up. Flow rate and heat regained in the transfer from the coils to the sample dewar determine the temperature obtained, and an evacuated transfer tube is recommended to approach −170°; liquid nitrogen also can be added directly into the sample dewar. A copper-constantan thermocouple embedded into the sample cell is useful over the 30° to −196° range. An even less expensive variation of this approach would be to encase the magnet gap with Styrofoam in place of a dewar. This might prove advantageous in experiments in which high-scattering samples (such as turbid membrane preparations) are being used since a light pipe could then be placed directly against the sample to collect a greater percentage of the scattered light.

Measurements of MCD from room temperature to liquid-nitrogen temperature are sufficient for establishing the paramagnetic origin of intense *C* terms since this changes the Boltzmann distribution of ground-state populations and MCD intensity by a factor of 3.8. However, for very weak *C*-term signals or cases in which spin-exchange coupling is suspected, it is

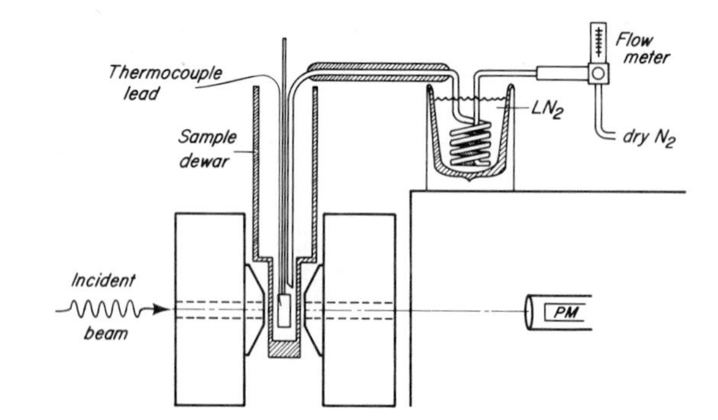

FIG. 4. Experimental apparatus for variable-temperature MCD measurements (20 to −196°C).

advantageous to obtain measurements at lower temperatures. Commercial cryostats using liquid-helium transfer or closed-cycle heat exchange capable of temperatures of 2°K and 10°K, respectively, are marketed by Air Products and Chemicals, Inc.

Several types of side effects can complicate the interpretation of low-temperature MCD measurements. Small increases in MCD intensity arise from solvent contraction on cooling, but for the above-mentioned glycol systems this results only in a volume change of 10–15% on going to −196°C and has generally been ignored. Changes in bandwidths of transitions, however, can lead to much larger effects. While the Soret band of most heme proteins is not appreciably narrowed at low temperature, the *Q* bands can become much sharper and more intense. Examples in which the MCD peak intensity of a diamagnetic reduced cytochrome increased due to narrowing almost as much as would be expected for a *C* term have been reported.[10,16] Such effects can be checked by recording the absorption and/or CD spectra at low temperature and by monitoring bandwidth or peak-to-trough splitting to the MCD curve. Further effects also can be observed with systems which are photosensitive, such as the carbon monoxide complex of heme proteins, since photodissociation may not be reversible at low temperatures or photostationary states may be established at high light intensities. Reversible conformational changes in heme proteins also can occur as the temperature is lowered.[7]

Results

The first measurements of the temperature dependence of the MCD of a heme protein establishing the presence of paramagnetic MCD effects were described by Briat *et al.* for cytochrome b_2.[17] Since then other reports have appeared on hemoglobin,[7,18,19] myoglobin,[9,19] cytochromes *c*,[10,18] b_5,[10] and *P*-450,[20] cytochrome *c* oxidase,[21] and some isolated heme derivatives.[22,23]

[16] A. M. Arutyunyan, A. A. Konstantinov, and Y. A. Sharanov, *FEBS Lett.* **46,** 317–320 (1974).

[17] B. Briat, D. Berger, and M. Leliboux, *J. Chem. Phys.* **57,** 76–77 (1972).

[18] M. A. Livshitz, A. M. Arutyunyan, and Y. A. Sharanov, *J. Chem. Phys.* **64,** 1276–1280 (1976).

[19] S. Yoshida, T. Iikuka, T. Nozawa, and M. Hatano, *Biochim. Biophys. Acta* **405,** 122–135 (1975).

[20] T. Shimizu, T. Nozawa, M. Hatano, Y. Imai, and R. Sato, *Biochemistry* **14,** 4172–4178 (1975).

[21] G. T. Babcock, L. E. Vickery, and G. Palmer, *J. Biol. Chem.* **251,** 7907–7919 (1976).

[22] H. Kobayashi, T. Higuchi, and K. Eguchi, *Bull. Chem. Soc. Jpn.* **49,** 457–463 (1976).

[23] T. Shimizu, T. Nozawa, and M. Hatano, *Bioinorg. Chem.* **6,** 77–82 (1976).

Typical results for both ferrous and ferric forms of heme proteins are discussed below.

Ferrous Case

The MCD spectra of reduced, Fe(II) heme proteins are the most straightforward to interpret since the low-spin ($S = 0$) form is diamagnetic and only the high-spin ($S = 2$) form can yield temperature-dependent C terms. For example, reduced cytochrome c exhibits no change in its Soret MCD spectrum from 22 to $-132°C$, and the visible region shows only effects due to band narrowing which are also apparent in the absorption spectrum.[10] High-spin ferrous forms of the above-mentioned heme proteins on the other hand exhibit MCD spectra in the Soret region which are extremely temperature sensitive and must be composed predominantly of C terms. The shape of the MCD curve of deoxyhemoglobin is very similar to that of deoxymyoglobin shown in Fig. 2, but the cytochrome oxidase case is more complex since this enzyme contains both high- and low-spin heme a. Figure 5 illustrates how the high-spin cytochrome a_3 ($S = 2$) portion of the spectrum can be resolved by a temperature-dependence study: the difference spectrum between any two temperatures yields only the C-term contribution of that chromophore while the low-spin cytochrome a ($S = 0$) component spectrum is not affected. The shape of the room-temperature MCD spectrum of high-spin reduced horseradish peroxidase[24] is also similar to these curves consistent with a C-term origin, but reduced cytochrome P-450 exhibits a Soret region MCD spectrum which is very different[20,25,26] even though it is thought to exist in an $S = 2$ state. The reason for the unusual MCD band shape in ferrous P-450 is not known, but the difference from other systems emphasizes the importance of temperature-dependence measurements for assignment of the $S = 0$ and $S = 2$ spin states. Triplet ($S = 1$) states are not generally considered to be populated in hemoproteins, but evidence for this spin state has been obtained for Fe(II) phthalocyanin in certain solvents (see Stillman and Thompson[27]). The paramagnetism of this spin state would also be expected to yield Faraday C terms.

Ferric Case

For Fe(III) heme proteins both the low-spin ($S = 1/2$) and high-spin ($S = 5/2$) states are paramagnetic and give rise to C terms. Experimentally it

[24] T. Nozawa, N. Kobayashi, and M. Hatano, *Biochim. Biophys. Acta* **427,** 652–662 (1976).
[25] P. M. Dolinger, M. Kielezewski, J. R. Trudell, G. Barth, R. E. Linder, E. Bunnenberg, and C. Djerassi, *Proc. Natl. Acad. Sci. U.S.A.* **71,** 4594–4597 (1974).
[26] L. Vickery, A. Salmon, and K. Sauer, *Biochim. Biophys. Acta* **386,** 87–93 (1975).
[27] M. J. Stillman and A. J. Thompson, *J. Chem. Soc., Faraday Trans. 2* **70,** 790–804 (1974).

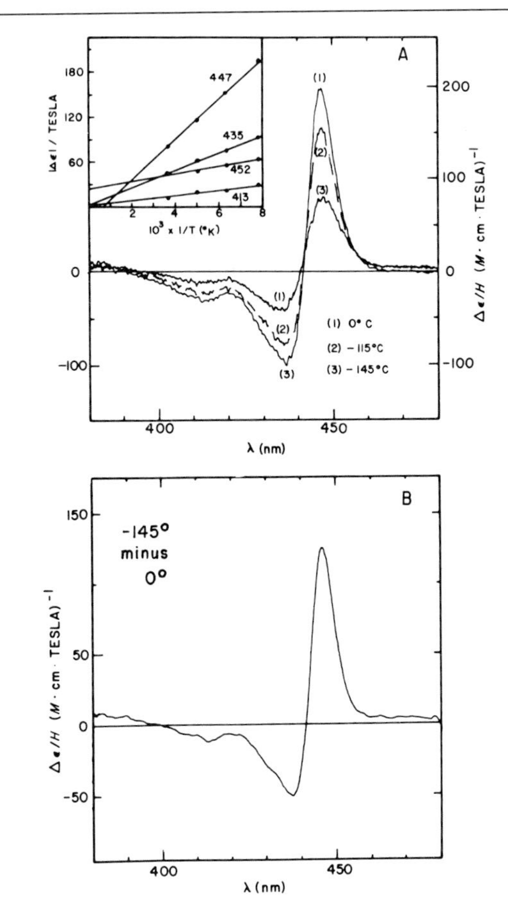

FIG. 5. Temperature dependence of the Soret MCD spectrum of reduced cytochrome oxidase.[21] [Reprinted with permission from *J. Biol. Chem.* **251** (1976).]

has been found, however, that in the Soret spectral region the MCD intensity of the low-spin form is much greater than that of the high-spin form. Figure 2 shows the MCD spectra of the completely low-spin cyanide derivative of ferrimyoglobin and of the completely high-spin fluoride form, and the intensity of the latter is seen to be neglible by comparison. In addition, it was found that other complexes of myoglobin which produce thermal spin-state mixtures give rise to MCD spectra whose intensities are approximately linearly proportional to the amount of low-spin form present in the equilibrium.[9] This is presumably due to the difference in the extent of spin-orbit coupling in the planar low-spin vs. the out-of-plane high-spin iron. Several other completely low-spin heme proteins differing in their axial coordination and even heme type (cytochromes b, b_5, c, c_1, and f and hemopexin complexes with proto- and deuteroheme) also have MCD intensities at 20°C similar to the ferrimyoglobin–cyanide complex. This suggests that as an initial approximation of spin state one can compare the room-temperature spectrum of a given ferriheme protein with myoglobin cyanide assuming that a long-wavelength extremum value of $\Delta\epsilon/H = -90$ to -100 $(M \cdot cm \cdot T)^1$ and a short-wavelength peak value of about $+70$ $(M \cdot cm \cdot T)^{-1}$ corresponds to the 100% low-spin state. The existence of an immediate spin state ($S = 3/2$) has been proposed for cytochrome c',[28] but the MCD behavior of such systems has not been investigated.

This simplified procedure of comparing single-temperature measurements is often sufficient for establishing spin-state differences within a single system, e.g., changing solvent conditions or addition of enzyme substrates, effector molecules, axial ligands, and denaturants, but should not be relied upon for accurate determinations of absolute spin states. Cytochrome oxidase and cytochrome P-450 are examples of systems where the myoglobin results cannot be applied directly. The low-spin ferric forms of each of these enzymes exhibit much weaker Soret MCD intensities than would be expected on the basis of the above arguments. In the case of cytochrome oxidase it was found that ferriheme a in a completely low-spin complex does not yield the same Soret C-term intensity as heme b or c, and hence a different standard must be used for a-type heme complexes. The cytochrome P-450 difference is not understood, but like a number of other unusual spectroscopic properties of the protein it may be a result of mercaptide coordination.

The MCD quantity that would be most valuable to correlate with spin state is the actual C-term intensity expressed as the ratio of the MCD intensity to the absorption intensity and determined from the change in the MCD spectrum with temperature. This may not be possible due to effects of temperature on spin equilibria (see below), but an example of measurements on a purely low-spin heme protein is present in Fig. 6. MCD spectra of cytochrome b_5 were recorded at six temperatures from 22 to -196°C, and the trough and peak extrema intensities, as well as the CD trough, are plotted against the reciprocal of the absolute temperature (ideally we would compare the integrated intensity changes with temperature). The slope of this plot represents the C-term contribution to the MCD spectrum. That weaker A and/or B terms also contribute somewhat to the spectrum is evidenced by the fact that the data at 419 nm do not extrapolate to zero at infinite temperature where C terms disappear; comparison of the intensity change between any two temperatures also shows that the effect

[28] M. M. Maltempo, *J. Chem. Phys.* **61**, 2540–2547 (1974).

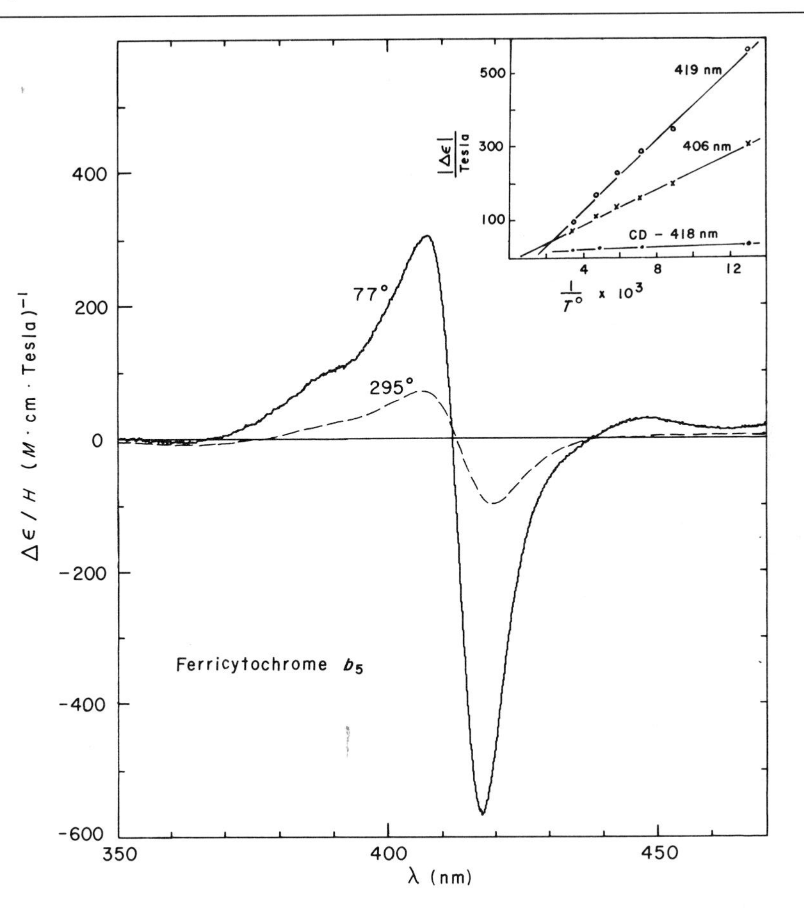

FIG. 6. Temperature dependence of the Soret MCD spectrum of oxidized cytochrome b_5.[10] [Reprinted with permission from *J. Am. Chem. Soc.* **98** (1976).]

is greater than would be expected on the basis of Boltzmann ground-state populations, thus indicating a temperature-independent component of opposite sign at this wavelength.

Spin-state Equilibria

Because of the large differences in shape, intensity, and temperature dependence of the MCD of heme proteins in different spin states, the technique can be used to investigate spin equilibria. In the ferrous state most heme proteins seem to be exclusively high or low spin with no clear examples of a $(S = 2) \rightleftharpoons (S = 0)$ thermal spin equilibrium. In the ferric state, however, the high- and low-spin forms often lie close in energy with the equilibrium a sensitive function of temperature as well as axial ligand field strength. By analysis of the temperature dependence of the MCD intensity of ferric spin-state mixtures one can extract the thermodynamic parameters for the interconversion between the two forms. Figure 7 presents hypothetical examples in which $(S = 5/2) \rightleftharpoons (S = 1/2)$ spin equilibria are poised at an equal mixture of both forms near room temperature. The broken lines represent typical MCD intensities for pure low-spin (upper curve) and high-spin (lower curve) ferric heme complexes; a linear temperature dependence which extrapolates to zero at infinite temperature (i.e., no A or B terms) has been assumed. The presence of a thermal spin equilibrium leads to nonlinearity and a failure of the curve to extrapolate to near zero at $1/T = 0$. Equilibria having an $S = 1/2$ ground state would approach the low-spin limit asymptotically, the curvature being most easily recognized in cases with a large negative enthalpic change; spin mixtures with an $S = 5/2$ ground state would actually exhibit an initial decrease in MCD intensity as the temperature is lowered. The actual ΔH and ΔS values for an experimental system can be determined by curve fitting.

Spin Coupling

Some heme proteins possess more than one heme and/or other paramagnetic centers which may be close enough for magnetic exchange to occur. The interaction may involve direct overlap of the orbitals containing

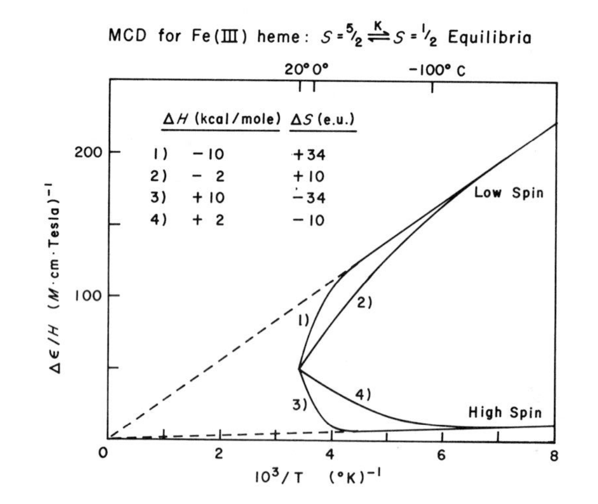

FIG. 7. Temperature dependence of the Soret MCD intensity for hypothetical cases of Fe(III) low-spin/high-spin equilibria.

the unpaired electrons or may occur through a superexchange mechanism involving orbitals of bridging diamagnetic species. A parallel alignment of spin results in a ferromagnetic interaction and an antiparallel alignment leads to antiferromagnetism; either of these states can exist in a thermal equilibrium with the noninteracting state, and the temperature dependence of the susceptibility can be used to determine the strength of the exchange interaction. The reader is referred to Mabs and Machin[29] for a quantitative treatment of the magnetic properties of polynuclear transition-metal complexes.

The effect spin coupling has on the MCD heme proteins has not yet been explored in any detail, but the sensitivity of MCD to the heme spin state suggests that the technique will prove useful for exploring the behavior of such systems in the future. In particular, the ability to obtain magnetic data over a broad temperature range may make it possible to investigate states lying at higher energies than is possible by EPR. This is illustrated by some results obtained for cytochrome oxidase,[21] a heme enzyme which contains two molecules of heme *a* and 2 g-atoms of copper per mole. In the fully oxidized enzyme a strong antiferromagnetic coupling between one cupric ($S = 1/2$) center and ferricytochrome a_3 ($S = 5/2$) to yield an $S = 2$ system has been proposed to account for the lack of an observable EPR signal for either species and for the linearity of the magnetic susceptibility (see Palmer *et al.*[30]). The MCD of the oxidized enzyme, however, does not show any paramagnetic effects attributable to the $S = 2$ state over the temperature range 20 to $-145°$, i.e., the MCD of cytochrome a_3 appears to be quite weak as expected for the noninteracting $S = 5/2$ state. The addition of cyanide to the enzyme converts cytochrome a_3 to the low-spin ($S = 1/2$) state but does not produce any new EPR signals, thus suggesting that cytochrome a_3 and the EPR-undetectable copper are still spin coupled at low temperature. The room-temperature MCD spectrum, however, exhibits an intensity expected for the $S = 1/2$ cyanide complex in an uncoupled system. While it is possible that a similar MCD curve could arise from the ferromagnetic $S = 1$ coupled state, the results are most simply interpreted in terms of a weak exchange interaction, J, so that spin coupling is manifest only in the lower-temperature range of the EPR measurements (the lack of an observable Cu(II) $S = 1/2$ EPR signal at higher temperatures could result from cross-relaxation effects with the iron $S = 1/2$ state without strong ferromagnetic coupling).[21] A temperature-dependence

[29] F. E. Mabs and D. J. Machin, "Magnetism and Transition Metal Complexes," pp. 170–203. Chapman & Hall, London, 1973.

[30] G. Palmer, G. T. Babcock, and L. E. Vickery, *Proc. Natl. Acad. Sci. U.S.A.* **73,** 2206–2210 (1976).

study of the MCD of this derivative should provide valuable insight into the mechanism of metal–metal interactions in this enzyme.

Acknowledgments

The author owes special thanks to Professors Kenneth Sauers, Todd Schuster, and Melvin Calvin and Dr. Melvin Klein for their encouragement during the course of the studies reviewed here. This research was supported in part by NSF Grant PCM-76-20041 and NIH Grant HL-17494 to T.S. and in part by the Division of Biomedical and Environmental Research of the U.S. Energy Research and Development Administration.

[18] Infrared Spectroscopy of Ligands, Gases, and Other Groups in Aqueous Solutions and Tissue

By JOHN C. MAXWELL and WINSLOW S. CAUGHEY

Infrared spectroscopy is one of two forms of vibrational spectroscopy. The other is Raman spectroscopy. In the infrared experiment, light of the energy associated with a given vibration is absorbed. The Raman experiment measures the energy of a vibration as a *difference* between the much greater energies of incident and scattered light. Both types of spectroscopy have recently enjoyed markedly expanded use in biological systems.[1,2] Infrared spectra of proteins have been used to explore relationships between the conformation and the frequencies of infrared bands, especially the amide bands I and II found in the region from 1680–1430 cm^{-1}.[3] Difference infrared spectroscopy of proteins has also been used to a limited extent to follow ionization equilibria of carboxylic acids[4,5] and hydrogen bonding of SH groups.[6] In this article we are primarily concerned with the methods employed to study small ligands bound to metals in heme, copper, and other metalloproteins. However, these methods can be applied to any observable vibration.

Infrared Spectroscopy in Aqueous Media: General Considerations

The presence of water in biological materials restricts the measurement of infrared absorption. Water absorbs strongly in the infrared region—

[1] J. C. Maxwell and W. S. Caughey, *Biochemistry* **15**, 388 (1976).
[2] T. G. Spiro, *Biochim. Biophys. Acta* **416**, 169 (1975).
[3] F. S. Parker, "Applications of Infrared Spectroscopy in Biochemistry, Biology and Medicine." Plenum, New York, 1971.
[4] S. N. Timasheff, H. Susi, and J. A. Rupley, this series, Vol. 27 [23].
[5] R. E. Koeppe II and R. M. Stroud, *Biochemistry*, **15**, 3450 (1976).
[6] G. H. Bare, J. O. Alben, and P. A. Bromberg, *Biochemistry* **14**, 1578 (1975).

ISBN 0-12-181954-X

more strongly in some regions than in others (Fig. 1). At least some radiation (e.g., 1%) must pass through the material under observation at the wavelength of interest if an accurate absorption measurement is to be made. As the absorption by water increases, the maximal thickness of sample (path length) must decrease and the minimal concentration of vibrating groups must increase. The path length of the CaF_2 cell used to obtain the spectra in Fig. 1 is only 0.030 mm and is even shorter in the KRS-5 cell (0.015 mm). Infrared measurements on solutes are therefore more readily made in regions of low absorption by water (the "windows") from 4000 cm^{-1} to 3800 cm^{-1}, 2800 cm^{-1} to 1750 cm^{-1}, and 1500 cm^{-1} to 900 cm^{-1} than in regions of strong absorption. If the solute has infrared bands of interest in the region not readily accessible in the presence of water, it is frequently possible to observe these bands by exchange of H_2O with D_2O which has "windows" at frequencies where H_2O does not (Fig. 1).

Since the band of interest may well appear amid a large number of other bands, its positive identification can be greatly facilitated by use of *difference* spectroscopy and by shifts in frequency detected as a result of the substitution of an isotope in the specific vibrator of interest. The bands of approximately Lorentzian shape exhibit several useful parameters. The

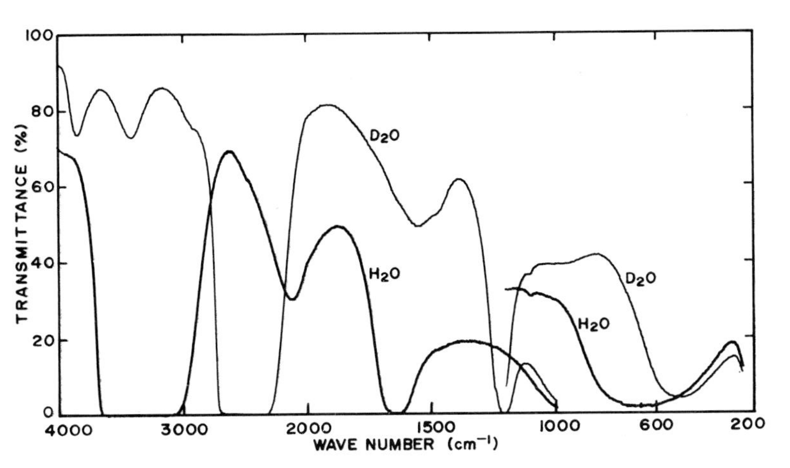

FIG. 1. Infrared spectra of H_2O vs. air (heavy trace) and D_2O vs. air (light trace) recorded in percent transmission mode. A cell with CaF_2 windows and 0.030-mm path length was used to record the 4000–1000 cm^{-1} region; CaF_2 begins to absorb strongly at 1200 cm^{-1} and is not useful below 1000 cm^{-1}. A cell with KRS-5 windows and 0.015-mm path length was used to record the spectral region from 1200–250 cm^{-1}. KRS-5 (thalium bromide-iodide) has a useful range at 4000–250 cm^{-1} but its high refractive index (2.38) creates some technical problems (see text). Both H_2O and D_2O have regions of strong absorption, but, since these regions do not overlap significantly in the 4000–650 cm^{-1} region, infrared studies can be carried out throughout this range by use of either H_2O or D_2O.

frequency at maximum absorption varies with changes in bonding between the vibrating atoms and, to a lesser extent, with the immediate environment (solvent and/or protein). The *band width* depends upon the stability or uniformity of the interactions between the vibrator and its environment. Other characteristics of bonding are the *intensity* and the shift in frequency due to isotope substitution. Intensities are also useful analytically. A wide variety of sampling conditions are satisfactory. Essentially physiological conditions (e.g., body temperature, neutral pH, and aqueous medium) may be studied or, if desired, wide deviations from these conditions may be used. Since at these long wavelengths scattering is far less important than it is in the visible region, intact tissues may be examined as well as clear solutions. Optimal sampling procedures and infrared cell construction will vary depending upon the nature of the sample and the region of the spectrum. Recent advances in the design of infrared spectrometers of both dispersive and interferometer types and in computer interfacing with the spectrometer have markedly increased sensitivity, resolution, and data-processing capabilities.

Spectral Regions of Special Interest

1750 to 2500 cm^{-1}. The region of the infrared spectrum between 1750 and 2500 cm^{-1} represents a "window" for both water and proteins (Figs. 1 and 2). Those bands that do appear in this region are therefore the most readily studied due to the lower and generally featureless absorption of water, protein, and other tissue components and the insignificant scattering for particulate samples. Here difference spectra are generally obtained satisfactorily with only water in the reference cell, and several cell window materials with high transmission in this region are available. Examples of vibrators that have been studied are the metal-bound ligands CO, N_3^-, CN^-, CNO^-, CNS^-, and $CNSe^-$ and the gaseous anesthetic N_2O.[7-9]

1000 to 1750 cm^{-1}. Water, protein, and other tissue components absorb strongly with many resolved or partially resolved bands in the region from 1000–1750 cm^{-1} (Figs. 1 and 2). For much of this region D_2O is a more useful medium than is H_2O. Obtaining well-isolated single bands for a ligand or other vibrator in this region will usually require very carefully measured difference spectra, e.g., with precisely matched reference and sample cells. Figures 3 and 4 illustrate the use of such difference spectra to

[7] J. O. Alben and W. S. Caughey, *Biochemistry* **7**, 175 (1968).

[8] S. McCoy and W. S. Caughey, *Biochemistry* **9**, 2387 (1970).

[9] J. M. Caughey, W. V. Lumb, and W. S. Caughey, *Biochem. Biophys. Res. Commun.* **78**, 897 (1977).

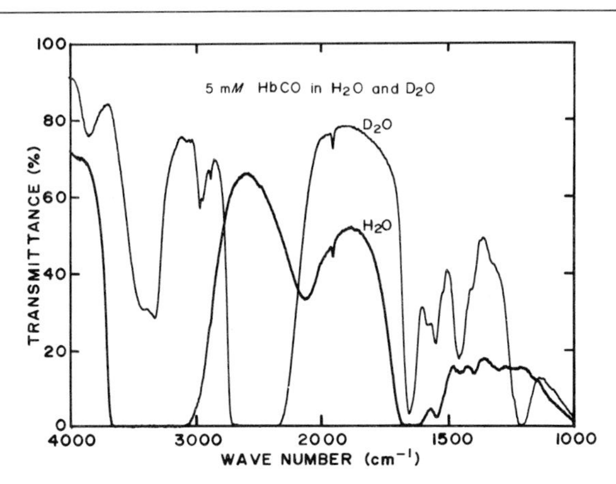

FIG. 2. Infrared spectra of ca. 5 m*M* human carbonyl hemoglobin in H_2O (heavy trace) and D_2O (light trace) recorded in the percent transmission mode. The cells had CaF_2 windows and path lengths of 0.030 mm. The sharp absorption at 1950 cm^{-1} is due to heme-bound carbon monoxide. Absorptions due to protein are seen more clearly in D_2O because strong H_2O absorption obscures protein vibrations.

isolate the NO stretch band from among the amide bands of nitrosyl hemoglobin A.[1] These spectra also illustrate the importance of isotopic substitution (in this case $^{14}N^{16}O$ vs. $^{15}N^{16}O$) for positive identification of the bands of interest. Similar techniques were utilized to obtain the O–O stretch for oxyhemoglobins[10-12] and oxymyoglobin[13] with difference spectra between $^{16}O-^{16}O$ and $^{12}C-^{16}O$ species, $^{18}O-^{18}O$ and $^{12}C-^{16}O$ species, and $^{16}O-^{16}O$ and $^{18}O-^{18}O$ species. Difference spectra between species which differ in ligand and/or oxidation state reveal numerous spectral changes of interest other than those of the ligand *per se*. However, these changes have not been well studied as yet. Some of these bands may arise from protein and others from the aromatic porphyrin ring.

Below 1000 cm^{-1}. Exploration of the spectral region below 1000 cm^{-1} is limited by strong absorption of both H_2O or D_2O (Fig. 1) and by the low intensity generally found for bands in this region. Nevertheless the region is of great interest. Most metal–ligand atom vibrations (e.g., Fe–N, Fe–O,

[10] C. H. Barlow, J. C. Maxwell, W. J. Wallace, and W. S. Caughey, *Biochem. Biophys. Res. Commun.* **55,** 91 (1973).

[11] J. C. Maxwell and W. S. Caughey, *Biochem. Biophys. Res. Commun.* **60,** 1309 (1974).

[12] W. S. Caughey, J. C. Maxwell, J. M. Thomas, D. H. O'Keeffe, and W. J. Wallace, *in* "Metal Ligand Interactions in Organic Chemistry and Biochemistry" (B. Pullman and N. Goldblum, eds.), D. Reidel Publishing Co., Dordrecht-Holland, pp. 131–152 (1977).

[13] J. C. Maxwell, J. A. Volpe, C. H. Barlow, and W. S. Caughey, *Biochem. Biophys. Res. Commun.* **60,** 1309 (1974).

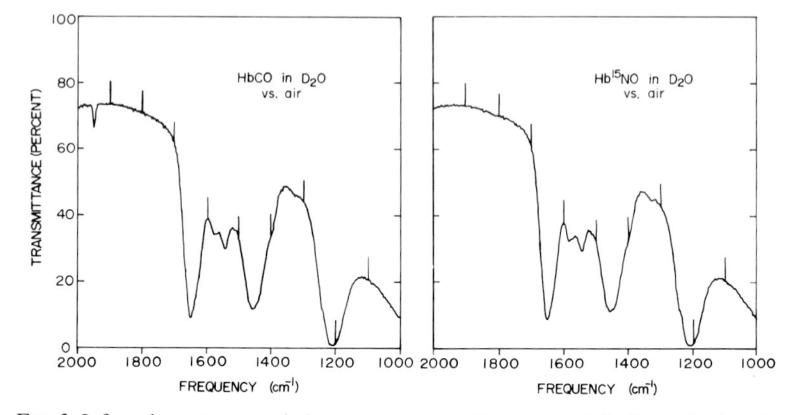

FIG. 3. Infrared spectra recorded as percent transmittance vs. air for hemoglobin solutions in D_2O 4.9 m*M* in heme: carbonyl species on the left; nitrosyl ($^{15}N^{16}O$) species on the right. Cells had CaF_2 windows and path lengths of 0.26 mm. Wave-number markers appear at 100 cm^{-1} intervals on each spectrum. [Reprinted with permission. Copyright by the American Chemical Society.]

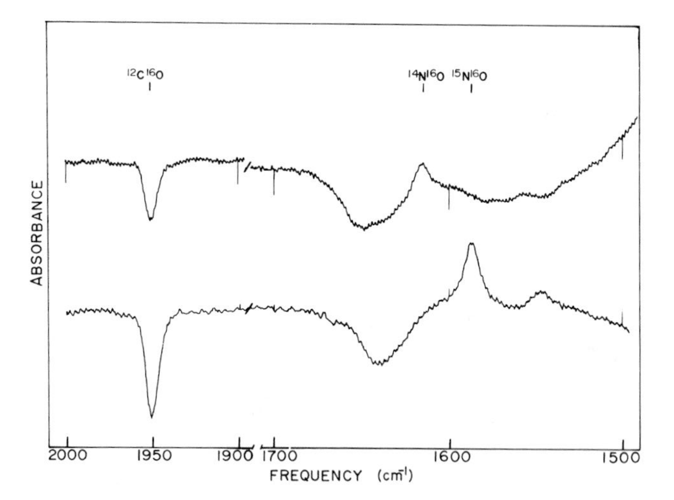

FIG. 4. Infrared difference spectra recorded in the absorbance mode. The upper trace shows the spectrum for hemoglobin solutions 2.7 m*M* in heme in 0.01 *M* sodium phosphate in D_2O buffer (pD 7.0) with the $^{14}N^{16}O$ species in the sample cell and the $^{12}C^{16}O$ species in the reference cell. The lower trace shows a similar spectrum for $^{15}N^{16}O$ vs. $^{12}C^{16}O$ species and was obtained from the same cells used to obtain the spectra of Fig. 3. Conditions were the same for both isotopes except for a difference in protein concentration. The negative band at 1951 cm^{-1} is due to bound CO, and the positive bands at 1615 and 1587 cm^{-1} result from bound $^{14}N^{16}O$, and $^{15}N^{16}O$ respectively. The broad negative band at ca. 1650 cm^{-1} is primarily due to a slight mismatch in cell path lengths. The HbCO solution was always placed in the longer path length cell so that any absorptions due to protein would appear as negative bands. [Reprinted with permission. Copyright by the American Chemical Society.]

Cu–O) occur between 200 and 1000 cm^{-1}, and the frequencies of these vibrations relate directly to the strength of metal–ligand bonding. Such bands are readily seen in infrared spectra for protein-free hemes and hemins in nonaqueous media,[14] but have not yet been observed for corresponding heme proteins. Other bands of interest also appear in this region, e.g., the O–O stretch for peroxides appears near 850 cm^{-1}.[15] Some progress in the study of bands in this region for aqueous systems of biological interest has been made with the Raman approach,[16–18] which receives less interference from the presence of water, but thus far, little has been done with infrared spectra below 1000 cm^{-1}. The recent improvements in technique do give promise that significant progress in this region of the infrared can be expected fairly soon.

Above 2500 cm^{-1}. Stretch bands that involve hydrogen appear above 2500 cm^{-1} and provide strong absorption from water, protein, and other tissue components due to O–H, C–H, S–H, N–H, etc. Also light scattering becomes more important for particulate systems at frequencies greater than 2500 cm^{-1}. This spectral region has long been of interest for protein studies but is less relevant to the study of ligand binding to metals. However, Alben *et al.*[6] have used the S–H stretch to study ligation effects on quaternary structures.

Band Parameters

A single nearly Lorentzian-shaped absorption band has several characteristics or parameters of value for interpretations of structure, bonding, and the environment about the vibrator as well as for the determination of how much vibrator is present. The object of the infrared experiment is to establish these parameters which are considered separately below. For ligand-binding studies, the infrared bands associated with the ligand provide a uniquely *direct* probe for the ligand, and the parameters associated with a ligand band provide important information about the ligand-binding site.

Frequency. The "band frequency" is that frequency where the band has reached maximum absorbance. For an accurate measurement of this

[14] N. Sadasivan, H. I. Eberspaecher, W. H. Fuchsman, and W. S. Caughey, *Biochemistry* **8,** 534 (1969).

[15] W. S. Caughey, C. H. Barlow, J. C. Maxwell, J. A. Volpe, and W. J. Wallace, *Ann. N.Y. Acad. Sci.* **244,** 1 (1975).

[16] J. B. R. Dunn, D. F. Shriver, and I. M. Klotz, *Proc. Natl. Acad. Sci. U.S.A.* **70,** 2582 (1973).

[17] J. S. Loehr, T. B. Freedman, and T. M. Loehr, *Biochem. Biophys. Res. Commun.* **56,** 510 (1974).

[18] A. Brunner, *Naturwissenschaften* **61,** 129 (1974).

frequency it is often helpful to locate the midpoints of lines drawn parallel to the base line at various heights from the base line to the point of maximum absorbance and to use the midpoints to determine the "band frequency." A band for a bound ligand tends to be extremely sharp (i.e., narrow bandwidth) which, in the absence of unresolved overlapping bands, permits the band frequency to be determined with high accuracy (i.e., ± 0.2 cm^{-1}). With some instruments, the lack of precision in the wavelength drive does not permit reproduction of the frequency within 0.2 cm^{-1}. In such a case, it may be advantageous to calibrate the frequencies recorded by recording a standard spectrum either just before or just after the band of interest without an interruption in the monochromater scanning drive. Such a spectrum is illustrated in Fig. 5 where the spectrum for the C–O stretch of hemoglobin carbonyl was recorded; the spectrometer was then converted to single-beam operation to record the H_2O vapor spectrum without stopping the scan. Spectra for H_2O and several other gases are well known and provide extremely accurate wavelengths standards. The National Bureau of Standards has provided a valuable source for spectral calibration standards.[19]

The frequencies of bound ligands may range quite widely as the protein differs. The wide range found for heme protein carbonyls is shown in Table I. Representative frequencies for other ligands are in Table II. The frequency may shift due to: (1) changes in bonding between the metal and the ligand; (2) steric restrictions about the ligand which prevent the ligand from achieving the preferred stereochemistry (e.g., bent Fe—C—O bonds can occur in hemoglobin[20,21] and myoglobin[22] carbonyls despite the preferred linear bonding stereochemistry normally found in metal carbonyls); (3) weak bonding interactions (usually of the dipole–dipole type) between the vibrator and its immediate environment (solvent, protein, etc.); and (4) Fermi or other vibronic interactions with neighboring vibrators.[10,11] For these reasons, the band frequency can serve as an exceedingly sensitive probe for differences or similarities among binding sites. A few of the many possible examples will be mentioned. The wide variety of frequencies found for CO in heme proteins (Table I) is an illustration of the wide variations in structure that occur in these sites. On the other hand, the consistency of normal hemoglobin CO stretch values (1951 ± 2 cm^{-1}) over a wide range of species—mammals, birds, reptiles, and fish—demonstrates a remarkable consistency in binding-site character over a broad span of evolution. Consequently, detection of a CO band at a different frequency in

[19] E. K. Plyler, A. Danti, L. R. Blaine, and E. D. Tidwell, *J. Res. Natl. Bur. Stand.* **64,** 1960.
[20] R. Huber, O. Epp, and H. Formanek, *J. Mol. Biol.* **52,** 349 (1970).
[21] E. J. Heidner, R. C. Ladner, and M. F. Perutz, *J. Mol. Biol.* **104,** 707 (1976).
[22] J. C. Norvel, A. C. Nunes, and B. P. Schoenborn, *Science* **190,** 568 (1975).

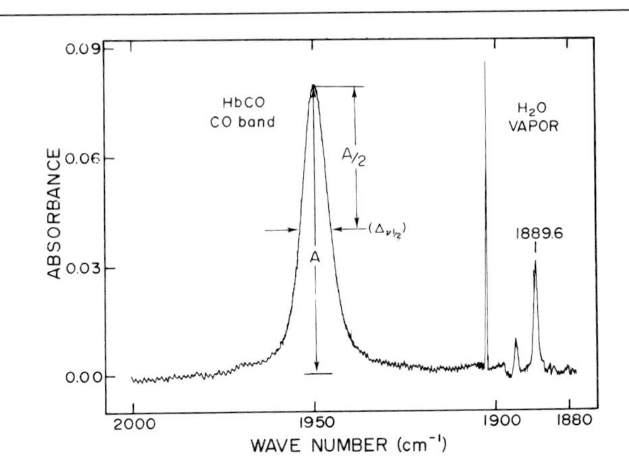

FIG. 5. An infrared spectrum recorded in the absorbance mode with highly expanded ordinate and abscissa for the difference spectrum of CO vs. O_2 human HbA reconstituted with deuteroheme from 2000–1903 cm^{-1}. Without interruption of the monochrometer drive, the samples were removed and the instrument converted for single-beam operation to allow the water vapor spectrum to be recorded (1903–1880 cm^{-1}). A water vapor peak such as the one at 1889.6 cm^{-1} provides a highly accurate standard for wave number calibration. For this spectrum the base line was chosen so as to intersect the spectrum ca. 30 cm^{-1} on each side of the peak maximum. The measurements needed for the calculation of the apparent maximal molecular extinction coefficient (ϵ) and bandwidth at half height ($\Delta\nu_{1/2}$) are indicated; for the computations required to go from the absorbance measurement to true ϵ values, see the text.

the blood represents a convenient and rapid method for the detection of abnormal hemoglobins.[23,24] The differences in frequency for CO bound to hemoglobin, myoglobin, and cytochrome *c* oxidase provide a means for detection of CO binding at these sites in heart and other tissue[25] as well as in solutions.[9,13,26] Similarly, on the basis of band frequency, CO binding at heme iron rather than copper(I) is indicated for cytochrome *c* oxidase,[24] and the high and low spin-states for azido methemoglobins and metmyoglobins can be differentiated.[8,27] Nitrous oxide (N_2O) represents a nonligand example in that at levels typical for anesthesia N_2O in brain exhibits ν_3 at more than one frequency in accord with N_2O molecules being located in both aqueous and nonaqueous media within the brain.[9]

[23] W. S. Caughey, J. P. Alben, S. McCoy, S. Charache, P. Hathaway, and S. Boyer, *Biochemistry* **8,** 59 (1969).

[24] W. J. Wallace, J. A. Volpe, J. C. Maxwell, W. S. Caughey, and S. Charache, *Biochem. Biophys. Res. Commun.* **68,** 1379 (1976).

[25] J. C. Maxwell, C. H. Barlow, J. H. Spalholz, and W. S. Caughey, *Biochem. Biophys. Res. Commun.* **61,** 230 (1974).

[26] W. S. Caughey, R. A. Bayne, and S. McCoy, *J. Chem. Soc. D.* 950 (1970).

[27] J. O. Alben and L. Y. Fager, *Biochemistry* **11,** 842 (1972).

TABLE I
INFRARED SPECTRAL PROPERTIES FOR CARBON MONOXIDE BOUND TO IRON(II) HEME PROTEINS

	ν_{CO} (cm^{-1})	$\Delta\nu_{\frac{1}{2}}$ (cm^{-1})	ϵ (M^{-1} cm^{-1} × 10^3)	B (M^{-1} cm^{-2} × 10^{-4})
Hemoglobin A	1951	8	3.7	3.4
Myoglobin (bovine heart)	1944[a]	12	2.0	2.8
Cytochrome *c* oxidase (bovine heart)[b]	1963.5	5.5	4.9	2.8
Cytochrome P 450_{cam} (substrate bound)[c]	1940	12.5	2.3	3.3
Cytochrome P 450_{cam} (substrate free)[c]	1940	~14		3.5
	1962			
Cytochrome P 420_{cam}[c]	1965	~25		
Cytochrome P 450_{LM}(PB)[c]	1948	~22		
Cytochrome P 450_{LM}(3MC)[c]	1954	~26		
Hemoglobin Zurich	1958[d]	8		3.2
	1951[d]	8		
Hemoglobin M_{Emory}[d]	1951	8		
	1970			
Rabbit hemoglobin[e]	1951	8		
	1928	10		
Hemopexin heme carbonyl[f]	1950	25		
Horseradish peroxidase[g]	1933	12		4.4
	1906	14		
Leghemoglobin[h]	1948	7		

[a] S. McCoy and W. S. Caughey, *in* "Probes of Structure and Function of Macromolecules and Membranes" (B. Chance, T. Yonetani, and A. S. Mildvan eds.), Vol. 2, p. 289. Academic Press, New York, 1971. A small but significant band at 1933 cm^{-1} is also always present. The origin of this band is still under investigation.

[b] S. Yoshikawa, M. G. Choc, M. C. O'Toole, and W. S. Caughey, *J. Biol. Chem.* **252,** 5498 (1977).

[c] D. H. O'Keeffe, R. E. Ebel, J. A. Peterson, J. C. Maxwell, and W. S. Caughey, unpublished observations. Cytochrome P-450_{cam} was isolated from *Pseudomonas putida* grown with d-camphor. Cytochrome P-450_{LM} was prepared from liver microsomes of rats pretreated with either phenobarbital (PB) or 3-methylcholanthrene (3MC).

[d] W. S. Caughey, J. O. Alben, S. McCoy, S. Charache, P. Hathaway, and S. Boyer, *Biochemistry* **8,** 59 (1969).

[e] N. A. Matwioff, P. J. Vergamini, T. E. Needham, C. T. Gregg, J. A. Volpe, and W. S. Caughey, *J. Am. Chem. Soc.* **95,** 4429 (1973).

[f] U. Muller-Eberhard, W. T. Morgan, J. C. Maxwell, and W. S. Caughey, unpublished observations.

[g] C. H. Barlow, P. I. Ohlsson, and K. G. Paul, *Biochemistry* **15,** 2225 (1976).

[h] C. A. Appleby, J. C. Maxwell, and W. S. Caughey, unpublished observations.

TABLE II
REPRESENTATIVE STRETCHING FREQUENCIES AND INTEGRATED INTENSITIES FOR HEMOGLOBIN- OR HEMIN-BOUND LIGANDS

Ligand	Stretching frequency (cm^{-1})	Integrated intensity (M^{-1} cm^{-2})
CO	1951[a]	3.4×10^4
NO	1615[b]	2.5×10^4
O_2	1107[c]	0.4×10^4
N_3	2025[d]	2.0×10^4
CN^-	2125[d]	0.5×10^4
$SeCN^-$	2075[d]	
SCN^-	2063[d]	
OCN^-	2167[d]	

[a] J. O. Alben and W. S. Caughey, *Biochemistry* **7**, 175 (1968).
[b] J. C. Maxwell and W. S. Caughey, *Biochemistry* **15**, 388 (1976).
[c] C. H. Barlow, J. C. Maxwell, W. J. Wallace, and W. S. Caughey, *Biochem. Biophys. Res. Commun.* **55**, 91 (1973).
[d] S. McCoy and W. S. Caughey, *Biochemistry* **9**, 2387 (1970).

Band Width. The width of an infrared band is conveniently expressed as the width in cm^{-1} at one-half peak height in absorbance (Fig. 5). The notation for this parameter—the "half bandwidth"—is $\Delta\nu_{1/2}$. For the measured $\Delta\nu_{1/2}$ to represent the true value, the instrumental resolution must be equal to, or less than, one-third of $\Delta\nu_{1/2}$. It is also important to establish the base line accurately. The location of the base line is not always obvious and is a frequent source of error. A useful approach is to consider that significant absorption starts at points 2.5 half bandwidths each side of the band frequency. Computer-assisted adjustments of nonlinear and nonflat base lines can be helpful but care must be taken not to introduce artifacts.

Observed bandwidths vary widely; data for heme protein carbonyls are in Table I. The binding of a ligand to protein usually results in a narrower band than is found for the ligand free in solution. Also conformational changes in protein may cause a large shift in both width and frequency. The bandwidth reflects the medium about the vibrator. Thus N_2O in water, methylene chloride, carbon tetrachloride, and *n*-hexane exhibit $\Delta\nu_{1/2}$ values of 11.2, 9.1, 7.9, and 14.5 cm^{-1} respectively.[9] Similar solvent effects are found with carbonyl and nitrosyl hemes.[1]

The bandwidth is determined by the nature of the population of vibrators of different energy that is seen by the incident infrared radiation. If the vibrating dipole interacts with its environment to give a wide range of energies, the band is broad, reflecting the energy distribution. On the other hand, a very narrow band is indicative either of little effect of the medium on the vibrating dipole or of a highly ordered environment in which the effects of the medium on the dipole are similar for the entire ligand population. Studies with different solvents reveal that solvation effects differ with the polarity and the shape of the solvent molecules. The dipole–dipole interactions between solvent molecules and the infrared vibrator affect the frequency as well as the intensity of the vibrating dipole seen in the infrared spectrum. A very narrow bandwidth as for CO in the carbonyl of cytochrome *c* oxidase ($\Delta\nu_{1/2}$, 6 cm^{-1}) indicates an unusually narrow range of C–O stretch energies which, in turn, requires a highly uniform environment of the CO ligand.[28] This could result from the complete absence of polar groups in the immediate environs of the CO since, with nonpolar groups, their positions with respect to the vibrating dipole of interest will have little effect on infrared energies. Alternatively, if the CO does in fact experience interaction with an adjacent dipole(s), then the relative position(s) of ligand and the polar group(s) must be remarkably uniform; within the population of CO ligands, there can be little variability in the steric positioning of CO with respect to the surrounding groups. The environment about CO at the oxygen binding in the $P_{450_{LM}}$ cytochromes, where $\Delta\nu_{1/2}$ is ca. 25 cm^{-1}, must be much less uniform than is the case with the oxidase carbonyl.

Band Height. The band intensity may be expressed in terms of height (extinction coefficients) or area (integrated intensities). The area is less sensitive to experimental conditions and is therefore the more reliable parameter for quantitative work. However, the heights can be very useful and are the more easily measured. The apparent molar extinction coefficient (ϵ_M) is defined as

$$\epsilon_M = \frac{A}{cl}$$

where A is the difference in absorbance between the base lines and the band maximum, c is the molar concentration, and l is the path length of the cell. The same requirements for resolution and base line location apply here as in the determination of $\Delta\nu_{1/2}$. Ligand band intensities have not been studied extensively; however, a few extinctions coefficients have been determined (Table I). For hemoglobin A and cytochrome *c* oxidase carbonyls, the ϵ values are independent of concentration.[28]

Integrated Intensity. The area of a band is conveniently determined as the apparent integrated intensity (B). The B value in M^{-1} cm^{-2} units can be computed from the band expressed in ϵ_M and cm^{-1} units either by computer-assisted integrations or manually by such procedures as triangu-

[28] S. Yoshikawa, M. G. Choc, M. C. O'Toole, and W. S. Caughey, *J. Biol. Chem.* **252,** 5498 (1977).

lation, planimetry,[29] and "weighing paper." Triangulation is a simple yet effective method whereby the half bandwidth, maximum absorbance, and an additional factor that accounts for the Lorentzian shape of the band are multiplied as follows:

$$B = (\Delta\nu_{1/2}) \times (\epsilon_M) \times (1.24)$$

Computer interfacing with the spectrometer makes integration a particularly easy procedure (Fig. 6). For all methods the accurate location of the base line is critical, whereas the resolution is not. The integrated intensity is useful for the measurement of how much vibrator is present as well as for correlations with structure, bonding, and the environment. However, only a few ligands and proteins have been studied in this way. Illustrative data appear in Tables I and II. In the most extensive study to date the B values for a single vibration (CO) remain quite constant over a range of heme proteins where ϵ and $\Delta\nu_{1/2}$ values vary widely (Table I).

Isotope Shift. The energy of a vibration is dependent upon the masses of the vibrating atoms. For an isolated diatomic oscillator the effect of a change in "mass" (isotope substitution) can be computed rather accurately. The frequency (ν) is found from the following expressions:

$$\nu = \frac{(k/\mu)^{1/2}}{2\pi}$$

where

$$\mu = \frac{m_1 m_2}{m_1 + m_2}$$

Here k is the harmonic force constant, μ is the reduced mass, and m_1 and m_2 represent the masses of the two atoms in the diatomic molecule. If k remains constant upon isotope substitution then from a knowledge of the frequency with natural abundance isotopes, the frequency expected with another isotope can be computed. Shifts in frequency observed for a few ligands are compared with values computed on the basis of a simple diatomic model in Table III. Of course, once bound, a diatomic ligand is no longer a simple diatomic system. An observed shift provides proof of the identity of the band, and its magnitude can give evidence on the nature of metal–ligand bonding. Because of the frequency difference the bands for two isotopes can experience rather different effects from Fermi resonance

[29] $$B = \frac{\text{Band area in cm}^2}{h \times w \times c \times b}$$

where h is cm/1.0 absorbance unit, w is the number of cm^{-1}/cm, c is the molar concentration, and b is the cell path length in cm.

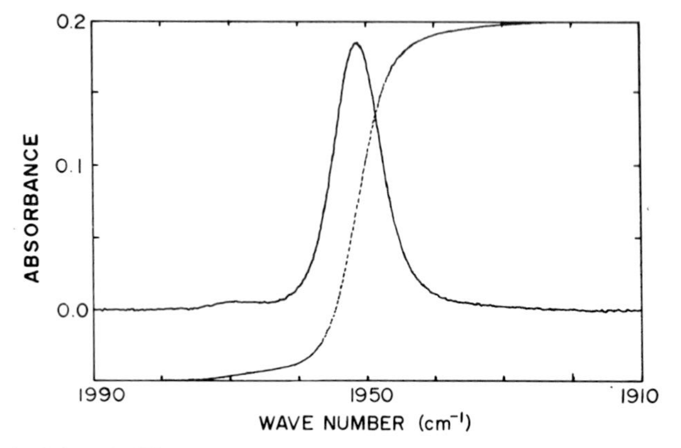

FIG. 6. An infrared difference spectrum of CO hemoglobin vs. O_2 hemoglobin plotted in absorbance (solid trace); the broken line represents the spectral integral from 1980–1920 cm^{-1}. Human hemoglobin solutions were 14 m*M* in Fe, and the CaF_2 cells had 0.038-mm path lengths. The low-intensity band at 1970 cm^{-1} is frequently seen but is of unknown origin. The integral plot shows the "wings" of the infrared band contribute very little to the integrated intensity. In our studies, we find insignificant absorption outside the region 2½ half bandwidths on each side of the absorption maximum. This contrasts with what is expected for a purely Lorentzian curve in which greater than 10% of the total integrated intensity would appear outside the region confined by 2½ half bandwidths on each side of the maximum.

and other vibronic coupling. Therefore isotopic substitution provides a means for detection of such effects.[10,11] Isotope substitution is not expected to influence $\Delta\nu_{1/2}$ (or intensity parameters) detectably since such an effect is proportional to the ratio of the two frequencies.

Cells and Sample Preparation

Cell Window Materials. The choice of water-insoluble window material depends primarily on three factors: (1) the transmission range, (2) the refractive index, and (3) the resistance to thermal and mechanical shock. It is advantageous to have the material as transparent as possible in the spectral region of interest. The refractive index (n) is important for two reasons. One is the loss through reflection at the surface of the window. The percent infrared energy reflected (R) from the surface can be approximated by the following expression:

$$R = \frac{(n - 1)^2}{(n + 1)^2}$$

and the total energy throughput for a typical cell with four surfaces (T_4) is about $(1 - R)^4$. Thus for a material like KRS-5 with an $n = 2.58$ at 2000

TABLE III
CALCULATED AND OBSERVED ISOTOPE SHIFTS IN INFRARED SPECTRA OF HEMOGLOBIN-BOUND LIGANDS

		$^{12}C^{16}O$	$^{13}C^{16}O$	$^{12}C^{18}O$
	ν^*/ν[b]		.9788	.9759
HbCO[a,c]	ν_{CO} observed (cm^{-1})	1951	1907	1907
	ν_{CO} calculated		1907.5	1904
		$^{14}N^{16}O$	$^{15}N^{16}O$	
	ν^*/ν		.9820	
HbNO[c]	ν_{NO} observed	1615	1587	
	ν_{NO} calculated		1586	
		$^{16}O_2$	$^{18}O_2$	
	ν^*/ν		.9428	
HbO_2[d]	ν_{O_2} observed	1107	1065	
	ν_{O_2} calculated		1044	

[a] J. O. Alben and W. S. Caughey, *Biochemistry* **7**, 175 (1968).
[b] ν^*/ν is the calculated ratio of the stretching frequencies where ν^* is the frequency of vibration for the isotopically labeled ligand and ν is the frequency for the natural-abundance ligand.
[c] J. C. Maxwell and W. S. Caughey, *Biochemistry* **15**, 388 (1976).
[d] C. H. Barlow, J. C. Maxwell, W. J. Wallace, and W. S. Caughey, *Biochem. Biophys. Res. Commun.* **55**, 91 (1973).

cm^{-1}, the losses due to reflection for an empty cell represent nearly 50% of the energy incident upon the cell. A second reason for concern about refractive index is the possible formation of interference fringes. If the windows in a cell are parallel and n for the sample between the windows is much lower than the n of the window, the spectrum of the sample will be recorded upon an interference fringe pattern with a periodicity related to the path length (see path length determination section). A solution to this problem is the use of a wedge-shaped spacer which makes the windows nonparallel, but path length matching is complicated by use of such a spacer. The resistance to thermal and mechanical shock is important if experimental procedures are used which require wide changes in temperature and pressure.

The transmission range and refractive index for some water-insoluble crystals are given in Table IV. Only three windows have been widely used for heme and other redox proteins: CaF_2, KRS-5, and AgCl. For most

TABLE IV
INFRARED SPECTRAL PROPERTIES OF SOME WATER-INSOLUBLE CELL WINDOW MATERIALS

	Low-frequency cutoff (cm^{-1})	Refractive index
KRS-5	250	2.38
AgBr	300	2.31
AgCl	450	2.00
Ge	600	4.01
Si	650	3.49
Itran-2	850	2.26
BaF_2	850	1.45
CaF_2	1100	1.40
Sapphire	1870	1.74
Polyethylene	(useful range 600–50 cm^{-1})	1.54

studies, CaF_2 is the window material of choice at frequencies greater than 1000 cm^{-1}. CaF_2 begins absorbing strongly near 1200 cm^{-1} and is not useful below 1000 cm^{-1}. It has a low refractive index and is very hard and strong. A good optical finish can be maintained without need for frequent repolishing, and long service can be expected if it is received without cracks from cleavage or drilling. A most useful property, especially for heme protein studies, is its transparency to UV–visible light. A spectrum from 300–700 nm may be recorded for the same cell for which an infrared spectrum was obtained. CaF_2 is not suitable for use with solutions containing ammonia in which it is soluble, but it is generally resistant to dilute acid or base. BaF_2 has a greater infrared transmission range but is weaker than CaF_2. KRS-5 which is composed of thallium bromide and iodide has a very wide transmission range (4000–250 cm^{-1}), but its high refractive index (2.38 at 2000 cm^{-1}) results in high reflection losses and persistent fringes in liquid cells. Also, etching of the surface occurs upon long exposure to aqueous solutions; KRS-5 is slightly soluble in water. However, the windows are easily repolished with silver polish or jewelers rouge to give satisfactory performance. With path lengths of 0.015–0.025 mm, H_2O may be used down to ca. 850 cm^{-1} and D_2O may possibly be used to 250 cm^{-1} although 650 cm^{-1} is often a more realistic cut-off. Special care should be exercised in handling KRS-5 because it is rather soft and will cold flow and because thallium is quite toxic but the toxicity is presumably not cumulative. The properties of AgCl resemble those for KRS-5; the transmission range extends only to ca. 450 cm^{-1}. Rods of AgCl are pressed to the desired thickness to prepare the windows. Since these plates are not crystalline, they are well suited for applications that involve low temperatures or sudden changes in temperature. AgCl is also quite soft and will cold flow. Isolation of AgCl by use of Teflon gaskets and spacers to separate the AgCl from metals prevents the corrosion of the metal that will occur when it is in direct contact with AgCl. Isolation from UV light is also necessary to prevent darkening of the windows. Frequently, when AgCl windows are first received, the smooth surface prevents aqueous solutions from satisfactorily wetting the surface with the result that air bubbles form as the cell is filled. The bubble formation can be reduced by "roughing up" the surface with jewelers rouge while keeping the AgCl plate flat.

Cells and Cell Matching. Infrared cells are available commercially in a wide range of styles and prices. Most have metal bodies that clamp a spacer of the appropriate thickness to provide the desired path length between the two windows. The cells usually have adapters for a syringe through which liquid samples are introduced to fill the cell. About 0.2 ml is required for most cells regardless of path length. Bubble formation often results from improper filling or drying of the cell or from an increase in sample temperature which can cause release of dissolved gas. Most serious studies will require cells that allow for temperature monitoring and control. Spacers and gaskets made from silver, lead, or Teflon can be used satisfactorily. Lead amalgum is not recommended for this use because it is difficult to separate from windows and thus complicates disassembly of the cell to clean the windows as is frequently necessary. With highly viscous materials where use of a syringe is unsatisfactory, such as with mitochondrial and microsomal pellets, the cell may be disassembled and a small amount of pellet simply squeezed between the two windows. Tissue slices must also be specially placed in the cell.

The pathlength of the cell is usually determined by one of two methods—the interference fringe and standard absorber methods. In the first, the interference fringe pattern is obtained with cells in which the window surfaces are parallel by recording the infrared spectrum of the empty cell vs. air. The pattern obtained is used for the accurate computation of the distance between the cell windows by one of the two expressions that follow:

$$L = \frac{nw_1 w_2}{2(w_2 - w_1)1000} \quad \text{or} \quad L = \frac{n(10)}{2(f_1 - f_2)}$$

where L = cell path length (in mm), w_1 = initial wavelength (in microns), w_2 = final wavelength (in microns), f_1 = initial frequency (in cm^{-1}), f_2 = final frequency (in cm^{-1}), and n = number of fringes between initial and final points.

For wedged cells or cells of poor optical quality the "standard absorber method" may be used to determine the overall effective pathlength. Here, the cell is filled with an absorber of known absorptivity such as benzene (ϵ_M, 1960 cm^{-1} = 100) or carbonyl hemoglobin A (ϵ_M, 1951 cm^{-1} = 3700), the spectrum is recorded, and the appropriate calculations made.[28,29]

It is necessary to know the exact path length if an extinction coefficient or integrated intensity measurement is to be carried out. However, for many studies, as in the recording of some difference spectra, the exact path length need not be known but rather it is important for both the sample and the reference to have the *same* path length. Equal path lengths are most readily accomplished through adjustment of the pressures on the spacers between windows to the point where the interference fringe patterns for the sample and reference cells are superimposable. It may be necessary to try a number of different spacers in order to find two that can be closely matched by adjustments in applied pressure.

As was pointed out above, the region between 1750 and 2500 cm^{-1} is made particularly convenient to study because a suitable reference cell may be one that contains only water. A variable-path length cell is useful for such a reference cell. Since this region is largely free of protein and buffer salt absorptions, the water in a variable-path length cell can be used to compensate for the water in the sample cell. The path length is simply adjusted until a flat base line is obtained in the region of interest. Light scattering from particulate samples (e.g., erythrocytes, microsomes) is not a problem due to the long wavelength of the radiation employed. However, high salt or protein concentrations may change the water spectrum slightly making achievement of a flat base line difficult, in which case the addition of salt or protein to the water in the variable-path length cell frequently makes it possible to get a satisfactory base line.

A problem that can arise when difference spectra are being observed in D_2O in the 1000–1750 cm^{-1} region relates to the extent of hydrogen exchange by deuterium. As mentioned above, infrared absorptions due to protein are not easily seen in H_2O solutions. The O–H deformation at about 1600 cm^{-1} obscures the Amide I band and complicates the detection of the Amide II band. On the other hand in D_2O these bands are readily observed since the deformation for O–D is shifted to near 1200 cm^{-1}. However, after H_2O is replaced with D_2O, there are some time-dependent changes in the spectrum that can be ascribed to slow H → D exchange in the protein. For example, the Amide II vibration, a combination of C–N stretch and N—H binding, is sensitive to the isotope (H or D) in the peptide bond; after addition of D_2O the Amide II band shifts slowly from ca. 1550 cm^{-1} to ca. 1450 cm^{-1} due to the progressive deuteration of peptide bands. These H → D exchange reactions make it necessary when carrying out difference spectroscopy in D_2O solutions to prepare the sample and reference materials with care to have the extent of deuteration the same in both cells.

Computers and Data Processing

Computers have been utilized for infrared problems only to a limited extent, generally for the identification of unknown compounds. For this, the computer is used to search spectra on file for a spectrum that most nearly matches the spectrum of an unknown. With biological samples the ability to express infrared spectra in digital form for storage and for computer processing not only greatly facilitates the manipulation of the spectral data but very significantly increases the power of the infrared technique. Computer interfacing has now been achieved for both dispersive and interferometer type[30] spectrometers and has been very satisfactorily applied to biological systems. An example with a dispersive instrument is in our laboratory where we recently interfaced a Tektronix graphic computing system (Model 4051) with a Perkin-Elmer Model 180 Infrared Spectrometer. The enhancement in capability was very significant indeed for intact tissue and aqueous protein studies. Programs have been written for the co-addition (averaging) of spectra to improve signal-to-noise ratios. These permit ready achievement of a 5-fold enhancement in sensitivity over a single scan of 70 min by the accumulation of 25–30 scans each of 6-min duration. In theory, the signal-to-noise level should increase in direct proportion to the square root of the number of scans that are co-added; so, with adequate sample stability, an even greater increase in sensitivity may be achieved by processing more scans. As important as sensitivity enhancement is the ability to store spectra on tape for later retrieval for analysis by any of several software programs which include:

1. Transmittance to absorbance calculations
2. Base-line slope correction
3. Peak integration
4. Band resolution
5. Subtraction of a ratioed spectrum B from spectrum A
6. 5 to 15 point smoothing routines
7. First derivative spectra

The integration routine is very useful; the integral curve can be plotted on the spectrum as shown in Fig. 6. If a valid base line has been chosen the integration curve should asymptotically approach its lower and higher range over the limits of the integration. Integrating in this way we have demonstrated that the C–O stretch band for carbonyl hemoglobin A is linear in integrated intensity vs. concentration from 0.004–20 m*M*.

The peak separation routines are also very useful. Overlapping bands may be resolved by the method of mirror image subtraction shown in Fig. 7 or by fitting Lorentzian curves to the observed spectrum (Figs. 8 and 9).

[30] L. Y. Fager and J. O. Alben, *Biochemistry* **11,** 4786 (1972).

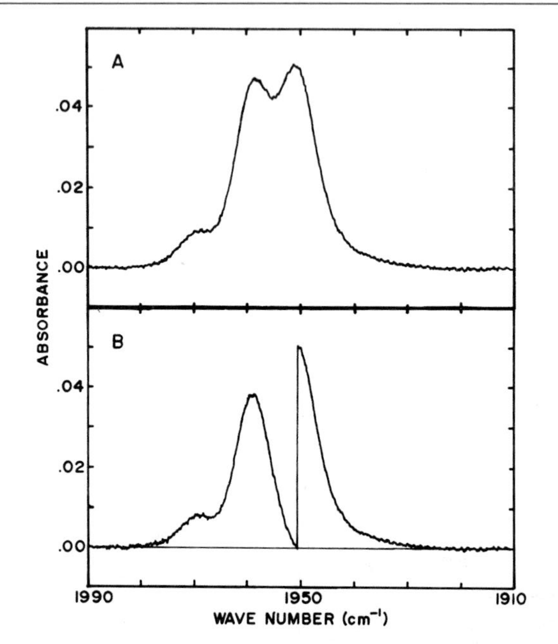

FIG. 7. (A) Infrared difference spectrum of CO hemoglobin Zurich vs. H_2O plotted in absorbance. (B) The same spectrum as (A) in which the mirror image of the spectrum from 1950.7–1925 cm^{-1} has been subtracted from the spectral region 1976.4–1950.7 cm^{-1}. If the overlapping peaks are separated by at least a half bandwidth, mirror image subtraction can provide a convenient method to partially deconvolute complex spectra.

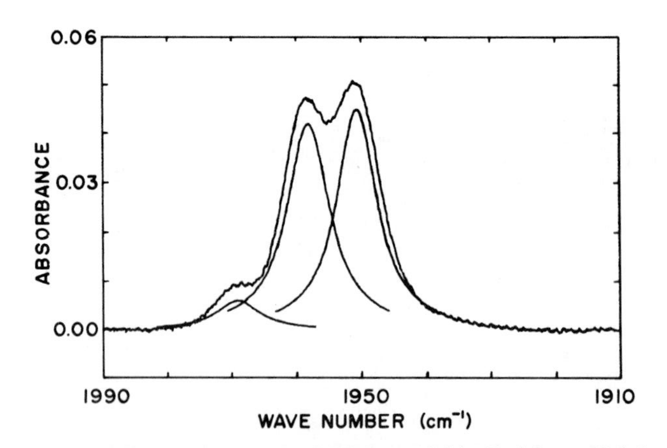

FIG. 8. Infrared difference spectrum of CO hemoglobin Zurich vs. H_2O in absorbance plotted with three computer-generated Lorentzian curves. The cumulative subtraction of these curves from the observed data yielded a nearly flat base line. The computer curves were generated with absorption maxima at 1969, 1958, and 1950.5 cm^{-1} with half bandwidths of 8, 8, and 7.9 cm^{-1}, respectively; maximal absorption intensities are 0.006, 0.042, and 0.045, respectively.

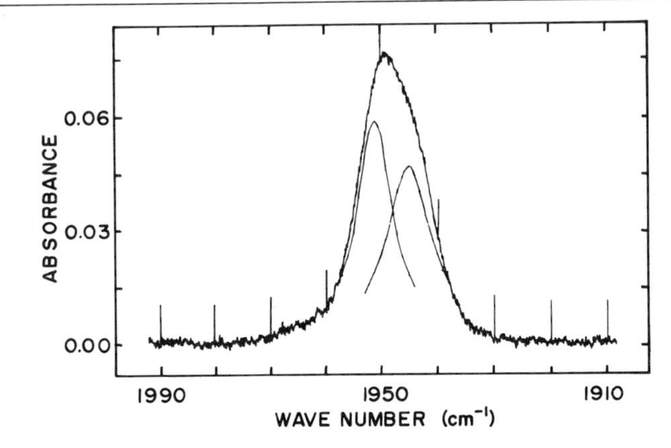

FIG. 9. Infrared difference spectrum of opossum CO hemoglobin vs. H_2O recorded in the absorbance mode. One subunit (α) of opossum hemoglobin differs from most mammalian hemoglobins in that a glutamine residue replaces the almost invariant distal histidine. This amino acid replacement causes a distinct change in spectrum (e.g., see Fig. 6). Lorentzian curves were hand-fitted to the recorded data. A "best fit" assuming equal areas for both bands was obtained with the following parameters: Beta chain $\nu_{CO} = 1951$, $\Delta\nu_{1/2} = 8$; and alpha chain $\nu_{CO} = 1945$, $\Delta\nu_{1/2} = 10$.

The first is a rather simple method in which the mirror image of the right half of peak B is subtracted from the left half of peak B thereby leaving only peak A.

A program whereby any percentage of spectrum B may be subtracted from spectrum A can be extremely useful as a means of obtaining a difference spectrum "electronically" without need for actually carrying it out "optically" on the spectrometer where very careful matching of sample and reference cells is required. However, "electronic" difference spectra may not be as accurate as the "optical" spectra in regions of high absorbance. High-precision detectors can be important if subtraction routines are used in regions where little radiation passes through the sample. The programs (in BASIC) used in our system are relatively straightforward to write (the data can be treated as a two-dimensional matrix).

Computer interfacing with the interferometers is an integral part of the system in that computer analysis is required for carrying out the Fourier-transform analysis. An infrared study of carbon monoxide binding to various hemocyanins using Fourier-transform interferometry has been reported by Fager and Alben.[30]

Techniques and Instrumentation

With double-beam dispersive spectrometers the spectrum is most accurately determined if the infrared radiation through the sample and reference beams impinges on the detector with nearly equal intensity in the region of interest. For example, a 1% difference in transmission when both beams are nearly balanced corresponds to a difference in absorbance of 0.004, whereas a 1% difference when the sample beam is only 10% of the reference beam intensity corresponds to a difference in absorbance of 0.10. The determination of the C–O stretch band for carbonyl hemoglobin may be taken as a typical experiment. A solution of freshly prepared oxyhemoglobin is injected into the reference cell. Another portion of the solution is treated with CO to form the carbonyl hemoglobin and then injected into the sample cell of the same path length as the reference cell. Upon recording the spectrum (e.g., from 2100–1800 cm^{-1}) only the absorption due to bound carbon monoxide is obtained because the water and protein concentration are the same in each cell (the dissolved CO is not seen under these conditions). The key in obtaining optimal spectra in the region of interest is an appropriate balance between instrument performance and the percent energy transmitted. For a given concentration, it is beneficial to increase the path length and thereby increase band height until the point is reached where any further increase in path length so reduces instrument performance (due to lowered light transmittance through the cell) that the overall sensitivity is not increased. Thus one must achieve the compromise between path length and energy transmission that gives an optimal sensitivity. The higher-quality spectrometers can generally be expected to perform better at lower light levels and therefore to provide greater sensitivity. It should also be recognized that most spectrometers do not perform well, if at all, if the sample beam intensity is greater than the reference beam intensity. Therefore care should be taken to maintain the sample beam intensity equal to or lower than the reference beam intensity.

Recent and Future Developments

The instrumentation for carrying out Fourier-transform infrared interferometry (FTIR) recently became available commercially. These instruments may be described most simply as rapid-scanning single-beam spectrometers. No slits or gratings are used since the "scanning" is accomplished by use of a Michaelson interferometer with two mirrors and a beam splitter. An interferogram is produced that must be Fourier-transformed into the frequency spectrum. The entire spectrum (dependent on the beam splitter) is obtained whether desired or not. Since computer assistance is required to perform the Fourier transform, data handling is an integral part of the system. These averaging and other data-handling capabilities as well as greater speed provide, in theory, a significant increase (e.g., 40-fold) in sensitivity over a dispersive instrument when the broad infrared region is under study. However, the high-quality dispersive spectrometers compete very well with the presently available FTIR instruments if only a small segment (e.g., 100 cm^{-1}) of the spectrum is observed. In some cases (i.e., in regions where little energy can pass), dispersive instruments can even outperform the FTIR systems. Since in FTIR the spectra are single beam, a high degree of digitization of detector input is required for accurate optical densities, and the environmental conditions (e.g., water content) in the sample compartment must be carefully controlled. It is anticipated that several technical advances in FTIR, such as the use of interference filters and cryogenic detectors, will further enhance the applicability of this relatively new technique to the study of biological systems.

It may also be anticipated that the attenuated total reflectance (ATR) or frustrated multiple internal reflectance (FMIR) techniques that have proven so widely useful in surface chemistry will find greatly expanded applicability in biochemistry. The technique involves the use of a crystal of high refractive index (KRS-5 generally) in intimate contact with a sample of lower refractive index. The infrared spectrum of the sample surface within 2–10 μm of the crystal is observed. An example of a potential use of ATR of great interest is the *in vivo* detection and quantitation of CO in tissue surfaces.

The potential for further applications of infrared spectroscopy in living systems is bright indeed. With present instrumentation, a wide range of biosystems await exploration. And the expected improvements in instrumentation and in sampling techniques will almost certainly expand the possibilities for application much further.

Acknowledgment

This work was supported by United States Public Health Service Grant HL-15980.

[19] X-Ray Absorption Spectroscopy of Metalloproteins

By SUNNEY I. CHAN and RONALD C. GAMBLE

Introduction

X-ray absorption spectroscopy is the measurement of the absorption of X-radiation by a material as a function of the incident photon energy. Two sources of information are contained in such a spectrum: the absorption edge fine structure (AEFS) and the extended X-ray absorption fine structure (EXAFS).

ISBN 0-12-181954-X

When applied to metalloproteins, the AEFS spectrum is often useful in ascertaining the oxidation state and site symmetry of the metal centers as well as the nature of the surrounding ligands. This information is derived from careful measurements of the energies and intensities of the electronic transitions which occur as the incident photon energy is scanned through an absorption edge, typically over 30 eV. From the EXAFS spectrum, the distance between a metal center and its ligand in a metalloprotein may be obtained. The small modulations in the absorption coefficient, which occur from 100–1000 eV above an absorption edge of the metal, are caused by variations in the probability of photoelectron emission. This probability reflects the extent to which the photoelectron wave back-scattered from a nearby atom to the metal center is in- or out-of-phase with respect to the emitted photoelectron wave. Theoretical methods have been developed to Fourier-transform the modulations in absorption amplitude with energy to give the distances between atoms.

X-ray absorption spectroscopy extends back to the 1930s, and much literature on various classes of compounds has ensued. The impact of this technique on coordination chemistry, of particular interest because of its application to metal-containing proteins, has been reviewed recently by Srivastava and Nigam.[1] Although the experimental techniques have improved with time, as reviewed by Thomsen and Cuthill,[2] the intensity of the X-ray sources, usually from Bremsstrahlung radiation, is not sufficient to accumulate data within reasonable time for systems of low metal content. The application of this spectroscopy to metalloproteins was not possible until the past several years.

A recent technical development has renewed interest in X-ray absorption spectroscopy, particularly for biological systems. The advent of synchrotrons and storage rings, intended originally for high-energy physics experiments, has made possible the production of synchrotron radiation with fluxes 10^5- or 10^6-fold more intense than those previously available. Thus data for dilute systems such as metalloproteins can be acquired within short periods.

Instrumentation

The X-Ray Source. Central to the X-ray absorption spectroscopy of dilute biological systems is a high-flux X-ray source. The use of synchrotron radiation has met this need. The general properties of synchrotron

[1] U. C. Srivastava and H. L. Nigam, *Coord. Chem. Rev.* **9,** 275 (1972–1973).

[2] J. S. Thomsen and J. R. Cuthill, *in* "X-ray Spectroscopy" (L. V. Azaroff, ed.), p. 26. McGraw-Hill, New York, 1974.

radiation and its application to X-ray spectroscopy are discussed by Madden.[3] The large machines needed to produce synchrotron radiation are the offspring of research in high-energy physics. There are only several such facilities throughout the world (cf. Table I). Not all facilities are capable of producing significant X-radiation, nor do all of those with the capability have ports which render access to this radiation. Within the United States, only the Cornell and Stanford facilities presently have instrumentation for X-ray absorption spectroscopy. This review will concern itself only with work which has emerged as part of the Stanford Synchrotron Radiation Project (SSRP). However, the principles and techniques discussed here are applicable to work done at other facilities. SSRP is funded by the National Science Foundation and is operated as a national research facility.[4] It provides and maintains the ports and monochromators on the storage ring. The storage ring, SPEAR, in turn is built and maintained by the Stanford Linear Accelerator Center.[5,6]

At SPEAR, synchrotron radiation is emitted by electrons which are accelerated to relativistic velocities. A single bunch of electrons orbits the storage ring under the control of horizontally deflecting and focusing quadrupole magnets. This orbit frequency is about 1 MHz, and the entire group passes a point in the orbit in about 0.1 nsec. At relativistic energies, each electron radiates a sharp cone of radiation in the forward direction, tangential to the electron trajectory and linearly polarized in the orbital plane. Typically, the storage ring operates at an energy of 3–3.5 GeV and a current of 40 mA. The synchrotron radiation emitted consists of a broad spectrum of collimated X-rays with a high energy cutoff (intensity down one decade) of 20–30 keV photons. Because of slightly different trajectories of the electrons, there is a vertical divergence of ~ 1 mradian in the emitted radiation. The horizontal divergence is determined by that fractional part of the orbit from which the X-rays are allowed to fall on the sample and is typically 2–10 mradians. The sample sees about 10^6 pulses/sec of radiation, corresponding to the successive passage of the electron bunch through that part of the orbit. A complete description of the properties of the orbiting

[3] R. P. Madden, *in* "X-ray Spectroscopy" (L. V. Azaroff, ed.), p. 338. McGraw-Hill, New York, 1974.

[4] An overview of SSRP is provided by H. Winick and I. Lindau, *in* "Synchrotron Radiation: Properties, Sources, and Research Applications," SSRP Rep. No. 76/03. Stanford Synchrotron Radiation Project, Stanford University, Stanford, California, 1976.

[5] The address for submitting proposals for experiments is: Director, SSRP, Bin 69, c/o Stanford Linear Accelerator Center, P.O. Box 4349, Stanford University, Stanford, CA 94305.

[6] In addition to the SSRP staff, much of the early and subsequent design and construction of monochromators and detectors were performed by researchers of the Bell Laboratories, Stanford University and University of California, Berkeley.

TABLE I
SYNCHROTRON RADIATION RESEARCH FACILITIES AS OF 1977[a]
AND THE APPROXIMATE FLUX AT THREE X-RAY ENERGIES

Location	Energy (GeV)	Current (mA)	Bending radius (m)	Critical energy (keV)	Photon flux (photons·sec^{-1}·keV^{-1}·(2π rad)$^{-1}$·10^{-17}) Photon energy 12 keV (1.03 Å)	6 keV (2.07 Å)	2 keV (6.20 Å)
Storage rings							
DORIS, Hamburg, Germany	4.0	~250	12.2	11.6	60	130	682
SPEAR, Stanford, USA	4.0	~100	12.7	11.1	24	91	280
VEPP-3, Novosibirsk, USSR	2.25	200	6.0	4.2	9.8	74	310
VEPP-2M, Novosibirsk, USSR	.67	100	1.22	.54[b]			
ACO, Orsay, France[c]	.54	100	1.1	.32[b]			
INS-SOR II, Tokyo, Japan[c]	.30	100	1.0	.059[b]			
TANTULUS I, Wisconsin, USA[c]	.24	100	.64	.048[b]			
SURF II, NBS, Washington, USA[c]	.24	10	.84	.036[b]			
DCI, Orsay, France	1.8	500	3.8	3.4	13	88	600
PACHRA, Moscow, USSR	1.3	300	4.0	1.1[b]			
IPP, Moscow, USSR[c]	1.35	100	2.5	2.2	0.31	5.6	70
DARESBURY, U.K.[c,d]	2.0	1000	5.55	3.2	17	210	1300
ADONE, Frascati, Italy[e]	1.5	60	5.0	1.5	0.012	0.93	42
PEP, Stanford, USA[d]	15	100	120	44	180	362	800
PETRA, Hamburg, Germany[d]	15	80		38			

Synchrotrons							
Cornell, USA	12	2	100	38	2.8	5.0	13
DESY, Hamburg, Germany	7.5	10–30	31.7	29.5	25	48	140
ARUS, Yerevan, USSR	7.0	20	24.6	19.5	13	27	65
NINA, Daresbury, U.K.[f]	5.0	40	20.8	13.3	17	49	120
BONN I, Germany	2.5	30	7.6	4.6	2.8	13	53
INS-SOR I, Tokyo, Japan	1.3	30	4.0	1.22[b]			
Frascati, Italy	1.1	10	3.6	.82[b]			
C-60, Moscow, USSR	.68	10	1.6	.44[b]			
BONN II, Germany	.5	30	1.7	.16[b]			
LUND, Sweden	1.2	40	3.6	1.06[b]			

[a] We gratefully acknowledge Dr. H. Winick at SSRP for compiling this information.

[b] The flux at these photon energies is lower than can be calculated using the tables of R. A. Mack, "Spectral and Angular Distributions of Synchrotron Radiation," Rep. No. CEAL-1027, Cambridge Electron Accelerator, Cambridge, MA 02138, U.S.A. (1966).

[c] Dedicated to synchrotron radiation.

[d] Operational 1979–1980.

[e] Synchrotron radiation beam lines under construction.

[f] Terminates operation end of 1977.

electron beam at SPEAR and the resulting X-ray spectrum is presented by Winick.[7,8]

The Monochromator. There are three X-ray beamlines presently set up at SSRP for absorption spectroscopy. Although each monochromator differs in geometry and the choice of crystals, there are common characteristics which will now be described.

Some of the considerations for interfacing a monochromator to a storage ring are given by Pianetta and Lindau.[9] The vacuum–atmosphere interface as well as the potential personnel hazards in utilizing synchrotron radiation require special shielding and safety equipment to be installed between the source and the monochromator. At SSRP, the monochromator is located some 15–20 m from the source. At this distance, the beam height is about 1 cm when electrons alone are circulating in the storage ring (electron energy 3–3.5 GeV, current 30–40 mA). Its width is determined by the planar angle of orbit allocated to the monochromator, and for a horizontal divergence of 2 mradians, it is about 3 cm.

The monochromator consists of a channel-cut crystal [e.g., Si (220)], which allows the beam to be reflected by two surfaces (cf. Fig. 1). With appropriate slits, the X-rays are monochromatized to about 1 eV bandwidth. The flux is typically of the order of 10^9 photons/sec/eV bandwidth. Energy selection within a range of 3–30 keV is set by a stepping motor, which is controlled by computer. The method of calibration of the monochromator will be discussed later.

One problem with using only a crystal for monochromation is that higher harmonic radiation, particularly $\lambda/2$ for the Si (220) reflection, still passes through the monochromator and can interfere with the absorption measurements at wavelength λ. In fact, the $\lambda/2$ incident photon intensity is comparable to the intensity of the fundamental frequency. For this reason absolute absorption coefficient measurements are difficult to make. However, since the absorption cross-section decreases as λ^3, absorption at $\lambda/2$ should not significantly affect the edge and extended fine structure observed at λ.

A focusing monochromator system has recently been completed, which provides 50- to 100-fold greater intensity than that from the unfocused monochromators. Focusing is accomplished by a metal-coated mirror

[7] H. Winick, "Stanford Synchrotron Radiation Users Handbook," SSRP, W. W. Hanson Laboratories of Physics, Stanford University, Stanford, California, 1974.

[8] H. Winick, "The Stanford Synchrotron Radiation Project (SSRP)," SLAC-PUB-1439. Stanford Linear Accelerator Center, Bin 69, Stanford, California, 1974.

[9] P. Pianetta and I. Lindau, "High Resolution X-ray Spectroscopy Using Synchrotron Radiation: Source Characteristics and Optical Systems," SSRP Rep No. 76/05. Stanford Synchrotron Radiation Project, Stanford University, Stanford, California, 1976.

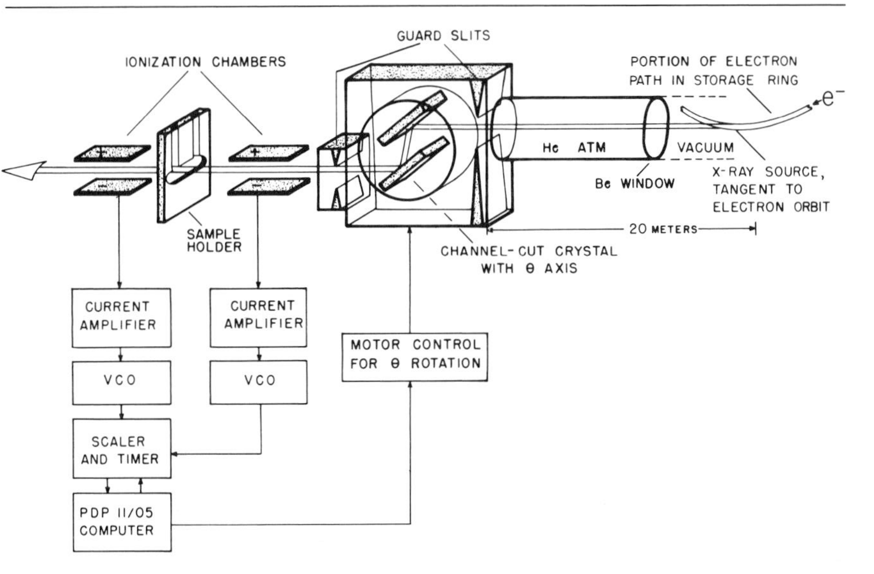

FIG. 1. Diagram of the instrumentation for X-ray absorption spectroscopy. The drawing is not to proportion, as the storage ring is much larger than shown.

(ground concave for horizontal focusing and bent for vertical focusing) which is placed upstream of the monochromator. The focus size is about 2 mm vertical by 4 mm horizontal. In addition to greater intensity and smaller beam size, interference by the above-mentioned higher harmonics is eliminated, because the mirror is set to an angle just below the critical angle of the next allowable harmonic wavelength. Accordingly the beam is highly monochromatic.

Calibration of the Monochromator. Energy calibration is accomplished by correlating a specific feature within an absorption edge to a setting on the stepping motor which controls the Θ rotation of the channel-cut crystal. Usually the sample is a metal foil (a few microns thick) whose K-absorption edge is well characterized and within the region of interest. A particular peak within the edge is assigned its known energy, thus defining a setting of the stepping motor. This energy is known usually to within 1 or 2 eV. Any incremental change in energy is then calculated from the crystal lattice parameter. Alternatively, two positions of Θ may be defined using two foils and subsequent energies determined by interpolation.

The reproducibility of a particular setting is important in edge absorption work. Coincidentally, minor variations on the crystal surface cause sharp changes in the efficiency of transmission of the crystal as it is stepped through certain angles. These "glitches" are very reproducible and thus provide an internal energy reference from one spectrum to another.

The Detectors. An absorbance measurement requires that both the incident and transmitted intensity be carefully measured. If A is the absorbance, I_0 the incident intensity, I the transmitted intensity, μ the absorption coefficient, ρ the density of the sample, and x the path length of the sample, then

$$A = \frac{\mu}{\rho} x = \ln \left(\frac{I_0}{I} \right)$$

The current method for measuring I_0 is by means of an ionization chamber, about 15 cm in length and containing a weakly absorbing gas, e.g., a He–Ne mixture. The use of this gas permits greater than 90% of the radiation to fall on the sample. When the ionization chamber is properly designed,[9a] the current produced across it is proportional to all the incident beam. With appropriate integration this current is amplified and applied to a voltage-controlled oscillator. The frequency is determined by scaling for the period of measurement at the given energy. The efficiency of the ionization chamber is smoothly varying with energy, provided no absorption edges for the gas are present in the energy range to be examined.

The transmitted intensity can be measured in a similar fashion, except that the completely absorbing Ar gas is used for greater efficiency. Because different gases are used for the measurement of I_0 and I, the different efficiencies of the two ionization chambers must be accounted for. This could be done by running a blank, i.e., a spectrum with no sample. In any event an accurate determination of A is usually difficult due to interference from higher harmonics. Fortunately, for edge spectroscopy, one typically only scans over an energy range of a few hundred eV at a photon energy of several keV. The interference, accordingly, is smoothly varying and relatively small, and can be eliminated by subtraction techniques to be discussed.

The sensitivity of the X-ray absorption spectroscopy may be increased by two orders of magnitude by measuring the fluorescence emission at and above the absorption edge.[10] Because the probability of radiative transition is directly related to the X-ray absorption cross-section, the fluorescence intensity (e.g., K_α radiation when the K-absorption edge is studied) can often be a more suitable indicator of the change in absorption. In these experiments, the plane of the sample holder is held at a 45° angle to the incident beam to observe fluorescence at 90°. A lithium-drifted germanium detector with an energy resolution of 200 eV, reduces the influence

[9a] For design considerations, see chapter by J. W. Boag, *in* "Radiation Dosimetry" (F. H. Attix and W. C. Roesch, eds.), Chapter 5. Academic Press, New York, 1968.

[10] J. Jaklevic, J. A. Kirby, M. P. Klein and A. S. Robertson, *Solid State Commun. (U.S.A.)* **23,** 679 (1977).

of elastic and inelastic (Compton) scattering. This discrimination between fluorescence and scattering is particularly advantageous when a dilute system is studied. Evidence for the sensitivity of the fluorescence detection method has been provided by the successful detection of Mn in the chloroplasts of a leaf at the concentration level of 10–50 ppm. A more recent development is the use of an array of nine scintillation detectors. With scintillation counters the energy resolution is reduced, the background scattering is accordingly greater, and the ensuing statistics are not as favorable. However, the larger solid angle for detection permits a much greater flux for detection. Data obtained using such a system will be presented later.

A comparison between the absorption and fluorescence detection methods has been made.[10] Although the EXAFS features are virtually the same, there are some differences in the edge structure.

Sample Considerations

Many of the same considerations for preparing materials studied by conventional X-ray absorption spectroscopy hold for biological samples. In the latter case the criteria are usually more stringent because greater signal-to-noise ratio (S/N) is required to detect low metal contents.

Sample Thickness. The optimal thickness of the sample depends on whether the edge fine structure or EXAFS region is to be investigated. Criteria for optimal thickness are discussed by Azaroff and Pease.[11] For studies through the edge, the optimal thickness should take into account the absorption coefficients above (μ_a) and below (μ_b) the edge. The optimal thickness is obtained by setting $d(I_a - I_b)/dx = 0$, where I_a and I_b are the intensities above and below the edge, and dx is the incremental thickness. Then,

$$x_{\text{optimal}} = \frac{\rho \ln (\mu_a/\mu_b)}{(\mu_a - \mu_b)}$$

For EXAFS the usual criterion is

$$x_{\text{optimal}} \sim \frac{\rho}{\mu_a}$$

Although the overall spectral features are not dependent on specimen thickness, some of the details are affected. Therefore a consistent criterion should be used in the choice of sample thickness.

[11] L. V. Azaroff and D. M. Pease, *in* "X-ray Spectroscopy" (L. V. Azaroff, ed.), p. 284. McGraw-Hill, New York, 1974.

For a liquid sample the concentration of the absorbing species can usually be adjusted to satisfy the above criteria given the cell thickness. However, difficulties often arise with a solid sample, particularly if it is highly absorbing. Here, this requisite thickness may be only several microns. In this case, a fine powder should be used wherein the grain size is smaller than the characteristic distance (ρ/μ_a) of the X-ray absorption.

With most biological samples which contain aqueous solution, one encounters a separate problem. Because the effective metal concentration in these samples is usually low, the solvent will dominate the absorption and therefore the optimal thickness is determined by the characteristic distance of water, which is about 1 mm for energies of several keV. A similar consideration would apply to solid samples in which the metal component is small, e.g., a metalloprotein powder.

Quantity of Material. The amount of sample for transmission measurements is determined by the cross-section of the incident beam and the optimal path length as estimated from the absorption coefficient. Since the beam (approximately 1-mm high and 15-mm wide) after monochromation moves a few millimeters vertically as the energy is scanned, one must take this into account in the design of the sample holder.

Usually a 100–300 mg sample is adequate for transmission measurements provided the ratio of metal atoms to nonmetal atoms is greater than about 10^{-3}. For example, the Mo–Fe component of nitrogenase (1.71 Mo/220,000 daltons) gave spectra of good S/N with 100 mg of lyophilized material.[12] For dilute metal systems, the fluorescence measurement has been shown to be more sensitive than the transmission measurement. Although initially only applied to EXAFS studies,[10] it has recently been applied to the AEFS experiments[13] as well. Concentrations as low as 1 m*M* bean plastocyanin (1 Cu/10,800 daltons) provide satisfactory data when several 1 sec/point scans are averaged using the nine-detector array described earlier.

Sample Preparation. For solid samples, compressed pellets may be used for transmission measurements. The EXAFS spectra of Cu-tetraphenylporphyrin (Cu-[TPP]), Ni[TPP], and methemoglobin[14] were obtained in this manner. Usually solid samples are pressed into a 10×20 mm milled hole in 0.2–2 mm thick Plexiglas plates. They are sealed on each side with 0.001″ Kapton for windows. The exact dimensions and material for

[12] S. P. Cramer, T. K. Eccles, F. Kutzler, K. O. Hodgson, and L. E. Mortenson, *J. Am. Chem. Soc.* **98,** 1287 (1976).

[13] V. W. Hu, S. I. Chan, and G. S. Brown, *Proc. Natl. Acad. Sci. U.S.A.* **74,** 3821 (1977).

[14] B. M. Kincaid, P. Eisenberger, K. O. Hodgson, and S. Doniach, *Proc. Natl. Acad. Sci. U.S.A.* **72,** 2340 (1975).

the sample holder differ depending on the nature of the sample and the particular experimental arrangement.

If absorption is high, then the solid, ground to a fine powder, may be diluted with another inert powder which is transparent to X-rays. Alternatively, the powder may be mixed with parlodian that is dissolved in some solvent to which the powder is insoluble, e.g., acetone or ethylacetate. After thorough mixing, this suspension is aspirated onto a warm glass plate. Following evaporation of the solvent, the plastic film can easily be peeled from the surface by exposure to moist air.

Molybdenum compounds have been prepared as fine powders dispersed in an epoxy matrix that is supported by aluminum foil backing.[12] Generally this method is useful only for elements whose edges appear at high energy, where absorption by the matrix and support is small.

Liquid samples for both transmission and fluorescence measurements have similar requirements. Syringe-type fittings are usually added for ease of filling the sample cell described above. For aqueous solutions, water absorption usually dominates and a path length of 1 mm for the 5–10 keV range is adequate. The total volume then is 0.2 cm^3. A more concentrated solution, of course, will provide better S/N for a given scan rate.

Other Considerations. Other factors concerning the sample include temperature control, elimination of oxidative atmosphere for sensitive samples, and radiation damage. Each experimenter must devise his own methods; however, there are some common features which may be mentioned here.

Spectra are normally run at room temperature. For studies at low temperature, a special dewar containing liquid nitrogen is positioned just above or below the sample holder and connected to the sample holder via a metal connection which transfers heat away from the sample. An evacuated chamber with Kapton windows surrounds the sample holder. Although the sample is at reduced temperatures, it is not necessarily as low as liquid nitrogen temperature.

Samples sensitive to oxidation are prepared under a nitrogen or other inert atmosphere and placed in a tightly sealed sample holder.

Although the intense X-ray flux can potentially damage a protein sample, there has been little evidence for this.[13,15] However, one group of investigators has noted that oxyhemoglobin transforms to its met form after a 2-hr exposure to X-radiation.[16]

[15] S. P. Cramer, T. K. Eccles, F. Kutzler, K. O. Hodgson, and S. Doniach, *J. Am. Chem. Soc.* **98,** 8059 (1976).

[16] Dr. George Brown, personal communication (1977).

Sensitivity

The S/N requirements for measuring the AEFS or the EXAFS are comparable and will be considered together.

For transmission measurements, the S/N is essentially limited by photon statistics at fluxes of 10^8 photons/sec or less. At higher fluxes, instrumental noise dominates, and the S/N is limited to around 10^4 for an integration time of 1 sec/point. Greater S/N can be obtained by a longer integration time or, preferably, by taking successive scans. In the event of failure of the storage ring, the latter has the advantage that some data are complete, which can be averaged with other data. Other factors, such as sample granularity or large intensity variations caused by momentary fluctuations of the electron beam, can also limit the ultimate S/N.

An example of how this S/N specification translates to obtaining useful data for an EXAFS experiment is discussed by Kincaid *et al.*[14] Several factors determine the time required for an experiment. Kincaid *et al.* have set the S/N for the modulations in the fine structure to be five for providing the minimum information necessary to interpret the spectrum. If one assumes that (1) these modulations are about 10% of the height of the absorption edge and (2) the flux is 10^8 or greater, then pure samples containing as little as 1 iron atom in 20,000–30,000 molecular weight can provide useful information with a 1-sec integration/point. A typical scan time would be 30 min or less.

Data Collection and Initial Reduction

The collection of data usually proceeds fairly rapidly. Data are placed on a mass storage device such as floppy disk or magnetic tape. This is necessitated because experimental time is at a premium. In fact, every part of the actual scan is automated to facilitate collection. The data stored are I_0, I, and the monochromator angle. Typically, 30 min are required for a single scan over 1000 eV. The incident photon energy is calculated for each absorption datum from the incremental increase in the monochromator angle.

For transmission measurements, I and I_0 may be used directly to compute the absorbance ($A = \text{In}/I_0/I$) as a function of energy. The background absorption and the difference in efficiency between the two detectors may be removed by subtracting a spectrum obtained without a sample. An example of an EXAFS spectrum for rubredoxin is given in Fig. 2A.

For fluorescence detection, scattering from processes other than the specific one monitored must be subtracted from the ratio F/I_0. This base-line correction is similar to that for absorption. The resulting ratio, then, is

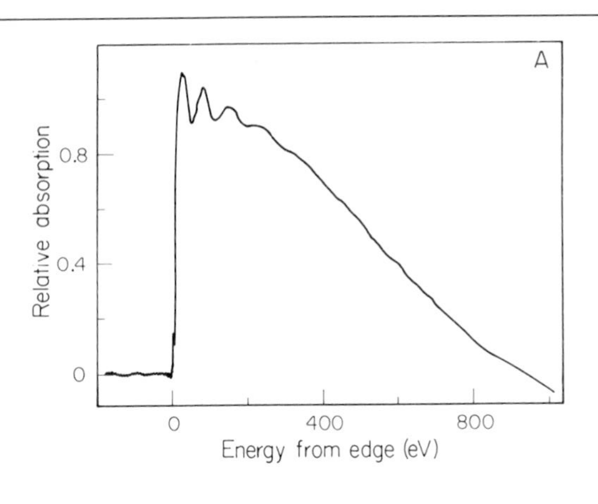

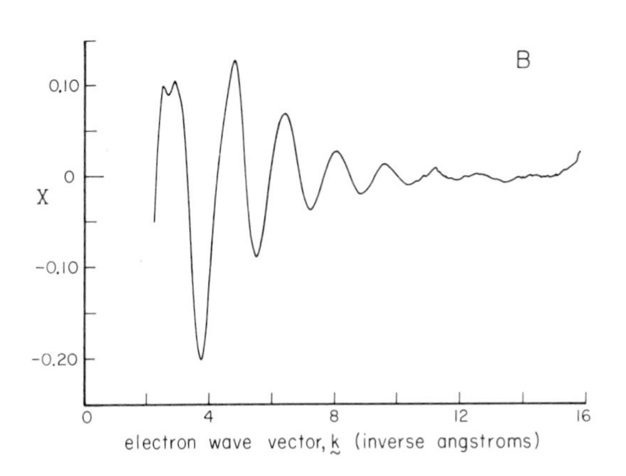

FIG. 2. Fe K-edge absorption spectrum of rubredoxin. In (A), the absorption vs. X-ray energy is presented. These data have been corrected for absorption from other components (see text for details). In (B) the unnormalized EXAFS contribution, χ, is shown as a function of the wave vector, **k**. [Courtesy of D. Sayers and B. Bunker, University of Washington, Seattle (1976) (unpublished).]

directly proportional to the absorbance. An example of a spectrum which has been obtained using this method of detection is shown in Fig. 3, where the Cu K-edge spectrum of cytochrome c oxidase is depicted.

The absorption spectrum contains contributions from several electronic transitions. When only the K-orbital contribution is desired, other absorption effects are removed by subtracting the data above the edge from an extrapolated curve from below the edge. A Victoreen formula, $A = C\lambda^3 - D\lambda^4$ (where λ is the wavelength of the photon and C and D are the

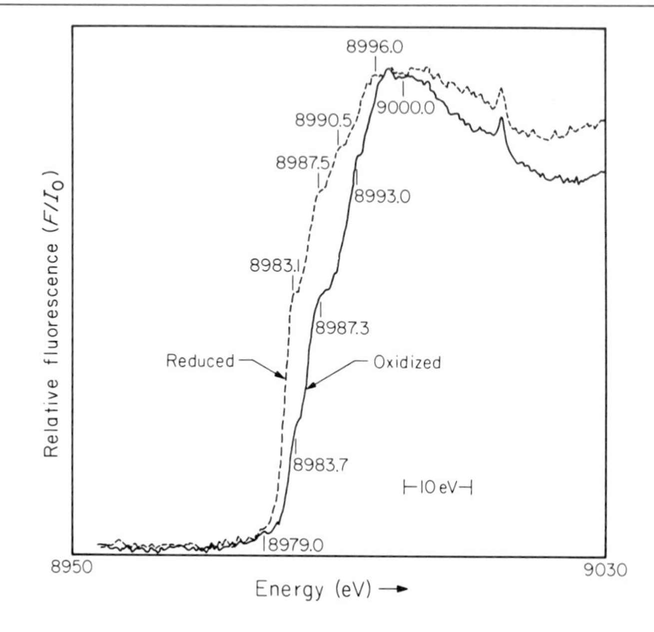

FIG. 3. Comparison of the Cu K-edge spectra of reduced and oxidized cytochrome c oxidase. The ratio F/I_0 is the fluorescence intensity, F, normalized to the incident intensity, I_0. Background scattering has been subtracted from these data. [V. W. Hu, S. I. Chan, and G. S. Brown, *Proc. Natl. Acad. Sci. U.S.A.* **74,** 3821 (1977).]

constants to be fitted) is often applied,[17] although other polynomials have been used.[15] The resulting K-edge contribution can then be normalized to an arbitrary value between 0 and 1[15] or to a per-atom absorption scale.[17]

Often the absorption data are manipulated to extract only the EXAFS, or oscillatory part, of the spectrum. Simple subtraction of a fitted curve from above the EXAFS region is not adequate because the EXAFS signal is a small part of the total signal. A Fourier filtering technique has been shown to be useful.[15,17] In this procedure the large sawtooth-like function from the K-edge absorption is first removed by subtracting a straight line, obtained from the first and last data points of the EXAFS region. The energy points on the abscissa are then converted to momentum, or k space (see later section on Fourier transform for details). The resultant data are then transformed back into frequency space. This frequency spectrum contains contributions from remnants of the smoothly varying K-edge (low frequency), EXAFS (intermediate frequencies), and random and instrumental noise (broad spectrum). The lower frequency and higher frequency com-

[17] F. W. Lytle, D. E. Sayers, and E. A. Stern, *Phys. Rev. B* **11,** 4825 (1975).

ponents may be eliminated upon back transforming into k space through appropriate selection of limits of integration in frequency. The result is a plot of the EXAFS contribution vs. **k** vector ($Å^{-1}$) (cf. Fig. 2B). Examples applying this technique are provided by Lytle *et al.*[17]

Analysis of Edge Spectra

X-ray absorption spectroscopy involves excitation of the core electrons of an atom. As the incident photon energy is increased there are certain critical values where a bound electron may absorb a photon and be excited into the continuum. Depending on the principal shell from which the electrons are excited, the transitions are designated K, L, or M edges. Because the absorption coefficient increases abruptly at the edge but decreases slowly and monotonically above the edge, the spectrum has the shape of a single sawtooth function. For atoms of intermediate atomic number, several absorption edges are usually located in the keV range. Only the K-edge will be considered in the following discussion.

Within a small energy range ($\sim$30 eV) that is slightly below each of the critical energies, there is an additional set of characteristic absorption peaks which correspond to electronic transitions from inner occupied orbitals to outer unoccupied orbitals. These peaks are closely spaced, and are typically only a few eV apart. These transitions give rise to the fine structure in the absorption edge. The energies and intensities of these transitions are usually sensitive to the charge state or valence of the atom in question. If in addition the atom is chemically bound, then the outer orbitals, and hence the energy levels, are affected by orbital overlap with neighboring atoms. Therefore, careful measurement of the edge spectra can provide information about the valence state of a metal atom and the nature of the surrounding ligands.

Although the edge fine structure of a free molecule can often be interpreted in terms of simple molecular orbital theory, the information is necessarily qualitative.[18] Nevertheless, the spectra of similar compounds, whose structures are known, can serve as an aid in the analysis of the spectrum from a metalloprotein. In fact, comparative measurements of the edge spectrum of a metal atom in a metalloprotein with those of model compounds with known valence states has been useful in determining the valency of the metal atom in the protein. We now illustrate two approaches for obtaining this information.

Shift of the Main Absorption Edge. If the main edge is essentially structureless, then its position can provide some charge information. For

[18] W. H. E. Schwarz, *Angew. Chem., Int. Ed. Engl.* **13**, 454 (1974).

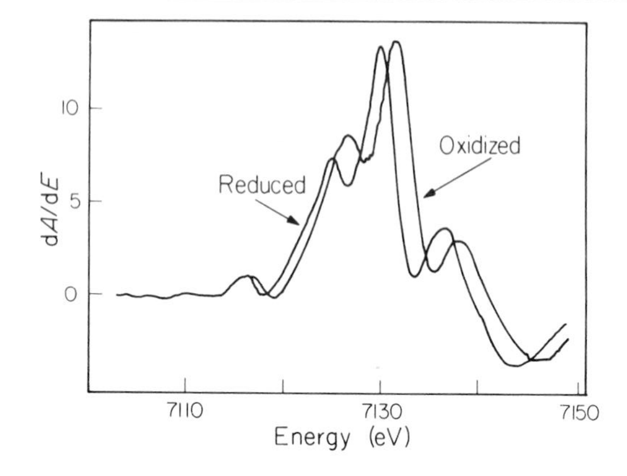

FIG. 4. Comparison of the Fe K-edge *derivative* spectra of oxidized and reduced cytochrome c at pH 7. [Courtesy of A. M. Labhardt, Stanford University, Stanford, California (1976) (unpublished).]

example, Cramer and co-workers[12] have estimated the "coordination charge" of Mo in nitrogenase from *Clostridium pasteurianium* by comparing the edge energy with those observed for several Mo compounds. In the model Mo compounds the edge energy was found to increase linearly with the increasing coordination charge calculated for each compound. A shift of approximately 10 eV (K-edge is ca. 20 keV) for a charge variation of 5 was noted. In this manner, the charge of the Mo in nitrogenase was estimated to be 2.3 ± 0.3.

The absorption edge differences between the oxidized and reduced forms of a metalloprotein have also been investigated. In Fig. 4, the Fe K-edge *derivative* spectra is shown for the two forms of cytochrome c. Although there are only minor differences in the detailed features, the edge of the oxidized protein is about 2 eV above that of the reduced form. A similar shift is noted between deoxy- and oxyhemoglobin.[19]

Assignment of Specific Transitions. If structure within the edge is present, then the absorption peaks may be assigned to specific transitions. Some of the principles reviewed by Srivastava and Nigam[1] can provide a useful starting point, particularly for ionic metal binding. When the bonding is strongly covalent as might be the case in some metalloproteins, peak assignments are more difficult to make. Shulman and colleagues[20] have considered this problem in the case of the Fe-containing proteins, rubre-

[19] P. Eisenberger, R. G. Shulman, G. S. Brown, and S. Ogawa, *Proc. Natl. Acad. Sci. U.S.A.* **73,** 491 (1976).

[20] R. G. Shulman, V. Yafet, P. Eisenberger, and W. E. Blumberg, *Proc. Natl. Acad. Sci. U.S.A.* **73,** 1384 (1976).

doxin and cytochrome *c*, and have provided a qualitative explanation for the Fe *K*-edge spectra for these proteins.

A recent study on cytochrome *c* oxidase[13] specifically demonstrates the usefulness of AEFS spectroscopy. This work yielded a direct determination of the oxidation state of a metal atom involved in electron transport. Cytochrome *c* oxidase is a membrane-bound protein which mediates the electron transfer from reduced cytochrome *c* to molecular oxygen in mitochondria. This protein contains two heme irons and two Cu atoms. Of particular interest is the role of the two Cu atoms in the electron-transport process. The Cu *K*-edge spectra for both the oxidized and reduced enzyme are shown in Fig. 3. Not only is the edge of the oxidized enzyme several electron volts higher than the reduced form, but several features of the fine structure are different. The spectra of several Cu compounds in the 1+, 2+, and 3+ state were also taken to aid the assignment of the various peaks. Representative compounds of different coordinating geometry, ligand type, and degree of covalency were included for each oxidation state. Figure 5 summarizes the energy ranges over which the various electronic transitions may occur for a given oxidation state. It is apparent that the ranges for several of the transitions overlap when Cu in different oxidation states is considered. For example, it is not possible to distinguish Cu(II) from Cu(III) compounds. However, the range for the Cu(I) $1s \rightarrow 4s$ transition is uniquely defined and, coincidentally, both forms of oxidase exhibit a peak within this range. If this assignment is correct, it provides

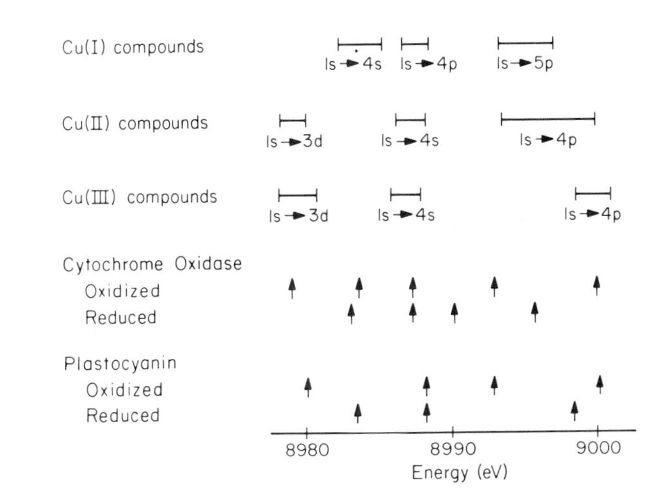

FIG. 5. Comparison of the ranges of *K*-edge transitions for Cu(I), (II), and (III) model compounds with those of the oxidized and reduced forms of cytochrome oxidase and plastocyanin.

evidence that both the oxidized and reduced enzymes contain at least one Cu(I). Because the integral under the peak for reduced oxidase is approximately twice that for the oxidized form, then two Cu(I) atoms must be present in the reduced state. Supportive evidence comes from the spectra of another Cu-containing protein, plastocyanin. On reduction, Cu(II) goes to Cu(I) and a peak also appears here within the range for the Cu(I) $1s \rightarrow 4s$ transition.

In summary, the utility of AEFS of a metalloprotein lies in its capability for monitoring specifically the outer electronic environment of a particular metal. Analysis of the spectra is still largely empirical, involving comparison to spectra of known structures. Clearly, the usefulness of the method can be extended greatly by a more complete theory.

Analysis of EXAFS Spectra

The Krönig fine structure or EXAFS has been known for some 40 yr, and although numerous attempts to explain the phenomenon have been made (see review by Azaroff[21]), only in the past several years has the theory been sufficiently developed to permit a molecular basis for its interpretation. One of the simpler, yet useful, approaches is presented by Sayers *et al.*[22] who calculated an EXAFS spectrum by a point-scattering theory. When absorption to the continuum occurs, a photoelectron is ejected with kinetic energy, $E = h\nu - E_0$, where $h\nu$ is the energy of the incident photon and E_0 is the critical energy necessary to free the bound electron. The photoelectron can be considered to be a spherical wave which expands in the atomic matrix. Neighboring atoms, assumed to be points, scatter part of the wave back to the emitting atom. The total wave received is simply the sum of the waves scattered by each atom. They showed that the extent of constructive or destructive interference varies with the momentum or de Broglie wavelength of the photoelectron. The EXAFS, or oscillatory part of the absorption coefficient, χ, above the edge, was shown to have the following general form:

$$\chi(k) = \frac{1}{k} \Sigma_j \frac{N_j}{R_j^2} S_j(k) \sin [2k R_j + 2\delta_j(k)]\, e^{-2\sigma_j^2 k^2}\, e^{-\gamma R_j} \qquad (1)$$

where $k = (0.263\, E\, (\mathrm{eV}))^{1/2}$ is the wave vector of the photoelectron; N_j is the number of atoms in the spherical shell, j, enclosing the absorbing atom at a distance, R_j; $S_j(k)$ is the amplitude of the back-scattering associated with each neighboring atom; and $\delta_j(k)$ is the phase shift for the outgoing

[21] L. V. Azaroff, *Rev. Mod. Phys.* **35,** 1012 (1963).

[22] D. E. Sayers, F. W. Lytle, and E. A. Stern, *Adv. X-Ray Anal.* **13,** 248 (1970).

electron wave relative to the 1s core state together with the phase shift on back-scattering. The first of the two exponential terms is the Debye-Waller factor, which takes into account thermal vibration and disorder; σ_j is the rms variation of an atom about R_j. The second exponential term describes the decay of the photoelectron with distance, and γ is the reciprocal of the mean free path for this decay. Stern[23] argues that effects due to short-range order, i.e., single scattering of the photoelectron back to the absorbing atom, dominate the EXAFS effect. Theories based on an extended lattice and long-range order have not been successful in accounting for the EXAFS spectrum, particularly for amorphous materials.

Multiple scattering effects have also been considered by Ashley and Doniach[24] for crystalline solids. The magnitude of the multiple scattering effect was calculated to be comparable to the second-shell single scattering contribution under certain circumstances. Lee and Pendry[25] have demonstrated similar effects on EXAFS from multiple scattering, for an atom in an outer shell which is shadowed by an atom closer to the absorbing atom. However, it appears that the effect of multiple scattering on the EXAFS spectrum is usually small.

Equation (1) can be used to determine the characteristic distances, R_j, if the back-scattering amplitude function, $S_j(k)$, the phase shift, $\delta_j(k)$, and the Debye-Waller factors, σ_j, are known. Unfortunately, this is seldom the case. Sayers *et al.*[26,27] have used the following Fourier transforms of the experimental $\chi(k)$

$$\phi(r) = (1/2\pi)^{1/2} \int_{k_{\min}}^{k_{\max}} e^{i2kr} \chi(k)\, dk \tag{2a}$$

$$\phi(r) = (1/2\pi)^{1/2} \int_{k_{\min}}^{k_{\max}} k^3 e^{i2kr} \chi(k)\, dk \tag{2b}$$

to obtain a profile in R space which peaks at certain characteristic distances (cf. Fig. 6). These peaks do not give R_j directly, but if the k-dependence of the phase shift $\delta_j(k)$ is known, R_j may be extracted from these characteristic distances. For example, if the phase shift varies linearly with k, i.e., $\delta_j(k) = \alpha_j k + \beta_j$, and there are reasonable theoretical arguments[25,27] for this assumption, then $\phi(r)$ will peak at $(R_j - \alpha_j)$. To obtain R_j, α_j must still be known. We shall discuss the determination of $\delta_j(k)$ later on in this section.

[23] E. A. Stern, *Phys. Rev. B* **10,** 3027 (1974).
[24] C. A. Ashley and S. Doniach, *Phys. Rev. B* **11,** 1279 (1975).
[25] P. A. Lee and J. B. Pendry, *Phys. Rev. B* **11,** 2795 (1975).
[26] D. E. Sayers, E. A. Stern, and F. W. Lytle, *Phys. Rev. Lett.* **27,** 1204 (1971).
[27] E. A. Stern, D. E. Sayers, and F. W. Lytle, *Phys. Rev. B* **11,** 4836 (1975).

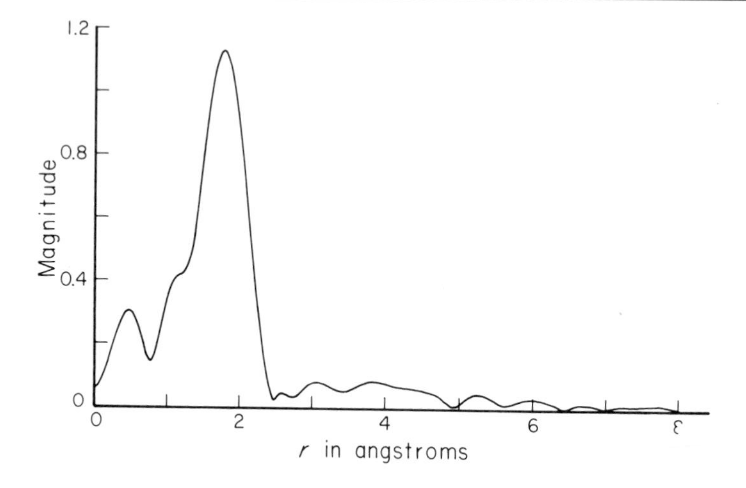

FIG. 6. Modulus of the complex Fourier transform of the EXAFS data vs. r for rubredoxin. The transform was taken over the k space range shown in Fig. 2. E_0 was chosen to be 7125 eV. [Courtesy of D. Sayers and B. Bunker, University of Washington, Seattle (1976) (unpublished).]

The Fourier-transform method has already demonstrated its utility in several investigations for determining nearest-neighbor distances around metal atoms.[19,26,28–32] In principle, the method has the potential of giving not only the position, R_j, but also the number, N_j, as well as the type of surrounding atoms which gives rise to $S_j(k)$. However, this additional information requires accurate amplitude data and precise accounting of the loss of EXAFS amplitude by other scattering processes. In addition, there are two other considerations which affect the quality of the Fourier-transformed data. The first is the limits of integration in k space. Stern and collaborators have shown that for the Eq. (2) to hold, k_{min} must be greater than 3 $Å^{-1}$.[23,27] For the effects of finite termination of the transform by k_{max} to be small, k_{max} must be much greater than k_{min}; the contribution to the width of each peak is $\Delta r \approx \pi/k_{max}$. The range of integration is typically 3–20 $Å^{-1}$. Hayes and colleagues have considered the effect of sharp cutoffs at k_{max} and k_{min}.[32] These limits essentially determine a window function in k

[28] F. W. Lytle, D. E. Sayers, and E. B. Moore, Jr., *Appl. Phys. Lett.* **24,** 45 (1974).
[29] D. E. Sayers, F. W. Lytle, M. Weissbluth, and P. Pianetta, *J. Chem. Phys.* **62,** 2514 (1975).
[30] R. G. Shulman, P. Eisenberger, W. E. Blumberg, and N. A. Stombaugh, *Proc. Natl. Acad. Sci. U.S.A.* **72,** 4003 (1975).
[31] D. E. Sayers, E. A. Stern, and J. R. Herriott, *J. Chem. Phys.* **64,** 427 (1976).
[32] T. M. Hayes, P. N. Sen, and S. H. Hunter, *J. Phys. C* **9,** 4357 (1976).

space which can cause a substantial interference of any residual background absorption with the actual structural peaks. As a technique to filter this modulation, they suggest the use of a square window convoluted with a Gaussian function, exp $(-2\,\sigma_w{}^2 r^2)$, where $\sigma_w{}^{-1}$ defines the width of this window. The second consideration is the choice of E_0, which determines the kinetic energy of the photoelectron and hence the zero point in k space. If the transform $\phi(r)$ is taken with the k^3 to enhance the contribution of data from higher k values where $\chi(k)$ is small, then the transform is relatively insensitive to E_0 as well as to k_{min}.[27]

We now turn to the question of the determination of $\delta_j(k)$. For the present, progress in the analysis of EXAFS spectra is mainly based on the use of phase shifts obtained empirically. The procedure is to use the phase shift which has been determined experimentally for a given central atom and its ligands from a similar system with known metal–ligand distances.[27] This approach may be of limited value to biological systems because precise analogues are usually not available. However, this requirement may be relaxed somewhat. Stern *et al.*[27] have proposed that the phase shift is the same for a given absorbing atom and that it is independent of the environment except for valence or charge effects. This principle of transferability is placed on firmer ground by Citrin *et al.*[33] who empirically determined phase shifts for atom pairs. Interatomic distances to within 0.02 Å could be determined using the phase shifts and EXAFS spectra for photoelectron energies above 100 eV. At least for simple compounds, this observation confirms the earlier postulates that phase shifts between atom pairs are essentially independent of the chemical surroundings.

The principle of transferability of phase shifts has been applied to a number of metalloproteins. Two groups of workers have independently measured the average Fe–S distance in rubredoxin. To determine the phase shift each group used a different model compound, FeS_2 by Sayers *et al.*,[31] and an iron-tris dithiocarbamate by Shulman *et al.*[30] [Sayers *et al.* assumed the same linear phase shift, α, for pyrite (FeS_2) and rubredoxin, whereas Shulman *et al.* applied a three-term polynomial expansion to $\delta_j(k)$.] Because there are four different Fe–S distances in rubredoxin, direct comparison of their results to crystallographic measurements of the metalloprotein is difficult, but there is excellent agreement between the two EXAFS measurements which used different model compounds, 2.30 Å ± 0.04 Å in the case of Sayers *et al.*[31] and 2.24 Å ± 0.1 Å in the case of Shulman *et al.*[30] These results argue in favor of the principle of transferability. In another experiment, the principle of transferability was used to determine the average Fe–N distances in various forms of hemoglobin.[19]

[33] P. H. Citrin, P. Eisenberger, and B. M. Kincaid, *Phys. Rev. Lett.* **36,** 1346 (1976).

Here the bisimidazole heme complex was used to estimate the phase shift. Again good agreement was found with crystallographic data.

It is apparent from the above discussion that the main difficulty in extracting structural information from EXAFS spectra is in determining the phase shift dependence on k. The assumption of a linear dependence by Sayers *et al.*[27] is satisfactory for the first shell, but others have shown that $\delta(k)$ is distinctly nonlinear,[25,34] an effect which may undermine the accuracy of determining the position of the outer shells. Experience suggests that it may be possible to improve on the precision of the transferability of phase shifts from atom pairs if the phase shifts observed for model compounds are expressed in terms of a power expansion in k, as Shulman *et al.*[30] had done in the case of rubredoxin. Cramer and co-workers[15] extended this approach by substituting the transform with a semi-empirical function which contains several parameters which may be varied to optimize the fit of the function to EXAFS spectra of known structure. Their proposed expression for $\chi(k)$ is

$$\chi(k) = \Sigma_j C_j k^{-\beta_j} \exp [-2\langle\sigma_j^2\rangle k^2) \sin (2R_j k + \alpha(k)] \quad (4)$$

with $\alpha(k) = a_0 + a_1 k + a_2 k^2$. The results of fits for certain organometallic compounds, including a number of iron porphyrins, indicate that distances out to the fourth coordination shell may be measured to within 0.1 Å. This curve-fitting method offers a convenient technique for applying tabulated phase shift and amplitude information on model compounds to metalloproteins of unknown structure and constituents.

There have been some efforts on *ab initio* calculations of the phase shifts.[34] These efforts have met with only limited success, since the calculated phase shifts produce EXAFS spectra which are only in qualitative agreement with experiment even for simple compounds such as Br_2 and $GeCl_4$. Recently a theoretical attempt to determine separately the contribution of the emitting and back-scattering atoms to the phase shift has been made, and this approach has met with fair success. Lee and Beni[35] calculated both the phase shifts and the amplitude functions, and applied these results to simple compounds using the Fourier-transform technique. The locations of the first shell peaks were of surprising accuracy (within 0.01 Å) when compared to crystallographic measurements. Part of their success was achieved by proper choice of E_0, which they permitted to vary slightly in order to get the first shell peaks of the imaginary component and the absolute value of the transform in R space to coincide. The floating of E_0 was rationalized in terms of small changes in the phase shifts at low k values

[34] B. M. Kincaid and P. Eisenberger, *Phys. Rev. Lett.* **34,** 1361 (1975).

[35] P. A. Lee and G. Beni, *Phys. Rev. B* **15,** 2862 (1977).

which depend on the chemical environment. Lee *et al.* have parameterized the amplitude functions and phase shifts[36,37] for elements with atomic numbers ranging from 5–36 (13–36 for back-scattered phases). The usefulness of these parameters in the analysis of EXAFS spectra for more complex compounds remains to be proven, but preliminary results on a number of simpler systems are certainly encouraging.

The Debye-Waller term, σ_j, in the expression for $\chi(k)$ includes both structural and thermal disordering. The latter contribution probably dominates for biological materials. The effect of disordering on the Fourier-transformed data is to broaden the peaks. (Except for finite termination of the transform integral, $\phi(r)$ would exhibit nearly delta functions at R_j for a perfectly ordered system.)

Finally, the mean free path, $1/\gamma$, has the effect of reducing the peak heights at increasing values of R_j.[27] It is energy dependent[24] and its value affects appreciably any estimate of the coordination number, N_j, of the outer shells. However, since shells beyond $j = 2$ cannot be accurately located at this time, a precise estimate of $1/\gamma$ is not important.

Note added in proof. A subsequent study [S. P. Cramer, K. O. Hodgson, W. O. Gillum, and L. E. Mortenson, *J. Am. Chem. Soc.* **100,** 3398 (1978)] has shown that the preliminary nitrogenase spectrum corresponds to air-oxidized Mo–Fe protein and that the edge position for the active material is about 20,010 eV. Thus, exposure to oxidation irreversibly oxidizes the Mo in nitrogenase. Detailed comparisons with model compound edges indicate that Mo in the active protein is in a high-sulfur environment, while the air-oxidized species contains one or two oxo groups bound to Mo.

Acknowledgments

We thank Drs. D. Sayers and B. Bunker, University of Washington, and A. Labhardt, Stanford University, for providing the unpublished data cited in Figs. 2, 4, and 6; Dr. H. Winick, Stanford University, for tabulating the information summarized in Table I; and Dr. P. Lee, Bell Laboratories, for sending us preprints of his recent calculations on phase shifts. Dr. G. Brown, SSRP, Stanford, reviewed part of this manuscript and offered a number of helpful suggestions. His assistance and advice are gratefully acknowledged. This work is partially supported by Grant GM-22432 from the National Institute of General Medical Sciences, U.S. Public Health Service, and Grant GP-38855X-3 from the National Science Foundation. R. G. is a recipient of a National Institutes of Health Postdoctoral Fellowship. This article is contribution No. 5600 from the Division of Chemistry and Chemical Engineering, California Institute of Technology.

[36] P. A. Lee, B.-K. Teo, and A. L. Simons, *J. Am. Chem. Soc.* **99,** 3856 (1977).

[37] B.-K. Teo, P. A. Lee, A. L. Simons, P. Eisenberger, and B. M. Kincaid, *J. Am. Chem. Soc.* **99,** 3854 (1977).

[20] Mössbauer Spectroscopy of Proteins: Electron Carriers

By ECKARD MÜNCK

I. Introduction

The basic principles of Mössbauer spectroscopy and its distinctive features have been thoroughly discussed by Moss[1] in Vol. 27 of this series. Adding at this time another comprehensive review serves no purpose unless a somewhat different point of view is taken. The aim of the present article is to discuss the contributions of Mössbauer spectroscopy to our understanding of iron-containing electron carriers. In addition we will attempt to point out some areas of future research on complex systems where the Mössbauer technique might be fruitfully employed. Throughout this article it is assumed that the reader is familiar with the basic principles as discussed by Moss.

The organization of this article is as follows: In Section II we outline the basic theory underlying the evaluation of magnetic Mössbauer spectra, the spin Hamiltonian formalism. This formalism is particularly suitable because it serves as a useful interface between experimental and theoretical work, and because it relates Mössbauer data to results obtained from complementary techniques such as EPR, ENDOR, far-infrared spectroscopy, and magnetic susceptometry.

In Section III we will discuss the salient features of the Mössbauer spectra observed for the most commonly occurring charge and spin states of iron. In particular we will stress, throughout this article, the difference between Kramers systems (Fe^{3+}, Fe^{+}, and spin-coupled clusters with half-integer electronic spin) and non-Kramers systems (Fe^{2+}, Fe^{4+}, and clusters with integer or zero spin).

A Mössbauer investigation of a protein containing only one iron atom yields in general a fairly complete set of parameters characterizing the electronic ground state of the iron atom. The complexity of protein Mössbauer spectra will not allow us to do a similar job on a protein containing many different iron centers. We can, however, shift emphasis and concentrate on more modest goals such as following the fate of the iron atoms as they undergo redox reactions. For such investigations we can utilize two distinctive features of the Mössbauer technique: (1) a Mössbauer resonance is observed regardless of the oxidation or spin state of the iron atoms; and (2) the spectral patterns of Kramers and non-

[1] T. H. Moss, this series, Vol. 27 [35].

ISBN 0-12-181954-X

Kramers systems are drastically different. We will illustrate the usefulness of Mössbauer spectroscopy for the investigation of complex proteins by considering a fictitious enzyme containing six metal atoms (Section V).

In Section VI we review the results obtained for some electron-transport carriers. We will focus on results which primarily, or exclusively, have been obtained from the Mössbauer technique.

In Section VII, finally, we discuss a variety of points the biochemist should consider in preparing samples for a Mössbauer investigation. This technique is still very much in the hands of physicists, and the most successful applications to biological problems have been the result of close interdisciplinary work. Moreover, many cooperations involve groups at different universities. Thus the list in Section VII might point the attention of the biochemist to problems, or solutions, easily overlooked in written correspondence or phone conversations.

II. Theoretical Background

A. *The Nuclear Hamiltonian*

In optical spectroscopy we measure transitions between the electronic ground state and excited states. The Mössbauer effect measures, as the synonymous expression "nuclear gamma-ray resonance" indicates, transitions between the ground state of a nucleus and its first excited state. In ^{57}Fe the quantum of energy exchanged is a γ-ray of energy $E_\gamma = 14.4$ keV and half-width $\Gamma = 4.7 \times 10^{-9}$ eV. (The width is given by the lifetime of the excited state.) The monochromatic ($\Gamma/E_\gamma = 3 \times 10^{-13}$) radiation emitted by a source (^{57}Co embedded in some appropriate matrix such as Cu or Rh metal) penetrates an absorber (our protein), and the transmitted radiation is detected by a proportional counter. The resonance energy spectrum is scanned by Doppler shifting the source relative to the absorber by an amount $\delta E = (v/c)E_\gamma$, where v is the source velocity and c the speed of light. The radiation transmitted through the absorber is recorded as a function of velocity (for ^{57}Fe, 1mm/s $\hat{=}$ 4.8×10^{-8} e V $\hat{=}$ 11.6 MHz). The resolution is adequate to resolve hyperfine interactions ($\simeq 5 \times 10^{-7}$ eV) of the ^{57}Fe nucleus with its electronic environment. The hyperfine splittings of the nuclear levels reflect electrostatic and magnetic interactions of the nuclear moments with (predominantly) the d electrons. The distribution of the d electrons, in turn, depends on the chemical and structural environment of the iron atom. Two facts make the Mössbauer effect an important spectroscopic tool. First, a Mössbauer resonance is observed regardless of the charge and spin state of the iron (there is no "Mössbauer-silent" iron). Second, the Mössbauer spectrum is characteristic of the spin and oxidation state of the iron. An obvious drawback for biological work is that the proteins have to be studied generally under nonphysiological conditions, either as single crystals, or lyophilized, or in frozen solution. Only a solid environment meets the prerequisite of the Mössbauer effect: recoilless emission and absorption of γ-rays.

A magnetic field acting on the ^{57}Fe nucleus removes the 2-fold degeneracy of the nuclear ground state (nuclear spin, $I_g = 1/2$) and the 4-fold degeneracy of the 14.4 keV excited state ($I_e = 3/2$). The nuclear excited state, in addition, has a quadrupole moment ($Q \simeq 0.2$ b) which can interact with an electric field gradient at the nucleus. This gives rise to a quadrupole splitting, an interaction only measurable by Mössbauer spectroscopy since the probing state is an excited nuclear level.

The energy splittings of the nuclear levels are described by the nuclear Hamiltonian

$$\hat{H}_n = -g_n\beta_n \vec{H}_{\text{eff}} \cdot \vec{I} + \frac{eQV_{zz}}{12}\left[3I_z^2 - \frac{15}{4} + \eta\,(I_x^2 - I_y^2)\right] \qquad (1)$$

where $g_n\beta_n\vec{I}$ is the nuclear magnetic moment ($g_n = g_g = 0.180$ for the ground state, and $g_n = g_e = -0.103$ for the excited state). $\vec{H}_{\text{eff}}$ is an effective magnetic field acting on the nucleus. The second term describes the interaction of the excited state quadrupole moment Q with an electric field gradient tensor [principal axes components V_{xx}, V_{yy}, and V_{zz}; $\eta = (V_{xx} - V_{yy})/V_{zz}$ is called the asymmetry parameter]. For the nuclear ground state, $Q = 0$. For $\vec{H}_{\text{eff}} = 0$ only the excited state is split, by an amount

$$\Delta E_Q = \frac{eQV_{zz}}{2}\sqrt{1 + \tfrac{1}{3}\eta^2}.$$

ΔE_Q is called the quadrupole splitting and is generally quoted in velocity units.

In addition to the quadrupole and magnetic interaction the Mössbauer spectrum yields another quantity, the isomeric or chemical shift δ. The latter results from the difference of the nuclear charge distributions for the ground and excited state. The isomeric shift is proportional to the s-electron density at the nucleus, and it manifests itself as an overall shift of the absorption pattern relative to zero velocity.

The basic Mössbauer spectra arising from Eq. (1) have been presented and discussed previously in this series; we will not repeat this discussion here and we refer the interested reader to the article by Moss.[1] Equation (1) looks rather simple, and it would appear that not too much information can be extracted from a Mössbauer spectrum. On the contrary, the spectra of proteins display a high degree of complexity. These intricate patterns result from a spatial anisotropy of $\vec{H}_{\text{eff}}$, its orientation relative to the field gradient tensor, and its dependence on the temperature and applied magnetic fields. In the following section we will discuss these matters by expressing $\vec{H}_{\text{eff}}$ in more familiar terms, and we will show that patterns arise which characterize the oxidation and spin state of the iron atom. We will discuss the spectra in the framework of the spin Hamiltonian since this formalism has been highly successful. Moreover, the same formalism is widely used to describe the data obtained by complementary techniques such as EPR, Endor, far-infrared spectroscopy, and magnetic susceptibility measurements.

B. The Spin Hamiltonian

The Mössbauer spectrum depends directly on the properties of the electronic states through the hyperfine interactions. Thus by studying the hyperfine splittings of the nuclear levels the nature of the lowest electronic states can be inferred. The character of these levels in turn gives structural and chemical information. It is useful to parameterize the fine structure (zero-field splitting) and hyperfine structure of the lowest electronic states in terms of a few coupling constants. This is achieved in the spin Hamiltonian description.

A spin Hamiltonian adequate for the description of our data may be written as

$$\hat{H} = \hat{H}_e + \hat{H}_{hf} \tag{2}$$

with

$$\hat{H}_e = D[S_z^2 - \tfrac{1}{3}S(S+1) + \frac{E}{D}(S_x^2 - S_y^2)] + \beta_e \vec{S} \cdot \tilde{g} \cdot \vec{H} \tag{2a}$$

and

$$\hat{H}_{hf} = \vec{S} \cdot \tilde{A} \cdot \vec{I} + \vec{I} \cdot \tilde{P} \cdot \vec{I} - g_n \beta_n \vec{H} \cdot \vec{I} \tag{2b}$$

Here $\hat{H}_e$ depends on electronic variables only. The first two terms in Eq. (2a) represent the zero-field or fine-structure splitting, and S_x, S_y, and S_z are the components of the electronic spin operator in a suitably chosen coordinate system fixed to the molecule. The last term in Eq. (2a) is the electronic Zeeman interaction that has to be specified, in general, by a g-tensor $\tilde{g}$. In EPR, magnetic susceptibility, and far-infrared work we are generally concerned with $\hat{H}_e$ only.

The terms in Eq. (2b) describe the hyperfine interactions. $\vec{S} \cdot \tilde{A} \cdot \vec{I}$ describes the magnetic interactions of the electronic cloud with the nucleus, and $\vec{I} \cdot \tilde{P} \cdot \vec{I}$ is the quadrupole interaction which when expressed in the principal axes frame of the electric field gradient tensor takes a form as specified in Eq. (1). The last term is the nuclear Zeeman term describing the interaction of the nuclear magnetic moment $g_n\beta_n\vec{I}$ with an applied field $\vec{H}$.

In Eqs. (2a) and (2b) the magnitudes of the various interactions are listed in descending order. The fine-structure parameters are in the order of a few wave numbers ($D \simeq 5$–$10\ \text{cm}^{-1}$ for heme proteins; $D = 1.5\ \text{cm}^{-1}$ and 8 cm^{-1} for ferric and ferrous rubredoxin, respectively; $D < 1\ \text{cm}^{-1}$ for siderochromes). For weak applied fields of about 10 G the electronic Zeeman interaction is comparable to the hyperfine terms; thus for applied fields larger than 100 G both electronic terms dominate the hyperfine interactions. Under these conditions it is possible to separate the nuclear and electronic coordinates, and Eq. (2b) can be manipulated into the form

$$\hat{H}_{hf} = \langle\vec{S}\rangle \cdot \tilde{A} \cdot \vec{I} - g_n\beta_n\vec{H} \cdot \vec{I} + \vec{I} \cdot \vec{P} \cdot \vec{I} \tag{3}$$

In Eq. (3) $\langle\vec{S}\rangle$ designates an appropriately taken expectation value of the electronic spin. Equation (3) can be understood as follows: If the fine-structure and electronic Zeeman terms are large compared to the hyperfine interactions, the quantization axis of the electronic spin is entirely determined by the electronic system and the effect of the nucleus on the orientation of the electronic spin is neglectible. This allows us to replace the operator $\vec{S}$ by its expectation value, which is determined by $\hat{H}_e$. Equation (3) may be rewritten as

$$\begin{aligned}\hat{H}_{hf} &= -\frac{-\langle\vec{S}\rangle \cdot \tilde{A}}{g_n\beta_n}\, g_n\beta_n\vec{I} - g_n\beta_n\vec{H} \cdot \vec{I} + \vec{I} \cdot \tilde{P} \cdot \vec{I} \\ &= -g_n\beta_n(\vec{H}_{\text{int}} + \vec{H}) \cdot \vec{I} + \vec{I} \cdot \vec{P} \cdot \vec{I}\end{aligned} \tag{4}$$

The quantity $\vec{H}_{\text{int}} = -\langle\vec{S}\rangle \cdot \tilde{A}/g_n\beta_n$ is called the internal magnetic field at the nucleus. The Mössbauer nucleus experiences an effective magnetic field which is the vector sum $\vec{H}_{\text{eff}} = \vec{H}_{\text{int}} + \vec{H}$. Equations (3) and (4) establish the relation between the spin Hamiltonian, Eqs. (2), and Eq. (1). Since $\langle\vec{S}\rangle$, and therefore $\vec{H}_{\text{eff}}$, depend through $\hat{H}_e$ on D, E, and $\tilde{g}$ these electronic parameters can be determined from the Mössbauer spectrum. Moreover, $\langle\vec{S}\rangle$ depends on the temperature T and on the magnitude and orientation of the applied magnetic field $\vec{H}$. Thus the absorption pattern of the Mössbauer spectrum can be changed by varying the experimental conditions, i.e., T and $\vec{H}$.

In the next section we will discuss the commonly occurring charge and spin states of iron compounds. We will separate the compounds in two classes: Kramers systems (complexes or iron clusters with an odd number of electrons, i.e., half-integral electronic spin) and non-Kramers systems (compounds or clusters with an even number of electrons, i.e., integer or zero electronic spin). The former, at low temperatures, yield generally magnetic Mössbauer spectra ($\langle \vec{S} \rangle \neq 0$) even in the absence of an applied magnetic field; the latter yield only quadrupole doublets at all temperatures unless a strong magnetic field is applied. We will discuss what is meant by "an appropriately taken expectation value of the electronic spin" and we will illustrate commonly occurring situations. For further information on the spin Hamiltonian formalism in connection with Mössbauer spectroscopy we refer the reader to the literature.[2-5]

III. Common Charge and Spin States

A. *Non-Kramers Systems*

All non-Kramers systems have one feature in common: $\langle \vec{S} \rangle$, and therefore $\vec{H}_{int}$, is zero unless the sample is studied in a strong applied field. Consequently, in zero field only the quadrupole interaction needs to be considered and the spectra consist of two absorption lines, a quadrupole doublet. Experimentally we obtain the magnitude, but not the sign, of the quadrupole splitting

$$|\Delta E_Q| = \frac{eQ\,|V_{zz}|}{2}\sqrt{1+\tfrac{1}{3}\eta^2}$$

and the isomeric shift δ. If low-lying orbital electronic states become populated, ΔE_Q will be temperature dependent. We may divide the non-Kramers system in two classes: diamagnetic compounds ($S = 0$) and paramagnetic specimens (integer, nonzero spin).

1. Complexes with $S = 0$. Typical representatives of diamagnetic iron compounds are the reduced forms of the cytochromes c and b_5, carbon monoxide complexes with reduced hemes, oxyhemoglobin, oxidized plant-type and bacterial-type ferredoxins, and the reduced high-potential iron protein. For these compounds $\vec{H}_{int} = 0$ under all experimental conditions. Application of a strong applied magnetic field results in magnetic splittings of the nuclear levels via the nuclear Zeeman term ($-g_n\beta_n\vec{H}\cdot\vec{I}$). For samples with randomly oriented molecules (powder) a characteristic spectrum results from which the sign of the quadrupole splitting and the asymmetry parameter η may be determined. Moreover, such spectra prove the diamagnetic nature of the metal center. Determinations of η for the

[2] G. Lang, *Q. Rev. Biophys.* **3,** 1 (1970).
[3] W. T. Oosterhuis, *Struct. Bonding (Berlin)* **20,** 59 (1974).
[4] E. Münck and P. M. Champion, *J. Phys. (Paris),* Colloque **35,** C6–33 (1974).
[5] P. G. Debrunner, *in* "Spectroscopic Approaches to Biomolecular Conformation" (D. W. Urry, ed.), p. 209. Am. Med. Assoc., Chicago, Illinois, 1971.

different iron atoms of a cluster (i.e, 4 Fe–4 S* proteins) are very difficult, if not impossible, since the iron atoms constituting the cluster may be dissimilar resulting in superposition of different spectra. The diamagnetic nature of the cluster, however, usually can be established reliably [even in the presence of some paramagnetic ($S \neq 0$) impurities].

2. Complexes with $S \neq 0$. The Mössbauer spectra of paramagnetic ($S \neq 0$) non-Kramers ions are rich in information since an internal magnetic field can be induced by application of a strong magnetic field. This allows the determination of zero-field splitting and magnetic hyperfine parameters. The Mössbauer effect is particularly useful here since paramagnetic non-Kramers ions are generally not amenable to EPR spectroscopy. High-spin ferrous ($S = 2$) proteins are typical representatives (reduced rubredoxin and cytochrome P450, deoxyhemoglobin, cytochrome *c*′). Fe^{4+} compounds fit into the same category (compound II of horseradish peroxidase and complex ES of cytochrome *c* peroxidase). A more complex situation occurs in fully reduced ferredoxin (spin-coupling between spin-coupled clusters).

As an example we briefly discuss some features pertinent to the high-spin ferrous state ($S = 2$). For symmetries lower than trigonal or tetragonal the 5-fold degeneracy of the spin quintet is totally removed by the zero-field splitting term [$D, E \neq 0$ in Eq. (2a)]. (The proteins studied so far exhibit a fine-structure splitting in the range from 10–50 cm^{-1}.) In the absence of an applied magnetic field a singlet state can have no magnetic moment ($\langle \vec{S} \rangle = 0$) and the Mössbauer spectrum consists only of a quadrupole doublet. An applied magnetic field, however, mixes the electronic states and produces a polarization, i.e., the expectation value of the spin $\langle \vec{S} \rangle \neq 0$ (this fact is used in magnetic susceptibility experiments). At sufficiently low temperatures only the lowest-spin singlet is populated. The appropriate spin expectation value appearing in Eq. (4) is $\langle \psi_g | S_i | \psi_g \rangle$, where ψ_g is the wave function of the spin ground state, and S_i stands for S_x, S_y, and S_z. At higher temperatures more than one spin level will be populated. Now two extreme situations may occur: The electronic spin may jump rapidly among the thermally accessible levels (fast relaxation), or the electronic relaxation rate may be slow (slow compared to the nuclear precession frequency, typically 10–30 MHz). In the latter case each spin level gives rise to a separate Mössbauer spectrum with an intensity given by the Boltzmann factor of the corresponding level. In the fast relaxation limit, on the other hand, the spin expectation value $\langle \vec{S} \rangle$ of Eq. (3) has to be taken as a thermal average over all five spin states, and only a single Mössbauer spectrum results.[6]

[6] P. M. Champion, J. D. Lipscomb, E. Münck, P. G. Debrunner, and I. C. Gunsalus, *Biochemistry* **14**, 4151 (1975).

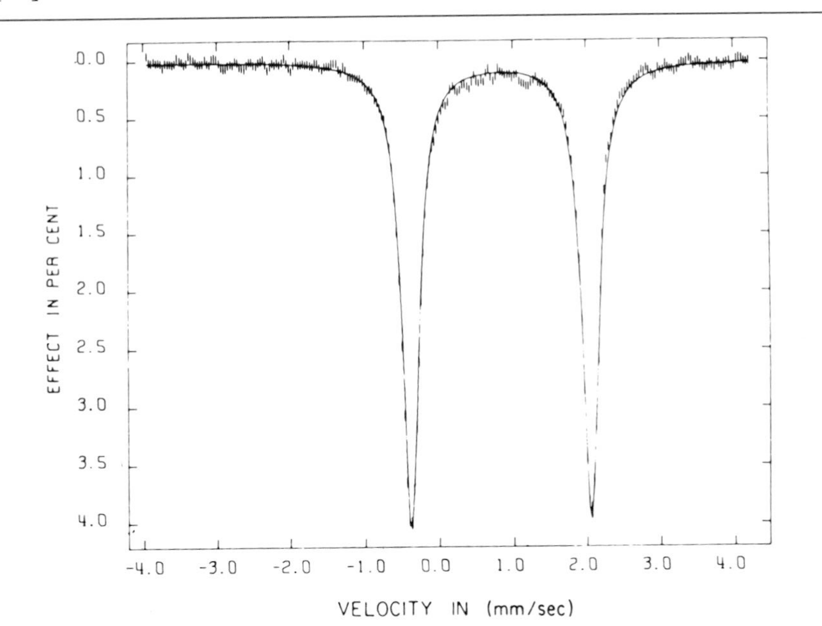

FIG. 1. Mössbauer spectrum of reduced cytochrome $P450_{cam}$ from *Pseudomonas putida*. The spectrum was taken at 4.2°K. Since the heme iron in reduced $P450_{cam}$ is in a high-spin ferrous ($S = 2$) configuration no magnetic hyperfine interactions are observed in the absence of a strong applied magnetic field. From the line positions the quadrupole splitting $\Delta E_Q = 2.42$ mm/sec and the isomeric shift $\delta = 0.82$ mm/sec (center of the spectrum) are obtained.

Mössbauer spectroscopy on paramagnetic non-Kramers ions using strong applied fields (up to 60 kG) has been applied successfully only recently.[6-9] The high-field spectra of these systems are exceedingly complex, and their understanding requires many measurements under different experimental conditions, and extensive computer simulations. On the other hand, such studies yield detailed information which cannot be obtained by other spectroscopic techniques. Moreover, the spectral patterns observed for various proteins are quite different. Thus high-field spectroscopy is sensitive to the details of the electronic structure and can therefore be used to elucidate structural details.

It is not the scope of this article to discuss the intricacies of high-field Mössbauer spectra. Without going into details we like to mention some relevant results from a study[6] of cytochrome $P450_{cam}$. In Fig. 1 we have

[7] P. M. Champion, R. Chiang, E. Münck, P. G. Debrunner, and L. P. Hager, *Biochemistry* **14,** 4159 (1975).

[8] P. G. Debrunner, E. Münck, L. Que, and C. E. Schulz, *Iron-Sulfur Proteins*, **3,** 381 (1976).

[9] T. Kent, K. Spartalian, G. Lang, and T. Yonetani, *Biochim. Biophys. Acta* **490,** 331 (1977).

displayed a zero-field spectrum, taken at 4.2°K, of reduced $P450_{cam}$. The quadrupole splitting $\Delta E_Q = 2.42$ mm/sec and the isomeric shift $\delta = 0.82$ mm/sec establish a high-spin ferrous configuration (values for $\delta > 0.7$ mm/sec are a characteristic marker for high-spin ferrous materials). The quadrupole splitting is almost independent of temperature indicating an isolated orbital ground state (but a spin quintet), separated in energy by at least 500 cm^{-1} from the next orbital state. Application of strong magnetic fields yields spectra showing paramagnetic hyperfine structure ($\langle \vec{S} \rangle \neq 0$). An example is given in Fig. 2; the solid line is the result of fitting Eqs. (2) to a series of spectra to extract a consistent set of parameters.[6] The relevant results are as follows: The fine-structure parameter D is positive, $D \simeq +14$ cm^{-1} and $0 \leq E/D \leq 0.15$. The data demand a field gradient tensor with a large asymmetry parameter, $\eta > 0.6$; moreover, the field gradient is rotated substantially away from the frame of the zero-field splitting tensor. This implies a very low symmetry (triclinic). Of further interest is the fact

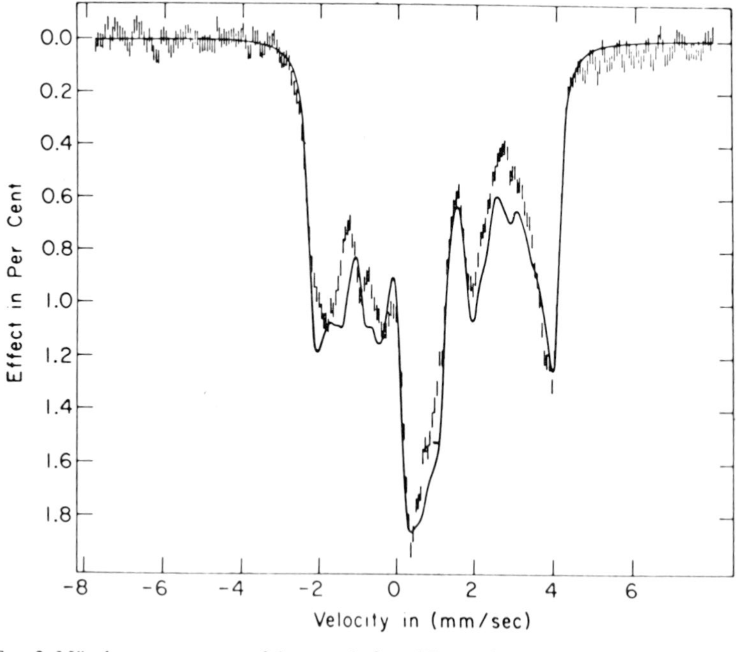

FIG. 2. Mössbauer spectrum of the sample from Fig. 1 taken at 4.2°K in an external field of 25 kG applied parallel to the transmitted γ-radiation. The applied field has induced an internal field H_{int} of about -150 kG. The intricate absorption pattern results from a complex interplay of magnetic and electric hyperfine interactions. The solid line was generated from Eq. (2) by means of a computer program. The data analysis and some methodology relevant to the high-spin ferrous state have been discussed in Champion *et al.*[6]

that the high-field spectra of reduced $P450_{cam}$ and reduced chloroperoxidase are practically identical[7] indicating similar, if not identical, ligand environments in the reduced form of the proteins. Moreover, the $P450_{cam}$ and chloroperoxidase spectra are quite different from the spectra of hemoglobin and horseradish peroxidase.[7] At present we cannot deduce ligand environments from these spectra. Nevertheless, high-field Mössbauer spectroscopy promises to be a tool capable of discerning fine details in electronic structure.

B. Kramers Systems

Kramers systems have one feature in common: In the absence of an applied field the electronic states are at least 2-fold degenerate; mostly they are doublets (Kramers doublets). In general, any system with an odd number of electrons, i.e., half-integer electronic spin, will give rise to Kramers doublets with observable magnetic properties such as magnetic hyperfine splittings in Mössbauer spectroscopy ($\vec{H}_{int} \neq 0$, $\langle \vec{S} \rangle \neq 0$) and magnetic resonance transitions in EPR spectroscopy. We may distinguish two situations. Systems with $S = 1/2$ yield generally an isolated Kramers doublet, and the zero-field splitting term in Eq. (2) need not be considered. For complexes with $S \geqslant 3/2$ the zero-field splitting term partly removes the $(2S + 1)$-fold degeneracy and $(2S + 1)/2$ Kramers doublets result. For small applied fields such that $\beta_e H \ll |D|$ (the doublets are well separated and the applied field does not mix the doublets), we can conveniently treat each doublet as a fictitious spin $S' = 1/2$ doublet; EPR and Mössbauer data are commonly analyzed in this way. For long electronic spin relaxation times ($T_e \geqslant 10^{-6}$ sec) each doublet gives rise to a separate Mössbauer spectrum.

1. Complexes with $S = 1/2$. In iron-containing proteins we encounter quite frequently complexes with $S = 1/2$. Many heme proteins in their resting state are low-spin ferric; examples are the cytochromes c, b_5, and P450. Complexes of reduced hemes with nitric oxide yield $S = 1/2$ states also; formally we can interpret them as low-spin Fe^{+}. Finally, iron-sulfur proteins present us with spin doublets which are the result of spin-coupling either two or four iron atoms to a resultant spin $S = 1/2$; examples are the reduced plant-type and bacterial-type ferredoxins and oxidized high-potential iron proteins.

We can describe the Mössbauer data of spin $S = 1/2$ complexes with the Hamiltonian [Eq. (2) with $D = E = O$]

$$\hat{H} = \beta_e \vec{S} \cdot \tilde{g} \cdot \vec{H} + \vec{S} \cdot \tilde{A} \cdot \vec{I} + \vec{I} \cdot \tilde{P} \cdot \vec{I} \quad (5)$$

As an example of an $S = 1/2$ system we discuss briefly some features of low-spin ferric hemes. These complexes have been discussed extensively in the literature;[2,4,10-12] the general features of their Mössbauer spectra are reasonably well understood. The method most commonly used to describe EPR and Mössbauer data follows the ligand-field treatment of Griffith[13] and of Oosterhuis and Lang.[10] Blumberg and Peisach[14] have evaluated the EPR data of a variety of compounds in the framework of ligand field theory and have classified the results in a "truth-diagram" which expresses the measured g-values in terms of ligand-field parameters describing the tetragonal and rhombic distortions of the heme iron. From a given g-tensor one can derive in this simple theory (restriction to t_{2g} orbitals and an isotropic covalency factor) the wave function of the electronic ground state. Except for scaling factors this wave function determines the A-tensor and the electric field gradient tensor.

Figure 3 shows Mössbauer spectra of oxidized cytochrome $P450_{cam}$ complexed to 2-phenylimidazole.[11] These spectra are typical for low-spin ferric hemes. A large quadrupole splitting (due to a missing d electron in an otherwise closed t_{2g} shell) in conjunction with a large and anisotropic hyperfine field gives rise to those broad spectra. The anisotropy of the A-tensor results from large orbital contributions which exceed the (isotropic) Fermi contact interaction.

We have chosen the spectra shown in Fig. 3 to indicate the promise of Mössbauer spectroscopy as a tool capable of probing the details of electronic structure—provided one can obtain the A-tensor from Endor spectroscopy to reduce ambiguities. The solid lines in Fig. 3 are the result of computer simulations by Sharrock *et al.* based on Eq. (5). These authors[11] first attempted to fit the data within the simple crystal field model using the experimentally known g-values. The model then yields theoretical spectra which account for the gross features of the experimental data. It turns out that one can improve the fits dramatically by choosing a value for the asymmetry parameter η of the field gradient tensor [Eq. (1)] which is quite different from that predicted by the model. (The fits to the cytochrome c and b_5 spectra can be improved in a similar way.) To rationalize this value

[10] W. T. Oosterhuis and G. Lang, *Phys. Rev.* **178,** 439 (1969).

[11] M. Sharrock, P. G. Debrunner, C. Schulz, J. D. Lipscomb, V. Marshall, and I. C. Gunsalus, *Biochim. Biophys. Acta* **420,** 8 (1976).

[12] P. M. Champion, E. Münck, P. G. Debrunner, P. F. Hollenberg, and L. P. Hager, *Biochemistry* **12,** 426 (1973).

[13] J. S. Griffith, *Nature (London)* **180,** 30 (1957).

[14] W. E. Blumberg and J. Peisach, *in* "Probes of Structure and Function of Macomolecules and Membranes" (B. Chance, T. Yonetani, and A. S. Mildvan, eds.), Vol. 2, p. 215. Academic Press, New York, 1971.

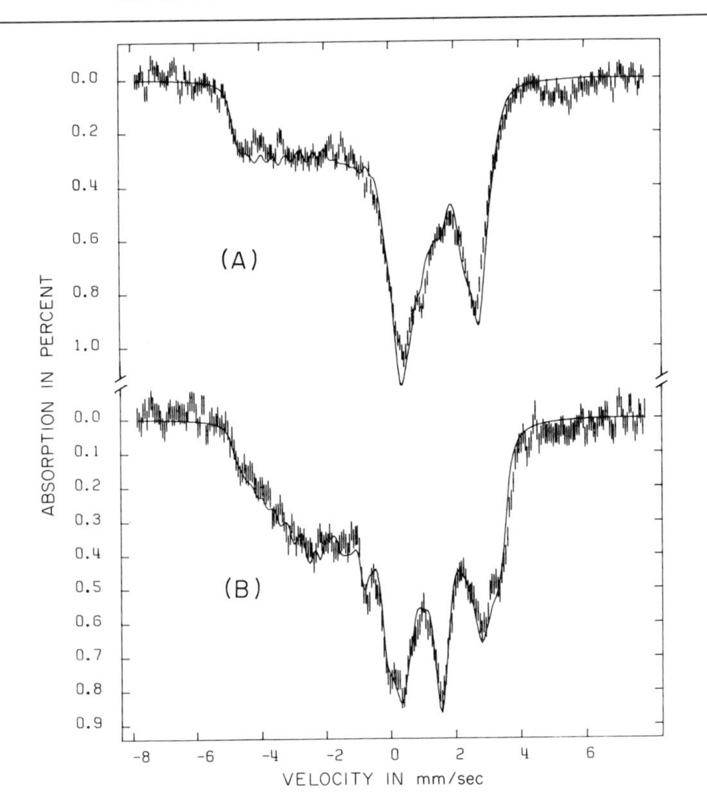

FIG. 3. Mössbauer spectra, measured at 4.2°K, of oxidized cytochrome $P450_{cam}$ complexed to 2-phenylimidazole. A magnetic field of 1600 G was applied parallel to the transmitted γ-rays in (A) and transverse in (B). The solid lines are computer-simulated spectra based on Eq. (5). The good fits were obtained by assigning different covalency factors to the out-of-plane orbitals, d_{xz} and d_{yz}, relative to the in-plane orbital, d_{xy}. At higher temperatures the electronic spin relaxation rate becomes fast, and the spectra collapse into a quadrupole doublet with $\Delta E_Q = 2.85$ mm/sec (at 200°K). For details see Sharrock *et al.*[11]

for η Sharrock *et al.* assigned different covalency factors N^2 to the out-of-plane orbitals, d_{xz} and d_{yz}, relative to the in-plane orbital, d_{xy}. Griffith[13] has pointed out that the d_{xz} and d_{yz} orbitals are able to participate in π-bonding with the ring nitrogens and suitable axial ligands, in contrast to the d_{xy} orbital. The fits shown in Fig. 3 result from $N^2_{xy} = 0.96$ and $N^2_{xz} = N^2_{yz} = 0.78$.

Although different covalency parameters are quite plausible the Mössbauer data do not unambiguously establish this interpretation. In evaluating the data it was assumed that the symmetry is not lower than D_2. In lower symmetries the three tensors in Eq. (5) will have different principal axes frames (adding as many as six more unknown parameters). Moreover, the A-tensor will not be symmetric any more for monoclinic and triclinic symmetries. Therefore, Endor measurements on low-spin ferric hemes are highly desirable. This technique allows precise determinations of the A-values and, equally important, it could give information about the relative orientation of the g- and A-tensor and thus provide crucial insight into the symmetry of the heme environment. Thus, Endor results could reduce the large number of unknowns in the Mössbauer data analysis. In any case, the analysis by Sharrock *et al.*[11] points in a very interesting direction.

2. Half-integer Spin $S \geqslant 3/2$. With the exceptions of an $S = 3/2$ cluster in the MoFe protein of the nitrogenase system, the biological systems to be considered in this section are high-spin ferric ($S = 5/2$). Many proteins contain the high-spin ferric ion in a variety of ligand environments (hemes, siderochromes, dioxygenases, and rubredoxin, to name only a few examples). Since heme proteins[2,4,15] and siderochromes[3] have been reviewed in considerable detail we will choose rubredoxin as an illustrative example.

For the high-spin ferric ion Eq. (2a) takes a particularly simple form since both the electron Zeeman interaction and the magnetic hyperfine interaction ($g_e\beta_e\vec{H} \cdot \vec{S}$ and $A_0\vec{S} \cdot \vec{I}$) are in general isotropic. The zero-field splitting term in Eq. (2a) partly removes the degeneracy of the spin sextet and three Kramers doublets result. For studies where $\beta_e H \ll |D|$ (the applied field does not mix the doublets to any appreciable extent) it is convenient to describe each Kramers doublet as a fictitious spin doublet $S' = 1/2$. The Mössbauer and EPR properties of each doublet can then be described by

$$\hat{H} = \beta_e \vec{S}' \cdot \tilde{g}_i \cdot \vec{H} + \vec{S}' \cdot \tilde{A}_i \cdot \vec{I} + \vec{I} \cdot \tilde{P} \cdot \vec{I} \qquad (6)$$

Equations (6) and (2) are made equivalent by a proper choice of the tensors $\tilde{g}_i$ and $\tilde{A}_i$ for each of the three doublets $i = 1, 2, 3$. For the high-spin ferric ion g_e can be taken as the free-spin value $g_e = 2.0023$; with this the g-tensors depend only on E/D, the rhombicity parameter of the fine-structure term. Conversely, the g-values obtained from EPR experiments determine only E/D. In this formalism it follows also that the components of the A_i tensors are proportional to those of tensors $\tilde{g}_i$. Specifically, we have

$$\frac{A_x}{g_x} = \frac{A_y}{g_y} = \frac{A_z}{g_z} = \frac{A_0}{g_e} \qquad (7)$$

for each of the three doublets.[16]

X-ray diffraction studies have firmly established that the iron in rub-

[15] E. Münck, *in* "The Porphyrins" (D. Dolphin, ed.). Vol. IVB, chapter 8, Academic Press, New York (1978), in press.

[16] H. H. Wickman, M. P. Klein, and D. A. Shirley, *Phys. Rev.* **152,** 345 (1966).

redoxin has a (distorted) tetrahedral environment of cysteine residues. Peisach *et al.*[17] and Blumberg and Peisach[18] have investigated the protein from *Pseudomonas oleovorance* with EPR spectroscopy. By taking into account both the temperature dependence of the observed EPR signals and five observed g-values these authors determined $D = 1.76\ cm^{-1}$ and $E/D = 0.28$. These values for the zero-field splitting parameters give rise to three almost equally spaced Kramers doublets, the middle doublet being 7.7°K above the ground state. The value quoted for E/D yields theoretical g-tensors $\tilde{g}_1 = (1.22, 9.52, 0.74)$ and $\tilde{g}_2 = (4.20, 3.97, 4.58)$ for the ground and middle doublet, respectively. The experimental data are $\tilde{g}_1 = (1.25, 9.24, 0.90)$ and $\tilde{g}_2 = (4.31, 4.02, 4.77)$. The agreement between theory and experiment is reasonable; some modifications in the spin Hamiltonian, however, are called for. We will return to this point after discussing the Mössbauer data.

Recently, Debrunner and co-workers[8,19] have investigated the protein from *Clostridium pasteurianum* in great detail. The low-temperature Mössbauer spectra of the ferric protein are displayed in Fig. 4. The features of these spectra are readily understood with the information we have from EPR spectroscopy.

At the lowest temperature, 1.5°K, only the ground Kramers doublet is populated. The Mössbauer spectrum associated with this doublet yields a simple six-line pattern (Fig. 4A). The direction of an applied magnetic field does not affect the intensities very much; this implies that the ground doublet has an easy axis of magnetization (see Section IV). This clearly must be the direction of the $g = 9.4$ component; i.e., the y direction of the zero-field splitting term. (The electronic spin, and therefore $\vec{H}_{int}$, tends to line up along the y axis, no matter how the molecule is oriented relative to the applied field. Thus the splittings of the nuclear levels are almost orientation independent and six fairly sharp lines result.) From the total magnetic splitting we obtain $H_{int,y} = -A_{1,y}\langle S'_y\rangle/(g_g\beta_n) = -365$ kG. The quadrupole interaction which is small compared to the magnetic splitting shifts the inner four lines (to the left) relative to the outer two lines (which move to the right), by an amount proportional to eQV_{yy}, the component of the quadrupole tensor along the prevailing internal field ($eQV_{yy} \cong 0.56$ mm/sec).

At 4.2°K both the middle and upper Kramers doublets become populated, their populations being given by a Boltzmann distribution. Our simple theory [Eq. (7)] predicts that the magnetic splitting of the middle doublet Mössbauer spectrum is about half of that for the ground doublet. The increase in absorption in Fig. 4B in the velocity range from -3 mm/sec

[17] J. Peisach, W. E. Blumberg, E. T. Lode, and M. J. Coon, *J. Biol. Chem.* **246,** 5877 (1971).
[18] W. E. Blumberg and J. Peisach, *Ann. N. Y. Acad. Sci.* **222,** 539 (1973).
[19] C. Schulz and P. G. Debrunner, *Int. Conf. Appl. Mössbauer Effect,* (1976); *J. Phys. Paris Colloque* **37,** C6–153 (1976).

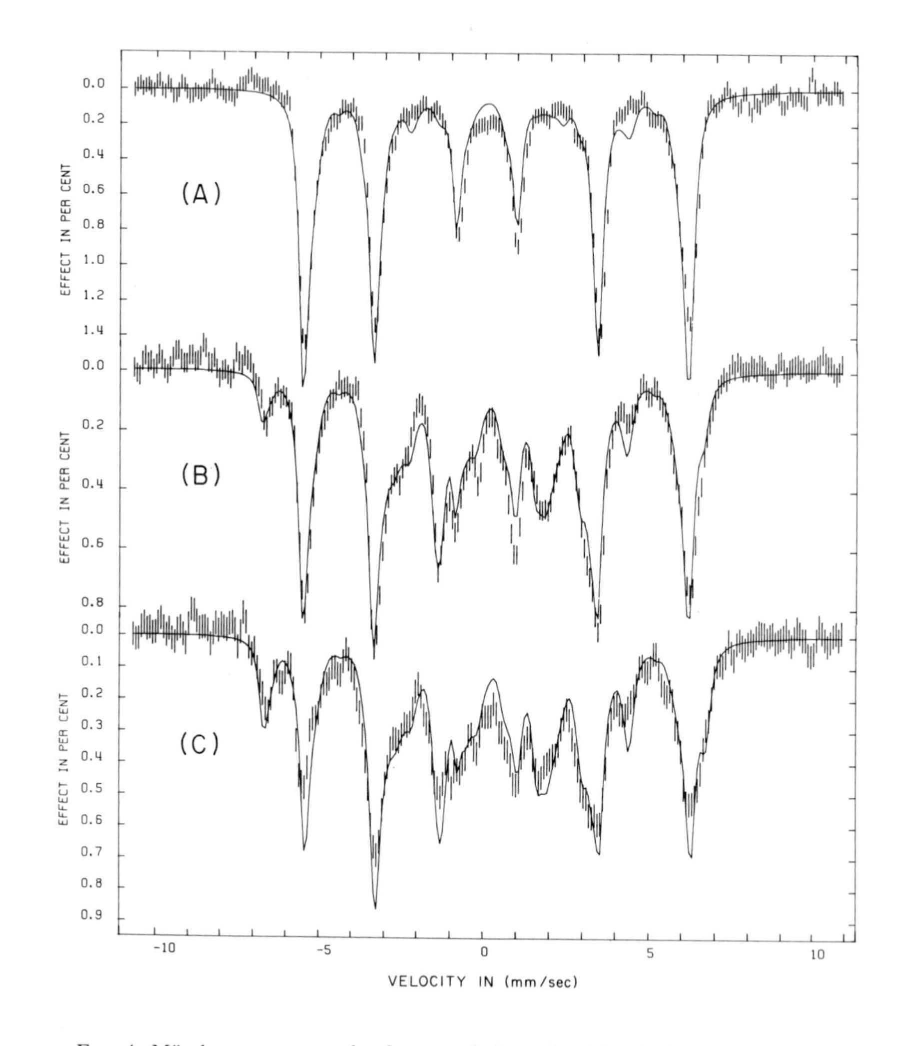

FIG. 4. Mössbauer spectra of a frozen solution of oxidized rubredoxin from *C. pasteurianum* taken in a transverse magnetic field of 1300 G at (A) 1.5°K, (B) 4.2°K, and (C) 10°K. The solid lines are computer simulations based on Eqs. (2), augmented by a fourth-order term [Eq. (8)] in the fine structure. The parameters used and their estimated uncertainties are as follows: $D = (2.5 \pm 0.5)$°K, $E/D = 0.17$, $\mu = 0.3 \pm 0.1$, $\Delta E_Q = -(0.5 \pm 0.1)$ mm/sec, $\eta = 0.2 \pm 0.1$, $A_x = -(22.7 \pm 1.5)$ MHz, $A_y = (-21.5 \pm 0.4)$ MHz, and $A_z = -(23.5 \pm 0.4)$ MHz. The g-tensors computed with these parameters are $\tilde{g}_1 = (1.10, 9.35, 0.75)$, $\tilde{g}_2 = (3.98, 4.09, 4.55)$, and $\tilde{g}_3 = (0.93, 0.73, 9.80)$. The rubredoxin work described here was done by P. Debrunner's group at Illinois; the rubredoxin samples were prepared by W. Lovenberg.

to +3 mm/sec is essentially due to the middle doublet. In addition, the spectrum in Fig. 4B shows new resolved peaks at −6.5 mm/sec and +5.5 mm/sec and a shoulder at +6.5 mm/sec; these lines result from the upper doublet. For this doublet, the spin Hamiltonian, for $E/D = 0.28$, predicts a g-tensor $\tilde{g}_3 = (0.65, 0.41, 9.77)$. These g-values predict for the upper doublet a Mössbauer spectrum similar to that of the ground doublet, with one major difference: the dominant g-value is now $g_z = 9.77$, i.e., the internal field has an easy axis along g_z and the spectrum measures V_{zz} ($eQV_{zz} = -1.0$ mm/sec). It turns out that the magnetic splitting for the upper doublet is 13% larger than that for the ground doublet. This finding violates Eq. (7), and the assumption of an isotropic magnetic hyperfine interaction (in the $S = 5/2$ representation) has to be dropped.

Figure 5 shows computer-simulated spectra for each of the three Kramers doublets. To account for the experimental data in Fig. 4 these component spectra have to be added according to the population of each doublet. This yields an interesting result. The spin Hamiltonian with $E/D = 0.28$ predicts an energy ratio $E_3/E_2 = 2.12$ of the upper and middle Kramers doublets relative to the ground doublet; the Mössbauer data demand $E_3/E_2 = 3.3 \pm 0.5$.

In order to fit the Mössbauer and EPR data of some siderochromes Spartalian and Oosterhuis[20] found it necessary to augment the spin Hamiltonian in Eq. (2b) by a fine-structure term of fourth order

$$\hat{H}_s = \frac{D\mu}{6}\left(S_x^4 + S_y^4 + S_z^4 - \frac{707}{16}\right) \quad (8)$$

Inclusion of this term allows one to vary the ratio E_3/E_2 while maintaining the features around $g = 9$ and 4.3. By choosing $E/D = 0.17$, $\mu = 0.3$, and $D = 1.76\ \mathrm{cm}^{-1}$ Debrunner's group was able to reconcile the EPR and Mössbauer data. The theoretical curves in Figs. 4 and 5 were computed with a set of parameters quoted in the caption of Fig. 4. Also included in the caption are the g-values resulting from the augmented spin Hamiltonian. Since the analysis of the rubredoxin data is still in progress the reader should view the results described here in the spirit of a progress report.

One further feature of rubredoxin is particularly noteworthy. The isotropic part of the magnetic hyperfine interaction $A_{av} = A_0 = -22.6$ MHz is considerably smaller than the −30 MHz typically observed for highly ionic Fe^{3+}. The value for A_0 compares well with that found for the ferric site in reduced 2 Fe−2 S proteins and it seems to be highly indicative of a tetrahedral sulfur environment. The low value for A_0 indicates that the d electrons are highly delocalized in rubredoxin.

[20] K. Spartalian and W. T. Oosterhuis, *J. Chem. Phys.* **59,** 617 (1973).

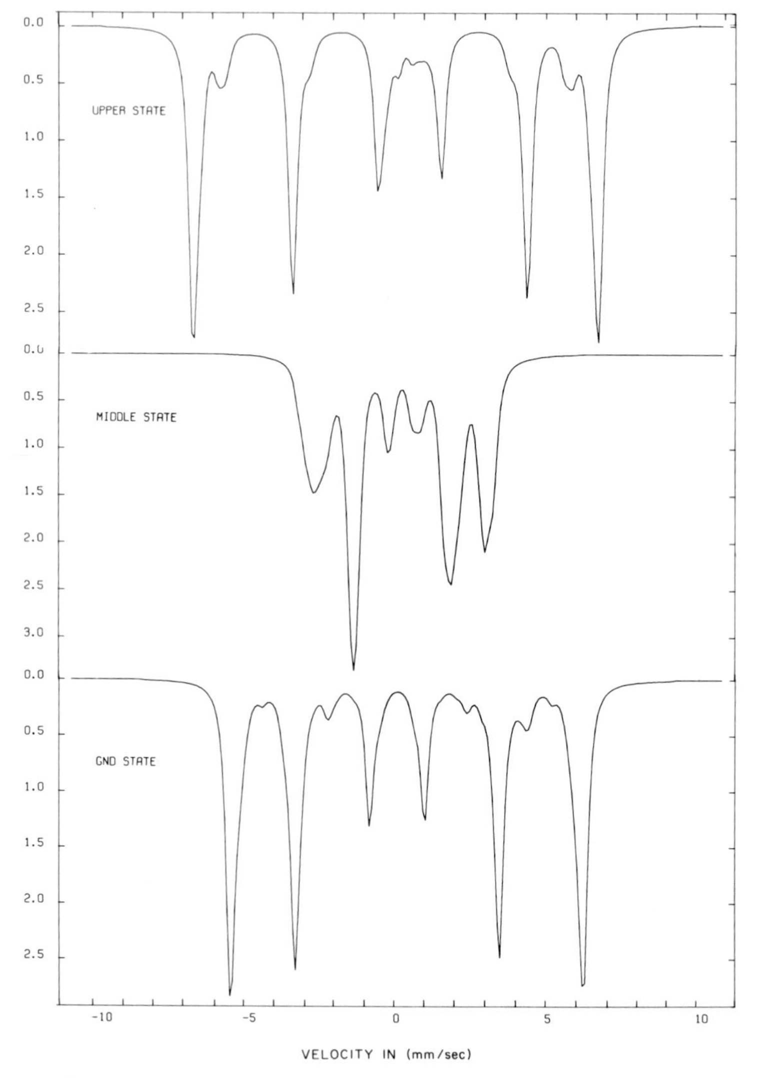

FIG. 5. Component spectra of the three Kramers doublets of oxidized rubredoxin. The spectra (powder averages over 15 × 15 different orientations) were computed with the parameters of Fig. 4. (From Debrunner *et al.*[8])

To summarize: The high-spin ferric system yields three Kramers doublets, each giving rise to a separate Mössbauer spectrum (this, of course, is only true if the electronic spin relaxation rate is slow compared to the nuclear precession frequency). At temperatures such that kT is comparable to the zero-field splitting we obtain a composite spectrum. The component spectra are weighted according to the population of the Kramers doublets with which they are associated. By properly decomposing spectra taken at different temperatures we obtain the energy separations of the three Kramers doublets, i.e., the zero-field splitting parameter D. Note also that the three spectra measure different components of the quadrupole tensor. For fast electronic spin relaxation rates (the spin is jumping rapidly among the thermally accessible states of the spin sextet) no magnetic hyperfine splitting will be observed. Application of strong magnetic fields, however, will induce an internal magnetic field which is proportional to A_0 and to the thermally averaged expectation value of $\vec{S}$. The latter depends on the zero-field splitting parameters and the temperature. Thus no matter whether the relaxation time is short or fast the magnetic hyperfine interaction and the fine-structure parameters can be determined.

IV. Connections with EPR Spectroscopy

The spin Hamiltonian Eq. (2) depends on many parameters. Therefore we should use all information which can be obtained from other techniques. For Kramers ions EPR spectroscopy generally can provide accurate values for the g-tensor and the parameter E/D. In favorable cases D can be determined with EPR also. In addition, there is a correlation between EPR and Mössbauer spectroscopy which we can use to our advantage: An EPR-active Kramers doublet implies a magnetic Mössbauer spectrum whose intensities depend on the orientation of the applied field (100 G is sufficient) relative to the transmitted Mössbauer radiation.

To elucidate this relation we consider two extreme situations for a spin $S = 1/2$ system. The system may be an isolated spin doublet as the ground state of the low-spin ferric ion or a Kramers doublet belonging to a spin multiplet. For an isotropic system the magnetic part of the spin Hamiltonian can be written as

$$\hat{H} = g\beta_e H_z S_z + A\vec{S} \cdot \vec{I} \tag{9}$$

where z is the direction of the applied field. For $H_z > 100$ G we can neglect off-diagonal terms in the magnetic hyperfine term, and we obtain for the internal magnetic field at the nucleus [Eq. (4)], $H^z_{\text{int}} = -\langle S_z \rangle A/g_n\beta_n$ and $H^x_{\text{int}} = H^y_{\text{int}} = 0$. Thus the internal field is parallel to the z axis, i.e., parallel to the applied field. If the external field is applied parallel to the observed Mössbauer radiation then all molecules of a randomly oriented sample will have their internal field aligned parallel to the direction of observation. Consequently, nuclear $\Delta m = 0$ transitions are forbidden, and the two middle lines of the six-line Mössbauer spectrum are suppressed (the 14.4 keV radiation is a magnetic dipole transition). On the other hand, if the field is applied perpendicular to the observation direction the nuclear $\Delta m = 0$ transitions are maximized.

The other extreme situation occurs if *two* g-values are vanishingly small compared to the third one, say $g_\zeta = g_\eta = 0$ and $g_\xi \neq 0$. In this case the electronic spin is quantized along the direction of g_ξ for practically all directions of the applied field. Consequently the internal magnetic field is directed along g_ξ, which is fixed to the molecule. Crudely speaking, the molecule does not care about the direction of the applied field, and the intensities of the Mössbauer spectrum are not affected by the orientation of the field.

The EPR transition amplitudes are proportional to g_ζ and g_η if the resonance is observed at g_ξ. An isotropic system therefore yields a strong EPR signal. An anisotropic doublet as defined above, on the other hand, is EPR-silent, since $g_\zeta = g_\eta = 0$. Even these extreme situations are encountered quite frequently. The high-spin ferric ion, for instance, exhibits in axial symmetry ($E/D = 0$) two Kramers doublets (the $\pm$ 3/2 and $\pm$ 5/2 sublevels) with $g_\zeta = g_\eta = 0$. The isotropic case occurs for $E/D = 1/3$ where the middle doublet has an isotropic g-value, $g = 4.3$. Most often we encounter situations between these extremes, and we find that doublets with small EPR amplitudes have associated Mössbauer spectra which depend only weakly on the direction of the applied field while doublets with large EPR amplitudes yield more drastic changes of the Mössbauer intensities.

This relation between the Mössbauer and EPR spectrum was usefully employed[21] in the investigation of the MoFe protein from nitrogenase (1–2 Mo and circa 30 Fe atoms). The MoFe protein yields a strong EPR signal with principal g-values at 4.3, 3.7, and 2.0 (it results from one Kramers doublet of a $S = 3/2$ system). It was not known whether this signal resulted from a molybdenum or from an iron center. Mössbauer spectroscopy can give an unambiguous answer: If the signal results from an iron center we have to observe, at low temperatures, a magnetic ($<\vec{S}> \neq 0$) Mössbauer spectrum whose intensities depend on the direction of an applied field in a way congruent with the anisotropy of the g-values. Such a component was observed in the Mössbauer spectrum. Moreover, quantitation of the

[21] E. Münck, H. Rhodes, W. H. Orme-Johnson, L. C. Davis, W. J. Brill, and V. K. Shah, *Biochim. Biophys. Acta* **400**, 32 (1975).

Mössbauer data showed that 12 iron atoms were associated with this component. A quantitation of the EPR signal (on the sample used for the Mössbauer study) yielded a spin concentration of 2 spins/molecule. Hence it could be concluded that the MoFe protein contains two identical EPR active centers ($S = 3/2$), each consisting of six iron atoms.

We have just discussed why it is useful to study Kramers ions in a weak applied field. In fact, it is almost mandatory to apply a magnetic field. This matter can be clarified by rewriting $\vec{S} \cdot \tilde{A} \cdot \vec{I}$ is $g_e\beta_e\vec{S} \cdot (\tilde{A} \cdot \vec{I}/g_e\beta_e) = g_e\beta_e\vec{S} \cdot \vec{H}_e$. The quantity $\vec{H}_e$ can be interpreted as a field acting on the electronic states ($H_e \simeq 10$ G for ^{57}Fe). If no external field were applied the spin Hamiltonian would be incomplete since we have ignored transferred hyperfine interactions. Such interactions arise when nuclear moments of ligand nuclei ($\vec{I}_j$) interact with the electronic spin $\vec{S}$ which is shared covalently to some extent between the iron atom and its ligands. The terms to be added to Eq. (5), or Eqs. (2), can be written as $\Sigma_j\vec{S} \cdot \vec{B}_j \cdot \vec{I}_j$ where $\vec{B}_j$ is the coupling tensor of the jth ligand nucleus. Transferred hyperfine interactions produce small random fields $[\vec{H}_j = \vec{B}_j \cdot \vec{I}_j/(g_e\beta_e)]$ which create a very complex situation. A weak magnetic field of a few hundred gauss decouples these interactions and defines a quantization axis for the electronic spin $\vec{S}$. This facilitates the data analysis appreciably. On the other hand, since zero-field Mössbauer spectra can be quite sensitive to transferred hyperfine interactions, we can use this sensitivity to identify ligands which are coordinated to the iron. This identification has to be done by comparison of isotopically substituted samples, i.e., ^{32}S versus ^{33}S, ^{16}O versus ^{17}O, or ^{95}Mo versus ^{98}Mo if a molybdenum–iron cluster is suspected. In many cases, the results may not justify the efforts and costs of preparing the samples. In a few cases, however, we might obtain crucial information. In any case, one should try EPR, Endor, or resonance Raman experiments first.

V. Complex Systems: What Can We Learn?

In Section III,B we have seen that the spectra of a protein with a single iron atom may be quite complex. Do we have a chance, then, to tackle a protein with many inequivalent iron centers, which often even consist of clusters with two or four iron atoms? Although the answer depends on the protein at hand, there are a variety of problems which can be elucidated by a Mössbauer study. It is quite clear that we will generally not succeed in obtaining a complete set of parameters. We should be able, however, to "follow" all or some of the iron atoms when they undergo redox reactions. We can do this by virtue of the fact that the low-temperature spectra of Kramers and non-Kramers systems are drastically different; the former yield magnetic spectra, the latter quadrupole doublets.

At this point let us consider a "fictitious" protein containing four nonheme iron atoms and two other metal atoms, e.g., molybdenum (the optical absorption spectrum is assumed to be uninformative). In the native form the protein exhibits an EPR signal with g-values at $g = 4.2, 3.8$, and 2.0. Such a signal can result from an $S = 3/2$ system (from the $\pm 1/2$ doublet) for $E/D = 0.03$. Does this signal result from Mo^{3+} or from a ferric ion in an intermediate spin state? Isotropic substitution experiments (with EPR) using ^{57}Fe and ^{95}Mo are inconclusive. We take a Mössbauer spectrum at 4.2°K and find a magnetic spectrum (MI) whose intensities depend on the orientation of a weak (500 G) applied magnetic field. This spectrum is superimposed by a quadrupole doublet DII, with $\Delta E_Q = 2.4$ mm/sec and an isomeric shift $\delta = 1.2$ mm/sec (see Fig. 6). The doublet contains 50% of the total absorption. Quadrupole splitting and isomeric shift unambiguously associate the quadrupole doublet with two high-spin ferrous ions, both residing in similar, if not identical environments. A Mössbauer spectrum taken at 1°K is identical to that taken at 4.2°K. This observation implies that the EPR active doublet of the $S = 3/2$ system is the ground state and that the $\pm 3/2$ doublet is separated by at least 10°K from the ground doublet; i.e., the zero-field splitting parameter $D > 0$. We now quantitate the EPR signal and find a spin concentration of 2 spins/molecule. These observations allow the following conclusions: The EPR signal results from two isolated ferric ions in an intermediate spin state $S = 3/2$; the other two iron atoms are high-spin ferrous. We now search for other stable oxidation states and oxidize the protein with some redox dye. We find (spectrophotometry on the dye) that four electrons can be removed (reversibly) from the protein. The $S = 3/2$ signal has vanished, and the sample exhibits no EPR signal. Apparently two of the four electrons are removed from $S = 3/2$ centers. Where did the other two electrons come from? Fe or Mo? We now take a Mössbauer spectrum at 4.2°K. We find a magnetic component (MII) with a magnetic splitting of 16 mm/sec and a quadrupole doublet (DI) with $\Delta E_Q = 0.7$ mm/sec and isomeric shift $\delta = 0.1$ mm/sec (Fig. 6, lower spectrum). The doublet quantitates again to 50% of the total Fe present. Since MII is magnetic it results from iron in an environment of half-integral electronic spin. The intensities of component MII are found to be independent of the orientation of a weak applied field. (If there were a field dependence we must have observed an EPR signal). We next subject the protein to a redox titration and find that two electrons can be removed from the proteins without affecting the intensity of the $S = 3/2$ EPR signal. A Mössbauer spectrum of a two-electron oxidized sample exhibits the magnetic components, MI and MII; no quadrupole doublets are observed. With this information we can make the following assignments: Doublet DI represents an oxidized form of the $S = 3/2$ centers; its Mössbauer parame-

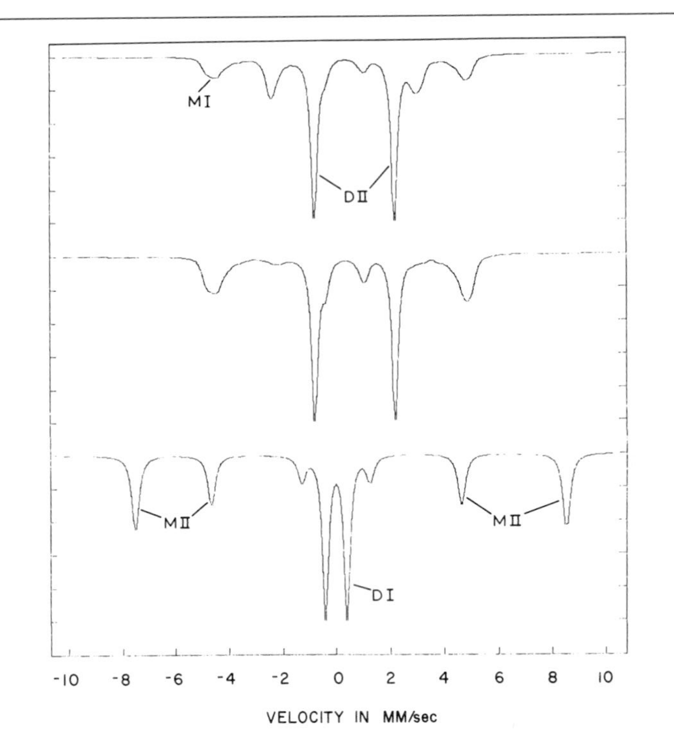

FIG. 6. Low-temperature Mössbauer spectra of the fictitious protein discussed in the text. The upper and middle spectra result from the native material showing a superposition of the magnetic spectrum MI (species I in a magnetic state) and the quadrupole doublet DII (species II appearing as a doublet). The four iron atoms divide in two classes. Two iron atoms, in identical environments, give rise to doublet DII with $\Delta E_Q = 2.4$ mm/sec and $\delta = 1.2$ mm/sec, establishing a high-spin ferrous configuration. The other two iron atoms, also in identical environments, give rise to spectrum MI. Note the intensity changes between a transverse (upper) and parallel (middle) applied magnetic field (500 G). The field dependence of MI establishes that the observed EPR signals at $g = 4.2$, 3.8, and 2.0 result from an iron center with $S = 3/2$ (MI results from the $\pm$ 1/2 doublet; we have assumed the $D > 0$ and that the $\pm 3/2$ doublet is not populated at the temperature the spectra were taken). The lower Mössbauer spectrum results when the sample is oxidized by four electrons. Both spectral components MI and DII have vanished, and the four iron atoms have appeared in MII and DI. Component DI is the conspicuous doublet ($\Delta E_Q = 0.7$ mm/sec and $\delta = 0.1$ mm/sec) centered around zero velocity. The area under DI is the same as the area under MII, i.e., both components represent two iron atoms. Note that species II is EPR silent in both oxidation states. The assignment of the species is: DI (Fe^{4+}) $\overset{e}{\rightleftharpoons}$ MI(Fe^{3+}, $S = 3/2$) and MII (Fe^{3+}, $S = 5/2$) $\overset{e}{\rightleftharpoons}$ DII (Fe^{2+}, $S = 2$).

ters suggest Fe^{4+} [hence $Fe^{4+} \overset{e}{\rightleftharpoons} Fe^{3+}$ ($S = 3/2$)]. The magnetic spectrum MII is an oxidized form of the high-spin ferrous ions. The magnetic splitting of 16 mm/sec strongly suggests high-spin ferric iron.

How can we explain that the fully oxidized protein is EPR-silent? The fact that we observe a magnetic Mössbauer spectrum at 4.2°K implies that the electronic spin relaxation time is no longer than 10^{-7} sec. Thus we cannot blame lifetime broadening to explain the missing EPR signal. We observe, however, that the magnetic spectrum MI collapses into a quadrupole doublet at 10°K implying that the relaxation time is now shorter than 10^{-9}. We can rationalize the missing EPR signal in the following way: The spectrum MII results from a high-spin ferric ion in an environment of tetragonal or trigonal symmetry ($E/D \simeq 0$); the zero-field splitting parameter D is large and negative, say $D = -8$ cm^{-1}. For these values of D and E Eq. (2) predicts a $\pm 5/2$ ground doublet with $g_x = g_y = 0$ and $g_z = 10$. This doublet is EPR-silent and yields a Mössbauer spectrum like MII. The $\pm 3/2$ doublet, the lowest excited state, is EPR-silent also. The uppermost doublet, the $\pm 1/2$ state, should yield intense EPR signals at $g = 6$ and $g = 2$. It is separated, however, by an energy $\Delta = 6D$ from the ground state and therefore only to 10^{-5}% populated at 4.2°K; at higher temperatures the $\pm 1/2$ doublet becomes populated but the EPR signal is not observed due to fast spin relaxation. (High-spin ferric hemes are made for EPR spectroscopists; the $\pm 1/2$ doublet is the ground state, since $D > 0$). The situation described here is certainly quite sneaky; it occurs, however, in the MoFe protein of the nitrogenase system. Obviously, the spectra DII and MII reflect iron atoms which are EPR-silent in two oxidation states. By measuring the intensity of the MII spectrum in a redox titration we can determine the redox potential. This might be a useful application of the Mössbauer technique when other methods are not applicable.

Our fictitious protein is actually an enzyme transforming substrate X into Y. The EPR signal is not affected upon substrate binding; the substrate binding site is not known. A Mössbauer spectrum of the enzyme–substrate complex reveals that the Mössbauer parameters of doublet DII change to $\Delta E_Q = 2.2$ mm/sec and $\delta = 1.3$ mm/sec. This observation suggests that the two ferrous ions are the substrate binding sites.

The reader may notice that we obtained appreciable information on our fictitious protein by a very limited Mössbauer investigation. We just took data at 4.2°K and distinguished between magnetic spectra (Kramers ions) and quadrupole doublets (non-Kramers ions). We could have obtained much more information by varying the temperature and by applying strong fields. The reader may have noticed also that the $S = 3/2$ centers were interpreted as isolated Fe^{3+} ions with intermediate spin $S = 3/2$. With the information given above we cannot exclude, however, the possibility that the $S = 3/2$ centers consist of a spin-coupled Mo–Fe pair, say Fe^{3+} ($S = 5/2$) and Mo^{4+} ($S = 1$). This question could possibly be settled by comparing the hyperfine parameters of MI with those of typical ferric compounds. Moreover, we could have taken Mössbauer spectra in zero magnetic field on samples enriched in either ^{98}Mo($I = 0$) or ^{95}Mo ($I = 5/2$). Spectra taken in zero-field allow us to observe transferred hyperfine interactions of nuclei in the vicinity of the ^{57}Fe nucleus (see Section IV).

VI. Results on Some Electron-Carrier Proteins

A. *Rubredoxin*

Since we have discussed recent results on ferric rubredoxin in some detail in Section III,B, we restrict ourselves here to the reduced protein. Debrunner and co-workers[8,19] have studied the high-spin ferrous form in strong applied magnetic fields. The zero-field splitting is quite large, $D \simeq +8\ cm^{-1}$, and it contains a sizable rhombic component, $E/D = 0.12$. As in the case of ferric rubredoxin, inclusion of a fourth-order term produced improved fits to the data. There is some evidence that a fourth-order term probably need not be considered: A model complex[22] for reduced rubredoxin yields Mössbauer spectra quite similar to those of the reduced protein. Moreover, a recent far-infrared study[23] on this model complex shows that the fine-structure splitting is well described with the terms in Eq. (2a). This information together with recent EXAFS studies on the nature of the "short bond" should give us new and detailed insight into the electronic structure of the "simplest" of the iron–sulfur proteins.

B. *2 Fe–2 S* Proteins*

Mössbauer spectroscopy has been an indispensable tool in elucidating the structure of the 2 Fe–2 S* proteins (plant-type ferredoxins). These proteins were discovered with EPR spectroscopy; a very unusual EPR signal—now referred to as the $g = 1.94$ signal—was observed in 1960 by Beinert and Sands.[24] These proteins contain two iron and two labile sulfur atoms, and they can exist in two oxidation states. In the oxidized form the proteins are diamagnetic ($S = O$); upon reduction the $g = 1.94$ resonance appears. Magnetic susceptibility experiments established that the EPR signal results from an $S = 1/2$ spin system. EPR experiments on samples

[22] V. Petrouleas, A. Simopoulus, and A. Kostikas, *Int. Conf. Appl. Mössbauer Effect,* (1976); *J. Phys. (Paris)* Colloque **37,** C6–159 (1976).

[23] P. M. Champion and A. J. Sievers, private communication.

[24] H. Beinert and R. H. Sands, *Biochem. Biophys. Res. Commun.* **3,** 41 (1960).

enriched with ^{57}Fe and ^{33}S revealed that the paramagnetic centers interact with iron, labile sulfur, and cysteine sulfur. Reductive titration experiments showed that one electron is required to fully reduce the protein. The first Mössbauer experiments confirmed the diamagnetic nature of the iron atoms in the oxidized proteins; moreover, the data showed that both iron atoms reside in equivalent environments. Most importantly, Mössbauer spectroscopy established that the spectra of both iron atoms changed drastically upon a one-electron reduction process. Thus, the iron atoms could not be isolated but had to reside in a cluster consisting of a pair of iron atoms. Further Mössbauer experiments performed under conditions where the electron spin relaxation time is fast (at temperatures around 200°K) proved that the iron atoms are dissimilar in the reduced protein. Quadrupole splitting and isomeric shift of one iron atom were typical of iron in a high-spin ferric state; the parameters of the second iron were found to be typical of a high-spin ferrous configuration. Moreover, the parameters of the second site resembled those of reduced rubredoxin indicating a tetrahedral environment of sulfur atoms. Mössbauer measurements taken at 4.2°K in strong applied magnetic fields showed unambiguously that the magnetic hyperfine interactions of the ferrous and ferric sites have opposite signs. This information, taken together with results from many different techniques, especially electron-nuclear double-resonance studies, finally established the nature of the active site: a spin-coupled pair of iron atoms, bridged by two labile sulfur atoms, and suspended into the protein matrix by cysteine residues.

For a detailed discussion and a comprehensive list of references the reader is referred to reviews by Palmer[25] and by Sands and Dunham.[26] Comprehensive Mössbauer studies and data analyses have been published by Dunham *et al.*[27] and Münck *et al.*[28]

C. 4 Fe–4 S* Proteins

For the 2 Fe–2 S* clusters we have accumulated a wealth of detailed, and consistent, information from a variety of spectroscopic techniques; crystallographic data, however, are not available yet. For the 4 Fe–4 S* proteins we have X-ray diffraction data and excellent model complexes; the spectroscopic data, however, have not yielded much detailed informa-

[25] G. Palmer, *Iron-Sulfur Proteins* **2,** 285 (1973).

[26] R. H. Sands and W. R. Dunham, *Q. Rev. Biophys.* **7,** 443 (1975).

[27] W. R. Dunham, A. Bearden, I. Salmeen, G. Palmer, R. H. Sands, W. H. Orme-Johnson, and H. Beinert, *Biochim. Biophys. Acta* **253,** 134 (1971).

[28] E. Münck, P. G. Debrunner, J. C. M. Tsibris, and I. C. Gunsalus, *Biochemistry* **11,** 855 (1972).

tion about the electronic structure of these clusters. In contrast to the centers of the plant-type ferredoxins the 4 Fe–4 S* clusters do not show the clear distinction between ferric and ferrous components. All Mössbauer investigations reported so far suggest that the four iron atoms are inequivalent to a certain extent, yet the components are not sufficiently resolved to allow an unambiguous decomposition of the spectra. There is ample evidence that spin coupling operates between the four iron atoms in all redox states; we are lacking, however, a theoretical framework such as that provided by Gibson and Thorneley for the 2 Fe–2 S* centers. (The spin-coupling problem is immensely more complex since four iron atoms are involved.)

From the results on the plant-type ferredoxins we might be tempted to assign formal oxidation states to the iron atoms constituting the 4 Fe–4 S* clusters. Formally, we can construct the cluster of the oxidized high-potential iron protein (HIPIP) from three ferric and one ferrous iron atom. The Mössbauer data,[29] however, show that the four iron atoms have the same (within rather narrow bounds) quadrupole splitting and isomeric shift. This suggests that we have to view these structures more in terms of cluster orbitals than in terms of an ionic picture (high-spin ferric or ferrous) which has been so successful in describing the 2 Fe–2 S* clusters. In the reduced form of the latter proteins the odd electron is localized highly on one iron atom; in the 4 Fe–4 S* proteins the electrons seem to be shared rather equally among all iron atoms.

At low temperatures magnetic hyperfine interactions are observed in the oxidized form of the HIPIP protein from *Chromatium*[29] and in the reduced form of the ferredoxin from *Bacillus stereathermophilus*.[30] Similar magnetic patterns (quite distinct from those of the 2 Fe–2 S* centers) have been observed for the Fe-proteins of the nitrogenase systems from *Klebsiella pneumoniae*[31] and *Azotobacter vinelandii*.[32]

The plot thickens when we consider the ferredoxins which contain two 4 Fe–4 S* clusters. Mathews *et al.*[33] have reported and interpreted a half-field resonance in the EPR spectra of the fully reduced protein. These proteins seem to present us with a double spin-coupling problem: Each individual cluster, in the fully reduced protein, is strongly spin-coupled to a spin $S = 1/2$ system, and both clusters are weakly coupled to yield a singlet

[29] D. P. E. Dickson, C. E. Johnson, R. Cammack, M. C. W. Evans, D. O. Hall, and K. K. Rao, *Biochem. J.* **139,** 105 (1974).

[30] R. N. Mullinger, R. Cammack, K. K. Rao, D. O. Hall, D. P. E. Dickson, C. E. Johnson, J. D. Rush, and A. Simpoulus, *Biochem. J.* **151,** 75 (1975).

[31] B. E. Smith and G. Lang, *Biochem. J.* **137,** 169 (1974).

[32] W. H. Orme-Johnson and E. Münck, unpublished results.

[33] R. Mathews, C. Carlton, R. H. Sands, and G. Palmer, *J. Biol. Chem.* **249,** 4326 (1974).

and a triplet state. This suggestion is supported by Mössbauer investigations;[34,35] the low-temperature spectra ($T = 0.45°K$) consist of a single quadrupole doublet (singlet and triplet states are non-Kramers systems and should yield nonmagnetic spectra in the absence of a strong applied field).

For further details the reader may consult a review prepared recently by C. E. Johnson.[36]

D. *The Cytochromes c and b_5*

The cytochromes *c* (axial heme ligands histidine and methionine) and b_5 (two histidines) are low-spin complexes in both stable oxidation states. The ferric proteins display quite similar EPR spectra and, not surprisingly, the Mössbauer spectra of both proteins have a close resemblance also. Lang, Herbert, and Yonetani[37] have reported a comprehensive Mössbauer study of cytochrome *c* from *Torula utilis*; Münck and Strittmatter[38] have investigated the b_5 protein from calf liver.

In the ferric form both proteins are low-spin ($S = 1/2$), and at 4.2°K broad spectra typical of low-spin ferric hemes are observed. At higher temperatures, $T = 195°K$, the electronic spin relaxation rate is fast enough so that well-defined quadrupole doublets are observed; $\Delta E_Q = 2.14 \pm 0.05$ mm/sec and $\delta = 0.21 \pm 0.03$ mm/sec for cytochrome *c*, and $\Delta E_Q = 2.27 \pm 0.03$ and $\delta = 0.23 \pm 0.03$ mm/sec for the b_5 protein. The quadrupole splitting, at 195°K, of lyophilized ferricytochrome *c* was found to be 10% smaller than that observed in frozen solution. The (magnetic) low-temperature spectra of cytochrome *c* have been evaluated by Lang *et al.* in the framework of the ligand field model alluded to in Section III,B. The model describes the data reasonably well although some refinement is in order.

The reduced cytochromes *c* and b_5 are in a diamagnetic, low-spin ferrous configuration. Thus in the absence of an applied field only a quadrupole doublet is observed at all temperatures. At 4.2°K, $\Delta E_Q = 1.17 \pm 0.05$ mm/sec and $\delta = 0.45 \pm 0.03$ mm/sec for ferrocytochrome *c*, and $\Delta E_Q = 1.04 \pm 0.03$ mm/sec and $\delta = 0.43 \pm 0.02$ mm/sec for cytochrome b_5. The quadrupole splittings of both proteins increase by about 0.03 mm/sec when the temperature is raised to 195°K. Both proteins have a positive quadrupole coupling constant, $eQV_{z'z'} > 0$, and $\eta \simeq 0.5$.

[34] L. Bogner, F. Parak, and K. Gersonde, *Int. Conf. Appl. Mössbauer Effect,* (1976); *J. Phys. (Paris)* Colloque **37,** C6–177 (1976).
[35] C. L. Thompson, C. E. Johnson, D. P. E. Dickson, R. Cammack, D. O. Hall, U. Weser, and K. K. Rao, *Biochem. J.* **139,** 97 (1973).
[36] C. E. Johnson, *Iron-Sulfur Proteins* **3,** 283 (1977).
[37] G. Lang, D. Herbert, and T. Yonetani, *J. Chem. Phys.* **49,** 944 (1968).
[38] E. Münck and P. Strittmatter, unpublished results.

All of the t_{2g} orbitals of a low-spin ferrous configuration are occupied and thus no quadrupole splittings would be expected in a first approximation. Covalency effects, however, will transfer some charge from the iron orbitals to the ligands. Lang and co-workers[37] have pointed out that the observed Mössbauer parameters of ferrocytochrome *c* would be consistent with a situation in which two of the three t_{2g} orbitals lost nearly equal amounts of their charge (about 15%) relative to the third. This suggestion fits possibly to the findings of Sharrock *et al.*[11] discussed above (see Section III,B. and Fig. 3). We have to emphasize, however, that the measurements on ferrocytochrome *c* (frozen solution) do not provide any information on the spatial correlation between the principal axes of the electric field gradient tensor and the heme group. Thus we should not view $V_{z'z'}$ as a component parallel to the heme normal and interpret the ferrocytochrome *c* results as evidence for preferential delocalization from the d_{yz} and d_{xz} orbitals (in theoretical work the heme normal is usually taken as the z direction). The situation is different in ferricytochrome *c*; here we know from single-crystal EPR work[39] that the largest principal g-value (g_3 = 3.06) is within 5° of the heme normal. With the EPR information the spatial correlation is established in the magnetic spectra even for measurements on frozen solutions.

VII. Considerations for the Biochemist

A. *Physical State of Samples*

Only a fraction of the γ-rays penetrating the absorber are capable of resonance absorption. This fraction is described by the Debye-Waller factor f. The f-factor is vanishingly small for liquid samples. To observe resonance absorption the protein has to be embedded in a solid, i.e., frozen solution, lyophilized materials (biologically not very desirable), or single crystals have to be used. The Debye-Waller factor is strongly temperature dependent. For a protein in frozen solution $f \simeq 0.7$ at 4.2°K, dropping by about a factor of 2 when the temperature is raised to 200°K. Thus at the temperature of liquid helium the signal-to-noise ratio is improved by a factor of about 4. The main reason, however, for working at 4.2°K, or even at 1.5°K, is that at low temperatures the electronic spin relaxation time is generally long enough to permit the observation of paramagnetic hyperfine structure (for Kramers ions) in the Mössbauer spectrum.

In the future we can expect an increased activity of Mössbauer spectroscopy towards the elucidation of proteins containing multiple iron centers. If the f-factors associated with different iron sites were the same, the

[39] C. Mailer and C. P. S. Taylor, *Can. J. Biochem.* **50,** 1048 (1972).

Mössbauer spectrum would provide a precise determination of the number of iron atoms associated with each observed component. (Such information would be very valuable for the determination of the structure of novel iron clusters, i.e., the iron components of the MoFe protein from nitrogenase.[21]) Although it is generally assumed that the *f*-factors are the same, hardly any precise information is available on this question.

It is possible to observe the Mössbauer resonance also in a "softer" environment, as in highly viscous liquids (for instance in glycerol or in the smectic or nematic phases of liquid crystals). Moreover, the Mössbauer resonance has been observed, at room temperature, for membrane-bound enzymes and for tightly packed bacterial cells. In the latter case[40] the Mössbauer lines were quite broad due to diffusional broadening.

B. Isotopic Enrichment

Since ^{57}Fe has a natural abundance of only 2.2% the signal can be increased by a factor of 40 by using isotopically enriched material. A Mössbauer investigation requires a sample with approximately 0.2–1 μM of ^{57}Fe per inequivalent iron site. A full Mössbauer investigation is rather time-consuming (typically 12–24 hr per spectrum) since the sample needs to be studied over a wide range of temperatures. Additionally, considerable information, often crucial, can be obtained by studying the samples in magnetic fields up to 80 kG, often at different temperatures. For such studies ^{57}Fe concentrations of 1 m*M* per inequivalent site are desirable. In many cases one can obtain useful information with as little as 0.1 μmol of ^{57}Fe (or 4 μmol of natural iron) present. A necessary requirement for such measurements is that the iron is in a non-Kramers state, i.e., only a quadrupole doublet is observed. For instance, the effects of perturbing solvents, or of pH, on the quadrupole splitting and isomeric shift of ferrocytochrome *c* could be studied on samples containing natural iron.

Isotopic enrichment can be achieved in many ways. (The prize of 90% enriched ^{57}Fe has jumped in 1978 from $3.30/mg to $9/mg.) In the most favorable case the iron is tightly bound in the active center, and the protein can be obtained from a microorganism grown on an ^{57}Fe-enriched medium. Researchers new to this endeavor should get some advice from experienced groups (selection of iron-free chemicals; iron content of malt extracts; iron exchange with stainless-steel components in carboys and fermentors). Direct chemical exchange of iron, where possible, offers an inexpensive solution (plant-type ferredoxins; heme proteins

[40] E. R. Bauminger, S. G. Cohen, I. Novik, S. Ofer, Y. Yariv, M. M. Werber, and M. Mevarech, *Int. Conf. Appl. Mössbauer Effect,* (1976); *J. Phys. (Paris)* Colloque **37,** C6–227 (1976).

which can be reconstituted with ^{57}Fe-enriched protoporphyrin; enzymes which are activated with iron). Reconstitutions, however, may be troublesome in that the samples may become contaminated with adventitiously bound iron or in that biologically less active species result as a consequence of irreversible processes associated with the biochemical manipulations. In many instances chemical reconstitutions have been very successful; but in some cases (cytochrome $P450_{CAM}$ from *Pseudomonas putida* for instance) quite some time and effort were wasted on these procedures. However, many systems offer no other choice but chemical reconstitution (*Ceratotherium simum* cannot be grown inexpensively on ^{57}Fe!).

For proteins with chemically different iron centers (e.g., hemes and iron–sulfur centers as in nitrite reductase) chemical reconstitutions might offer distinct advantages if some metal components could be chemically exchanged. This would offer a selective enrichment of either component and therefore a substantial simplification of the Mössbauer spectrum.

C. *Sample Thickness*

A γ-beam penetrating a protein sample is attenuated by a variety of processes (electronic absorption and scattering, mainly photo-effect and Compton scattering). The useful signal increases linearly with the ^{57}Fe concentration (for concentrations up to about 100 μg $^{57}Fe/cm^2$) while the counting rate decreases exponentially with the sample thickness ($I = I_0 e^{-\sigma\rho t}$, where σ is the linear extinction coefficient for 14.4 keV γ-rays, ρ the density of the sample, and t the sample thickness). The optimum thickness for a protein in frozen solution is approximately $\rho \cdot t \simeq 0.6$ gm/cm^2, i.e., $t \simeq 6$mm.

The electronic absorption increases rapidly with atomic number Z. The cross-sections for 14.4 keV γ-radiation of the elements from oxygen to krypton are well described by $\sigma = 0.003 Z^3$ cm^2/gm. Thus solvents containing sulfur and chlorine severely restrict the sample thickness. For instance, the sulfur component of 2-mm dimethylsulfoxide ($\rho = 1.1$ gm/cm^2) absorbs 70% of the 14.4 keV γ-radiation. For the same reasons chlorinated plastics and glass should be avoided as sample cells. Water has a cross-section of about 1.7 cm^2/gm and a typical protein 1.3 cm^2/gm. If samples are treated with iodoacetamide to reduce protein aggregation the iodine should be removed by dialysis.

D. *Sample Cells*

The typical Mössbauer cell is a cylindrical cup (about 10–15 mm diameter and 8–15 mm length) into which the protein is filled and then frozen. The cell is typically made of nylon or delrin (Lucite can crack upon freezing). Cheap polyethylene cuvettes, on the other hand, are often satisfactory (i.e., 2-ml vials from Olympic Plastics, 5/8″ OD); however, they easily deform upon freezing, which might be troublesome if the sample holder assembly fits fairly snugly. If odd-shaped cells are required one may use polyethylene cuvettes which can be molded rather cheaply from appropriate dies.

Often one might want to cap the sample cell with a lid. Some caution is necessary when the samples are to be immersed in liquid helium or nitrogen. The seal must either be perfect or a bleeding hole should be drilled into the lid (if it is only a protective or thermal shielding device) since accumulation of liquid inside the sample cell might destroy the cell and its content when the sample is warmed up.

It might be necessary to prepare specimen under anaerobic conditions (for O_2 permeability of various plastics consult Quispel[41]) and to freeze them immediately after some biochemical manipulations, i.e., the Mössbauer cell should be connected to some glass apparatus which allows purging with inert gases and insertion of syringes through serum caps. After freezing the sample it can be a problem to disconnect the Mössbauer cell from the glass. This problem can be solved by fitting a brass adapter snugly over the top of the Mössbauer cell. Since brass and plastic materials have drastically different thermal expansion coefficients the Mössbauer cell can be disconnected reasonably well after freezing. This procedure can also be used to prepare samples with the rapid-freeze technique. The place of the glass apparatus is taken by a funnel which collects the material after it has been injected into a cold isopentane bath.

E. Sample Temperatures

In Mössbauer experiments the sample is studied over a wide range of temperatures. With few exceptions the most interesting regions are from 1.5–25°K and from 100–250°K. At low temperatures a variable-temperature study allows one to determine zero-field splittings of spin multiplets either by varying the temperature to achieve population changes of the electronic levels, or by mixing the states in strong magnetic fields. At higher temperatures one may follow the temperature dependence of the quadrupole splittings. At higher temperature there usually is no need for great precision in temperature measurements, and sensors attached to the sample holder (usually copper) will do the job. At low temperatures the

[41] A. Quispel, *in* "Biology of Nitrogen Fixation" (A. Quispel, ed.), p. 499. North-Holland Publ., Amsterdam, 1974.

problem is more serious, and temperature measurements to within $\pm 0.2°K$ require some attention, especially when the sample is cooled in a gas stream and its temperature is controlled by a heater coil. For critical applications it is advisable to embed the sensor into the protein solution prior to freezing. Low-temperature thermistors such as the HE-T2 from Keystone Carbon Company offer a good solution; they are rugged, small (2-mm diameter), quite insensitive to applied magnetic fields, and reasonably inexpensive. To avoid heat influx through the thermistor leads, which would falsify the temperature reading, one can solder a foot of thin manganin wire to the sensor leads. The manganin wires can be wound on a 3-mm plastic sleeve slipped over the thermistor leads, and the whole unit can be embedded into the protein solution.

Accidents leading to the destruction of a sample during a Mössbauer experiment happen very rarely. One note of caution regarding the control of temperature might be useful. Many sensors, like thermistors and carbon resistors, have negative temperature coefficients, i.e., the resistance increases at low temperatures. An open circuit due to a broken sensor wire or a bad contact is interpreted by the heater control circuitry as a too-low sample temperature and maximum current is applied to the heater coil. Some sample chambers are so well isolated that the sample temperature can reach 100°C resulting in the destruction of the protein. Therefore it is advisable to incorporate a circuit into the control electronics which senses such abnormal conditions. Although such accidents happen very rarely it would be a pity to destroy a precious enzyme such as cytochrome *c* oxidase from *Nessiteras rhombopteryx,* grown on an ^{57}Fe-enriched medium.

F. Effects of Freezing

The validity of extrapolating data obtained on frozen solutions to physiological conditions has been the subject of much concern. It is quite clear that the protein is affected by locking it into a solid matrix. The question is whether the structure of the metal center is modified sufficiently by freezing so that we arrive at the wrong conclusions regarding its electronic environment. In discussing low-temperature measurements on frozen solutions we have to differentiate between a variety of effects.

First, in virtually all cases investigated the proteins regained their full biological activity when thawed. Second, at low temperatures we are investigating essentially the orbital ground states. Many molecules, however, have excited states whose energies are separated only by a few hundred wave numbers from the ground state. Occupation of these states may be essential to the functioning of the molecule under physiological conditions. A temperature-dependent quadrupole splitting signals the presence of such states; in that respect Mössbauer spectroscopy is extremely useful. Third, for many heme proteins we observe thermal mixtures of high- and low-spin ferric species. These transitions are usually observed well below the freezing point. The question then is whether these transitions are caused by freezing or whether the low-temperature measurements just signal us the presence of low-lying excited states. It has been observed in both Mössbauer[42] and magnetic susceptibility[43] experiments that the spin populations may depend on the freezing rate, pH, and ionic strength of the protein solution. An interesting spin transition is observed for chloroperoxidase from *Calderiomyces fumago*. The Mössbauer data[4] show a well-resolved transition from high-spin to low-spin in the temperature range from 250–100°K. The same phenomenon is observed when the protein is mixed with dimethylformamide using the techniques developed by Douzou;[44] the spin transition (observed with optical spectroscopy) occurs now in the liquid phase between 280 K and 220°K.

Although freezing effects should require our constant attention, not much Mössbauer work has been published addressing this point. The reason is obvious: Mössbauer samples are always frozen on a slow timescale regardless of whether the sample is plunged into liquid N_2 or cooled in a stirred isopentane bath. The rapid-freeze technique, well developed in EPR spectroscopy, injects (by means of a syringe ram) a fine spray of protein solution into cold isopentane. The problem is that the sample cannot be recycled for further biochemical studies. Researchers are quite reluctant to sacrifice an ^{57}Fe-enriched protein for such an experiment. It is probably more economical to employ EPR (if applicable), magnetic susceptibility, or resonance Raman experiments to explore these matters. Recent low-temperature resonance Raman experiments on cytochrome *c* are encouraging; the positions of the vibrational bands were found to be the same in the frozen state (down to 4.2°K) as in solution.[45]

G. Combining Mössbauer and EPR Studies

Iron complexes with half-integral electronic spin generally exhibit an EPR signal. As discussed above, Mössbauer and EPR spectroscopy are highly complementary, and a combined EPR and Mössbauer study can yield more information than either technique alone.[21] In complex situations it is therefore advisable to study the same sample with both techniques. For stable species it should be sufficient to fill a Mössbauer cell and an EPR tube

[42] G. Lang, T. Asakura, and T. Yonetani, *J. Phys.* C **2,** 2246 (1969).

[43] T. Iizuka and M. Kotani, *Biochim. Biophys. Acta* **194,** 351 (1969).

[44] P. Douzou, *Mol. Cell. Biochem.* **1,** 15 (1973).

[45] P. M. Champion, D. W. Collins, and D. B. Fitchen, *J. Am. Chem. Soc.* **98,** 7114 (1976).

from the same batch. For species with lifetimes comparable to the time necessary to freeze a sample (4–10 sec) this procedure is unreliable since it is practically impossible to freeze EPR and Mössbauer samples in identical fashions. Therefore it is necessary to conduct both investigations on the Mössbauer sample. These matters are presently explored in W. H. Orme-Johnson's laboratory at Wisconsin using a Large Sample Accessory Cavity (Varian). This cavity should be suitable to accommodate a helium dewar and a 12-mm diameter Mössbauer cell. As we probe more deeply into intricate systems such as cytochrome *c* oxidase or nitrogenase it will be mandatory to use such combined investigations on the same specimen.

H. Short-lived Species

In the future the Mössbauer technique will be used increasingly to investigate short-lived intermediates. Such species can for instance be prepared with rapid-freezing techniques. Alternatively, one might be able to prepare samples by increasing the lifetime of the species in appropriate antifreeze solvents using the techniques developed by Douzou.[44] Short-lived species can pose one problem about which the Mössbauer spectroscopist should be aware: the lifetime of an intermediate, even when kept in the frozen state, is not necessarily long compared to the running time of a Mössbauer experiment; this is particularly true when the samples are studied in the temperature range from 190–240°K. In past years we have lost two samples in our laboratory by overlooking these matters. Compound I of chloroperoxidase, prepared with the rapid-freezing technique, decayed within 12 hr while a spectrum was taken at 200°K; the ternary complex of 3,4-protocatechuate dioxygenase with substrate and O_2 decayed substantially within 6 hr at 240°K.

[21] Magnetic Susceptibility Applied to Metalloproteins

By THOMAS H. MOSS

Introduction

The magnetic state of metals in biological systems is often a key chemical parameter. Both the valence and type of metal coordination can be indicated by magnetic measurements. Though magnetic resonance or other spectroscopic data can frequently be interpreted in terms of the magnetic properties of a metal system, the only direct measurement is that of the magnetic susceptibility. In some cases, as in those with integer ionic spin

ISBN 0-12-181954-X

states, resonance methods cannot provide data but susceptibilities may be clearly measurable. On the other hand, it must be remembered that magnetic susceptibility is generally measured as a bulk property of a sample, representing a possible average of several quite different component magnetic species. In this sense, it can be deceptive or much less informative than spectroscopic data, where individual species may be resolved and separately quantified. In general, magnetic susceptibility is a necessary measurement for complete characterization of a metal-containing biological system, but it is not sufficient without ancillary data.

The theory and techniques of magnetic susceptibility measurement have been described previously in this series,[1] and many of the basic points of that report remain timely. This discussion will aim at complementing that and other reviews[2,3] by stressing new areas and subtleties of application which were not included in the earlier papers.

The Physical Basis of Susceptibility Measurements

Magnetic susceptibility is defined simply as the characteristic parameter which relates the magnetization of a sample to the strength of an applied magnetic field. It is, therefore, the sum of the orientational or distributional responses to an applied field by the individual molecular, atomic, and subatomic magnetic moments within the sample. It measures the number of such moments and their magnitude, which in metalloproteins translates in part to the number of paramagnetic sites, and the number of unpaired electrons at each site. Other contributions, discussed below, may have to be subtracted from the sum to separate the paramagnetic contribution. In some cases, the orientations of molecular or atomic magnetic moments are linked to that of the electrostatic crystal or molecular field and may not respond as free magnetic moments would in an applied field. In these situations the energy levels of the magnetic system will have a complicated dependence on the applied field magnitude and direction. Because of the link between magnetic and electrostatic properties, analysis of these measurements can yield additional chemical-structural information beyond that inferred from the number and strength of individual atomic or magnetic moments.

For a collection of individual paramagnetic moments, such as metal ions in a biological sample, the tendency to align along the field will be balanced against a tendency to thermal disorder. The former tends to

[1] T. Iizuka and T. Yonetani, Vol. 26, p. 682.
[2] T. Iizuka and T. Yonetani, *Adv. Biophys.* **1,** 157 (1970).
[3] L. N. Mulay, *Anal. Chem.* **42,** 325R (1970).

increase the net magnetization of the sample, the latter to reduce it. At high temperatures, therefore, net magnetization in an applied field (and susceptibility) will tend toward zero; at low temperatures it can be very large. Application of straightforward Boltzman statistics for the fraction aligned leads to the simple Curie law expression for the susceptibility. In the most basic case of a two-level system (a spin 1/2 ion aligned parallel or antiparallel to the applied field) the fraction in the upper and lower levels will be:

$$\frac{N_u}{N} = \frac{e^{-\mu H/kT}}{e^{\mu H/kT} + e^{-\mu H/kT}}$$

$$\frac{N_l}{N} = \frac{e^{\mu H/kT}}{e^{\mu H/kT} + e^{-\mu H/kT}}$$

Here N_u and N_l are the number of ions in the upper and lower states and $N = N_u + N_l$ is the total. With a magnetic moment μ for the ion, the energy of interaction with the applied field H is just μH. It is that value related to kT which determines the exponential power and indicates whether the sample will be highly polarized, with most of the ions in the lower state, or disordered, with equal populations in each level.

The difference in population between the two orientation levels, $N_l - N_u$, multiplied by the magnitude of the projection of the magnetic moment along the field direction, $g\mu_\beta/2$ for spin 1/2, gives the net sample magnetization:

$$M = \frac{g\mu_\beta N}{2}\,\frac{e^{\mu H/kT} - e^{-\mu H/kT}}{e^{\mu H/kT} + e^{-\mu H/kT}} = \frac{Ng\mu_\beta}{2}\tanh x \quad \left(x = \frac{gJ\mu_\beta H}{kT}\right)$$

For $\mu H \ll kT,\ e^{\mu H/kT} \sim \mu H/kT$ and:

$$M \sim \frac{Ng\mu_\beta}{2} \cdot \frac{g\mu_\beta H}{2kT}$$

For an atom with spin J greater than 1/2, the results can be generalized to:

$$M = NgJ\mu_\beta B_J(x)$$

$B_J(x)$ is the Brillouin function:

$$B_J(x) = \frac{2J+1}{2J}\operatorname{ctnh}\left(\frac{(2J+1)x}{2J}\right) - \frac{1}{2J}\operatorname{ctnh}\left(\frac{x}{2J}\right)$$

For $x \ll 1$ this again simplifies by approximation of the exponential to:

$$M = \frac{NJ(J+1)\,g^2\mu_\beta^2}{3kT}\,H$$

and:

$$\chi = \frac{NJ(J+1)\,g^2\mu_\beta^2}{3kT}$$

The $1/T$ temperature dependence is the characteristic Curie law for susceptibility, and it generally holds over most of the experimentally accessible temperature and field range. For instance, for valence-two copper ions and a typical large laboratory magnetic field of 10 kG,

$$\frac{g\mu_\beta H}{2kT} \simeq \frac{2}{2.8T} \ll 1 \quad \text{for } T \gg 10^\circ\text{K}$$

At either high fields or very low temperatures the more complex Brillouin functional dependence must be used to calculate the susceptibilities. At these limits the Brillouin function depends less strongly than linearly in H or $1/T$ so that the susceptibility is sometimes said to show "saturation" characteristics. Figure 1 illustrates the functional dependence of $B_j(x)$ for H or $1/T$ sufficiently large so that $x \gg 1$ and $B_J(x)$ is nonlinear.

The square root of the number multiplying the quantity $N\mu_\beta^2/3kT$ is called the "Bohr magneton number." For a spin J system this is obviously $g\sqrt{J(J+1)}$ and has only a few possible discrete and characteristic values given by the allowed values of J. However, the term is also often used to characterize inhomogeneous systems or systems in which there is a contribution to the magnetization from both spin and orbital angular momenta; in those cases the Bohr magneton parameter may take on any value in a given range and lose its significance as an indicator of a particularly defined ionic state.

In principle, for biological samples the susceptibility should reflect only the magnetic properties of unpaired electron spins. This would limit the

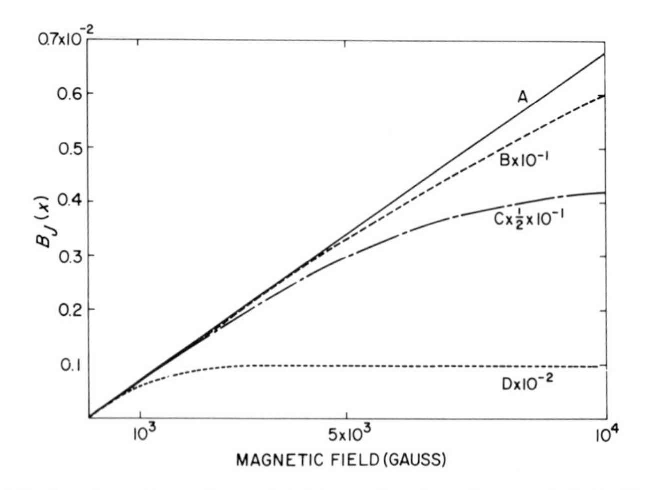

FIG. 1. The functional dependence of $B_j(x)$ as a function of magnetic field, H, where $x = gJ\mu_\beta H/KT$. T is represented as a parameter; curve A is for $T = 10$°K; curve B, $T = 1$°K; curve C, $T = 0.5$°K; curve D, $T = 0.1$°K.

information available to only that concerning the paramagnetic transition metals or relatively stable free radicals. The reason for this is encompassed in the theory of "quenching" of atomic angular momentum[4] in the presence of an electrostatic field. Such a field will exist in any but a gaseous sample so that the "quenching" applies to any biological sample measurement. The significance of the "quenching" is that the angular momentum of the orbital motion of the atomic electrons is not measurable. However, in the presence of a magnetic field a small amount of the orbital motion does become measurable ("unquenched") and perturbs the spin-only properties. Since nearly all magnetic measurements including susceptibility techniques depend on applied fields, this perturbation is always potentially present. The extra angular momentum can be expressed in a deviation of the electronic g-value from the spin-only magnitude of 2.0023. In fact, since the orbital angular momentum of the electronic orbitals can vary with the spatial properties of the orbitals, the g-value becomes spatially anisotropic and must be expressed as a tensor. In most biological magnetic measurements polycrystalline or other randomly oriented samples are used so that the measured value becomes a spatially averaged scalar. However, if a single crystal is used the measured value will depend on relative orientations of the applied field and the molecular axis.

In all cases where the orbital angular momentum plays a role in the atomic magnetic properties, the interactive coupling of the spin and angular momenta must also be considered as well as their individual coupling to the applied field. This coupling can cause the spin moments to respond to an externally applied field as if they were not freely orientable, and make the total magnetization of the sample a complex function of the applied field. In extreme cases the energy levels of the magnetic system may be more strongly perturbed by the internal electrostatic field than any attainable laboratory magnetic field. "Zero-field splitting" is said to apply in these situations. The magnetic properties of the spin 5/2, Fe(III) heme proteins provide a classic example of this behavior.[5]

The important point of the "orbital moment" contributions for biological applications is that the measurements can provide data on more than the simple number and magnitude of individual spin magnetic moments. The g-value deviation and other differences from spin-only susceptibility values can give extra information on the orbital configurations and the ligand field.

The small contribution of the orbital atomic moments to the susceptibility has a diamagnetic as well as paramagnetic component. This can be

[4] C. P. Slichter, "Principles of Magnetic Resonance," p. 65. Harper, New York, 1963.
[5] A. Tasaki, J. Otsuka, and M. Kotani, *Biochim. Biophys. Acta* **48,** 266 (1961).

understood simply as a reflection of Lenz's law, which states that a magnetic field applied to an electrical circuit will induce a current in the circuit which will oppose the flux change. The induced current will set up a field opposite to that of the applied field. An electronic orbital can be thought of as a resistance-free circuit so that the opposing current will persist as long as the applied field persists. This is true for closed-shell electronic configurations as well as partly filled ones, so that all atoms in a biological sample contribute to the diamagnetism. Though the diamagnetic effect is small, there are usually many more diamagnetic carbon, nitrogen, hydrogen, and oxygen atoms than paramagnetic ones, so that most biological samples show a net diamagnetic effect at room temperature. In fact, since the diamagnetic species are so diverse and predominant it would be difficult to separate out the interesting paramagnetic terms if it were not for the fundamental fact that the diamagnetic properties are temperature independent. They depend purely on the "induction" effect rather than on thermally determined state occupation. By making susceptibility measurements over a wide temperature range the paramagnetic contribution can, then, be easily separated from the constant "background" diamagnetism. Historically, that background proved troublesome when it was not possible to study biochemical samples over wide temperature ranges. Though in small inorganic molecules it was not hard to calculate the diamagnetism for subtraction from the overall result, this was impossible for proteins with thousands of diamagnetic atoms. Only by comparing two chemically similar forms, one with a paramagnetic center and one without, was it possible to work at all with one-temperature measurements.[6]

In most applications, the accuracy of magnetic susceptibility measurements has not been such as to require very great refinement of theory. For systems in which free radicals provide the measurable magnetism, the g-values are so close to the spin-only value that the susceptibility can be interpreted directly from the Curie law and the experimentally determined parameter becomes the number, N, of such radicals providing the measured result.

For systems involving paramagnetic ions, g-values can deviate sufficiently from spin-only values so that the measured values become an important indicator of electronic state. In theory and practice, ionic states with no intrinsic angular momentum—half-filled transition shells, like Mn(II) for instance[7]—give spin-only susceptibilities. The exact correspondence of measured and Curie law values then becomes a clear indicator of such a state. For ions like $S = 1/2$ Fe(III) or Cu(II) the orbital contribution perturbs the g-values by roughly 10% of the free-spin value,

[6] L. Pauling and C. Coryell, *Proc. Natl. Acad. Sci. U.S.A.* **22,** 210 (1963).

[7] J. H. Van Vleck and W. G. Penney, *Philos. Mag.* [7] **17,** 961 (1934).

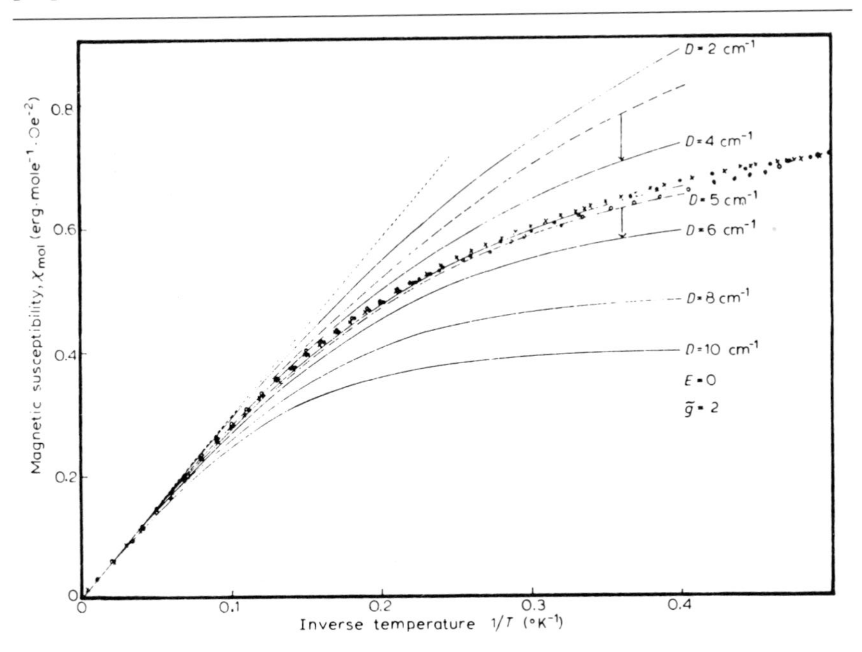

FIG. 2. Experimental and calculated values of the temperature dependence of ferric hemoglobin and myoglobin susceptibility as a function of inverse temperature. The various values of D indicate theoretical curves for various zero-field splittings of the S_z^2 states of the $S_z = 5/2$ iron ion. ×, ○, and ● indicate experimental values for myoglobin solution, hemoglobin solution, and hemoglobin polycrystals, respectively.

and that 10% deviation is reflected in, and characteristic of, the susceptibility of that state.[8] In a very few cases single crystals of biological materials have also been measured with sufficient accuracy to detect the spatial anisotropy of the susceptibility,[9] and this has proven interpretable in terms of the g-value anisotropy.

The "zero-field splittings" of various ions have also proven measurable (Fig. 2), both by measurements over wide temperature range[5] as well as over wide applied field range.[10] In some cases, the temperature-dependent measurements have also provided important confirmation of theory relating to thermal mixtures of electronic states,[11] in which it was postulated and confirmed that states of differing magnetic properties lay close enough so that the susceptibility should reflect a thermally excited mixture of them. A case will be discussed in the application section.

[8] A. Ehrenberg, *Sven. Kem. Tidskr.* **74,** 3 (1962).

[9] H. Uensyama, T. Iizuka, H. Morimoto, and M. Kotani, *Biochim. Biophys. Acta* **160,** 159 (1968).

[10] N. Nakano, J. Otsuka, and A. Tasaki, *Biochim. Biophys. Acta* **236,** 222 (1971).

[11] T. Iizuka and M. Kotani, *Biochim. Biophys. Acta* **181,** 275 (1969).

Limitations of the Theoretical Basis for Susceptibility Studies

In general, magnetic susceptibility studies have not been limited by problems in theoretical development. In contrast, the accuracy of measurements and the difficulty of obtaining homogeneous or well-defined samples have proven to be the limitation in analysis. With improvement in these areas considerably more sophisticated theoretical interpretation could be attempted. However, because the susceptibility almost inevitably will measure only time, directional, and chemical "average" properties of a sample, it will never lend itself to the detailed theoretical analysis of the high-resolution spectroscopic techniques.

Experimental Methods

There are a wide variety of means for the experimental determination of magnetic susceptibility. Force methods depend on the vertical or torsional force on a sample in an inhomogeneous applied magnetic field. Induction methods rely on the current induced in a coil as a magnetically polarized sample is moved in and out of the coil. Nuclear resonance techniques utilize the local field inside a paramagnetic sample to cause measureable shifts in the resonance frequency of probe nuclei. These are but a few of the often ingenious approaches which have appeared in the nearly 200-year history of biological magnetic susceptibility studies.

Among the force methods, the Faraday technique has been used most often in biological experiments (Fig. 3). In this technique the magnetic poles are shaped to give, over a finite region, a linear field gradient in the vertical direction, superimposed on a constant applied field. The sample is hung from a sensitive balance in the linear field gradient region. The measurement is made by measuring the force on the sample as a function of applied field. The susceptibility can be taken from the relation:

$$F = \chi H \frac{dH}{dz}$$

This is obtained directly by taking the derivative of the system energy, $\frac{1}{2} MH = \frac{1}{2} \chi H^2$, as a function of position. To avoid sample movement in the field gradient region a "null method" is often used to measure the force indirectly. Typically, a current loop is wound about the sample holder through which a current is passed to create a moment opposite in sign to that induced in the sample by the applied field. The magnetic moment of the sample is thus proportional to the current needed to nullify it.

In a simple spin $= 1/2$ system the magnetization M is linear in field at low field, and the constant slope of force vs. $H(dH/dz)$ yields the constant

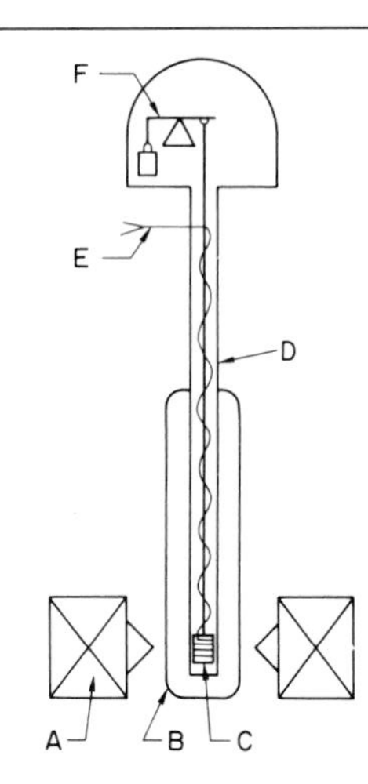

FIG. 3. Schematic diagram of a Faraday balance. A is the magnet for the polarizing field, B the dewar, C the sample, D the suspension tube for the sample (usually containing heat-exchange gas), E the wires for sample heating or for applying a "nulling" current around the sample to oppose the sample magnetic moment, and F is the balance.

susceptibility. To distinguish Curie paramagnetism from temperature-independent diamagnetism, or other magnetic effects, the susceptibility measurement is normally made over a range of temperatures. For this purpose the sample is hung in a dewar placed between the pole pieces, with the sample isolated from the cryogenic fluid to allow temperature variability. Resistive heating may be included in the sample holder, while low-pressure exchange gas provides a coupling to the cooling bath. Temperature is changed by varying heater current and/or exchange-gas pressure, and measurements are made at appropriate points. In some cases, in the absence of exchange gas, radiative or conductive heating and cooling are at a rate such that desired susceptibility-temperature points can be measured as the sample slowly heats or cools.

Torsion methods are very similar, with the exception that optical methods are normally used to detect either a field-induced sample torque or torque null as created by a current loop around the sample holder.

In the Gouy (force) technique the sample is in the form of a long cylinder which extends from a region of strong homogeneous magnetic field to a region of neglible field. The sample is then pulled in or pushed out of the high-field region depending on its paramagnetism or diamagnetism. The force on the sample is obtained as in the Faraday method and is:

$$F = \tfrac{1}{2}\,\mu_0\,(\chi - \chi_{\text{air}})AH^2$$

A is the cross-sectional area of the sample, H the field strength, and χ_{air} the volume susceptibility of the surrounding air.

The most common induction or flux apparatus generally operates in a vibrating mode (Fig. 4), with the unknown sample being moved in and out of a coil, or between two halves of a split coil.[12] A steady polarizing field must also be imposed to give the paramagnetic sample a net moment with which to induce the current. In some arrangements the polarizing field is applied by a superconducting coil in which the sample is positioned. The coil can then be center-tapped and used as the pick-up coil as well, as the sample is moved from one half to the other. In other cases the polarizing field may be a large electromagnet with the pick-up coils being entirely separate.

The AC signal derived from the pick-up coil is rectified and recorded as a DC voltage proportional to the sample moment. The moment is recorded as a function of polarizing field as in the force methods and thereby yields the susceptibility. Enclosing the sample in an appropriate dewar allows measurements as a function of temperature.

Superconducting flux measurement devices, such as the SQUID, are similar to other flux devices in that they rely on measurement of a current induced in a coil by a flux change.[13] However, they make use of the fact that the current induced in a superconducting loop will be persistent and can be measured at any convenient time and in a variety of indirect ways by coupling to another coil. In principle, the sample need be moved in or out of the detecting coil but once. As in vibrating-sample devices, a separate polarizing field must be provided, and temperature variation must be accomplished through use of a sample chamber capable of variable thermal coupling to a cryogenic system.

Nuclear magnetic resonance techniques have been used to measure the susceptibilities of soluble materials in liquid solution.[14] Two separate solutions are usually placed in concentric tubes. One contains a magnetic sample and a "probe" substance—a substance with a strong, unsplit narrow resonance. The second contains the probe substance alone. The presence of the magnetic sample creates an internal field in the solute in addition to the polarizing field, and the probe resonance line is shifted with respect to the probe line from the tube without magnetic solute. The shift, Δf, between the two resonances is a measure of the solute susceptibility as given by

$$\chi = \frac{3\Delta f}{2\pi f m} + \chi_0 + \frac{\chi_0(d_0 - d_s)}{m}$$

[12] A. G. Redfield and C. Moleski, *Rev. Sci. Instrum.* **43,** 760 (1970).
[13] E. J. Cukauskas, D. A. Vincent, and B. S. Deaver, *Rev. Sci. Instrum.* **45,** 1 (1974).
[14] K. D. Bartle, B. J. Dale, D. W. Jones, and S. Maricic, *J. Magn. Reson.* **12,** 286 (1973).

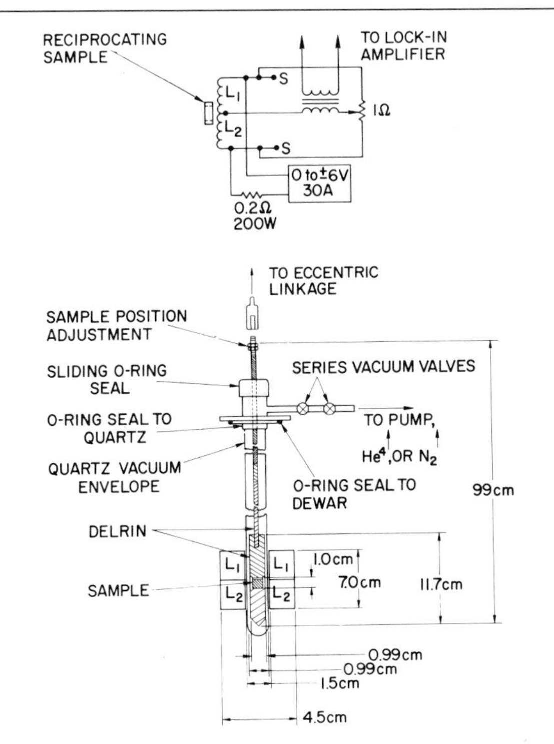

FIG. 4. A detailed schematic of a simple vibrating-sample susceptibility device, with simplified circuit diagram and physical dimensions.

Δf is the shift in frequency of the probe nucleus, from its normal value, f, due to the presence of the surrounding paramagnetic material; m is the mass of the dissolved material in 1 ml of solution; d_0 is the density of the solvent and d_s is that of the solution with the dissolved paramagnetic material; and χ_0 is the mass susceptibility of the solvent.

Problems and Limitations of the Experimental Methods

The sensitivity of the balance used is one obvious limitation on the accuracy and sensitivity of the Faraday or Gouy measurements. There are a number of other less obvious difficulties, however. A particularly troublesome one is the effect of gas convection or condensation–evaporation in the suspension chamber of the balance. As was mentioned previously, for sensitivity and delineation of complicated energy level splittings, biological samples must often be studied as a function of temperatures in ranges as low as attainable in ordinary cryogenic apparatus; 1–10°K or 1–100°K ranges are commonly needed. As temperature is varied in this region gases may be either condensed or evaporated on and from the sample and suspension itself or on and from the sample chamber walls. Though the heat-exchange gas is normally helium at low partial pressure, it is relatively difficult to insure that no other gases are present, adsorbed or absorbed, in parts of the system. This is especially true with biological samples which may be solutions frozen in relatively large sample containers which cannot be baked or "cleaned" in a physical sense. The condensation or evaporation on and from a sample holder can make a very large spurious change in sample weight as temperature is changed. Thus the common method of measuring a weight difference with and without applied magnetic field must often be done against a rapidly changing base line of zero-field weight. With the experimental chamber walls also part of the condensation–evaporation surface system, gaseous convection currents may also be set up along inevitable temperature gradients in the experimental chamber. These, too, can cause rapid and erratic changes in zero-field base-line weight. These problems can usually be controlled by taking great care with the exchange-gas volume and composition, by outgassing the systems prior to use, and by making temperature changes during the measurement process relatively slow and uniform. Nonetheless, they are quite common problems and should be planned for in both experimental design and analysis of results.

Another common problem in biological Faraday measurements, that of sample positioning, is also directly related to the typically low concentration of paramagnetic centers. To achieve the largest measurable effects, the amount of sample weighed should, of course, be maximized. However, many biological preparations, in addition to being inherently low in paramagnetic center density, are conveniently handled only in solution. Both of these factors imply that a sample with reasonable numbers of paramagnetic centers will be large in volume. Depending on the strength of the individual paramagnetic centers, on the order of a micromole of them may be needed for a reliable measurement. For a protein with one center per molecule, this implies a cubic centimeter of millimolar solution. Many magnet systems are designed so that the region of constant magnetic field gradient, dH/dz, is smaller or barely comparable to this. Small positioning errors can then introduce systematic inaccuracy. This is especially true if the device is calibrated with samples of different size or shape than those actually measured. Even with geometrically equivalent calibration and measurement samples, errors can be large if the calibration sample position is not reproduced exactly.

In addition, in most devices sample movement will occur in response to the applied field. If the sample is in a region where the field gradient changes with position, an error can occur proportional to this change. The null techniques used to overcome this difficulty can introduce problems of their own; the current loop wire may introduce paramagnetic impurities to the sample holder, and the leads from it to a voltage and measurement system may compromise the balance sensitivity.

A third basic problem with the Faraday method is common to other techniques. The magnetic properties of the sample holder and suspension must be known very well and should be as nearly independent of temperature as possible. At low temperatures, even trace impurities or contaminants of the holder and suspension can become strongly paramagnetic or ferromagnetic. The standard difference measurement of sample in holder minus empty holder then becomes a small difference of relatively large numbers with strong temperature dependence and limited accuracy. Maintaining strict cleanliness and material purity control over sample holder and suspension is thus essential to reliable base-line measurements.

These factors make estimates of instrumental sensitivity difficult to define. In a Faraday system the sensitivity in principle is determined by the balance properties and can be defined as that value of sample susceptibility which gives a measurement at least comparable to noise. This is generally quoted as $\sim 10^{-9}$ cgs susceptibility units for Faraday devices. However, the exact method of measurement, including the averaging process used in analyzing individual measurements made as a function of field or temperature, affect this value and make it difficult to apply directly to actual experimental accuracy. Experiments on biological materials have not actually achieved this accuracy, and it is doubtful that sample purity alone can be controlled to this degree. Actual Faraday susceptibility experimental accuracy on biological samples is probably at least an order of magnitude less than the 10^{-9} cgs figure.

The Gouy technique has all the balance sensitivity and convection–condensation problems of the Faraday technique. However, because it relies on the total field gradient from between the poles to the outside region, it avoids the difficulties of accurate sample positioning inherent in the Faraday method where the gradient is constant over only a small spatial region. However, it has the overwhelming problem of requiring a very large amount of sample material. For a large electromagnet applying fields of several thousand gauss, it may require a sample holder 20 cm or more in length to span the region of high field to the areas where the field is negligible. For high-purity biological samples, providing several milliliters of concentrated material for each measurement may make preparative time prohibitively long. In addition, because of the large sample size it is difficult to design a temperature-control system for measurements of temperature-varying paramagnetism. This severely limits the utility of the technique.

Because vibrating-sample induction devices are mechanically rigid, they are free of many of the problems in handling high-sensitivity balances.

In general, however, it is difficult to attain the very high sensitivity of a balance measurement. The problem of sample holder cleanliness is important as it is in all measurement methods. A unique problem to the vibrating-sample technique is that of spurious mechanical vibrations. If these occur at frequencies near that of the measurement motion they will induce a signal in the pick-up coil which will be recorded as a sample measurement. This problem is best dealt with by operating at frequencies far from natural vibrations and by using filtering techniques.

Sample geometry and positioning are also potential problems in vibrating-sample devices. As in the Faraday measurements, sensitivity considerations generally lead to maximizing sample size. The induction pick-up and polarization coils, however, will inevitably have nonuniform end effects which can be accurately calibrated only if experimental and calibration samples are of exactly similar geometry and oscillate precisely about the same position. In practical biological applications the sensitivity of vibrating-sample symptoms is probably somewhat less than that of a Faraday balance, perhaps by as much as an order of magnitude. The ease of use and reproducibility may mitigate this disadvantage.

Superconducting magnetic flux measuring devices are new enough in biological application so that routine practical problems are not yet completely predictable. Up to the present, it has been found that small vibrations of sample or component parts, which introduce small flux changes, have been most troublesome. The large volume and the need to use a container for frozen liquid samples aggravate that problem, relative to that for small inorganic solid samples, by putting a large mass of material in or near the flux change pick-up system. It remains to be seen whether this problem will remain as more sophisticated construction techniques are devised. In principle, some of the highest sensitivities should be attainable by this method, in the 10^{-10} cgs susceptibility unit range.

The nuclear magnetic resonance technique is flawed chiefly by being restricted to temperatures where high-resolution NMR measurements can be made, specifically those where liquid solutions can be maintained. This precludes the information needed about low-lying energy levels, but, on the other hand, may allow measurements at interesting high temperatures where antiferromagnetic uncoupling of spins can be observed. The sensitivity of this technique is lower than most others are in principle, because it involves measurement of line separations which may be only a few times line width even for strongly magnetic samples. It is probably limited in most NMR systems to 10^{-7} cgs susceptibility units. A further complication involves data analysis itself. Most analyses calculate the field at the probe nuclei embedded in a paramagnetic medium by assuming simple point dipole interactions. The situation with anisotropic paramagnets in solution, as would be the case for many biologically interesting molecules, has not been fully analyzed.

It should be noted also that the NMR susceptibility measurement depends on a knowledge of the difference in density between the solute and the solution of the magnetic material. This can be a small difference of numbers known with only limited accuracy, and this puts a limit on the accuracy of the susceptibility determination.

Applications

An account of the two centuries of magnetic susceptibility applications in biology would fill many reviews. We mention a few applications here only to further illustrate the range and limitation of the technique. That usually cited as the first was a measurement by Faraday in 1845 on arterial blood.[15] He remarked that he should follow that up with a measurement of venous material. Had he done so successfully he would have had at his fingertips one of the more important indicators of the mechanism of O_2 binding in the protein.

Pauling's measurements in the 1930's[16,17] were the first real follow-up to Faraday's idea. These were done by the Gouy method but were sufficiently sensitive to detect and quantify the change in spin state of hemoglobin iron among the $S = \frac{5}{2}$, $S = \frac{1}{2}$, $S = 2$ and $S = 0$ forms. They stimulated the first real thought concerning the unusual versatility of hemoglobin iron–ligand bonding. They were also noteworthy in overcoming the difficulty of single-temperature measurement in estimating diamagnetic contributions to the overall magnetism. This was done by using the diamagnetic CO form as a "blank" and measuring the paramagnetic effects in other forms by their difference from the CO–hemoglobin result.

The first meaningful single-crystal biological susceptibility measurement was done using another heme protein, myoglobin.[18] Because of the anisotropy of g-values in the high-spin ferric forms ($g_x = g_y = 6, g_z = 2$), the susceptibility varies greatly with the angle between applied field and molecular axis. This variation has been used to map the orientation of protein heme groups relative to the crystal axes, and it has been used as a valuable constraint on interpretation of X-ray diffraction patterns in determining total molecular structure.

The magnetic measurements on hemoglobin have been extended to

[15] Quotation from P. W. Selwood, "Magnetochemistry." Wiley (Interscience), New York, 1956.

[16] C. Coryell, F. Stitt, and L. Pauling, *J. Am. Chem. Soc.* **59,** 633 (1937).

[17] C. Coryell and F. Stitt, *J. Am. Chem. Soc.* **62,** 2942 (1940).

[18] F. R. McKim, *Proc. R. Soc., Ser. A* **262,** 281 (1962).

many heme proteins. The linkage between iron spin state and the nature of the ligands bonded perpendicular to the heme plane made the physical susceptibility measurements a standard chemical tool.[19,20] The concept was broadened to include the notion of mixed spin states, in which chemical bonding of an intermediate character, between those typified as "ionic" or "covalent," was indicated by susceptibilities intermediate between those defined by the alternative pairings of the electron spins.[21] Measurements over wide temperature ranges showed that the spin states could exist in mixtures in thermal equilibrium,[22] and that the exact details of the thermal mixtures could become a fine indicator of the chemical bond. The notion was broadened further with the realization that some ionic spin states could exist in "quantum mixtures,"[23] in which the individual ionic wave function was not a pure spin state in thermal mixture with other ions, but was itself a quantum mixture of wave functions characterizing distinct spin states. These quantum mixtures have quite distinct temperature dependence from those of the thermal mixtures, as well as different implications for the chemical bond character,[24] and thus they have become another sensitive indicator of chemical bonding and structure.

One further dimension to the studies of temperature variation of heme protein susceptibility was pointed out in a short but extremely important paper by Otsuka.[25] The author noted protein susceptibility temperature dependences which could not be explained by the simplest Boltzman statistics for spin state thermal populations. He realized, however, that change of spin state for a heme iron was linked to a structural movement of the iron in and out of the porphyrin plane. These movements could involve degrees of freedom such as weak electrostatic and/or van der Waals bonding with many protein sites. Such degrees of freedom would associate a magnetic state change with an entropy change which could be very different from that indicated by the spin multiplicity alone. Temperature-varying measurements in a variety of cases proved this concept to be valid[26] and allowed the use of susceptibility data for a characterization of protein state which is far broader than that of ionic spin state alone.

Measurement of field in addition to temperature dependence of susceptibility has proven to be another means to detect fine details of metal-ion

[19] M. Kotani, *Prog. Theor. Phys. Suppl.*, 174 (1961).

[20] P. George, J. Beetlestone, and J. S. Griffith, *Rev. Mod. Phys.* p. 441 (1964).

[21] J. S. Griffith, *Proc. R. Soc., Ser. A* **235,** 23 (1956).

[22] T. Iizuka, M. Kotani, and T. Yonetani, *Biochim. Biophys. Acta* **167,** 257 (1968).

[23] M. Maltempo, T. H. Moss, and M. A. Cusanovich, *Biochim. Biophys. Acta* **342,** 290 (1974).

[24] M. Maltempo and T. H. Moss, *Q. Rev. Biophys.* **9,** 181 (1976).

[25] J. Otsuka, *Biochim. Biophys. Acta* **214,** 235 (1970).

[26] T. Yonetani, T. Iizuka, T. Asakura, J. Otsuka, and M. Kotani, *J. Biol. Chem.* **247,** 863 (1972).

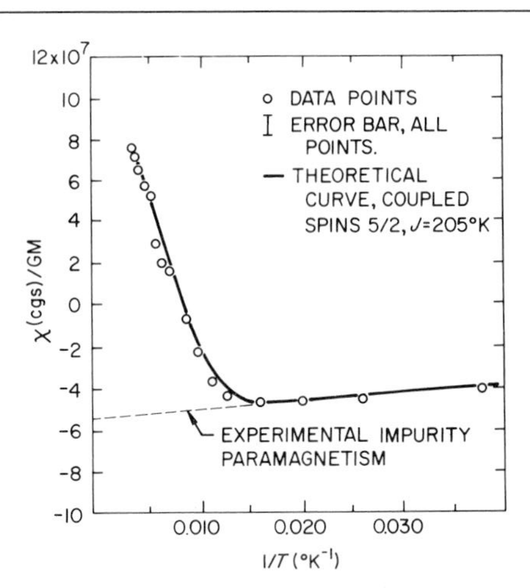

FIG. 5. Temperature dependence of the magnetic susceptibility of an iron–porphyrin dimer showing antiferromagnetic behavior of the coupled iron atoms. The solid curve is that expected theoretically for a pair of high-spin Fe(III) ions coupled antiferromagnetically by a term $2J\overline{S}_1 \cdot \overline{S}_2$ in the Hamiltonian for the system. This type of coupling has been postulated as responsible for some of the abnormal magnetic properties of cytochrome oxidase.

electronic state.[27] Heme proteins have energy-level splittings of only a few degrees Kelvin in some states, which can be clearly identified by the susceptibility behavior at high fields of 5–50 kG.

Though biological susceptibility measurements were historically catalyzed by interest in heme protein chemistry, they have played a strong role recently in other areas. The unusual magnetic properties of the nonheme iron proteins of the ferredoxin class suggested possible linkages between the two iron atoms in a single protein. Susceptibility measurements confirmed antiferromagnetic coupling,[28,29] and the notion of "metal clusters"[30] as an active center in enzymic and other functions stimulated many magnetic measurements. The "cluster" idea of metal function extended to copper and other metal-containing proteins, with susceptibility measurements of antiferromagnetic behavior playing a key role.[31,32] The very weak magnetic behavior of antiferromagnetic pairs showing only slight uncoupling at higher temperatures (Fig. 5) has placed

[27] P. Champion, E. Munck, P. G. Debrunner, T. H. Moss, J. D. Lipscomb, and I. C. Gunsalus, *Biochim. Biophys. Acta* **376,** 579 (1975).

[28] T. Moss, D. Petering, and G. P. Palmer, *J. Biol. Chem.* **244,** 2275 (1969).

[29] G. Palmer, W. R. Dunham, J. A. Fee, R. H. Sands, T. Iizuka, and T. Yonetani, *Biochim. Biophys. Acta* **245,** 201 (1971).

[30] T. H. Moss and J. A. Fee, *Biochem. Biophys. Res. Commun.* **66,** 799 (1975).

[31] T. H. Moss and T. Vanngard, *Biochim. Biophys. Acta* **371,** 39 (1974).

[32] T. H. Moss, D. C. Gould, A. Ehrenberg, J. S. Loehr, and H. S. Mason, *Biochemistry* **12,** 2444 (1973).

new demands on the sensitivity of biological susceptibility measurements and has stimulated interest in improvements. In fact, the first contributions of the SQUID concept in biological questions have been in this area.[33]

Conclusion

Susceptibility measurements, though limited in detailed information content, are a basic characterization of metal state in biological molecules. A wide variety of experimental techniques exist, with trade-offs among them related to sensitivity vs. ease of use and versatility. Measurements over a wide temperature and field range are generally most productive and are often necessary for meaningful interpretation. In some cases the temperature dependence may have implications for protein structure far beyond that of the simple indication of the number of impaired metal-ion electrons.

[33] E. I. Solomon, D. M. Dooley, R. Wang, H. B. Gray, M. Cerdonio, F. Mogno, and G. L. Romani, *J. Am. Chem. Soc.* **98,** 1029 (1976).

[22] Determination of Oxidation-Reduction Potentials[1]

By GEORGE S. WILSON

Of particular significance in recent years has been the coupling of electrochemical and spectroscopic techniques (e.g., EPR, fluorescence, NMR, and UV-visible) to identify and characterize electroactive centers in proteins. With present methodology it is possible to study proteins containing multiple centers as well as electron-transport-coupled species as found, for example, in mitochondria. This article will concentrate primarily on techniques involving the union of electrochemistry and UV-visible spectroscopy in the study of *in vitro* redox systems.

Establishment of Equilibrium

As a fundamental thermodynamic parameter, the importance of establishing equilibrium between all potential-determining species cannot be overemphasized. As indicated in Fig. 1, meaningful results will be obtained only if *chemical* equilibrium is achieved between the various species in

[1] This work supported in part by National Science Foundation Grant MPS 73-08683.

ISBN 0-12-181954-X

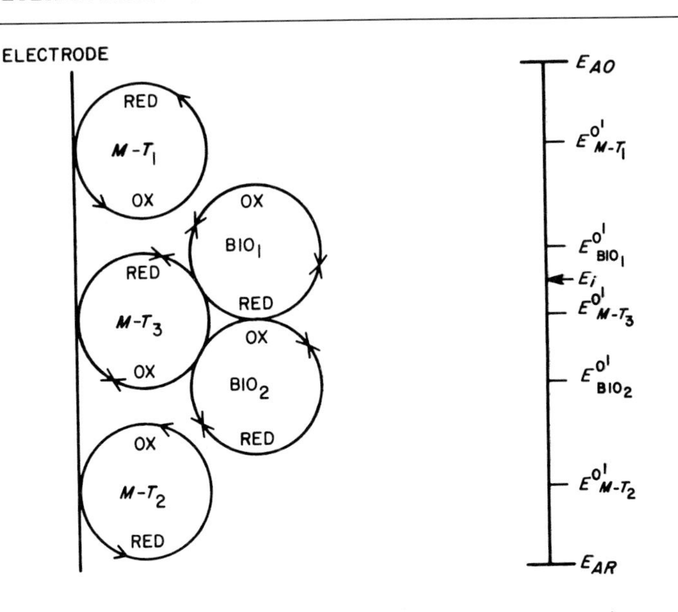

FIG. 1. Schema for mediator-titrant coupled electrochemical reactions.

solution and *electrochemical* equilibrium is established at the electrode solution interface. Under the usual null current potentiometric conditions, stable and meaningful potentials will be obtained only if the rates of the reaction at the electrode surface

$$\mathrm{Ox} + ne^- \underset{k_b}{\overset{k_f}{\rightleftharpoons}} \mathrm{Red} \tag{1}$$

are rapid in both directions. Electron-transfer rates are an intrinsic property of a particular redox state (k_f, k_b), its concentration, and the characteristics of the electrode surface. It follows that rapid electrochemical equilibrium might be established for Reaction (1) if Ox and Red are at approximately equal and *suitably* high concentrations. Even at relatively high concentrations most proteins do not establish stable potentials because the heterogeneous charge-transfer (electrochemical) rates are low. The mediator-titrants shown in Fig. 1 will make electrochemical measurements possible if both heterogeneous and all associated homogeneous electron-transfer rates are sufficiently high. This means that for potentiometric measurements the Ox/Red ratio for a mediator should vary from about 0.1–10.0 so that its formal potential spans the range of ± 60 mV ($n = 1$) or ± 30 mV ($n = 2$) around the expected midpoint potential of the biocomponent. It must be established that the mediator-titrants themselves are chemically stable and merely transfer electrons without complexing or otherwise interacting with the components so as to affect their electrochemical behavior. Such interactions are not easy to identify, but variation of the initial concentration of the mediator over a factor of 5–10 sometimes reveals diagnostic spectral or potentiometric information. Where possible, mediators of different structure should be employed to see whether the same results are obtained. Finally, it should be established that the component(s) under study obeys an appropriately constituted Nernst equation for a wide range (0.1–10.0) of redox ratios. The data should be time independent and also independent of the direction in which the redox ratio is varied. Hysteresis effects are not uncommon and result from a variety of causes including denaturation and/or precipitation of protein.

Acquisition of Nernst Equation Data

Spectrophotometry is often used to determine the extent of reduction of a biocomponent. Ideally a wavelength can be chosen at which the absorbance varies monotonically between the fully oxidized and fully reduced states of a particular prosthetic group. It is a good practice to verify the Ox/Red ratio at several different wavelengths. Such studies and a knowledge of the overall redox stoichiometry will often reveal the existence of spectrally invisible redox components as found for cytochrome oxidase.[2]

Method of Mixtures. The equilibrium constant, K, for the redox reaction involving the *one*-electron reduction of B_O

$$B_O + {}_nM\text{-}T_R \rightleftharpoons B_R + {}_nM\text{-}T_O \tag{1}$$

is given by

$$K = \frac{[B_R]}{[B_O]} \cdot \left(\frac{[M\text{-}T_O]}{[M\text{-}T_R]}\right)^n \tag{2}$$

where B_O and B_R and $M\text{-}T_O$ and $M\text{-}T_R$ are the oxidized and reduced forms of the biological molecule and mediator-titrant respectively. The redox ratios $[B_O]/[B_R]$ and $[M\text{-}T_O]/[M\text{-}T_R]$ are determined spectrophotometrically. As Minnaert[3] has shown, a plot of log $([B_O]/[B_R])$ vs. log $([M\text{-}T_O]/[M\text{-}T_R])$ yields a straight line of slope n and intercept $-\log K$. In turn $\log K$ is directly proportional to the difference in formal potentials ($\Delta E^{0\prime}$) between the biological redox couple and the mediator-titrant:

$$\Delta E^{0\prime} = E_B^{0\prime} - E_{M-T}^{0\prime} = \frac{RT}{nF} \ln K \tag{3}$$

[2] J. L. Anderson, T. Kuwana, and C. R. Hartzell, *Biochemistry* **15,** 3847 (1976), and references therein.

[3] K. Minnaert, *Biochim. Biophys. Acta* **110,** 42 (1965).

If a linear plot of the correct slope results for $-1.0 \leq \log([M\text{-}T_O]/[M\text{-}T_R] \leq 1.0$, then the formal potential of the B_O/B_R couple may be estimated if that of the mediator-titrant is known. The latter is selected to have a formal potential comparable to that of the biological system. Nonlinear plots have also recently been resolved.[2] This technique has the advantage of greater simplicity because potentiometric measurements may not be required, and it can therefore be used in situations where such measurements are difficult or impossible to make.[4] The price paid for simplicity is a lack of possibly valuable potentiometric or stoichiometric data regarding specific interactions between the mediator and biological component or other system peculiarities.

"Polarographic" Method. The so-called polarographic method involves the measurement of a current proportional to concentration which results from the application of a potential between a reference and indicator electrode. The latter is usually a platinum microelectrode which is often vibrated or rotated to increase sensitivity and reproducibility. This method is presently restricted to monitoring of small molecules such as mediator-titrants, some coenzymes, and such species as oxygen and ascorbate. It is often possible to measure concentrations significantly lower than permitted by either potentiometry or spectrophotometry.[5]

Potentiometric–Spectrophotometric Titration

Titrant is added from a microburet so that the redox state of the biocomponent can be varied in a step-by-step fashion. Since very few proteins have been observed to establish well-defined potentials at an indicator electrode, it will be necessary to add mediators with formal potentials within 30–60 mV (*vide supra*) of the measured potential. For a single redox center, one mediator is ordinarily sufficient for a given set of measurements, whereas in multiple-centered complexes several may be required. During a titration it is important to establish that the redox ratio as measured spectrophotometrically is consistent both with the measured potential and with the degree of conversion expected from the reaction stoichiometry. Verification of this latter point is often neglected because of the experimental difficulties associated particularly with oxygen-sensitive titrants. Many of the mediator-titrants commonly employed have spectra in the visible region which overlap extensively with those of biocomponents. This, of course, requires corrections which can introduce considerable

[4] G. S. Wilson, J. C. M. Tsibris, and I. C. Gunsalus, *J. Biol. Chem.* **248,** 6059 (1973).
[5] A. H. Caswell and B. C. Pressman, *Anal. Biochem.* **32,** 396 (1969).

uncertainty into the measurements and argues for minimizing overlap where possible.

Potentiometric–Spectrophotometric Titration (Coulometric)

It was observed independently by Wilson[6] and Kuwana[7] that the indirect titration of biocomponents could be accomplished by coulometric generation of mediator-titrants at controlled current or controlled potential. Thus as indicated in Fig. 1 the redox state of the mediator-titrants can be varied by an electrochemical perturbation applied to the electrode according to the reaction:

$$M\text{-}T_O \pm ne^- \rightleftharpoons M\text{-}T_R \tag{4}$$

If the reaction of the mediator-titrant (*M-T*) with the biocomponent involves complete, well-defined stoichiometry and if the *M-T* is generated at a known current efficiency, then the charge passed can be directly related to the extent of biocomponent conversion. For several systems thus far examined[2,8,9] in detail the presence of proteins and/or detergents does not appear to interfere seriously either with static or dynamic measurements involving *M-T* redox state perturbations. It is clear from Eqs. (2) and (3) that the *M-T* cannot perform both mediator and titrant functions simultaneously if Reaction (1) is to proceed quantitatively to the right. Consequently, the titrant(s) is chosen so that $\Delta E^{0'}$ is large and the mediator is chosen according to previously described criteria. If two titrants are employed ($\Delta E^{0'} >> 0$ and $\Delta E^{0'} << 0$) then the biocomponent can be sequentially cycled between oxidized and reduced states. This technique has the advantage that titrant can be "added" electrochemically in precisely and accurately controlled amounts without dilution of the biocomponent solution. This greatly simplifies studies of oxygen-sensitive species since the titration cell can be completely sealed off for the duration of the experiment. If the charge Q (coulombs) required to titrate the protein to a certain redox level is measured as a function of the absorbance change of the redox chromophore (ΔA) one obtains:

$$\frac{\Delta A}{\Delta Q} = \frac{b}{FV} \cdot \frac{\Delta \epsilon}{n} \tag{5}$$

where b is the optical path length (cm), F the Faraday, V the volume of the cell (liter), $\Delta\epsilon$ the difference in molar extinction coefficient between the

[6] D. B. Swartz and G. S. Wilson, *Anal. Biochem.* **40,** 392 (1971).
[7] M. Ito and T. Kuwana, *J. Electroanal. Chem.* **32,** 415 (1971).
[8] M. D. Ryan and G. S. Wilson, *Anal. Chem.* **47,** 885 (1975).
[9] J. S. Ranweiler, Ph.D. Thesis, University of Arizona, Tucson (1975).

oxidized and reduced forms, and n the number of equivalents per mole. Under the conditions of Reaction (1) for $K \gg 0$, a linear variation of ΔA with charge should result if: (1) the stoichiometry remains constant and (2) $\Delta\epsilon$ is unaffected by the state of oxidation or reduction. From the total charge required to titrate the chromophore the concentration of the protein can be calculated:

$$C = \frac{Q}{nFV} \tag{6}$$

Clearly if a biocomponent contains more than one active center these can be identified even if spectrally "invisible" because a nonlinear relationship between ΔA and Q will result. In other words, consumption of oxidant or reductant does not produce a corresponding spectral change. The method of mixtures and the potentiometric/spectrophotometric method rely on absorbance ratios to determine redox states of *all* biocomponent species at each point in the titration. This is impossible if some species are spectrally invisible. The absorbance–charge approach has been applied to studies of cytochrome oxidase.[2] Simultaneously obtained potentiometric data can also provide valuable complimentary information.

Interpretation of Nernst Equation Data

The formal potential, $E^{0\prime}$, is defined by Eq. (7) below for *unit concentrations* of the oxidized and reduced forms of Reaction (1).

$$E = E^{0\prime} + \frac{RT}{nF} \ln \frac{[\mathrm{Ox}]}{[\mathrm{Red}]} \tag{7}$$

The formal potential is ionic-strength dependent, and for highly charged redox systems like ferri/ferrocyanide careful control of electrolyte concentrations is essential.[10] If a form of the Nernst equation consistent with expected reaction stoichiometry can be established, then the formal potential should be evaluated. A more common practice is to operationally evaluate the midpoint potential, E_m, which is observed at the half-titration point. For simple redox systems E_m and $E^{0\prime}$ are equivalent. Clark,[11] however, cites several examples where this is not the case.

In general n values which deviate from the expected stoichiometry by more than 10% should be reevaluated. As Wilson and Dutton[12] have pointed out, two independent, closely spaced one-electron redox compo-

[10] J. E. O'Reilly, *Biochim. Biophys. Acta* **292,** 509 (1973).

[11] W. M. Clark, "Oxidation-Reduction Potentials of Organic Systems." Williams & Wilkins, Baltimore, Maryland, 1960.

[12] D. F. Wilson and P. L. Dutton, *Biochem. Biophys. Res. Commun.* **39,** 59 (1970).

nents can yield a sigmoidal Nernst plot. Unless this situation is recognized, apparent n values between 0.5 and 1.0 can easily be obtained. Attention should also be called to the redox behavior characterized by flavins and quinones in which the n value determined from a Nernst plot can be nonintegral. Although the overall n value is two, the limiting value can vary between one and two depending upon the relative stability of the intermediate "semiquinone."[11] For this reason quinone or quinone-like mediator-titrants should be used with caution especially with the method of mixtures.

If the redox ratio is measured spectrophotometrically as a function of the indicating electrode potential, multicomponent systems may be resolved from the sigmoidal titration curve which results.[13] We have used a similar program[14] to resolve a two-component system assumed to contain two chromophoric centers with equal extinction coefficients but different concentrations. Using a method of steepest descent calculation it is possible to adjust the percent composition and potentials of the two components to obtain the best fit of the experimental data.

Valuable information can be obtained from the pH dependence of the formal potential. The apparent ionization constants obtained from an $E^{0\prime}$ vs. pH plot[11] do not necessarily correspond directly to the values expected for amino acid functional groups even when the local hydrogen ion activity is considered. Ligation, for example, usually causes an acidic shift in the functional group's apparent pK_a. The noncorrespondence of spectral and electrochemical transitions is well known and can occur for any of the following reasons: (1) Simultaneous transitions occur in both oxidized and reduced forms such that no net change in $E^{0\prime}$ results. (2) One ligand is replaced with another of equal "strength" but different spectral properties. (3) Observed spectral transitions are totally unrelated to perturbations in the redox active group. (4) Spectral measurements are insensitive to perturbations such as conformational changes affecting the redox group. These questions are discussed in more detail elsewhere.[9]

Experimental Techniques and Apparatus

Titration Cells. Numerous titration cells[15-17] have been described for potentiometric/spectrophotometric titrations. Because of its inherent simplicity we prefer a design similar to that described by Hawkridge and

[13] P. L. Dutton and J. B. Jackson, *Eur. J. Biochem.* **30,** 495 (1972).
[14] M. A. Cusanovich, personal communication (1972).
[15] B. D. Burleigh, Jr., G. P. Foust, and C. M. Williams, Jr., *Anal. Biochem.* **27,** 536 (1969).
[16] P. L. Dutton, *Biochim. Biophys. Acta* **226,** 63 (1971).
[17] J. A. Culbert-Runquist, R. M. Hadsell, and P. A. Loach, *Biochemistry* **12,** 3508 (1973).

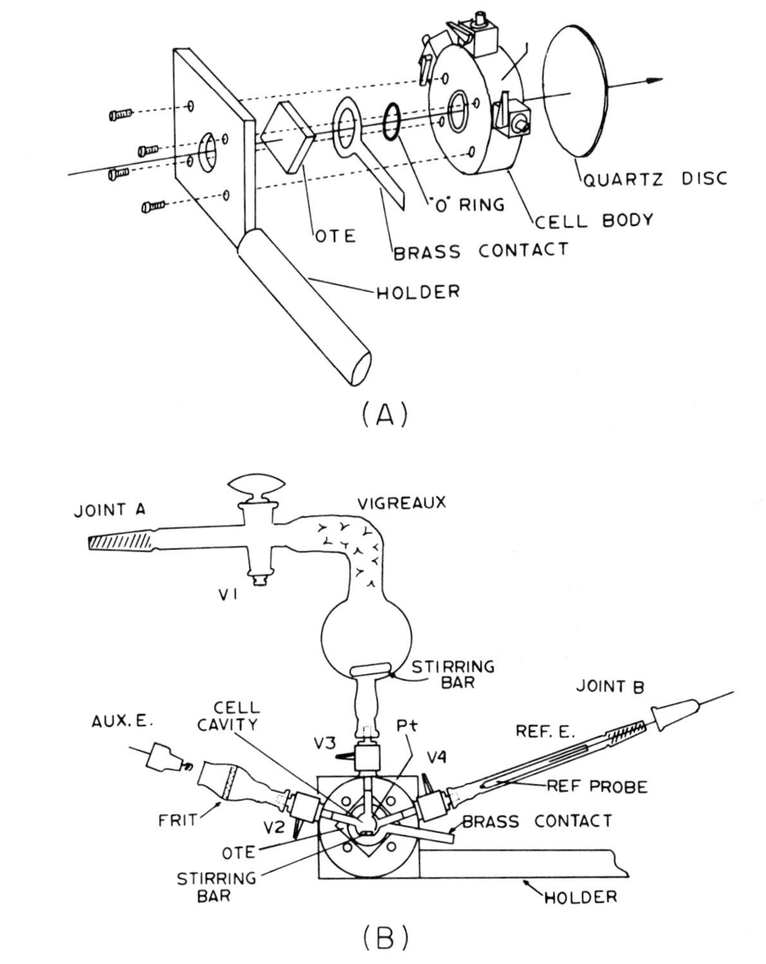

FIG. 2. Spectroelectrochemical cell. [*Anal. Chem.* **45,** 1021 (1973). Reprinted with permission from *Anal. Chem.* **45.** Copyright by the American Chemical Society.]

Kuwana.[18] (See Fig. 2A and B.) This cell is oxygen tight ($<5 \times 10^{-7} M\ O_2$), requires a small volume ($\sim$2 ml) of solution, can be readily adapted to volumetric or coulometric titrant addition, and is compatible with a variety of complimentary experiments (EPR, flash photolysis, preparative-scale electrochemistry, etc.). We construct the cell body of Kel-F which is sturdier and more resistant to organic solvents than Lucite. The cell holder

[18] F. M. Hawkridge and T. Kuwana, *Anal. Chem.* **45,** 1021 (1973).

and side arms have been modified to fit more easily into a Cary 14 cell compartment. A further simple but useful modification is to replace V3 with a Fisher Model 13-639-92 microcombination pH electrode mounted in a 10/18 $ Teflon thermometer adapter (Kontes K-179800). The reference electrode attached to V4 is replaced with a serum cap, and a Gilmont Model S3200 ultraprecision micrometer syringe is inserted through it. With this arrangement the reduced form of an autooxidizable protein can easily be titrated with acid, base, or ligand. The desired redox level can be maintained by coulometric generation of mediator-titrant. The pH reference electrode can be used in combination with either the glass or Pt indicator electrodes.[19]

The cell preparation procedure involves filling the reference probe with NaCl (saturated with AgCl) and attaching the reference electrode arm to V4. An identical auxiliary electrode arm containing a Pt wire sealed into a 7/25 $ joint placed on V2. Valves V1–V4 are opened so that the reference and auxiliary electrode arms can be carefully degassed in 3–5 vacuum (0.1 Torr)/purified nitrogen (3 psi) cycles. The cell is returned to a slight positive nitrogen pressure and valves V2–V4 are closed. About 2.5 ml of a previously degassed analyte solution (protein, mediators, and buffer/supporting electrolyte) are placed in the reservoir above the cell. The cell is again tilted to the vertical position and the solution forced slowly under nitrogen pressure into the cell and electrode arms. V3 is closed and the degassing reservoir removed. The cell is now placed in a Cary 14 spectrophotometer with a special cell holder/shielded magnetic stirrer. (Fields generated by permanent magnets can cause undesirable fluctuations in photomultiplier response.) The light beam should be masked so that light is not scattered off the stirring bar which is stationary during spectral measurements. Details of cell construction and modifications are described elsewhere.[9,18] The minimum optical path length possible with this design is about 0.75 cm, typical cells being 1 cm.

Purge Gas Purification. In handling autooxidizable species we have found properly purified nitrogen quite satisfactory. The gas is passed through a 3 × 30 cm column containing a Cu-Ni pellicular solid reductant (BTS catalyst R 3-11, BASF Wyandotte Corp.). We have found this material to be superior to all other common reductants[11] due to its high efficiency and room-temperature operating conditions. The catalyst is regenerated with hydrogen, and a heating tape is wrapped around the column to drive off the water produced. The catalyst also quantitatively removes traces of CO in the purge gas. All connections in the nitrogen train should be fabricated of glass or 1/8″ copper tubing.

[19] J. S. Ranweiler and G. S. Wilson, *J. Bioelectrochem. Bioenerg.* **3**, 113 (1976).

Reference Electrode. The AgCl/Ag electrode is preferred as it is readily adapted to a variety of configurations. A 16–18 gauge silver wire cleaned in 1 *M* HNO_3 is anodized in 0.1 *M* HCl to produce a smooth tan-plum colored deposit. The wire is then inserted into a reference probe solution saturated with both AgCl and NaCl. This electrode has a potential of 0.200 ± 0.002 V vs. NHE (normal hydrogen electrode) to which all measurements should be referred. The reference electrode potential is checked daily against a saturated calomel electrode before use. In anaerobic applications conducting gel salt bridges should be avoided in general. Otherwise a polyacrylamide gel (1 *M* electrolyte) similar to that used for electrophoresis[6] has proven superior to agar.

Indicator Electrodes. For null current measurements (potentiometry) platinum has been the most widely used. In the absence of a well-defined electrochemical equilibrium, this electrode will often assume potentials characteristic of platinum oxides formed on the surface (high-potential region) or surface catalytic reduction (low-potential region). In general, poorly established electrochemical equilibria will be characterized by a potential dependence on the solution stirring rate not reflected in the simultaneous spectral measurements. In some instances gold gives improved results. It is increasingly evident, however, that neither of these electrodes is inert as widely assumed. One should be especially cautious in working with sulfur-containing species.

Electrochemical measurements involving nonzero currents are much more sensitive to blockage of the electrode surface by surfactants (lipids, proteins, electrode reaction products, etc.). Nevertheless, with proper choice of electrode and conditions this difficulty can, in some instances, be circumvented. Greatly facilitating cell design (Fig. 2) is the optically transparent electrode (OTE) contacted with a brass or copper ring. Thin metal films of gold and platinum as well as *n*-type semiconductors (Sb-doped tin oxide or Sn-doped indium oxide) can be employed. The latter are available from PPG Industries (Pittsburgh, PA) under the trade names of NESA and NESATRON, respectively. Surface resistivity should be 5–35 Ω per square. An optical window from about 300–700 nm can be achieved. The base line can be improved by depositing the film on quartz and by using fluoride-doped tin oxide.[20] At pH 7 tin oxide (Fig. 2) may be used at applied potentials in the range $+1.5$ V to -0.65 V vs. NHE. Details concerning the spectral and electrochemical properties of these semiconductor electrodes are presented elsewhere.[21] Neutral-density gold, nickel, and mercury-coated nickel minigrids have also been used as OTE's.[22] These have the

[20] G. S. Wilson, unpublished results (1975).

[21] N. Winograd and T. Kuwana, *in* "Electroanalytic Chemistry" (A. J. Bard, ed.), Vol. 7. Marcel Dekker, New York, 1974.

[22] W. R. Heineman, B. J. Norris, and J. F. Goelz, *Anal. Chem.* **47**, 79 (1975).

advantage that the optical characteristics are determined by the spacing of the grid holes and not by the thickness or condition of the electrode surface. They have been extensively employed in optically transparent thin-layer cells (optical path length 50–200 μm, volume 20 μliters) and have advantages in some applications.[23] Vitreous or glassy carbon (Tokai Electrode Mfg. Co., Tokyo, Japan)[6] and carbon paste[24] also have been successfully used in biological voltammetric applications.

Instrumentation—Potential Measurements. A high-impedance (>10 MΩ) input digital voltmeter or pH meter will serve this purpose. Some pH meters have a recorder output which is useful in establishing electrode drift.

Instrumentation—Coulometric Titrant Generation. Figure 3 shows the design of a potentiostat/integrator suitable for measuring the charge injected in generating the requisite amount of mediator-titrant at controlled potential. To reduce a biocomponent the potential of the generating electrode is stepped from E_i to E_{AR} at which point $M\text{-}T_{2R}$ is generated in a reproducibly stirred solution at a diffusion-controlled rate (see Fig. 1) for the desired period of time. The accumulated charge is due to both the Faradaic current and the charging of the electrode double layer. The data are corrected for the latter contribution by stepping the potential from E_i to E_{AR} in the absence of a Faradaic reaction (buffer/electrolyte only) for the specified time. After generating $M\text{-}T_{2R}$ the potential is returned to E_i (an open-circuit potential, i.e., no potential applied in the region of $E^{0\prime}{}_{BIO_1}$). The charge, potential, and spectrum are now measured. If E_i or E_A is changed during the experiment then a new charging-current correction must be established. The integrator has a sensitivity (volt/coulomb) nominally equal to 0.2 × CF sensitivity (volt/ampere) as indicated in Fig. 3. It must be calibrated against a known current and checked for drift. $M\text{-}T_{2R}$ is generated in steps until the titration is complete. The Faradaic charge will have to be corrected for any mediators also titrated. To reoxidize BIO_1, the potential is changed from E_i to E_{AO} in an analogous fashion. One reducing cycle is usually necessary to remove traces of oxygen in the cell. The constant-current titration procedure described elsewhere[6] is useful for maintaining a controlled solution-potential level.

Mediator-Titrants. The choice of M-Ts is critical to successful measurements yet little is known about the chemistry of these redox couples and still less about their specific interaction with proteins. Table I gives a list of mediators that have been commonly used for redox potential determinations. This list is by no means comprehensive but is intended rather to illustrate characteristics we are aware of which may influence results.

[23] J. G. Lanese and G. S. Wilson, *J. Electrochem. Soc.* **119,** 1039 (1972).

[24] R. L. McCreery, R. Dreiling, and R. N. Adams, *Brain Res.* **73,** 23 (1974).

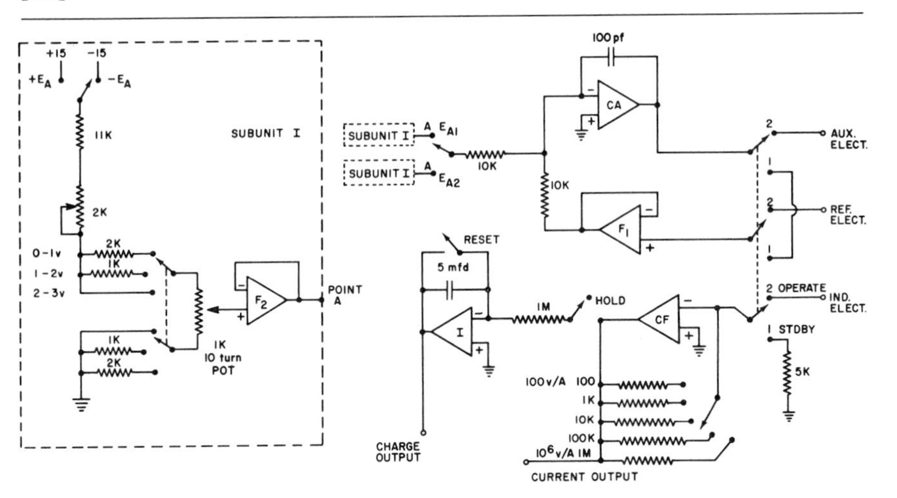

FIG. 3. Circuit diagram for potentiostat-coulometer. F_1 = follower (LF356H), F_2 = follower (LM741); CA = potentiostat [Burr-Brown 3582J (± 90 V P.S.)]; CF = current follower (LF356H); I = integrator (Burr-Brown 3293/14). All resistors 1%. Power supply and amplifier offset connections not shown.

Certain classes of compounds [naphthoquinones (especially ortho), aminoindophenols, and sulfhydryl] have in general been eliminated because of their instability or irreproducible potentials. We have also excluded some dyes where there is confusion about the correspondence of names and properties in the 1930s literature with that of presently available compounds.

Most of the M-Ts can be reduced by catalytic hydrogenation. The M-T is placed in an anaerobic vessel containing a gas dispersion tube or frit. A small amount of 5% platinized asbestos (J. T. Baker) is introduced, followed by bubbling with hydrogen. The reduced solution is then transferred after filtering out the asbestos to a microsyringe (see above) in which it can be stored for several days without oxidation if proper precautions are taken. It is often useful to estimate the formal potential of a biocomponent by using a series of graded potential reductants to establish when reduction occurs. Potentials can usually be estimated within ±100 mV so that further, more exact choices of mediator can be made. The following mediator characteristics might be noted:

A. Unstable in one or more oxidation states (pH 7). Fortunately, the M-Ts presented are usually sufficiently stable at physiological pH to permit measurements over a period of several hours. Mediators such as compounds 17 and 20 have been observed to be photochemically unstable and have influenced the course of photophosphorylation reactions.[25] The

[25] M. A. Cusanovich and M. D. Kamen, *Biochim. Biophys. Acta* **153,** 418 (1968).

TABLE I
CHARACTERISTICS OF MEDIATOR-TITRANTS

No.	Name	Em_7 (vs. NHE)	Comments (see text)	Reference
1.	$Mo(CN)_8^{-4}$	0.798	E	a
2.	Ferrocene 1,1’ dicarboxylic acid	0.644	A	a
3.	Ferrocenyl methyl trimethyl ammonium perchlorate	0.627	A	a
4.	Ferrocene 1,1′ dicarboxylic acid	0.530	A	a
5.	1,1′ Bishydroxymethyl ferrocene	0.465	A	a
6.	Ferrocene	0.422	D, Tween 20, A	b
7.	$K_3Fe(CN)_6$	0.408	$\mu = 0.1\ P_i$, C, E	c
8.	Hydroxyethyl ferrocene	0.402	A	a
9.	Ferrocene acetic acid	0.365	A	a
10.	1,1′ Dimethyl ferrocene	0.341	D, Tween 20, A	a
11.	*N,N,N′,N′* Tetramethyl *p*-phenylenediamine	0.276	A	e
12.	Disodium 2,6 dibromobenzeneone-indo-3′ carboxyphenol	0.250		d
13.	2,6 Dichlorophenol indophenol	0.217	B	f
14.	1-Naphthol 2-sulfonate indophenol	0.135		f
15.	Toluylene blue	0.115		f
16.	Fe EDTA	0.096	B	g
17.	Phenazine methosulfate	0.080	A	f
18.	Thionine (Lauth’s violet)	0.056		f
19.	Phenazine ethosulfate	0.055		f
20.	Methylene blue	0.011	A, D	f
21.	Fe $(oxalate)_3^{-3}$	0.005	0.15 *M* $K_2C_2O_4$, A, B	h
22.	5-*N*-methyl l-oxyphenazine (pyocyanine)	−0.034		f
23.	Indigotetrasulfonic acid	−0.046		f
24.	2,8 Dihydroxyphenoxazine (resorufin)	−0.051		f
25.	Indigodisulfonic acid	−0.125		f
26.	2-OH-1,4 Naphthoquinone	−0.152		f
27.	Anthraquinone 1,5 disulfonate	−0.174		f

28. Anthraquinone 2,6 disulfonate	−0.184	A	[f]
29. Flavin mononucleotide	−0.219	A	[f]
30. Anthraquinone 2 sulfonate	−0.225	C	[f]
31. Phenosafranine	−0.252	B	[f]
32. Safranine T	−0.289	B	[f]
33. Neutral red	−0.329		[f]
34. 1,1′ Dibenzyl 4,4′ bipyridylium dichloride (Benzyl viologen)	−0.358	B, F	[i]
35. 1,1′ Ethylene 2,2′ bipyridylium dichloride (Diquat)	−0.361	F	[i]
36. 1,1′ Bis(hydroxyethyl)4,4′ bipyridyl dihalide	−0.408	F	[a]
37. 1,1′ Dimethyl 4,4′ bipyridylium dichloride (methyl viologen)	−0.449	C, F	[i]
38. 1,1′ Propylene 2,2′ bipyridylium dibromide	−0.556	F	[i]
Titrants (Not Suitable as Mediators)			
39. Ascorbic acid	(0.058)	F	[f]
40. NADH		"oxidase" present	[f]
41. Sodium dithionite		B, F	[j]

[a] R. Szentrimay, P. Yeh, and T. Kuwana, *Am. Chem. Soc. Symp. Ser.* **38,** 143 (1977).
[b] P. Yeh and T. Kuwana, *J. Electrochem. Soc.* **123,** 1334 (1976).
[c] J. E. O'Reilly, *Biochim. Biophys. Acta* **292,** 509 (1973).
[d] M. A. Cusanovich, personal communication (1976).
[e] J. A. Friend and M. K. Roberts, *Aust. J. Chem.* **11,** 104 (1958).
[f] W. M. Clark, "Oxidation Reduction Potentials of Organic Systems." Williams & Wilkins, Baltimore, Maryland, 1960.
[g] I. M. Kolthoff and C. Auerbach, *J. Am. Chem. Soc.* **74,** 1452 (1952).
[h] W. B. Schaap, H. A. Laitinen, and J. C. Bailar, *J. Am. Chem. Soc.* **76,** 5868 (1954).
[i] E. Steckhan and T. Kuwana, *Ber. Bunsenges. Phys. Chem.* **78,** 253 (1974).
[j] W. H. Orme-Johnson and H. Beinert, *Anal. Biochem.* **32,** 425 (1969).

oxidized form of compound 6 (ferricinium ion)[26] or compound 11 (Wurster's Blue)[27] appear sufficiently stable to permit use over short periods of time, especially where the unstable state can be generated electrochemically at low concentrations.

B. Complicated chemistry at nonphysiological pH. A number of the M-Ts become increasingly unstable as the pH is raised. In most cases, the reactions and products are unknown, but base-catalyzed hydrolysis and condensation reactions are good possibilities. Potentiometric/spectral titration of the mediator alone is a good preliminary way to detect such anomalies. Compound 34[28] shows anomalous behavior in basic solution. Others showing some base instability are compounds 13, 31, and 32. The metal complexes (compounds 16 and 21) should not be used above about pH 8, and compound 41 will rapidly form colloidal sulfur below pH 5.5.

C. Interacts with other mediators. Dithionite is often used as a reductive titrant. Its oxidation product (sulfite) is quite reactive and can form adducts with quinones and other species. It is also known to interact with flavoproteins and inhibit enzymic activity.[29] Spectral evidence suggests interactions between compound 37 and ferrocyanide[28] as well as the reduced form of compound 37 and reduced compound 30.[30]

D. Limited solubility/detergent solubilized. Sometimes poor solubility leads to drifting potentials—behavior typified by thiazines such as compound 20 in which the oxidized form is considerably more soluble than the reduced. It has recently been demonstrated that mediators can be detergent solubilized to yield well-defined redox couples.[26] A series of ferrocenes (some detergent solubilized) ranging in potential from 0.34–0.64 V have recently been reported[31] which should extend M-Ts into a range where suitable systems are lacking.

E. Inhibits or affects biological activity. Inhibition of O_2 uptake activity by ferrocyanide in cytochrome oxidase is well known.[32] The presence of excess ligands such as EDTA in compound 16 can affect biological activity especially where metals are involved as co-factors.

F. Desirable spectral window. For reductions the viologens (compounds 34–38) have the advantage of being strong reductants in which the oxidized form (usually in excess) is practically transparent in the visible region. At low concentrations compounds 39 and 41 do not interfere in either oxidation state. The latter, however, exhibit very complicated redox chemistry.

[26] P. Yeh and T. Kuwana, *J. Electrochem. Soc.* **123,** 1334 (1976).

[27] R. N. Adams, "Electrochemistry at Solid Electrodes." Dekker, New York, 1969.

[28] A. H. Corwin, R. R. Arellano, and A. B. Chivvis, *Biochim. Biophys. Acta* **162,** 533 (1968).

[29] B. E. P. Swoboda and V. Massey, *J. Biol. Chem.* **241,** 3409 (1966).

[30] A. Nakahara and J. H. Wang, *J. Phys. Chem.* **67,** 496 (1963).

[31] R. Szentrimay, P. Yeh, and T. Kuwana, *Am. Chem. Soc. Symp. Ser.* **38,** 143 (1977).

[32] W. W. Wainio, *Oxidases Relat. Redox Syst., Proc. Symp., 1964* p. 622 (1965).

[23] Redox Potentiometry: Determination of Midpoint Potentials of Oxidation-Reduction Components of Biological Electron-Transfer Systems

By P. LESLIE DUTTON

Introduction

Oxidation-reduction (abbreviated to redox) potentiometry has become a routine technique in several laboratories interested in understanding the energetics of biological electron-transfer processes and the reactions coupled to electron transfer. The information it provides is an important part of the description of biological oxidation-reduction systems. It is as equally applicable to the components involved in the ultra-rapid primary events of photosynthesis as it is to respiratory systems and many of the enzymes involved in metabolism and drug detoxification. The basic practical approach is the same for all the different systems. With the appropriate simple equipment and when certain guidelines are followed the measurements can be as straightforward as pH determination.

Simple Theory and the Nature of the Information Obtained from Oxidation-Reduction Potentiometry

A redox component or "couple," A_{red}/A_{ox}, can be represented as undergoing the following reaction:

$$A_{reduced} = A_{oxidized} + ne^- \quad (1)$$

where n is the number of electrons (e^-) which leave the reduced component as it goes oxidized and are incorporated as it goes reduced. The n value is usually one or two, but it can be more. Such an equation represents only half the complete oxidation-reduction reaction; it is called a half cell. Another half cell is required to complete the reaction because electrons are not free entities identifiable in the chemical sense. The complete cell is:

$$A_{red} + B_{ox} = A_{ox} + B_{red} \quad (2)$$

In this case as written, B obviously has an n value that is the same as that of A. If B and A had different numbers of electrons to exchange per mole (i.e., different n values) then the ratios of A and B would be adjusted appropriately to account for the number of electrons exchanged. The equilibrium constant ($K_{eq} = [A_{ox}][B_{red}]/[A_{red}][B_{ox}]$) of Eq. (2) will depend on the individual affinities for electrons of the A and B couples. Thus if B has a higher affinity for electrons than A under standard conditions (unit activities) the reaction will tend to the right, and it would be understood that B is a stronger oxidizing agent or a weaker reducing agent than A. Since in this

ISBN 0-12-181954-X

case the standard free energy of the reaction (i.e., ΔG°, the free energy of the reaction maintained under standard, unit-activity conditions) will from the usual expression and sign convention, $\Delta G^\circ = -RT \ln K_{eq}$, be negative, K_{eq} will be expected to be > 1. The general expression to account for nonstandard conditions is:

$$\Delta G = \Delta G^\circ + RT \ln \frac{[A_{ox}]\,[B_{red}]}{[A_{red}]\,[B_{ox}]} \tag{3}$$

Common to all redox reactions between two half cells is electron transfer. Because of this we can avoid having to list equilibrium constants or ΔG° of the reactions between innumerable redox couples, and instead we can refer the electron affinities of individual redox half cells to one half cell which is chosen as a standard. This is usually the standard hydrogen half cell which is usually written as $\frac{1}{2} H_2 = H^+ + e^-$. This half cell is made up of hydrogen gas at 1 atm in equilibrium with a solution of H^+ of unit activity (i.e., pH = 0). Because of the nature of the measurement and the reactions, it is appropriate to use electrical potential units throughout. This is achieved by using the equation $\Delta G = -nF\ \Delta E$ where ΔE is the redox potential difference between two half cells and F is the Faraday which is the chemical to electrical potential "conversion factor" (i.e., F = 96,493 C/chemical equivalent, or, since 1 J = 1 C/V, F = 96,493 J/V equivalent). The negative sign on the right of the equation is a matter of convention; that is, a positive redox potential change represents a low to high redox potential change for which, therefore, ΔG is negative. Once again n is the number of electrons transferred in the reaction; an n value of 2 provides twice as much $-\Delta G$ per mole as an n = 1 reaction. Thus substituting $\Delta G = -n\ \Delta E\ F$ in Eq. (3):

$$\Delta E = \Delta E_0 - \frac{RT}{nF} \ln \frac{[A_{ox}]\,[B_{red}]}{[A_{red}]\,[B_{ox}]}$$

and rearranging

$$\Delta E = \Delta E_0 + \frac{RT}{nF} \ln \frac{[B_{ox}]}{[B_{red}]} - \frac{RT}{nF} \ln \frac{[A_{ox}]}{[A_{red}]}$$

or

$$E_{(B)} - E_{(A)} = E_{0(B)} - E_{0(A)} + \frac{RT}{nF} \ln \frac{[B_{ox}]}{[B_{red}]} - \frac{RT}{nF} \ln \frac{[A_{ox}]}{[A_{red}]} \tag{4}$$

Since there is no absolute value for redox potentials, the chosen standard hydrogen half cell E_0 is given the value of 0 V at any temperature. If we identify A as the standard hydrogen half cell, the differences of Eq. (4) disappear (i.e., $[A_{ox}] = [A_{red}]$ and therefore $RT/nF \ln [A_{ox}/[A_{red}] = 0$; $E_{(A)} = E_{0(A)}$ which for the standard hydrogen half cell = 0 by definition), and it is accepted henceforth that the redox potential of the redox couple or half cell described here by B is referred to the hydrogen electrode. The expression becomes the well-known Nernst equation:

$$E_h = E_0 + \frac{RT}{nF} \ln \frac{[B_{ox}]}{[B_{red}]} \quad (5)$$

To denote that the reference half cell is the standard hydrogen half cell the subscript h is added to the redox potential form. E_0 is the standard redox potential of the B couple (i.e., for the reaction $B_{red} = B_{ox} + ne^-$, E_0 is that E_h at which $[B_{ox}] = [B_{red}]$ with $[B_{ox}]$ and $[B_{red}]$ maintained at unit activities at pH = 0). In general for work done at ~29°C the RT/F term together with a change to $\log_{10}$ simplifies the expression to:

$$E_h = E_0 + \frac{0.06}{n} \log \frac{[\text{ox}]}{[\text{red}]} \quad (6)$$

At 24°C the $2.303\,RT/F$ becomes very nearly 0.059, at 10°C it is 0.058, and at 10°C it is approximately 0.056.

In biology several liberties have to be taken with standard conditions. The activities of components in membranes are unknown, and even using concentrations it is not possible to approach standard conditions either in concentration, or if the H^+ is involved, in pH. The E_0, if it was measurable directly, would be a value obtained using concentrations much less than standard activities or concentrations; second, determinations at pH zero are not often very profitable with biological materials, and in most cases there is insufficient data to make an extrapolation back to pH zero from higher more physiological values. The symbol introduced by Michaelis (see Clark[1]) for "standard" half reduction or midpoint potential point under conditions other than pH zero and under conditions less than unit activities is E_m (pH = x) or E_{mx} where m stands for midpoint and x is the ambient pH of the determination. E_0', the original symbol of Clark,[1] still appears in textbooks where it signifies the midpoint at physiological pH values; however, since certain determinations may not be done exactly at pH 7.0, the other symbol, also later adopted by Clark, is better.

Figure 1 shows a typical Nernst curve for $n = 1$ and $n = 2$ plotted as the state of oxidation-reduction against E_h. These are the two parameters measured in redox potentiometry as will be described later. At the half reduction point, where [ox] = [red] (i.e., [ox]/[red] = 1), the $E_h = E_m$. On the right in Fig. 1 the data are plotted as a semilog plot to yield a straight line.

Obviously the proton in redox reactions cannot be ignored. The proton

[1] Clark, W. M. (1960). "Oxidation-Reduction Potentials of Organic Systems." Waverly Press, Baltimore, Maryland.

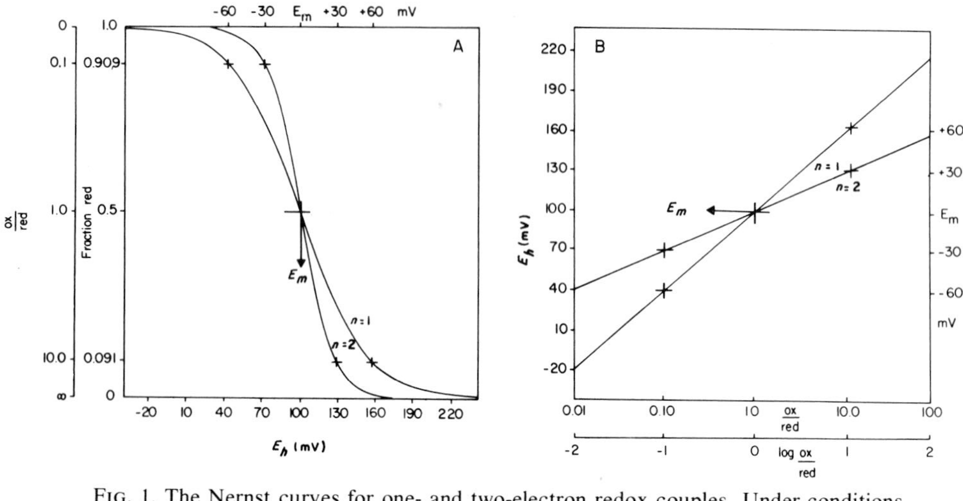

FIG. 1. The Nernst curves for one- and two-electron redox couples. Under conditions when the measured extent of reduction is 50% the redox couple is at its midpoint. In these examples, the $n = 1$ or 2 redox component is given at E_m of 100 mV. The same Nernst curves are plotted in (B) on a logarithmic scale. The slope of the straight line depends on the n value and the temperature (see text); at 29.25°C, the slope of an $n = 1$ or 2 line is 60 mV or 30 mV, respectively, per factor of 10 change in [ox]/[red].

is an integral part of the physical chemistry of many redox reactions. The underlying reason is that the oxidized form and the reduced form of a redox couple may be chemically very different as far as proton affinity is concerned. Thus, with one or two odd exceptions, the oxidized form of a couple is a stronger acid than the conjugate reduced form. In other words, from the Henderson-Hasselbalch expression (pH = pK + log [base]/[acid]) which is the proton analog of Eqs. (3) and (6), the pK (i.e., the pH when [base]=[acid]) is lower for the oxidized form than it is for the conjugate reduced form. This is intuitively easy to appreciate since the reduced form possesses n negative charges more than the oxidized form. The Nernst expression for the redox reaction in which the ambient pH is *in between* widely separate values for pK_{ox} and pK_{red} can be derived as follows.

We shall use a simple $n = 1$ redox reaction as an example:

$$\text{red H} = \text{ox} + e^- + \text{H}^+ \tag{7}$$

$$E_h = E_{m0} + 0.06 \log \frac{[\text{ox}][\text{H}^+]}{[\text{red H}]}$$

$$E_h = E_{m0} + 0.06 \log \frac{[\text{ox}]}{[\text{red H}]} + 0.06 \log [\text{H}^+]$$

$$E_h = E_{m0} + 0.06 \log \frac{[\text{ox}]}{[\text{red H}]} - 0.06 \text{ pH} \tag{8}$$

This expression tells us that the Nernst curve as a whole and hence its E_m will shift -0.06 V every pH unit increase as shown in Fig. 2. The hydrogen half cell adheres to this behavior; this is included in Fig. 2B. In the figure we have given the example of Eq. (7) an $E_{m7.0}$ of 0.10 V or 100 mV, and with each unit change in pH the E_{mx} is different by 0.06 V. In such cases in which the pH range studied encounters no pKs for ox or red, E_m at a specified pH, (E_{mx}) permits calculation of other E_m values at different pH values according to

$$E_{mx} = E_{my} - 0.06\ (\text{pH}\ x - \text{pH}\ y) \tag{9a}$$

If this condition extends to pH zero as in Fig. 2 then E_{mx} values could be related to the E_{m0} (E_0) of the half cell

$$E_{mx} = E_{m0} - 0.06\ \text{pH} \tag{9b}$$

So far we have dealt with redox reactions in which there is a net release and binding of a proton during oxidation-reduction. For most redox couples this proton exchange would be limited at lower pH values by a pK on the oxidized form (i.e., ox^+ H/ox) and at higher pH values by the pK on the reduced form (i.e., red H/red). In fact an increasing number of redox couples of biological interest are being revealed to have pKs in the physiological range and hence cannot be ignored. The pKs are determined by performing redox titrations to determine E_m values at different pH values. Thus extending the simple redox reaction of Eq. (7), and its uninterrupted E_m dependency of -0.06 V/pH unit, Fig. 3A shows experimental data which reveals pK_{ox} and pK_{red} values; the pKs are shown more clearly in the E_m/pH plot in Fig. 3B. The pKs are revealed because above and below the

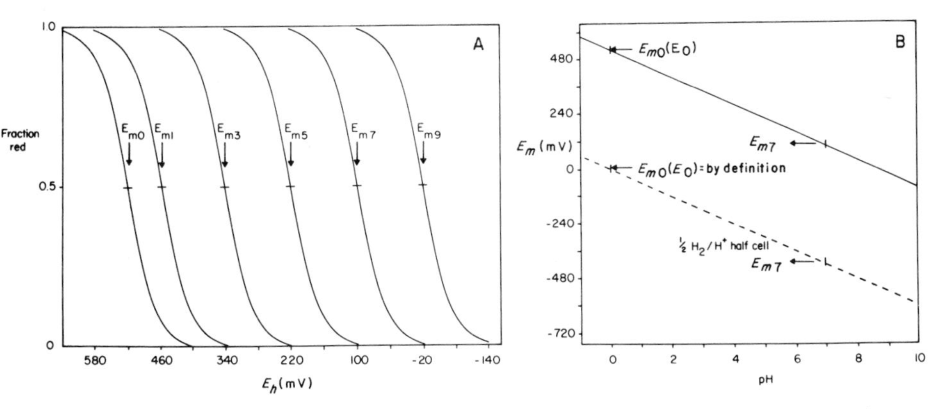

FIG. 2. The E_m/pH relationship of a redox couple. The Nernst curve ($n = 1$) of (A) of a redox couple (red H = ox + e^- + H^+) shifts upward 60 mV for every unit decrease in pH. (B) shows the E_m values (E_{m7} = 100 mV) changing as a function of pH. A hydrogen half cell is shown for comparison.

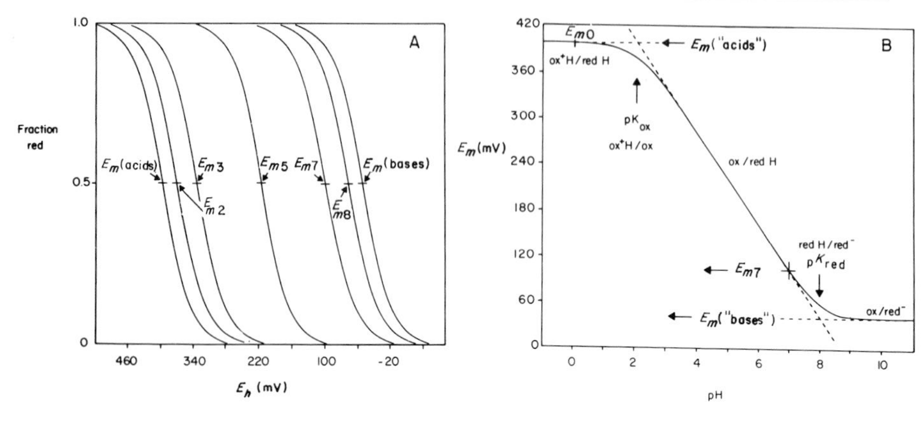

FIG. 3. The E_m/pH relationship when the redox couple has a pK on its oxidized form at 2 and a pK on its reduced form at 8. (A) shows what deviation one would observe experimentally as pH varied; a deviation from the 60 mV/pH unit dependence as shown in Fig. 2B, until the E_ms became totally pH-independent. (B) plots E_m vs. pH as in Fig. 2B. Three regions are shown: (pH $<$ pK_{ox}), where E_m approaches pH independence; (p$K_{ox} <$ pH $<$ pK_{red}), where E_m varies approximately 60 mV/pH unit; and pH $>$ pK_{red}, where E_m again becomes pH-independent; the predominating redox species are indicated in each region; dashed lines extrapolate to the respective pKs.

pKs the reaction in Eq. (7) becomes insignificant and there is no net proton exchange and the E_m value loses its dependency on the ambient pH. In this simple case there are three distinct regions where different redox reactions predominate. At pH values $\gg$pK_{ox} and pK_{red} the redox reaction is:

$$\text{red}^- = \text{ox} + e^- \tag{10a}$$

At pH values between pK_{ox} and pK_{red} the predominant reaction is as described for the reaction in Eq. (7), and at pH values $\ll$pK_{red} and pK_{ox} the reaction is

$$\text{red H} = \text{ox}^+\text{ H} + e^- \tag{10b}$$

The detailed description of the E_m/pH relationship of Fig. 3 can be depicted and derived as follows:

$$\begin{array}{ccc} \text{red H} & \xleftrightarrow{\quad E_{m(\text{acids})} \quad} & \text{ox}^+\text{ H} + e^- \\ K_{\text{red}} \updownarrow & & \updownarrow K_{\text{ox}} \\ \text{red}^- + \text{H}^+ & \xleftrightarrow[\quad E_{m(\text{bases})} \quad]{} & \text{ox} + e^- + \text{H}^+ \end{array}$$

We shall define K_{ox} and K_{red} as dissociation constants:

$$K_{\text{ox}} = \frac{[\text{ox}]\,[\text{H}^+]}{[\text{ox}^+\,\text{H}]} \qquad \text{and} \qquad K_{\text{red}} = \frac{[\text{red}^-]\,[\text{H}^+]}{[\text{red H}]} \tag{11}$$

In this system as a whole, the E_{mx} is measured at that E_h where the *total* oxidized species equals the total reduced species of the redox couple:

$$E_h = E_{mx} + 0.06 \log \frac{[\text{ox}] + [\text{ox}^+\,\text{H}]}{[\text{red}^-] + [\text{red H}]} \tag{12}$$

Substituting for [red$^-$] and [ox] from Eqs. (10) and (11):

$$E_h = E_{mx} + 0.06 \log \frac{[\text{ox}^+\,\text{H}]\,K_{\text{ox}}/[\text{H}^+] + [\text{ox}^+\,\text{H}]}{[\text{red H}]\,K_{\text{red}}/[\text{H}^+] + [\text{red H}]}$$

$$E_h = E_{mx} + 0.06 \log \frac{[\text{ox}^+\,\text{H}]\,(1 + K_{\text{ox}}/[\text{H}^+])}{[\text{red H}]\,(1 + K_{\text{red}}/[\text{H}^+])}$$

$$E_h = E_{mx} + 0.06 \text{ by } \frac{[\text{ox}^+\,\text{H}]}{[\text{red H}]} + 0.06 \log \frac{(1 + K_{\text{ox}}/[\text{H}^+])}{(1 + K_{\text{red}}/[\text{H}^+])}$$

Since, at pH values low enough to essentially protonate both red and ox (i.e., pH $\ll$ pK_{red} and pK_{ox} or $[\text{H}^+] \gg K_{\text{red}}$ and K_{ox}):

$$0.06 \log \frac{[\text{ox}^+\,\text{H}]}{[\text{red H}]} = E_h - E_{m(\text{acids})}$$

Therefore,

$$E_{mx} = E_{m(\text{acids})} - 0.06 \log \frac{(1 + K_{\text{ox}}/[H^+])}{(1 + K_{\text{red}}/[\text{H}^+])} \tag{13}$$

A different view of the same relationship can be derived from the $E_{m(\text{bases})}$ standpoint by substitution in Eq. (12) for [red H] and [ox$^+$ H] of Eqs. (10) and (11). This is:

$$E_{mx} = E_{m(\text{bases})} - 0.06 \log \frac{(1 + [\text{H}^+]/K_{\text{ox}})}{(1 + [\text{H}^+]/K_{\text{red}})} \tag{14}$$

Equations (13) and (14) as written are for the proton. Substituting "L" as any ligand to the redox couple makes the equation general for the differential binding of the oxidized and the conjugate reduced form to a ligand. L can be anything from a membrane phospholipid or protein to a small chemical group. If both [H$^+$] and [L] are interacting with the same redox couple then another term for the binding characteristics of [L] similar to that for the proton is added onto Eqs. (13) and (14). Equations for the measured E_m for redox couples with multiple n values and the binding of multiple protons and ligands can also be derived using the above approach.

There are numerous examples listed in the book by Clark.[1] Information from biological sources is also increasing regarding pKs and ligand binding constants (see, for example, Dutton *et al.*[2-7]).

Parameters Measured in Redox Potentiometry

The experimental goal is to measure the E_h and a corresponding state of oxidation or reduction of a redox couple: E_h is measured by electrodes, while the state of oxidation-reduction of a redox component is measured by some physical technique, usually some form of spectrometry (spectrophotometry of all kinds such as absorbance, fluorescence, phosphorescence, Raman; and magnetic approaches such as EPR, NMR, magnetic susceptibility, etc.). The character of the biological redox component usually dictates the chosen technique; the character also will dictate whether the oxidized or reduced form is measured. Several such correlations of E_h and state of oxidation or reduction should be taken at E_h values that extend over the range that encompasses the central part of the expected Nernst curve. To make the analysis straightforward it is preferable to obtain readings for the state of oxidation-reduction as nearly fully oxidized and reduced as possible (i.e., $E_h = E_m \pm 120$ mV). The information here will describe the Nernst curve of the redox couple and provide the E_{mx} value and the n value.

The Choice of Oxidant and Reductant to Change E_h

The E_h and the redox state of the redox system are changed routinely by addition of solutions of freshly prepared, well-buffered sodium dithionite as a reductant and potassium ferricyanide as an oxidant. Obviously other oxidants and reductants can be used but dithionite and ferricyanide react promptly and provide a large E_h range which usually goes well beyond the scope of the most experiments. Thus it is an easy matter to obtain near 100% reduced and oxidized values for the redox state assay. Sodium dithionite can take the E_h down to the hydrogen half cell potentials (i.e., E_m at pH 7.0, -420 mV) where water is seen to be reduced into hydrogen.

[2] Dutton, P. L., and Wilson, D. F. (1974). *Biochim. Biophys. Acta* **346,** 165–212.
[3] Dutton, P. L., and Prince, R. C. (1978). "The Photosynthetic Bacteria" (R. K. Clayton and W. R. Sistrom, eds.). Plenum Press, New York, in press.
[4] Prince, R. C., and Dutton, P. L. (1976). *FEBS Lett.* **65,** 117–119.
[5] Petty, K. M., and Dutton, P. L. (1976). *Arch. Biochem. Biophys.* **172,** 346–353.
[6] Lindsay, J. G., Owen, C. S., and Wilson, D. F. (1975). *Arch. Biochem. Biophys.* **169,** 492–505.
[7] Wilson, D. F., Lindsay, J. G., and Brocklehurst, E. S. (1972). *Biochim. Biophys. Acta* **256,** 277–286.

Care has to be exercised with dithionite since its oxidation products are strongly acidic. Dithionite solutions should be freshly prepared and well buffered (e.g., 0.5 *M*). There is no need to make up solutions of specified concentration. We tend to make up solutions strong enough so that small but controllable (microliter) volumes can be added to the redox system to elicit the desired E_h change. Dithionite solutions are more stable in the air at higher pH values, and keeping the solution cold and in a narrow tube also increases the time that it can be used. A "typical" time to give a dithionite solution would be 20–30 min, but this depends on the nature of the system and the E_h range being studied. If long-term stability is required, the solution can be kept in an anaerobic vial fitted with a septum.

On the oxidizing side, potassium ferricyanide can take the E_h up to a stable 530 mV. Typically, microliter additions of 100 m*M* or 10 m*M* solutions would be used to change the E_h. Higher E_h values with the ferri/ferrocyanide couple can be obtained if a little (<1 μM) cytochrome *c* peroxidase is added, followed by small additions of H_2O_2; the enzyme has a very low K_m for ferricyanide, which on oxidation to ferricyanide can take the E_h up to over 600 mV.

Practical Aspects of Correlating E_h and Redox State

The physical methods chosen to measure the fraction of a redox couple reduced or oxidized at a measured E_h value are usually dictated by the character of the redox couple. In many cases, E_h and the state of oxidation-reduction can be measured simultaneously and continuously. This is especially the case for room-temperature spectrophotometric determinations or if the couple displays a change in EPR signal or any other physically determinable parameter at room temperature. In certain cases, however, it is necessary to assay the state of reduction of a component at low temperature. In these cases biological material in which the redox constituents are poised at a known E_h value at room temperature can be removed, frozen to liquid nitrogen temperature, and assayed later.

Figure 4 shows basic equipment for redox potentiometry.[8] Figure 4A illustrates an anaerobic vessel for simultaneous measurement of redox potential and the state of reduction by optical means. This is suitable for a dual wavelength spectrophotometer. The light path is typically 1 cm, but cuvettes with light paths of as little as 2 mm are successful if the stirrer is turning directly over the slot at the top of the cuvette (see Fig. 4A). Stirring is as necessary in redox potentiometry as it is in measuring pH. It is convenient to have the magnetic bar suspended and rotated on a flat part of

[8] Dutton, P. L. (1971). *Biochim. Biophys. Acta* **226,** 63–81.

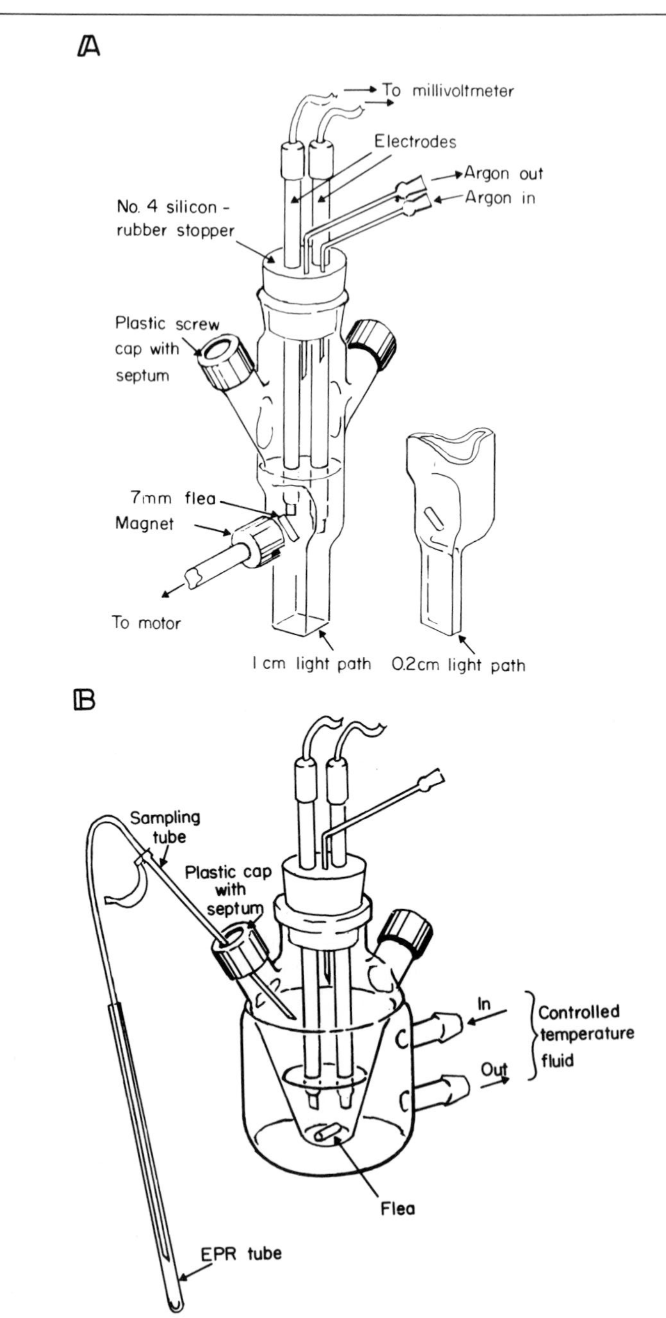

FIG. 4. (A) Spectrophotometric cuvettes for redox potentiometric titrations at ambient temperatures. (B) A vessel for redox potentiometric titration in which samples are taken anaerobically and frozen for analysis at low temperatures.

the vessel by a magnetic stirrer. Anaerobic conditions are required because oxygen is a strong oxidizing agent; by "anaerobic" we mean that gas (argon; for example Airco grade 5;99.999% minimum purity) with very low O_2 levels is continuously flushed over the biological suspension. The gas train need not be elaborate for normal work with mitochondria or photosynthetic material; all that is needed is a bubbler between the gas tank and the anaerobic redox vessel (this is to see if the gas is flowing and to "wet" the gas so that it does not evaporate the biological suspension). A bubbler for the gas exhaust is used to create a slight back pressure in the system and as an indicator of leaks. We usually use stainless-steel or copper tubing to carry the gas to the anaerobic cuvette. The stopper is made of silicone rubber. The side arms are capped with septums. Two side arms are not mandatory, but for further probes such as thermocouples or pH electrodes a second one can be useful.

Figure 4B shows a vessel for measuring the E_h of biological suspensions which is designed so that samples of known E_h can be transferred anaerobically to other vessels in which the redox state of the biological sample is to be measured. This approach is necessary for sample transfer into cuvettes for low-temperature spectrophotometry (see Dutton[8] for a design) or into EPR tubes or other vessels which can be frozen for assay. Figure 4B shows how the receiving vessel (in this case an EPR tube) can be flushed free of air and the transfer can be effected. The gas supply enters the redox vessel via a needle in the stopper but comes out through a tube inserted through the septum of one side arm which is directed into the receiving vessel. At the sampling point, after the receiving vessel is flushed free of air, the stainless-steel needle or tube is dipped into the biological solution or suspension. Since there is now no exit for the incoming gas, a pressure builds up (extent governed by gas tank regulator) and pushes the suspension down the capillary and into the EPR tube. When sufficient sample has been taken the sampling tube is removed from the biological suspension and the EPR tube is withdrawn and promptly frozen for EPR analysis.

Practical Notes About Electrodes

a. The Measuring Electrode. This is usually a shiny strip of platinum fused into a glass rod. Other metals such as gold also can be used. The metal is in direct contact with the aqueous solution containing the redox system under assay.

b. The Reference Electrode. So far we have talked about the standard hydrogen half cell or "electrode." In practical terms this half cell is cumbersome. Other more convenient "secondary" standard half cells are available and of these the standard calomel or the silver/silver chloride half cells or electrodes are most commonly used. They are calibrated against the standard hydrogen electrodes and so the E_h of a redox system can still be worked out by adding the potential of the secondary standard to the measured redox potential. Table I shows some standard potentials of the calomel and silver/silver chloride electrodes, taken from references 9 and 10, showing their temperature dependence relative to the standard hydrogen half cell which is deemed zero at all temperatures. For further information about the secondary and the hydrogen electrode, see Hills *et al.*[9-11] These reference electrodes of the calomel or silver/silver chloride kind are made up of a metal in contact with an insoluble salt in turn in contact with a solution containing the chloride. Contact with the solution containing the redox system under assay is made via a salt bridge (made of either KCl-agar or a porous plug). Figure 4 shows a typical arrangement of commercial electrodes in redox measuring equipment. We usually use a Radiometer K401 saturated calomel standard reference calomel electrode and Radiometer P101 platinum measuring electrode.

c. Routine Care of the Electrodes. The saturated KCl calomel electrode should be kept topped up with saturated KCl and when not in use maintained dipped in a similar KCl solution. The porous plug at the bottom should be kept free of precipitated protein and other organic matter, which can be removed by gently rubbing the end of the electrode on a flat sheet of extra-fine sandpaper. The platinum electrode should be kept shiny with occasional rubbing with fine abrasive. A new electrode system should be checked for proper function against other standard half cells.

Redox Potential Measurement of Biological Redox Systems: The Requirement for Redox Mediators

Unlike many small inorganic and organic redox couples, biological redox molecules and complexes, even when isolated from the membrane, are inert as far as the measuring electrodes are concerned. A predominant reason for this is that the redox center is often shielded by protein and so does not gain proper contact with the electrode surface. Redox mediators are required to act as go-betweens between the measuring electrode and the biological redox couple. Redox mediators are small organic or inorganic

[9] Hills, G. J., and Ives, D. J. G. (1961). *In* "Reference Electrodes, Theory and Practice" (D. J. G. Ives and G. J. Janz, eds.), pp. 127–178. Academic Press, New York.

[10] Janz, G. J. (1961). *In* "Reference Electrodes, Theory and Practice" (D. J. G. Ives and G. J. Janz, eds.), pp. 179–230. Academic Press, New York.

[11] Hills, G. J., and Ives, D. J. G. (1961). *In* "Reference Electrodes, Theory and Practice (D. J. G. Ives and G. J. Janz, eds.), pp. 71–127. Academic Press, New York.

TABLE I
SOME STANDARD POTENTIALS OF THE CALOMEL AND SILVER/SILVER CHLORIDE ELECTRODES AT DIFFERENT TEMPERATURES

Temperature	Calomel electrode (V) Saturated KCl	1 *N* KCl	0.1 *N* KCl	Ag/AgCl electrode (V)
10	0.2541	0.2839	0.3343	0.2314
20	0.2477	0.2815	0.3340	0.2256
25	0.2444	0.2801	0.3337	0.2224
30	0.2411	0.2786	0.3332	0.2191
Electrode reaction:	Pt,H_2 | H^+ unit activity | KCl solution | $Hg_2Cl_{2(S)}$ | Hg			Pt,H_2 | H^+Cl^- unit activity | $AgCl_{(S)}$ | Ag

redox agents; some are listed in Table II. There are several criteria that mediators must meet before redox potentiometry can be usefully employed:

a. The Redox Mediator Must React Effectively and Reversibly with the Electrode and the Biological Redox Component. In the presence of a mediator, addition of oxidant or reductant to a biological suspension should, in a matter of seconds, elicit both a change of E_h registered at the electrode, and a corresponding change in the state of oxidation-reduction of the biological redox component. Simple as this may seem, for a variety of

TABLE II
SOME REDOX MEDIATORS

Component	Approx. E_{m7} (mV)	*n* value
Potassium ferro/ferricyanide	430	1
2,3,5,6,-Tetramethyl phenylenediamine ("diaminodurol" or DAD)	260	2
N,N,N',N'-Tetramethyl phenylenediamine (TMPD)	260	1
N-Methyl phenazonium methosulfate (PMS)	80	2
N-Ethyl phenazonium ethosulfate (PES)	55	2
N-Methyl-1-hydroxyphenazonium methosulfate (pyocyanine)	− 34	2
2-Hydroxyl-1,4-naphthaquinone	−145	2
Anthraquinone-26-disulfonate	−185	2
Anthraquinone-2-sulfonate	−225	2
N,N'-Dibenzyl-4,4-bipyridinium dichloride (benzyl viologen)	−311	1
N,N'-Dimethyl-4,4,-bipyridinium dichloride (methyl viologen)	−430	1

reasons many redox mediators fail in one or both of these requirements when used in the concentration range (10^{-6}–10^{-3} M) that is useful for biological experimentation.

Failure of a redox couple to react effectively with the electrode is not confined to membrane- or protein-bound biological couples. For example, many organic redox couples that are candidates to serve as mediators are at equilibrium "two-electron" redox couples and have n values at or approaching 2.0. One rule of thumb regarding the strength of reaction with the electrode rests on the "stability" of the intermediate one-electron redox state or "semiquinone" analogue, which, being a free radical, interacts well with the electrode surface (see Chapter 7 in Clark[1] for a discussion of semiquinone analogues). Those $n = 2$ redox agents for which the semiquinone analogue is very "unstable" (e.g., duroquinone) are only weakly felt by the electrode in the 10^{-5}–10^{-4} M concentration range. Those agents which have a relatively stable (e.g., the phenazines) or fully stable "semiquinone" (i.e., are visibly $n = 1$ redox agents such as tetramethyl phenylenediamine, methyl and benzyl viologen) are much more active and are effective in the 10^{-6}–10^{-5} M concentration range. Nevertheless, the effectiveness of relatively weakly interacting mediators can be improved by increasing their concentration.

Redox mediators should be able to act not only in the aqueous phase of biological suspensions to make contact with the electrode, but should also be capable of transferring oxidizing and reducing equivalents into and through the membrane to make contact with membrane-bound redox components. This means that a redox mediator should not be so hydrophobic that it is partitioned almost entirely into the membrane, leaving insufficient mediator in the external aqueous medium to react effectively with the platinum electrode. Several quinones fall into this category. At the other extreme, polyionic mediators may not be hydrophobic enough. For example, potassium ferri/ferrocyanide is not a good membrane penetrant, and although a strong redox contact is made with the platinum electrode at low concentrations of the redox couple, direct interaction with membrane-bound biological redox components is slow and the direct contact weak. Again, the problems encountered with redox mediators that react only weakly with the biological redox components, for whatever the reason, can in many cases be overcome by simply increasing the concentration of the redox mediator if this does not adversely affect the experiment. However, a subtler way of dealing with the unfavorable partition of redox mediators between membrane and water is to use combinations of hydrophilic and hydrophobic redox mediators.

The rates of the electrode-mediator–biological component reaction must be rapid enough to achieve a true equilibrium (i.e., one where all redox complexes in the biological electron-transfer system are at the same E_h) and not a quasi-equilibrium that is established with other redox events that may tend to shift the system out of true equilibrium. Such a shift can arise from the presence of oxygen in the system, which can vigorously interact with mediators and individual components of the biological electron-transfer system at widely differing rates. Alternatively or additionally, the presence of an active "endogenous reducing system" (i.e., the unspecified flux of reducing equivalents from stored substrates into the electron-transfer system) may also serve to act differentially with the mediators and the individual biological redox components. If the rates involving oxidation or reduction in these two examples are faster than the interaction with the added redox mediators, then the E_h measured at the electrode will diverge from the proper level of reduction exhibited by the various biological redox components. Again the problem can be combatted by increasing the concentration of the redox mediators, together with the use of multiple mediators to hold or buffer the ambient E_h. However, use of the anaerobic conditions eliminates the oxygen problem, and, if there is a choice, the use of a biological material that does not have a high endogenous reducing system avoids the other problem; rat liver mitochondria are especially difficult in this respect, while pigeon heart mitochondria are low enough in endogenous reductant that this is of no deleterious significance; photosynthetic systems are also generally not adversely affected by an endogenous reducing system.

b. Redox Mediators Must Not Decompose Immediately on Oxidation or Reduction. Although this may seem an obvious point, many useful organic dyes used as mediators in redox potentiometry do not have an indefinite lifetime and, under certain conditions of pH or at E_h values that are far away from their midpoint, may undergo additional, irreversible oxidation or reduction reactions. Measuring beam light may also lead to the decomposition of certain redox mediators. Little systematic work has been done in this area, and our current knowledge is based on personal observations made during the course of many redox titrations on mitochondria and on photosynthetic bacterial chromatophores. It should be noted that some irreversible reactions may be catalyzed by the organelles rather than being intrinsic to the mediators themselves. Although this is not normally a major concern, the experimenter should be aware of this possibility and take steps to avoid the problem. The useful lifetime of a mediator can be tested under experimental conditions for periods much longer than the experimental time by checking for the promptness of the response of a redox component to a change in E_h. Correction for mediator decomposition can be made by adding fresh mediator(s) at the appropriate time. Alternatively the experiment can be performed in segments using new material and mediators for each segment.

c. The Mediator Must Not Chemically Modify the Biological Redox Component. If the mediator acts as a substrate to a redox enzyme, or chemically interacts with one form of the redox component (directly or through other agents), then the E_m of the component being measured will be modified [i.e., the redox mediator becomes a ligand to the redox component as discussed in Eqs. (13) and (14)]. An example of this has been revealed (B. Cohen and D. F. Wilson, personal communication) in the reaction of methyl or benzyl viologen or other analogues with the detoxification enzyme cytochrome P-450. This problem can be revealed by performing redox titrations with chemically distinct redox mediators and by drastically varying the redox mediator concentration.

d. The Mediator Should Not Interfere with the Accompanying Measurements of the State of Reduction of the Redox Couple Under Assay. Clearly, for example, if the state of reduction of a component is being measured optically, then a redox mediator itself undergoing optical changes concurrent with those of the agent being measured adds complications to the measurements. Similarly if the oxidized or reduced form of a redox component is being measured by a free radical EPR signal, complications arise from use of redox mediators which have similar signals in one or another of their redox states. It is an unfortunate fact that the best mediators are colored and undergo large absorbance changes on oxidation or reduction. They also often display significant free radicals in one of their oxidation states.

This problem can be overcome by performing "control" redox titrations without the biological material to check for mediator interference. Caution should be exercised, however, in recognizing that a mediator may behave differently in the presence or absence of a biological sample (e.g., different "semiquinone" stability and hence different levels of free-radical formation). Redox mediator contribution *in vivo* can be deduced by repeating the titration with increasing amounts of redox mediators to establish the systematic contributions.

General Rules about Reliability, Redox Mediators, and Redox Titrations

As much information as possible should be obtained in order to increase confidence that titrations have been successfully performed at equilibrium. With the considerations outlined in the preceding section it is a common practice to:

1. Perform redox titrations in both oxidative and reductive sequences. Identical results should be obtained; hysteresis implies disequilibrium.

2. The redox titration should be repeated with as wide a variation in concentration of mediators as possible in the 10^{-6}–10^{-3} M range. Use of different chemical structures of redox mediators and various combinations of different redox mediators adds confidence to results.

3. Ideally redox mediators should have E_m values that are similar (<60 mV difference) to the biological redox couple being measured. Reversibility becomes increasingly strained as the E_h is moved away from the E_m and the [ox]/[red] ratio becomes too large or too small. For example, in titrating one biological redox couple it is best, although not always possible, to have three mediators of differing E_ms so that together they more than cover the central part of the Nernst curve of the biological redox couple. If a redox titration is performed over a wide redox potential range, a string of redox mediators are used. Such extended redox titrations are common in exploratory work, and for the titration of certain respiratory or photosynthetic redox components which are present in electron-transfer systems as multiple electrochemical species of widely differing E_m values but similar spectrometric properties.

Redox Mediators and the Different Kinds of Poise They Establish with Biological Redox Systems

Redox "poise" is an index of the buffering capacity afforded by a redox mediator in the same way that pH buffers operate with acid–base change. Hence, the use of high concentrations of redox mediators will lead to a resistance to E_h change on the addition of oxidants or reductants such as ferricyanide or dithionite. As is well known for the pKs of pH buffers, the E_h region of greatest resistance is near to the E_m value of the redox mediator. In performing a redox titration over a wide range with a string of different redox mediators the extent of E_h change for the same addition of oxidant or reductant will be different depending on how close the E_h value is to the E_m values of the mediators, and on how much of the mediator is present. It is a matter of experience to know how much oxidant or reductant to add to elicit the required change in E_h at different E_h values.

The widely different rates with which redox mediators can react with biological electron-transfer components provides a means of obtaining both equilibrium thermodynamic information as well as pulsed activated kinetic data which is an expression of the equilibrium redox state before activation. This approach is applicable to photosynthetic systems but could be applied to the study of certain reactions in mitochondria, hemoglobin, and other respiratory systems. However, it is especially applicable to photosynthetic bacterial chromatophores, since their electron (proton) transfer system is cyclical and therefore requires no net input of reductant or oxidant; in this case an E_h can be established, and the chromatophores activated by one or more short single-turnover saturating flashes to induce a state of disequilibrium. The time course of the perturbation and relaxation can then be followed using rapid spectrometric techniques (see Dutton and Wilson[2]).

In this "poise and pulse" approach, several variants of the character of the redox poise are possible; the kind chosen depends on the material being studied. Two examples will serve to illustrate.

1. An amount of redox mediator is added to biological suspensions sufficient to provide a reliable equilibrium E_h reading, but insufficient to interfere with *in vivo* kinetics after pulsed activation.

With photosynthetic systems, particularly bacterial, redox potentiometry has been used in combination with kinetic analysis of flash-induced reactions from 10^{-12} sec to the seconds time range. Clearly the slower the time range studied, the more likely that redox mediator interference will be encountered, so the experimental conditions required to achieve equilibrium before activation and yet generate no kinetic contributions after activation become more stringent. In this kind of experiment the amount of redox mediator (i.e., the degree of buffering) is not of primary importance. The important condition is the absence of kinetic interference from the redox mediators, which will attempt to reestablish the equilibrium state by interacting with the redox components that are transiently out of equilibrium with the redox mediators. This problem can be tested and overcome by repeating kinetic analysis of components involved over the E_h range studied with a wide variation of redox mediators, and selecting the appropriate redox mediators. With care we have been able to obtain kinetics of cytochrome and reaction center oxidation-reduction up to 1 sec after microsecond flash activation, without significant interference starting from a reasonably reliable E_h value.

A special case of this kind of poise and pulse approach is the establishment of a reliable E_h value at room temperature, followed by anaerobic sampling as in Fig. 4 and freezing for light-induced analysis at temperatures down to liquid helium temperatures. At these temperatures kinetic contributions from redox mediators are negligible.

2. A catalytic quantity (i.e., concentration less than the biological redox components) of rapidly interacting redox mediator is used to provide a reliable E_h reading before activation, but after activation it contributes no buffering capacity to tend to hold the system at the starting E_h. This is applicable to multiple turnover activations that in number are in significant excess of the redox equivalents in the mediator.

Some Redox Titrations and Use of Redox Mediators

This section deals generally and briefly with some examples of redox titrations to illustrate the use of redox mediators.

Redox Potential Titrations of Cytochromes Analyzed in Combination with Room-temperature Dual Wavelength Spectrophotometry. Cytochromes are most often analyzed spectrophotometrically as the reduced form by their Soret or γ-bands (in the 400–460 nm range) or their α-bands (540–620 nm). Reduced minus oxidized difference extinction coefficients are much larger for the γ-bands as compared with the α-bands. In spite of this our current collection of redox mediators is best suited for work in the α-band region. Difference spectroscopy for the analysis in the α-band region of cytochromes *a*, *b*, and *c* is essentially free of absorbance interference from the redox mediators themselves (i.e., redox mediators up to the 10^{-5}–10^{-4} μM concentration range with reduced minus oxidized absorbance changes for the cytochromes in the region of 0.01–0.02 ΔA and above). This statement is true from high E_h values down to about 0 mV at pH 7.0. Thus, for example, the redox mediators potassium ferricyanide, DAD, PMS, PES, and duroquinone (and many other quinones) are particularly good in that their own redox-associated absorbance changes are not registered by wavelength pairs typically used for cytochrome assay. Below 0 mV (at pH 7.0 or from this value −60 mV/pH unit), however, precautions have to be taken against possible interference from pyocyanine, 2-hydroxy-1,4,-napthaquinone and the anthraquinones and the viologens (see next section).

In the γ-band region caution has to be exercised for most of the redox mediators since virtually all the mediators undergo some absorbance changes of their own on oxidation and reduction. The problem can be minimized or eliminated by the appropriate choice of wavelength pairs used in the measurement. However, a rather serious problem in the γ-band region is encountered at high E_h values when the yellow potassium ferricyanide assumes significant concentrations; this leads to major distortions of the redox titrations of such high midpoint potential cytochromes as cytochrome a_3 (E_{m7} = 380 mV; see Dutton *et al.*[12,13]) from respiratory systems, or the high-potential *c*-type cytochromes from photosynthetic bacterium *Chromatium vinosum*, or cytochrome *f* and the high-potential form of cytochrome b_{559} of chloroplasts. Unfortunately, the interference from ferricyanide is greatest at the sensitive E_h values beyond the E_m value where its contributions prevent the *direct* measurement of the near-100%-

[12] Wilson, D. F., and Dutton, P. L. (1970). *Arch. Biochem. Biophys.* **136,** 583–584.
[13] Dutton, P. L., and Storey, B. T. (1971). *Plant Physiol.* **47,** 282–288.

oxidized absorbance value of these kinds of high-potential cytochromes. Careful subtraction of the ferricyanide contribution is necessary.

I have purposely avoided giving any numbers regarding expected difference extinction coefficients of the redox mediators; these should be established experimentally for any particular set of determinations. The relative significance of mediator contributions is entirely dependent on the experimental conditions and the level of sophistication at which the work is pitched. For basic work it is no more serious than spectrophotometric interference from endogenous absorbance changes and often it is an easier matter to eliminate.

For general exploratory work, apart from following the guidelines described in the earlier sections of this article, it is a good idea to perform dual wavelength spectrophotometric analysis of redox titrations in both γ- and α-band regions and to use different reference wavelengths (e.g., analysis of cytochrome *b* at 560 nm in the α-band region can be done with the reference wavelength at 540 nm or 575 nm); this will detect interference from color changes from agents with broad spectra (redox mediators or endogenous redox components such as flavin). Alternatively, redox titrations can be done at different measuring wavelengths which permits the construction of E_h-resolved spectra of the redox components involved (e.g., Dutton and Jackson[14]). However, with the advent of the scanning dual wavelength spectrophotometer the need to perform several determinations is eliminated, and all the information can be obtained in one redox titration. A titration of this kind is shown in Fig. 5; it illustrates the almost complete[15] freedom of the titration from interference by redox mediators.

Experimental Tactics Regarding Use of Redox Mediators

a. Degas Supporting Medium. Before adding the aliquot of concentrated biological material or the redox mediators to the redox vessel/cuvette, degas the supporting aqueous medium by *bubbling* with argon for a few minutes. Bubbling is necessary since stirring under an atmosphere of argon takes a long time to degas. The degassing avoids any damage of the redox mediators by the interaction of the biological material and oxygen. It also saves having to wait for "anaerobiosis" induced by the biological respiratory system which may be weak in the absence of substrates. In such a case degassing therefore eliminates the need to unnecessarily add substrates or sodium dithionite.

b. An Example of a Redox Titration Requiring E_h Values above and below 300 mV. Mitochondrial cytochrome *a* ($E_{m7.2}$ = 210 mV) and cyto-

[14] Dutton, P. L., and Jackson, J. B. (1972). *Eur. J. Biochem.* **30,** 495–510.

[15] Erecinska, M., Oshino, R., Oshino, N., and Chance, B. (1973). *Arch. Biochem. Biophys.* **157,** 431–445.

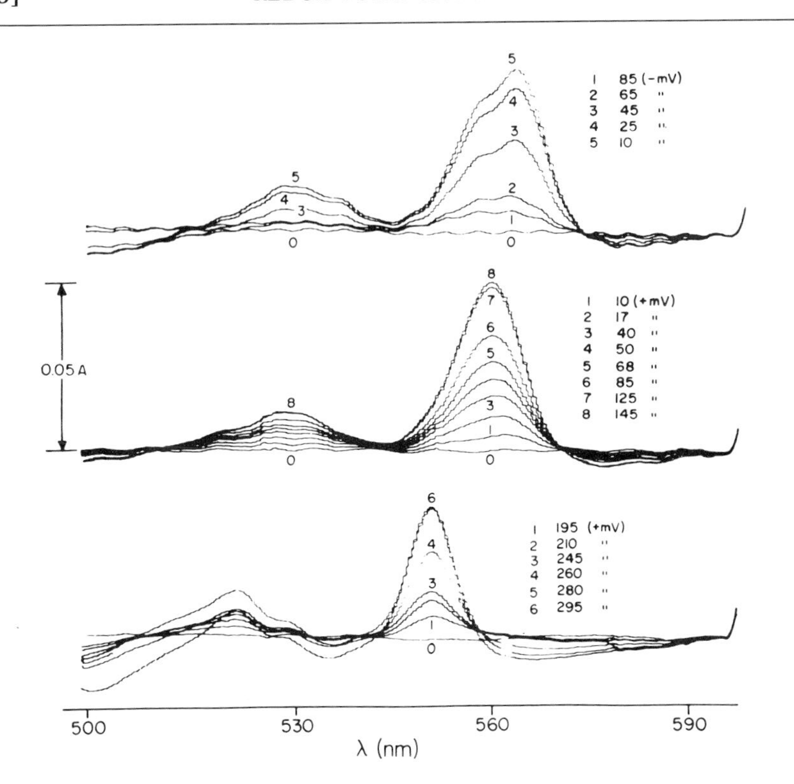

FIG. 5. Difference spectra of the cytochromes of succinate–cytochrome c reductase obtained during anaerobic potentiometric titration of the preparation. The succinate–cytochrome c reductase (2 mg protein/ml) was stirred under an argon atmosphere in 0.1 M phosphate buffer, pH 7.0 (0.008% Triton X-100, 0.008% deoxycholate). The redox mediators used were: 20 μM diaminodurol; 40 μM each of phenazine methosulfate, phenazine ethosulfate, and duroquinone; 5 μM pyocyanine, and 15 μM 2-hydroxyl-1,4-naphthoquinone. The figure shows an oxidative titration with potassium ferricyanide. The reference wavelength was 595 nm, and the scanning speed was 1.3 min/100 nm. Upper traces represent the absorbance changes taken in the potential range: −100 to −10 mV showing principally cytochrome b_{566}; middle: −10 to +145 mV showing principally cytochrome b_{560}; bottom traces: +145 to +275 mV showing principally cytochrome c_1.

chrome a_3 ($E_{m7.2}$ = 380 mV) are a good example. (1) Degas medium and add mitochondria. (2) Set wavelengths, for example, at 605–630 nm. The α-band region usually precludes the use of TMPD, which undergoes absorbance changes here, but does permit the use of DAD which has no absorbance contributions. (3) Add 5–100 μM DAD, chosen because it has an E_m value similar to that of cytochrome a; add 5–50 μM PMS (also no absorbancy contribution) to permit reliable E_h measurements at about 100 mV for the near-100% reduced cytochrome a/a_3 absorbance value. PMS is not obligatory but it does increase the reliability of the titration. (4) Add dithionite until E_h is ~100 mV; this is the maximum absorbance value of the titration. (5) Establish that the absorbancy measurement is not drifting with time for reasons other than oxidation-reduction; this base line is the 100% reduced value, and all numbers are referred to it. If it drifts during the experiment there will be significant errors. (6) Start adding ferricyanide, taking E_h readings at each new level of decreased absorbance. At each reading make sure values are steady; this may take 0.25–2 min and might be generally slowly drifting back to lower E_h values aided by the endogenous reducing supply. (7) As the 300 mV E_h is passed the stability of the DAD decreases. There is, however, no particular concern for this because there will be sufficient ferro/ferricyanide accumulated to take over as a mediator; if disequilibrium is indicated by instability of E_h with respect to the absorbance reading, some ferrocyanide (e.g., 50 μM) can be added. The titration is completed by taking the E_h as high as possible to a point where little ΔA occurs on changing the E_h. This should be about 500 mV; however, the addition of extra ferrocyanide may make it difficult to achieve 500 mV without adding enormous amounts of ferricyanide (e.g., with 50 μM ferrocyanide present it will take 50 μM to get the E_m and 500 μM to achieve an E_m value 60 mV beyond that). (8) If a reduction titration is done on the same material, then at E_h values just below 300 mV more DAD and PMS should be added to allow reliable E_h values to the completion of the titration; if a formal reductive experiment is *not* done, dithionite should still be added to check that the original starting absorbance has not changed during the titration since this is vital to calculation. A reductive titration can be done in a separate experiment. In this case, after the system is anaerobic a small amount of DAD (e.g., 10 μM) can be added (no ferrocyanide or PMS) to aid membrane mediation of ferricyanide. The E_h can then be promptly taken up as high as possible to establish the near-100% absorbance level. The E_h is then lowered by adding ferrocyanide until about 50–100 μM has been added (this increases the ferro/ferricyanide mediation concentration) and then by adding dithionite or by allowing endogenous reductant to act if this is a suitable rate (the rate can be adjusted by the amount of mediator present). Again at 300 mV DAD and PMS are added and the titration taken to completion at 100 mV. (9) If the titration is repeated in the γ-band region (445–455 nm), then the DAD can be replaced by TMPD again acknowledging that TMPD irreversibly decomposes at E_h values above 300 mV. As discussed before, ferricyanide unavoidably contributes in this region and requires a full accounting for the additions of ferri- and ferrocyanide and control experiments. PMS, not used in the published titrations, would have to be checked also and if necessary accounted for or replaced, or since it is not obligatory, eliminated.

c. Wide E_h Range Titrations. The procedure outlined above is the basis for all titrations. In the α-band region there are areas below 0 mV at pH 7.0 where redox mediators that we have used contribute significant absorbance changes if used in large amounts. Pyocyanine is blue (neutral pH) or purple (acid pH) in the oxidized form; this is as good a mediator as PMS or PES so it can be used in small (5 μM) amounts. To increase the poise in the 0 to -100 mV E_h range, duroquinone, itself ineffectual as far as its interaction with the electrode, is usually added with it; together they satisfy many of the criteria required of redox mediators. 2-Hydroxyl-1,4-naphthaquinone and the anthraquinone sulfonates can contribute so these should be kept at low levels (<20 μM) or otherwise be controlled for. Thus by adding anthraquinone-2-sulphonate, anthraquinone-2,6-disulphonate, 2-hydroxyl-1,4-naphthaquinone, pyocyanine and duroquinone, PES, PMD, DAD, and ferrocyanide an E_h titration can be performed at pH 7.0 from -250 to $+500$ mV.

Often wide-range E_h titration such as has been encountered in cytochrome *b* titrations of FeS proteins (see, e.g., Dutton *et al.*[14-17]) takes quite a long time to perform, depending obviously on the number of E_h/absorbance correlations taken. They can take hours if equilibration with the redox components is slow. It is essential in continuous-absorbance titrations to periodically check the near-100% reduced or oxidized level to make sure it has not shifted on the chart paper. Alternatively, the titration can be done in separate pieces paying attention to different E_h ranges on each determination.

It is important to repeat titrations with differing redox mediator concentrations. The same results should be obtained each time not only with respect to the E_m but also with regard to the shape of the titration curve and its overall absorbance change. Systematic distortion of redox curves or systematic increases in the overall absorbance change of the titration indicate mediator interference.

With examination of the state of reduction by EPR (e.g., cytochromes or iron–sulfur proteins) over a wide E_h range there is usually no limitation on the redox mediators added. This freedom applies to EPR free-radical signals not at g 2; otherwise it becomes very important to use dyes with little or no interfering g 2 signal of their own in the E_h range being examined. If possible the problem can be circumvented by the different microwave power properties of the biological redox couple and the mediators (see Ingledew *et al.*[18]).

[16] Dutton, P. L., Wilson, D. F., and Lee, C. P. (1970). *Biochemistry* **9**, 5077–5082.

[17] Ohnishi, T. (1975). *Biochim. Biophys. Acta* **387**, 475–490.

[18] Ingledew, W. J., Salerno, J. C., and Ohnishi, T. (1974). *Arch. Biochem. Biophys.* **177**, 176–184.

d. Low-temperature Analysis. Questions are often asked regarding the relevance of redox potential at low temperature and the possible shifts in E_m as the temperature is lowered. Although the E_m values of redox components almost certainly change as the temperature is lowered, it is unlikely that a sample of known E_h and state or reduction at room temperature will change its state of reduction if promptly (<1 sec) frozen to liquid nitrogen temperatures. Even if redox mediator and the measured biological redox component differentially change their E_m values and they are both in a partly reduced state at the point of freezing, activation energies for electron transfer prohibit electron transfer from occurring on a time scale that hinders the measurement. Although it cannot be ruled out, there has been no report of such interference, and there are now many examples of redox determinations done at both room and cryogenic temperatures. Most E_m values agree within 15 mV (see, e.g., Dutton and Wilson[2]).

Analytical Procedures for Resolving into Component Parts Redox Titrations Containing Multiple Electrochemical Species

In almost all cases redox titrations of biological electron-transfer systems involve more than one redox component. This arises from the multiple components having the same spectrometric properties but differing E_m values [e.g., the multiple iron–sulfur proteins which have similar EPR line shapes and g values, or multiple cytochrome species such as the two or three (or more) b cytochromes evident in many systems, cytochrome c and c_1, or cytochrome a and a_3]. For two components this makes plots such as those in Fig. 1B sigmoidal, approaching linearity and with the slope for the n value only at the extremities (see Wilson and Dutton[19]). For example, cytochrome oxidase would involve the ratio $[a_{ox}][a_{3ox}]/[a_{red}][a_{3red}]$. If there are more components, as is often the case with the b cytochromes or Fe–S proteins, this Nernst plot becomes a series of steps. The "height" of the steps depends on the separation in the E_m values of the components. With separations of above 150 mV as in the case of cytochrome a and a_3 for which the relative contributions to the overall titration are equal, the resolution is simple. It is a matter of taking a point between the two Nernst curves and calling the value 100% oxidized for cytochrome a. This of course is an approximation but usually the error is much less than the experimental uncertainty. When the E_m difference is smaller than 100 mV or if there is a large amount of one component compared to the other, the above simple approach leads to errors that may be significant. Computer techniques are then applicable to completely resolve the curves. Simple

[19] Wilson, D. F., and Dutton, P. L. (1970). *Biochem. Biophys. Res. Commun.* **39**, 59–64.

examples have been devised by M. Pring (see Ohnishi[17], and Dutton *et al.*[20]), with other ones given by Denis[21] and Hendler *et al.*[22]

Acknowledgments

I would like to express my gratitude to the National Science Foundation (grant nos. GB 28125 and later PCM-76-14902) and to the U.S. National Institutes of Health (grant no. PHS GM 12202) and to acknowledge my USPHS Research Career Development Award (1-K04-GM 70771) for financial support during the development of this technique. Thanks are also due to Roger Prince and Kenneth Wells and to Margaret Mosely and Barbara Bashford for their suggestions and help with the manuscript.

[20] Dutton, P. L., Erecinska, M., Sato, N., Mukai, Y., Pring, M., and Wilson, D. F. (1972). *Biochim. Biophys. Acta* **267,** 15–24.

[21] Denis, M. (1973). Ph.D. Thesis, University of Marseilles.

[22] Hendler, R. W., Town, D. W., and Shrager, R. L. (1975). *Biochim. Biophys. Acta* **376,** 42–62.

[24] Micro Methods for the Quantitative Determination of Iron and Copper in Biological Material

By HELMUT BEINERT

Introduction

There are many different methods available for the determination of iron and copper, chemical and spectroscopic; and for the chemical determination a great number of reagents have been proposed. No single investigator can be expected to have experience with all these methods, and it is reasonable that an investigator use the method that he has set up and working and in which he has gained confidence. The description that follows is therefore aimed at those who do not have a method set up and reliably working in their laboratory and who prefer to resort to chemical determinations rather than purely spectroscopic ones. However, irrespective of the final determination of the metal, which can be carried out by various means, the preparation of the samples for the final determination shares common features in most instances. In this respect also the precautions to be taken with the two metals, iron and copper, are very similar, since they are both ubiquitous contaminants. The main problem can be stated as this: The difficulty does not lie in finding the metal that is there, but rather in not finding metal that does not belong there.

ISBN 0-12-181954-X

The essential features of the method to be described[1] are wet ashing, evaporation of excess acid, reduction, neutralization with excess sodium acetate, development of color with a suitable bathophenanthroline, and extraction with a small quantity of organic solvent followed by spectrophotometric determination.

For Iron

Reagents

(1) Iron standard in dilute H_2SO_4
(2) H_2SO_4, concentrated
(3) HNO_3, concentrated
(4) $HClO_4$, 60%
(5) HCl, 2 *N*
(6) Mercaptoacetic acid, 1% (w/v) solution
(7) Sodium acetate, saturated solution
(8) 3-Methyl-1-butanol
(9) 4,7-Diphenyl-1,10-phenanthroline solution in 3-methyl-1-butanol

Comments to Procurement and Preparation of Reagents

(1) A concentrated primary standard, containing approximately 1 mg-atom Fe/100 ml, is prepared first. From this a working standard of 0.1 μg-atom/ml is derived by dilution with dilute H_2SO_4. The use of HCl is avoided, as iron chloride is volatile. Analytical-grade iron wire (56 mg) is degreased, placed in a 100-ml volumetric flask, and dissolved in a solution containing 2 ml concentrated HCl and 2 ml concentrated HNO_3 in 10–15 ml. When all the iron is dissolved the flask is made to volume with water. This standard is stable indefinitely. Iron(II)-ethylenediammonium sulfate is also a suitable stable primary standard.

(2–5) Ultrapure acids of low metal content are available from various sources (J. T. Baker Chemical Co.; EM Laboratories, Inc.; British Drug Houses, Ltd.).

(6) Also called thioglycolic acid. The aqueous solution has to be prepared approximately once a month since oxidation destroys it. The original sample, once opened, is also not stable for more than a few months. Disulfide formation can generally be recognized by the appearance of a precipitate. Low color yield in the reaction with phenanthroline is generally due to deterioration of this reagent.

(7) Sodium acetate may be obtained as ultrapure reagent. However,

[1] M. Van De Bogart and H. Beinert, *Anal. Biochem.* **20,** 325–334 (1967).

even this reagent, when used in the quantity needed here, contributes to the blank color. Its use in the required quantities also becomes very costly. Commercial reagent-grade acetate must be purified before use, and it is also advisable to apply the following procedure to the ultrapure reagent. A saturated solution of sodium acetate is prepared. It is advantageous to let this solution sit undisturbed for a few days, as some ferric hydroxide will settle out which then does not have to be removed by the laborious extraction procedure described below. However, extraction has to be carried out in any event, since some iron always remains in solution. After standing, the solution is filtered and collected in an Fe-free bottle. (The filter paper is prepared by washing with one of the Versenes and thereafter with sufficient water to completely remove Versene. The complete removal of Versene is essential or interference with reduction and binding of Fe by the phenanthroline may occur.) To the filtered solution a few milliliters of glacial acetic acid are added (about 6 ml/liter) to lower the pH to about 6, and some undiluted thioglycolic acid (1 ml/liter acetate) is added to reduce the contaminating iron left in solution. Iron is not reduced by thioglycolic acid at a pH higher than 6. Each liter of acetate solution is then shaken out with several 50-ml portions of phenanthroline/isoamyl alcohol reagent. The pink color of the isoamyl alcohol layer increases when the two phases are left standing together for several hours after each extraction. When after several extractions and standing periods only a faint pink color appears in the organic phase, the acetate solution is extracted twice with isoamyl alcohol to remove any phenanthroline from the acetate.

(8) Also called isoamyl alcohol; some lots of the reagent-grade alcohol were found satisfactory, while others contained iron. Isoamyl alcohol specially purified for iron analysis can be obtained commercially (G. Frederick Smith Company, Columbus, Ohio), or the ordinary reagent can be redistilled.

(9) Bathophenanthroline (G. Frederick Smith Company, Columbus, Ohio) (molecular weight 332), approximately 200 mg dissolved in 250 ml isoamyl alcohol. The reagent dissolves slowly and should preferably be left standing overnight. Warming the suspension in a water bath also speeds dissolution. This reagent slowly develops a yellowish color, which adds to the blank color. It is, therefore, advisable to prepare it in small lots fresh.

Glassware

Conical 12-ml graduated centrifuge tubes with ground-joint stoppers (e.g., Pyrex #8144) were found most suitable for ashing and extraction, as the approximate fluid volume can be continuously monitored. Micropipettes are used for all additions under 1 ml. In most cases absolute accuracy of these additions is not required, merely repeatability. The pipettes are stored in tubes covered by larger tubes and all bottles are covered, since laboratory dust usually contains appreciable amounts of iron or copper. Glassware must be carefully cleaned and water of low metal content must be available. Procedures for achieving this have been published.[2] In our laboratory we found it satisfactory to redistill via a Pyrex glass condenser previously deionized or distilled water from a large metal still. Particular attention has to be given to the cleaning of new glassware. It should be kept in mind that all glassware that has been used in the analytical procedure, although possibly not "clean" with respect to other uses, is probably still the cleanest for reuse in the same procedure. Thus, the tubes used for ashing are flushed with water, ethanol, and again water and briefly boiled with 2 *N* HCl. After rinsing with water and draining but not necessarily drying, they are ready for reuse. The tubes are touched only at the lower end and care is taken not to contaminate the upper rim since on ashing, through refluxing, as well as during extraction, liquid does wet the upper parts of the tubes. We found it most convenient and safest to store the tubes upside down between analyses, on a rack provided with appropriate clamps (e.g., commercial broom-clips). After the tubes are dry, they are capped with Parafilm. After storage for several weeks the tubes have to be cleaned with acid, and after even longer storage it is advisable to carry them through a blank ashing procedure with the reagents used in the procedure. This is absolutely necessary for new tubes or tubes used for other work. The stoppers are degreased, if necessary, boiled in 2 *N* HCl, flushed with water, and stored in a suitable container. It is recommended that all containers used for storage of stoppers, pipettes, or reagents have covers with lids such as Petri dishes which prevent settling out of dust or dirt on the rim (e.g., a beaker covered with a watch glass is unsatisfactory). Pipettes are cleaned with chromic or hot nitric acid, flushed, and dried before being used in the analytical procedure. When stored for more than a few days after this treatment they should be flushed before use with 2 *N* HCl, followed with water and then with the reagent to be pipetted. Two samples of reagent are drawn up and expelled into a waste container. After this treatment a pipette is usually used to make additions to a whole series of samples. If the addition of the reagent is made carefully and in order of increasing iron content of the samples, no significant contamination can result. After use, pipettes used for aqueous solutions not containing protein are again flushed with 2 *N* HCl and water. Pipettes used for pipetting organic solvents are flushed with ethanol immediately before and after use

[2] R. E. Thiers, *Methods Biochem. Anal.* **5**, 301 (1957).

and otherwise with 2 *N* HCl and water as above. Pipettes used for protein solutions are first cleaned with detergents and thereafter as just described. To prevent contamination of the stock reagents it is necessary to have a second set of smaller reagent containers (e.g., glass-stoppered tubes), in which the reagent is frequently renewed. All pipetting is done from this second set of containers.

Apparatus

One adjustable microburner for ashing is sufficient. Occasionally a second burner for expelling condensed water at the upper end of the tube is helpful. The time required to set tubes up properly in a digestion rack, and the chance that spattering may occur when the tubes are unattended, probably outweigh the inconvenience of the labor involved when the contents of each tube are ashed by hand. With the small quantities of material this requires only 5–10 min/tube. A shaking or swirling device is required for extraction. A buzzer according to Bessey *et al.*[3] is satisfactory or a mixer of the type called "Vortex." Brief, low-speed centrifugation is necessary for separating the phases after extraction. Spectrophotometry may be performed in any instrument that permits the use of volumes of the order of 0.3 ml.

Procedure

See also "Comments to Procedure" below.

(1) Align the tubes in a rack according to increasing iron content and cap with Parafilm. Remove caps only when additions are made, but do not place Parafilm on hot tubes.

(2) Add "blanks," standards, and samples to appropriate tubes.

(3) Add 50 μl concentrated H_2SO_4 (reagent 2).

(4) Add 50 μl concentrated HNO_3 (reagent 3).

(5) Ash over microburner for approximately 6 min.

(6) Add 25 μl $HClO_4$ (reagent 4) to tubes after contents have cooled down.

(7) Ash over microburner an additional 3 min, but avoid boiling.

(8) Add, after cooling, 250 μl of water.

(9) Add 125 μl 1% thioglycolic acid (reagent 6).

(10) Add 750 μl sodium acetate (reagent 7).

(11) Add 500 μl bathophenanthroline solution (reagent 9).

(12) Stopper tubes with glass stoppers.

(13) Extract by buzzing or swirling.

[3] O. A. Bessey, O. H. Lowry, M. J. Brock, and J. A. Lopez, *J. Biol. Chem.* **166**, 177 (1946).

(14) Centrifuge to separate layers.

(15) Withdraw ~0.3 ml of the organic phase and place in dry microcuvette.

(16) Read at 535 nm against blank containing isoamyl alcohol (reagent 8).

Comments to Procedure

(2) The samples should not contain excessive amounts of interfering ions. Of metals to be expected in biological material, copper should not be present in a more than 5-fold excess over iron. Interference from molybdenum can be controlled by choosing a pH above 5.5.[4,5] Of anions to be expected in biological material, phosphate, pyrophosphate, and cyanide interfere. When the sample is dissolved in 0.1 *M* phosphate, for instance, the use of a 20-μl sample may be safe, while it is not advisable to use 0.1–0.2 ml of a 0.1 *M* phosphate solution. Tris buffers, such as Tris acetate, which decomposes on ashing, are very suitable. It is also advisable to avoid large concentrations of reagents such as sucrose or bile salts, as ashing may become very cumbersome in this case; multiple additions of $HClO_4$ may be necessary, and blanks may rise correspondingly. In general, samples must be prepared in a way suitable to remove contaminating iron. For proteins this may be accomplished by dialysis against chelating agents. After equilibrium has been reached, a sample of the dialyzate, equal to the protein sample, is used as "dialyzate" blank. If such a blank is expected or known to contain not significantly more iron than the "reagent blank" (containing all reagents only), dialysis fluid may be added to the standards as well, thus eliminating the need for running separate reagent blanks. The sample and standards should contain not more than 10 μg-atoms of iron (0.5 μg Fe) and 1 mg of organic matter, optimally about 5 μg-atoms of iron and up to 0.25 mg of organic matter. A complete standard curve need not be run, but it is advisable to run standards at two levels, above and below that expected for the sample. If different samples are run with different liquid volumes, it is also helpful to run these same volumes of "dialyzate" blanks. It may be desirable, when a set of carefully calibrated micropipettes is not available, to use the same pipette for blanks, standards, and samples, in this order, with appropriate cleaning of the pipettes (see above) between sampling.

(5) With the conical tubes used, care should be taken never to heat the tip. The tubes are held in a nearly horizontal position so that most of the

[4] W. B. Fortune and M. G. Mellon, *Ind. Eng. Chem., Anal. Ed.* **10,** 60 (1938).

[5] E. B. Sandell, "Colorimetric Determination of Traces of Metals." Wiley (Interscience), New York, 1944.

liquid is in a shallow layer at the place where the tapered and the straight portion of the tube meet. The fumes are driven up the barrel of the tube. Water should not be allowed to condense near the top of the tube, as this may cause spattering should the water run back into the hot tube. A second burner may be used to advantage to expel water condensed in the upper part of the tube. Tubes are capped again with Parafilm as soon as they have cooled down sufficiently.

(7) $HClO_4$ usually clears up the digest readily, and only when an excess of organic matter is present is the addition of more $HClO_4$ necessary. The presence of undigested organic matter is generally indicated by the persistence of a yellow-brownish color even after cooling of the digest. Toward the end of the ashing period a bright yellow color may appear, which usually disappears on cooling or further ashing. This color does not stem from organic matter. Ashing may also be effectively accomplished with H_2SO_4 alone by addition of H_2O_2. After heating with the acid and cooling, 10-μl portions of 30% H_2O_2 are added until the digest is clear and colorless; the effect of H_2O_2 is similar to that of $HClO_4$.

(10) This amount of acetate was found to buffer the amount of acid which could maximally be left over, to a pH of about 4.2. The pH should be within the range of 4–6 as otherwise reduction and color formation are not optimal. The reductant (reagent 6) should always be added before the buffer (reagent 7) in order to ensure rapid and complete reduction of iron.

(11) This solution is viscous and drains slowly. It should be slowly and uniformly delivered into all tubes, even if a quantitative delivery is not possible. Only the relative amount delivered to each individual tube is of importance, not the absolute quantity.

(13) Vigorous swirling or buzzing is necessary to extract all the iron into the organic phase. The tubes are buzzed or swirled for 2 min, then centrifuged to clear the walls, buzzed or swirled for 1 more minute, and centrifuged again. The stoppers may be lubricated very slightly with silicone grease. Poor color yield is often due to incomplete extraction.

(15) The same transfer pipette and cuvette are used for all samples. The samples are again processed in the order of increasing iron content. Between readings in the spectrophotometer the pipette and cuvette are flushed 3 times with ethanol and dried in a stream of filtered air. If difficulty is experienced in withdrawing a sufficient amount of organic phase uncontaminated by aqueous phase, it may be helpful to tilt the tube slightly so that the organic phase breaks from the wall on one side and contracts on the other side, thus providing a thicker layer to draw from.

(16) The reagent blank should give an absorbance reading not in excess of about 0.03 (1-cm light path). This blank value is mainly determined by the success in the removal of Fe from the acetate solution. Five μg-atoms of Fe

give a net absorbance value of 0.190–0.210 under the conditions described. It should be kept in mind that the small quantity of reagent soluble in isoamyl alcohol can extract only a limited amount of Fe. It is stated by Peterson[6] that 1 ml of the reagent can extract 2 μg of Fe. The absorbance should therefore never exceed a value of 0.6. Absorbance and iron content are proportional below this absorbance value. If the amount of iron present in a sample has been misjudged, so that too high an absorbance is observed, the quantity of reagent 9 added in step 11 may be increased. It is not sufficient to dilute with solvent after extraction. The amount of reagent should of course then also be increased in all other samples of the series including standards and blanks.

It often occurs that the isoamyl alcohol layer becomes turbid after transfer into the spectrophotometer cuvette. This is probably due to formation of an emulsion of water in the alcohol layer. Brief warming of the cuvette in a water bath to a few degrees above room temperature regularly clarified the solution for a time sufficiently long to take readings.

Possible Reasons for Difficulties in Procedures

(a) Erratic results with samples, blanks, and standards: contaminated glassware, random contamination during improper handling, or insufficient extraction.

(b) Erratic results in samples only: insufficient ashing, not liberating all the iron or leaving organic iron chelators behind; presence of iron complexing salts such as phosphate or pyrophosphate.

(c) Low values for standards: deteriorated mercaptoacetic acid or incomplete extraction.

(d) High but reproducible values for blanks and standards: contaminated bathophenanthroline reagent either from contaminated isoamyl alcohol or reagent itself; contaminated acetate or other reagents. The phenanthroline reagent and the acetate are the most likely sources.

(e) High values of standards only, not blanks: dilution of secondary working standard is either incorrect or this standard has increased in strength by contamination or evaporation. It is advisable to replace the secondary standard frequently. Significant contamination of the primary standard of high iron content is less likely.

Simplified Procedure for Acid-extractable Iron

In cases where it is certain or can be expected that the iron which is to be determined is released by acid (or some treatment other than ashing), the

[6] R. E. Peterson, *Anal. Chem.* **25,** 1227 (1953).

ashing procedure can be eliminated. The following procedure was found to be convenient and more precise than the procedure described, when applied, e.g., to purified iron sulfur proteins: Reagents 1, 2, 5, and 7–9 from above are used, but mercaptoacetic acid is used as a 5% solution. The iron is released with sulfuric acid since this acid can be obtained with very low iron content, so that blank readings are <0.008 absorbance units at 1-cm light path. The volume of sample plus buffer or water or of standard plus buffer or water is made up to 150 μl. Then 20 μl of 98% sulfuric acid are added and the contents are all mixed. Thereafter, 50 μl of 5% thioglycolic acid are added with renewed mixing. After 10 min 400 μl of saturated acetate are added followed by 500 μl of reagent 9. The extraction is performed as described. With the samples, which now contain precipitated protein, very thorough extraction is necessary, since the denatured protein appears to impede the exchange of the reagent between the aqueous and organic phase. However, after extraction and centrifugation, the organic solvent layer is clear. In this procedure the blank readings are as low as 0.003–0.008 and 5 ng-atoms of iron give a net absorbance of 0.215–0.220.

For Copper

The method for the determination of copper is in many ways similar or identical to that used for iron. Details will therefore be given only when differences exist; otherwise reference is made to the iron method by "(see Fe)."

Reagents

(1) Copper standard in dilute H_2SO_4
(2) (See Fe)
(3) 2 *N* HCl
(4) H_2O_2, 30%
(5) Hydroxylamine hydrochloride, 1% (w/v) aqueous solution
(6) Sodium acetate, saturated solution
(7) 1-Hexanol
(8) 2,9-Dimethyl-4,7-diphenyl-1,10-phenanthroline solution in 1-hexanol

Comments to Procurement and Preparation of Reagents

(1) Since the water content of most copper salts is not well defined, it is advisable to prepare the standard solution from metallic copper, e.g., copper shot of sufficient purity. A concentrated primary standard is prepared by dissolving 63.5 mg metal in a 100-ml volumetric flask (see Fe). The resulting standard contains 10 μg-atoms/ml. A secondary working

standard is prepared by diluting 1 ml of this standard to 100 ml with dilute H_2SO_4, so that 1 ml contains 0.1 μg-atom or 6.35 μg Cu.

(2,3) (See Fe.)

(4) Reagent-grade H_2O_2 is satisfactory.

(5) Reagent-grade hydroxylamine hydrochloride is satisfactory. This solution should be prepared fresh every 2–3 months.

(6) The acetate solution is prepared similarly to that described for Fe but the solution is shaken with approximately 1/15 its volume of a 0.01% solution of dithizone (diphenylthiocarbazone) in chloroform. The acetate solution is extracted until the purple color of the copper–dithizone complex is no longer visible in the chloroform layer. The chloroform is then shaken out with *n*-hexanol (reagent 7). Remaining *n*-hexanol, dissolved in the aqueous phase, does not interfere. In order to avoid the formation of stable emulsions shaking has to be done judiciously. Centrifugation in copper-free tubes may help in breaking emulsions when formed.

(7) *n*-Hexanol; it is advisable to redistill the commercial products. *n*-Hexanol purified for copper determination is commercially available (G. Frederick Smith Company, Columbus, Ohio).

(8) Bathocuproine (G. Frederick Smith Co.) (molecular weight 360); approximately 25 mg are dissolved in 250 ml *n*-hexanol (see Fe). This solution should be made up fresh every few weeks.

Glassware

Similar considerations apply here as for the iron method. After cleaning with acid, glassware may be treated with dithizone in chloroform (0.01%) and then rinsed with chloroform. The acid-cleaned glassware need not be dried before dithizone treatment, if an acetone rinse is included.

Procedure

(1–2) (See Fe.)
(3) Add 100 μl concentrated H_2SO_4 (reagent 2).
(4) Ash over microburner for approximately 6 min.
(5) Add 20 μl of reagent 4 to tubes after contents have cooled down.
(6) Ash over microburner an additional 3 min, but avoid boiling.
(7) Repeat step 6 if necessary.
(8) Add, after cooling, 500 μl of water.
(9) Add 0.25 ml hydroxylamine hydrochloride solution (reagent 5).
(10) Add 1.5 ml acetate solution (reagent 6).
(11) Add 0.5 ml bathocuproine solution (reagent 8).
(12–14) (See Fe.)

(15) Withdraw 0.3 ml of the organic phase and place in dry microcuvette.

(16) Read at 479 nm against blank containinq *n*-hexanol (reagent 7).

Comments to Procedure

(2) Similar considerations apply here as for the iron procedure, and only a few comments specific for the copper procedure will be made. Heavy metals in amounts expected in biological materials do not interfere. The amount of phosphate present should be kept at a minimum. When copper-binding organic anions such as EDTA are present, it is particularly important to ash the samples thoroughly so that these chelators are destroyed. Otherwise complex formation and extraction into the organic phase is interfered with.[7] The samples and standards should not contain more than 15 ng-atoms of copper ($\sim 1\ \mu g$ Cu) and 2 mg of organic matter.

In the original description of the method,[1] the alternative method of ashing with H_2SO_4, HNO_3, and $HClO_4$ was also recommended. Since copper perchlorate may be volatilized under certain conditions, ashing should be carried out with H_2SO_4–H_2O_2 as described.[7]

(16) The reagent blank should give an absorbance reading not in excess of 0.02 (1-cm light path). Five ng-atoms of Cu give a net absorbance value of 0.105–0.115 under the conditions described. Since the solubility of the reagent is limited and therefore the quantity of copper that can be extracted, the total absorbance should not exceed 0.4. Absorbance and copper content are proportional below this value.

Difficulties likely to arise in the procedure for copper and their possible sources are analogous to those discussed above for iron.

[7] H. Beinert, C. R. Hartzell, B. F. van Gelder, K. Ganapathy, H. S. Mason, and D. C. Wharton, *J. Biol. Chem.* **245,** 225 (1970).

[25] Atomic Spectroscopy in Metal Analysis of Enzymes and Other Biological Material*

By CLAUDE VEILLON and BERT L. VALLEE

Introduction

The analytical determination of trace metals in biological matter is becoming increasingly important due to growing knowledge of the very critical role played by minute concentrations of various metals in complex biological systems and processes. Of primary interest in this regard are the mechanisms whereby trace metals serve catalytic, structural, conformational, and regulatory functions in metalloproteins,[1,2] their critical importance and essentiality in human nutrition, and their value as indicators of pathological lesions in various clinical conditions.[3] Knowledge, appreciation, and understanding of these mechanisms have progressed approximately in parallel with the ability to measure ever-decreasing concentrations of trace metals in biological systems by improved analytical techniques.

This field is in a period of very rapid development, both in terms of methodology and application, and, understandably, it is undergoing growing pains. Biochemists, nutritionists, and clinicians are continually offered a bewildering array of instrumentation with ever-increasing analytical capabilities, by-and-large developed by analytical chemists who usually work with relatively simple, inorganic matrices. In the words of Alan Walsh, they perceive success as a new "world's record" in low detection limits of some element or elements. Their enthusiasm is developed and conveyed by instrument manufacturers, and the products are then sold to users who can seldom be fully aware of their limitations when employed for samples of rather complex composition and matrices. Without ill intent, the consequent problems are but instances of the developers/suppliers (i.e., analytical chemists and manufacturers) providing instrumentation and techniques to users (i.e., biochemists, nutritionists, etc.) without the requi-

* Specific manufacturer's products are mentioned herein solely to reflect the personal experiences of the authors and do not constitute their endorsement nor that of the Department of Agriculture or Harvard University.

[1] Anderson, R. A., Bosron, W. F., Kennedy, F. S., and Vallee, B. L. (1975). *Proc. Natl. Acad. Sci. U.S.A.* **72,** 2989.

[2] Sytkowski, A. J., and Vallee, B. L. (1976). *Proc. Natl. Acad. Sci. U.S.A.* **73,** 344.

[3] Li, T. K., and Vallee, B. L. (1978). "Modern Nutrition in Health and Disease" (R. S. Goodhart and M. E. Shils, eds.), 6th ed., Chapter 8, Lea & Febiger, Philadelphia, Pennsylvania (in press).

ISBN 0-12-181954-X

site mutual knowledge and appreciation of problems inherent to their specialties.

This treatise is directed toward users faced with the problem of trace-metal determinations in biological matrices who have not been exposed previously to the nuances of analytical atomic spectroscopy. It is our objective to update such users regarding current capabilities and to point out some of the pitfalls. We will concentrate on trace-metal determinations in enzyme preparations: the hows, whys, and wherefores.

The methods most widely employed now for determinations of traces of metals are based on atomic spectroscopy—the interactions of analyte atoms with electromagnetic radiation. Broadly, these procedures fall into three categories: atomic emission, atomic absorption, and atomic fluorescence spectroscopy. All of these have developed over the years into highly sensitive, specific, and rapid means of chemical analysis. Instrumentally, all three techniques share many similar features. Each method has unique advantages and disadvantages, dependent in most cases on the application, type, number, and amount of samples, sensitivity required, and information desired. These three instrumental categories, including their virtues and limitations, and recent state-of-the-art developments will be examined briefly. The discussion will be limited to those techniques and instrumental arrangements that have been shown clearly to be applicable to analysis of biological materials, especially enzymes, much as they may not yet be developed commercially; it will neglect those for which only the potential has been demonstrated but which have not been reduced to readily applicable practice.

Atomic Emission Spectrometry

Instruments which measure the emission of electromagnetic radiation by sample atoms fall into several classifications, depending on their design configuration and the type of source used to atomize the sample and excite the sample atoms, thereby producing the desired emission. Instruments are designed either to perform single-element or—alternatively—multi-element determinations. Multi-element determination refers to instruments for simultaneous measurement at numerous wavelengths or rapid-scanning (or slew-scanned) instruments designed to make multi-element determinations on a single sample. Various atomization/excitation sources are employed for both types of instruments. These include chemical flames, high-voltage spark discharges, high-current arc discharges, and various plasma discharges produced by constricted arcs, radio-frequency fields, or microwave fields (a "plasma" being simply a highly ionized, electrically conducting gas).

Basis of the Method

The basis of atomic emission can be illustrated as:

$$M + \text{energy} \rightarrow M^* \rightarrow M + h\nu$$

where M represents the analyte atom in the gas phase, usually in its lowest or electronic ground state. In emission spectrometry the high-temperature gaseous environment of the source serves a dual function: dissociation of the sample introduced into the source into its component atoms (M); and imparting sufficient energy to the atoms by collisions with energetic species within the source to cause them to undergo a transition to a higher electronic energy level. This results in an excited atom, M^*, which, as one possible means, can then lose the acquired energy by a radiational process, i.e., emission of a photon, $h\nu$, and return to the original state. The energy of the emitted photon, $h\nu$, is equal to the energy difference between the electron energy levels involved in the transition. Emission is not the only means of energy loss, of course, but it is the only one of analytical importance.

Other transitions from higher to lower energy levels are possible also and do not necessarily end in the ground state, resulting in more than one emission wavelength for an element.[3a] Since atoms have only quantized electronic energy levels, and no vibrational or rotational levels, the wavelengths emitted by atoms are highly discrete and monochromatic, and their spectral width is of the order of 10^{-3} nm. This circumstance, and the fact that the energy level structure of any two atoms is not exactly the same, renders techniques based on atomic spectroscopy so specific and free of spectral interferences, despite the thousands of wavelengths emitted or absorbed by the elements.

These discrete wavelengths are commonly observed as lines when photographed in the conventional manner and are referred to as "lines" for that reason even though modern recording devices no longer visualize them as such. In most instruments, radiation from the source is passed through a narrow slit, dispersed according to wavelength by a prism or diffraction grating, and focused onto an image plane. Thus, visually or photographically the discrete emission wavelengths appear at the focal plane as narrow lines which are the images of the entrance slit at each discrete wavelength. Usually one or more of the transitions ending in the electronic ground state is most probable, resulting in characteristic lines which are also of greatest intensity. Transitions ending in the electronic ground state are referred to as "resonance lines."

The basic requirements for analytical determinations by atomic emis-

[3a] ν is the frequency of the emitted radiation and is related to the wavelength, λ, by $\nu = c/\lambda$, where c is the velocity of light.

sion spectroscopic methods are as follows: (1) the sample must be atomized and the resulting atoms excited; in atomic emission methods, these two functions are both accomplished by the source; (2) the wavelengths of the resulting characteristic line emissions must be spectrally separated and their relative intensities measured; and (3) the resulting intensities must then be compared with those from suitable standards of known concentration of elements measured under identical conditions in order to determine the content of the desired element(s). The majority of the problems encountered in an analysis, particularly a direct determination performed on biological material, are associated with the choice of a suitable standard, a subject which will be discussed (see below).

The fraction of sample atoms excited under any given source conditions varies exponentially with the source temperature, the Boltzmann distribution being a good approximation in instances where the source attains or approaches thermodynamic equilibrium. Thus, the higher the temperature of the excitation source, the greater the emission intensity for a given analyte atomic concentration in the source. Coupled to the increased intensity as a function of increased excitation temperature, the atomic concentration of the sample also usually increases in cases in which atomization is not complete at lower temperatures. One might at first expect that the analytical sensitivity (i.e., the decreased detection limit) can be increased almost without restriction, simply by choosing excitation sources of higher temperature, but in actuality this approach encounters several limitations. At higher temperatures, ionization of the sample atoms becomes important; the spectrum becomes more complex as more upper-level lines are excited; and, perhaps most important, the background emission of the source also increases rapidly.

The principal advantages of atomic emission are high sensitivity, i.e., low analytical detection limits for many elements, relatively simple instrumentation, good specificity and speed of analysis, and ready adaptability to simultaneous determinations of many different elements, henceforth referred to as multi-element analysis for the sake of simplicity. The principal limitations are centered about the type of excitation source used and the inseparability of the atomization and excitation processes. With relatively low-temperature sources, many elements with high excitation energies are not excited adequately, while with relatively high-temperature sources high background and complex spectra may require a spectrometric system of high resolution.

Excitation Sources

Analytical atomic emission spectrometry relies upon the source, both to atomize the sample and to excite the resultant analyte atoms. Generally, emission sources fall into two groups: flames and electrical devices. The former include chemical flames resulting from the combustion of various fuel/oxidant combinations, while the latter include electrical discharges of various types, such as arcs, sparks, and plasmas. The primary function of these sources is the same—atomization and excitation—but there are important differences in the manner in which each brings them about.

Flames. The chemical flame is one of the oldest of emission sources, dating to 1860, the time at which Kirchoff and Bunsen began their work; it remains in wide use to the present. Flame emission sources offer good analytical sensitivity for a number of elements, and their simplicity, convenience, and inexpensive operation are unsurpassed. They are used widely in clinical laboratories for routine determinations of sodium, potassium, calcium, magnesium, and other elements. Most chemical flames have temperatures in the 2000–3000°K range and provide good analytical sensitivity for elements having excitation energies—for resonance lines—of less than about 5 eV, and which are atomized to an appreciable extent, since they do not form stable, refractory oxides at these temperatures.

Flame emission sources have a number of limitations which reduce their suitability for many applications: (1) temperatures above about 3000°K cannot be obtained with the usual fuel/oxidant combinations which result in relatively poor analytical sensitivity for many elements whose principal resonance lines arc below about 300 nm (excitation energies >5 eV); (2) at these temperatures many elements form stable monoxides in the highly reactive flame environment, effectively removing them from the atomic emission process; and (3) in certain spectral regions considerable background emission of flames leads to reduced analytical sensitivity for elements whose primary resonance line(s) are in these regions.

Because of the decreased stability of compounds as a function of increased temperature, and because atomic emission increases exponentially with temperature, flames of higher temperature would prove valuable as emission sources. In fact, this has been demonstrated for the cyanogen/oxygen flame[4] having temperatures well above 4000°K. Good analytical sensitivity is observed for many elements, yet this source has not gained acceptance, primarily due to the relatively high cost of cyanogen.

Burner systems popular in atomic emission flame spectrometry generally are of two basic designs: the "total-consumption," and the "chamber-type." The so-called total-consumption burners are of the familiar Beckman type (Beckman Instrument Co.), designed initially by P. T. Gilbert, Jr.; they survive virtually unchanged to this day. The burner is quite safe, since fuel and oxidant are mixed externally. Sample solution is

[4] Fuwa, K., Thiers, R. E., and Vallee, B. L. (1959). *Anal. Chem.* **31,** 1419.

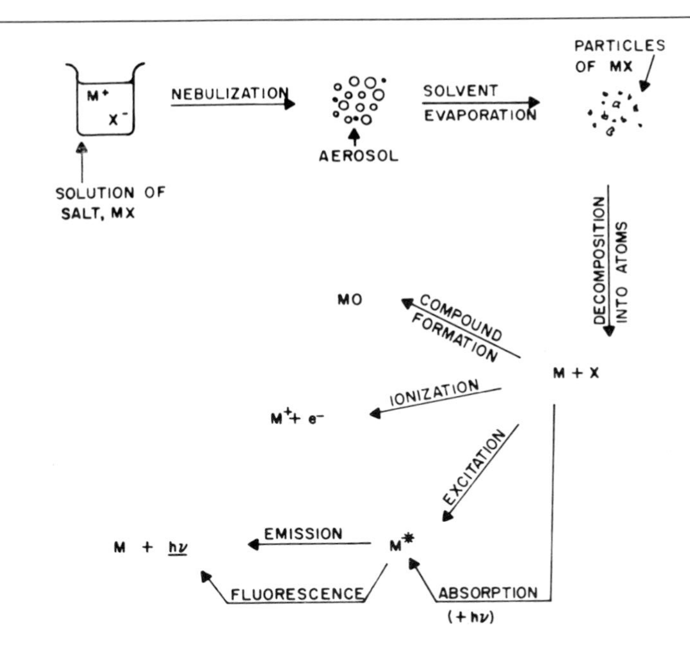

FIG. 1. Atomization/excitation process.

nebulized and sprayed into the flame directly. The "chamber-type" burners nebulize the sample solution in a separate chamber, where only a fraction of the finest aerosol droplets are carried into the flame. Fuel and oxidant are premixed in the chamber and conducted along with sample aerosol to a burner head, usually as employed in Meker-, Bunsen- or slot-type burners. The latter configuration is widely used today as an atomization cell for atomic *absorption* spectrometry (*vide infra*), although it has been shown to be a very good system for flame atomic emission, particularly with the nitrous oxide/acetylene flame.[5]

Rather than detail the many nuances of burner and nebulizer designs, considerations of burning velocity, flame composition, etc., the reader is referred to the chapter on emission flame spectroscopy by Vallee and Thiers[6] and the treatise by Mavrodineanu and Boiteux.[7]

No matter what burner system is employed, the series of events which occur in rapid succession in the atomization/excitation processes are similar, as illustrated in Fig. 1. The sample solution, usually aqueous, is dispersed into a fine mist which enters the flame. Here the droplets are desolvated and atomized rapidly, permitting excitation of the atomic

[5] Picket, E. E., and Koirtyohann, S. R. (1968). *Spectrochim. Acta, Part B* **23,** 235.

[6] Vallee, B. L., and Thiers, R. E. (1965). *Treatise Anal. Chem.* **6,** Part I, 3463.

[7] Mavrodineau, R., and Boiteux, H. (1965). "Flame Spectroscopy." Wiley, New York.

species. Atomic absorption and atomic fluorescence are also illustrated, to be discussed later; only the mode of excitation and/or measurement for these two processes differ.

The conversion of element(s) in the sample to gaseous atoms depends heavily on the thermodynamics and kinetics of the events of vaporization-dissociation. In this sense, the sample *matrix* may have a dominant effect on the final, steady-state, population of free atoms. This is often the case when dealing directly with biological matrices and emphasizes the importance of choosing standards carefully and verifying given procedures by independent means. This important aspect is overlooked quite often and will be reiterated in subsequent sections.

In addition to the analytically useful pathways of the analyte atoms (Fig. 1), two additional and undesirable processes are shown—compound formation and ionization—both of which occur to a greater or lesser degree depending upon the element and the flame conditions. Once an atom is ionized or becomes part of a molecule, it can no longer contribute to the *atomic* emission being measured. In general, ionization of the analyte atoms does not occur to a significant degree, except for elements with low ionization potentials in relatively hot flames. Formation of compounds is usually far more important.

Many elements form quite stable monoxides in flames:

$$M + \mathrm{O} \rightleftarrows M\mathrm{O}$$

Since this reaction is exothermic, an increase of the flame temperature would shift the equilibrium toward the free metal atom. Having little control over the flame temperature, a better approach is to lower the partial pressure of oxygen in the flame, thus shifting the equilibrium in the desired direction. By making the flame fuel-rich, especially for hydrocarbon fueled flames, the free oxygen content is lowered. Usually the flame temperature does not fall rapidly as the fuel/oxidant ratio increases above stoichiometry; thus, a net gain is realized. Similarly, the fuel-rich condition increases the partial pressures of highly reducing species within the flame, such as carbon monoxide and atomic carbon, further altering the equilibrium:

$$M\mathrm{O} + \mathrm{C} \rightleftarrows M + \mathrm{CO}$$
$$M\mathrm{O} + \mathrm{CO} \rightleftarrows M + \mathrm{CO_2}$$

Note, however, that this process will be applicable only in those cases where at the temperature of the flame, for example, CO is more stable than *M*O.

It must be stressed that processes like ionization and compound formation, or any other process which reduces the atomic concentration, because of matrix effects, *will equally affect atomic emission, absorption, and fluorescence.*

It was indicated earlier that for gaseous atoms in thermal equilibrium, as approximated in the flame, the observed emission intensity depends exponentially on the absolute temperature. Consequently, in flame atomic emission spectrometry it is of the utmost importance to control carefully all of the variables affecting the flame temperature. These include primarily the rates of fuel and oxidant flow, and that of sample introduction. Correct procedures for gas flow control and monitoring have been described.[8]

In choosing an analytical technique for trace element analysis, one very important criterion is the smallest amount of an element that can be determined, the so-called "detection limit." The terms "sensitivity" and "detection limit" are frequently employed for many techniques. Both are important criteria in evaluating the potential of an instrument or technique for a particular determination by different workers, and their semantic connotations frequently are very different.

In atomic emission spectrometry, flame or otherwise, *sensitivity* is a measure of the increase in the analytical signal as the concentration of the analyte element increases. In a plot of the emission signal versus the concentration of the element, the slope of the linear portion of the curve will be a measure of the sensitivity for that line of the element under given analytical conditions. It is not an "invariant" number, because a change in the source operating conditions, e.g., temperature, causes it to change. Furthermore, if the line emission is superimposed on a continuum of the background emission from the source, a change of the spectral bandwidth will alter the sensitivity. Narrowing the spectral bandwidth causes the line intensity to decrease, and the continuum background to decrease but not in equal proportion. The measured background intensity decreases more rapidly than that of the line, i.e., by the square root of the spectral bandwidth. Thus, reducing the spectral bandwidth by a factor of 4 improves the sensitivity by a factor of 2. This may or may not improve the detection limit, because decreasing the spectral bandwidth decreases detector signal and, since ultimately both noise and signal determine the detection limit, it may not improve.

The term *detection limit* is somewhat of an enigma. It implies the smallest amount of an element that can be *detected* reliably. In practice, it refers to the smallest concentration or amount of analyte that can be measured. There is no sharp boundary here between purely qualitative detection and quantitative determination: as the concentration of the sample is raised above the detection limit the analytical precision improves

[8] Veillon, C., and Park, J. Y. (1970). *Anal. Chem.* **42,** 684.

rapidly of course. Some criterion had to be established and accepted generally. Presently most workers define a signal-to-noise ratio of 2 as establishing the point which denotes the detection limit. In the case of emission, this usually represents "that concentration or quantity of analyte where the signal (above background or blank) is equal to twice the fluctuations of the background or blank." Oher criteria and definitions have been proposed and are used, but this is perhaps the one accepted most widely. It, too, is not an "invariant" number, because it may depend on several factors, such as sensitivity, type of signal processing, and time constant of the electronic measurement system.

Frequently one can "trade off" the duration of the time interval over which the measurement is performed for an improved signal-to-noise ratio; the "trade-off" is not linear, but rather a square root function, analogous to the spectral bandwidth sensitivity relationship mentioned earlier.

While on the subject of detection limits it might be appropriate to point out that these are usually expressed in concentration units. This is acceptable in cases in which there is no limitation of sample, by reasons of sensitivity or whatever. A more objective approach, allowing a more direct comparison of data and analytical techniques is to express the detection limits on an absolute basis such as weight: What is the smallest amount of an element that can be determined?

The preceding is intended to point out that sensitivity and detection limit values are good indications of the capabilities of an instrument or technique for trace analysis, but small differences may not be meaningful. While differences of an order of magnitude are certainly significant, a factor of 2 probably is not.

Detection limits for a number of elements by N_2O/C_2H_2 flame atomic emission spectrometry using a slot burner and taken from the work of Pickett and Koirtyohann[5] are shown in Table I. This system consumes about 3 ml min^{-1} of sample solution; thus, if 20 sec (i.e., 1 ml) are sufficient for a measurement, on an absolute basis, the minimum detectable amounts of each element, in micrograms, will be as shown in Table I. About a dozen of these elements can be determined at levels of 10 ng or less, about 30 can be determined at levels of 1 μg or less, and only a few have detection limits above the 1 μg level. While this study did not report on all the elements, it gives a good idea of the capabilities of flame atomic emission spectrometry, i.e., the capacity to determine perhaps half of the elements in the periodic table at microgram levels and below.

Spark Sources. High-frequency, high-voltage spark discharges have been used for many years as atomic emission sources, particularly with spectrographs and multichannel spectrometers for simultaneous, multielement analysis. This source was developed as an alternative to the early

TABLE I
DETECTION LIMITS BY EMISSION IN THE N_2O/C_2H_2 FLAME[5]

Element	Wavelength (*A*)	Detection limit (μg ml^{-1})
Al	3962	0.01
Ag	3281	0.02
Au	2676	0.5
Ba	5536	0.002
Be	2349	40.
Ca	4227	0.0001
Cd	3261	2.
Co	3454	0.05
Cr	4254	0.005
Cu	3274	0.01
Fe	3720	0.05
Ga	4172	0.01
Ge	2651	0.5
In	4511	0.005
La	5791	2.
Li	6708	0.00003
Mg	2852	0.005
Mn	4031	0.005
Mo	3903	0.1
Nb	4059	1.
Ni	3415	0.03
Pb	4058	0.2
Pd	3635	0.05
Pt	2660	2.
Re	3461	0.2
So	4020	0.03
Sn	2840	0.3
Sr	4607	0.0002
Ti	3999	0.2
Tl	5351	0.02
V	4379	0.01
W	4009	0.5
T	4077	1.
Zr	3601	3.

DC are sources which were rather unstable and not suitable for quantitative work. The individual discharges consist of a high-voltage pulse, followed by a series of oscillatory, arclike discharges. These individual discharges usually occur at radio frequencies and have a total duration on the order of 10 μsec. Approximately 1000 discharges occur per second, so the spark discharge occurs for only a relatively small fraction of the analysis time, approximately 1% in the above example, resulting in little heating of the bulk sample. This permits the direct analysis of solutions. The spark source is quite reproducible, permitting quantitative measurements, although the small amount of sample consumed leads to lower analytical sensitivity compared to the DC arc source. The *effective excitation temperature* in the discharge is well over 10,000°K which means that virtually any element will be excited.

Solutions are usually determined using a porous cup electrode or a rotating disc electrode. The former consists of a porous-bottom graphite cup containing the sample solution, with the counter electrode beneath and discharging to the wet bottom of the porous cup.

The rotating disc electrode consists of a rotating graphite disc, the lower edge of which dips into the sample solution and carries it to the spark discharge region at the top of the disc. Numerous other electrode arrangements have been used, but these are the most popular ones.

The spark source, when coupled to a spectrograph or multichannel, direct-reading spectrometer, represents a very powerful analytical measurement system. Quantitative determinations of many trace metals can be made simultaneously on a single small sample. This system has proven to be invaluable in the investigation of trace metals in biological materials, especially in the identification and characterization of new metalloenzymes.[9]

Arc Sources. The DC arc discharge is widely used as a spectrochemical excitation source, and is almost always employed in conjunction with a spectrograph or multichannel spectrometer. A relatively high-current, low-voltage discharge is maintained between two electrodes, usually graphite, one containing the sample. The effective excitation temperatures observed in the arc are typically in the 5000–6000°K range, and since all of the sample is vaporized by the discharge, this source exhibits very good analytical sensitivity for virtually all of the elements. However, as indicated previously, the early arc sources exhibited poor reproducibility limiting these to semiquantitative measurements, and they were not readily adaptable to the analysis of solutions.

The development of the Gordon arc[10] appears to have eliminated virtually all of the problems associated with quantitative analysis using the DC arc, such as reproducibility, matrix effects, and handling liquid samples. A novel, molten-bead tantalum cathode, controlled atmosphere, and pointed carbon anode serve to eliminate arc "wandering," the major source of trouble in earlier arc configurations. A more important characteristic is the reduction of matrix effects, achieved with the use of carbon anodes impregnated with silver chloride at an AgCl/sample weight ratio of 400.

[9] Vallee, B. L. (1955). *Adv. Protein Chem.* **10,** 317.

[10] Gordon, W. A., and Chapman, G. B. (1970). *Spectrochim. Acta, Part B* **25,** 123.

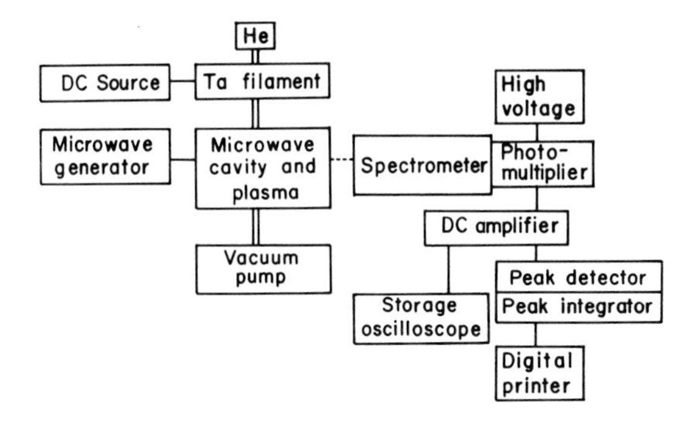

FIG. 2. Block diagram of apparatus.

Sample volumes of 10 μl are used, and unity-slope calibration curves are obtained for those elements determined simultaneously over a range of 10 ng to 10 μg. Hambidge has documented the use of this system for trace-metal determinations at nanogram levels in a variety of biological materials.[11] He obtained sensitivities, in nanograms, of 0.5 for Zn, 1.0 for Cr, Mn, and Mg, 2.5 for Cu, 4.0 for Fe, and 13 for Mo. Clearly, this system warrants further study.

Microwave Plasmas. The low-pressure, microwave-induced plasma, an emission source developed recently, has been shown to be one of the most sensitive yet designed for the determination of trace elements in aqueous solutions.[12] Basically, the system operates as follows: A helium flow at low pressure is contained in a quartz discharge tube. The tube passes through a microwave resonant cavity, into which is fed about 50 W of microwave power at a frequency of 2450 MHz. A plasma discharge is initiated in the tube and is sustained by the microwave field. Above the plasma, in the helium carrier gas stream, is a small tantalum filament onto which microvolume samples are placed and dried. The filament is then pulse-heated by capacitive discharge, vaporizing the sample which is then swept into the plasma. The resultant atomic emission is then measured using a spectrometer. Sample volumes of only 5 μl are used, detection limits (on an absolute basis) are in the 1–50 pg range, and little or no sample preparation is required.

A block diagram of the instrumental arrangement is shown in Fig. 2. The instrument is very compact, electronically sophisticated, and highly automated in its operation. The plasma source is shown schematically in Fig. 3.

This analytical system was designed primarily for the determination of *picogram* quantities of metals in metalloenzyme preparations. To verify its

[11] Hambidge, K. M. (1971). *Anal. Chem.* **43,** 103.
[12] Kawaguchi, H., and Vallee, B. L. (1975). *Anal. Chem.* **47,** 1029.

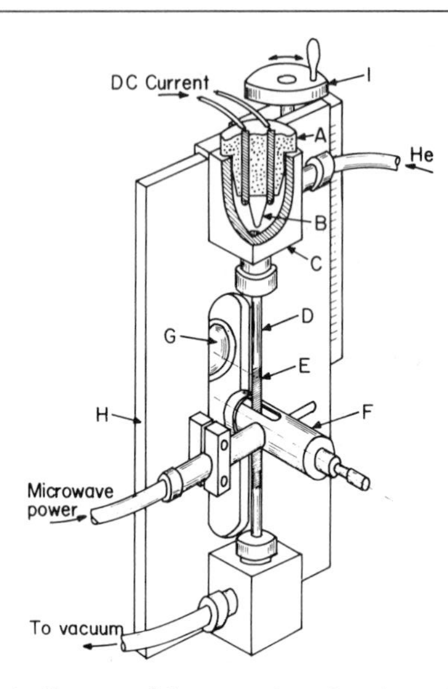

FIG. 3. Schematic diagram of the mounting of cavity and discharge tube.

utility for this purpose, Kawaguchi and Vallee[12] determined the metal stoichiometry in highly purified, well-characterized preparations of several zinc metalloenzymes, and compared these data to those obtained for the same material by atomic absorption spectrometry. The enzymes used as standard reference materials were bovine carboxypeptidase A, human carbonic anhydrase, horse-liver alcohol dehydrogenase, and *E. coli* alkaline phosphatase. In addition to complete agreement between the two independent methods, it is interesting to note that the atomic absorption method required about 100 μg of enzyme per determination, while the microwave-induced emission measurements required only about 0.1 μg of enzyme. Subsequently, they were able to demonstrate that reverse transcriptase from avian myeloblastosis virus—available only in miniscule amounts—contained stoichiometric amounts of zinc, while Cu, Fe, and Mn are absent.[13,14] Following this, the reverse transcriptases from mammalian tumor viruses (murine and simian leukemic viruses) have been shown similarly to contain stoichiometric amounts of zinc, employing this instru-

[13] Auld, D. S., Kawaguchi, H., Livingston, D. M., and Vallee, B. L. (1974). *Biochem. Biophys. Res. Commun.* **57**, 967.

[14] Auld, D. S., Kawaguchi, H., Livingston, D. M., and Vallee, B. L. (1974). *Proc. Natl. Acad. Sci. U.S.A.* **71**, 2091.

mental system,[15] as have been RNA polymerases I and II from *E. gracilis*, the RNA polymerase I from yeast, and methionyl-tRNA transferase.

Following these initial studies, the system has been utilized further in investigations of mercury substituted[16] and arsenic-labeled metalloenzymes.[17]

This emission source is unique in that it operates at a low pressure in an atomic gas having rather unusual metastable energy levels. Efforts have been made to understand and explain the abnormal excitation observed in this plasma.[18,19] Extraordinary enhancement effects have been observed for several elements when samples are placed in a KCl matrix.[12] This has also been studied by Busch and Vickers[20] in a similar source, and revolves mainly around the incongruity of a relatively low gas (i.e, kinetic energy) "temperature" and an extremely high effective excitation "temperature" in the plasma discharge. The KCl matrix alters this, and apparently serves as a spectroscopic and matrix buffer, analogous to that observed with the AgCl matrix in the Gordon arc (*vide supra*).

Emission Spectrometers and Spectrographs

General

Once a suitable emission source has been selected, a means of isolating or separating the atomic lines and measuring their intensities is required. Normally, this is achieved with either a spectrometer or spectrograph, the primary difference between the two being the means of detection and intensity measurements. Both consist of a dispersing element (monochromator), e.g., a prism or grating to separate the various radiations emitted according to wavelength, and detector(s) of some type. A *spectrometer* employs a photoelectric device, such as a photomultiplier, to convert the radiation into an electrical signal, while a *spectrograph* relies on photographic plates or film for detection. The basic design of spectrographs, the pertinent physical principles, and a suitable historical perspective can be found in the classical treatise of Harrison, Lord, and Loofbourow.[21]

[15] Auld, D. S., Kawaguchi, H., Livingston, D. M., and Vallee, B. L. (1975). *Biochem. Biophys. Res. Commun.* **62,** 296.

[16] Atsuya, I., Kawaguchi, H., and Vallee, B. L. (1977). *Anal. Biochem.* **77,** 208.

[17] Atsuya, I., Alter, G. M., Veillon, C., and Vallee, B. L. (1977). *Anal. Biochem.* **79,** 202.

[18] Kawaguchi, H., Atsuya, I., and Vallee, B. L. (1977). *Anal. Chem.* **49,** 266.

[19] Atsuya, I., Kawaguchi, H., Veillon, C., and Vallee, B. L. (1977). *Anal. Chem.* **49,** 1489.

[20] Busch, K. W., and Vickers, T. J. (1973). *Spectrochim. Acta, Part B* **28,** 85.

[21] Harrison, G. R., Lord, R. C., and Loofbourow, J. R. (1948). "Practical Spectroscopy," p. 605. Prentice-Hall, Englewood Cliffs, New Jersey.

Spectrographs. The use of a photographic emulsion as the detection system has several advantages: All the spectral information is recorded at once over a wide and continuous wavelength range; the emulsion is an integrating device, recording the product of intensity and time, and the permanent record of all of this information is available for study in the future. The primary disadvantages of photographic emulsions as detectors are associated with the image development process and the information readout. Both are time consuming and less accurate than photoelectric devices, and photography is complicated further by the fact that the emulsions of films or plates are inherently nonlinear devices, both in terms of intensity and wavelength, mandating careful calibration, one of the most time-consuming features of these systems.

Single-channel Spectrometers. The single-channel spectrometer is by far the one used most widely in flame atomic emission spectrometry, atomic absorption, and atomic fluorescence spectrometry. These spectrometers usually consist of a plane grating monochromator in an Ebert or Czerny-Turner mounting, a mirror focal length between 0.25 and 1 m, equal and bilaterally adjustable entrance and exit slits, a photomultiplier detector, and usually provisions for electrical wavelength scanning. Signal-processing electronics vary from simple electrometer circuits to sophisticated lock-in amplifiers, photon counting or integration, and readout systems vary from simple meters to direct concentration printout. Similar instrumentation is also used with certain high-energy excitation sources, plasmas.

Multichannel Spectrometers. These instruments are usually spectrographs in which the photographic detector has been removed and several exit slit–photoelectric detector units substituted; the slits are placed at appropriate points on the focal curve of the instrument to monitor specific wavelengths of elements. These direct-reading, multichannel instruments are usually arranged in the smaller instruments to analyze simultaneously for a few elements—to as many as 20 to 30 elements in the larger ones. The excitation sources are usually of the electrical discharge variety (arc, spark, plasma, etc.). While not very versatile in terms of being able to change easily the elements to be determined, they are fast, reproducible, and very well suited to the analysis of large numbers of samples of a given type. Some of the larger instruments are quite sophisticated electronically, with built-in computers to process the data, control the source, and print out the analytical results. The only real limit to the number of elements that can be determined simultaneously is the physical space behind the focal curve for the mounting of detectors. Despite the level of sophistication of modern multichannel emission spectrometers, they still represent a "brute force" approach and generally fall short of detection limits required for biological samples such as enzymes, largely due to the source employed. Some promising new approaches to simultaneous multi-element analysis by atomic emission spectrometry are mentioned in the following section.

New Developments. Several new instrument designs, combining suitable sources, dispersing devices, and detectors have appeared in recent years and show promise of improving the sensitivity, versatility, and multi-element capabilities of emission spectrochemical analysis.

In the area of dispersing devices, holographically ruled gratings have been developed and interest in the echelle grating has resurged recently. Holographic gratings, while not yet perfected in terms of groove shape, are perfect with respect to groove spacing, freedom from periodic errors, and so on, resulting in spectra of exceptional purity. Stray light, scatter, and ghosts are essentially absent, and very high resolving powers can be achieved. This should aid in the development of high-resolution, high-aperture monochromators which are important in emission spectrometry.

Renewed interest in the echelle grating has also been generated in recent years. These are relatively coarse gratings which depend on use at high diffraction orders to achieve high dispersion. The problem of overlapping orders is solved by cross-dispersion with a prism. The result is a more-or-less square array of wavelengths, i.e., spectral lines, with an order-overlap-free region (free spectral range).

In recent years, perhaps the greatest interest in multi-element emission techniques has centered on multichannel detectors. These include such devices as television camera tubes and arrays of photodiodes, phototransistors, or photoresistors (vidicons). Clearly, the photographic emulsion is a multichannel detector, but we are here referring to electronic devices which can replace the emulsion to good advantage. The use of these devices was first suggested by Margoshes.[22] Photodiode and phototransistor arrays were studied as detection devices for multichannel emission spectrometry by Boumans and Brouwer.[23] They found phototransistors to be superior to photodiodes. Busch, Howell, and Morrison[24] and Mitchell, Jackson, and Aldous[25] used silicon vidicon tubes for simultaneous multi-element flame emission spectrometric determinations. The advantages of these systems are that a compact, low-dispersion monochromator can serve to cover a wide spectral region, closely spaced lines can be measured, and they are fast. The main disadvantage, at least at present, is low sensitivity which is low relative to that of a photomultiplier. This appears to

[22] Margoshes, M. (1970). *Spectrochim. Acta, Part B* **25**, 113.

[23] Boumans, P. W. J. M., and Brouwer, G. (1972). *Spectrochim. Acta, Part B* **27**, 247.

[24] Busch, K. W., Howell, N. G., and Morrison, G. H. (1974). *Anal. Chem.* **46**, 575.

[25] Mitchell, D. G., Jackson, K. W., and Aldous, K. M. (1973). *Anal. Chem.* **45**, 1215A.

be a state-of-the-art limitation, and future developments with intensified tubes may overcome this limitation.

Atomic Absorption Spectrometry

Scientific history will no doubt record the development of atomic absorption spectrometry as one of the more significant ones in this period of phenomenal scientific growth. First proposed as a means of chemical analysis by Walsh[26] in 1955, the method received little attention for several years. Then the chain reaction began, no doubt aided to some extent by the burst of scientific advancement that occurred in the 1960s. Near the end of this decade, a plot of the number of atomic absorption instruments sold, assuming each occupied one square meter of space, versus time, when extrapolated to the year 2000, indicated that the entire land mass of the earth would be covered with atomic absorption spectrometers! While this plot was made with "tongue-in-cheek," it indicates the usual growth rate of an idea whose time has come. Much of our current knowledge of and interest in trace metals in biological systems—including the rapid increase in the number and diversity of metalloenzymes—is due to the development of this analytical technique.

While many of the basic advances came from investigators in trace-metal analysis, much of the refinement, perfection, sophistication, and "selling" of these ideas and the method are due to the efforts of scientists associated with the manufacturers of these instruments. Massive efforts by companies like Perkin-Elmer, Jarrell-Ash, and Instrumentation Laboratories (in the U.S.) and Techtron (in Australia; now a subsidiary of Varian), to name but a few, are to a great extent responsible for the wide acceptance and utilization of atomic absorption in numerous fields. Early zeal and enthusiasm promised to the scientific community that this technique could perform almost all analyses virtually without interferences; in short, it was the greatest thing since sliced bread! Such exaggerations have largely died away as the method was applied more broadly in various fields, and most of the virtues and limitations are now appreciated widely. Most, but not all.

Atomic absorption spectrometry has evolved into a sensitive, highly specific and widely used method-of-choice for trace-metal determinations. Significant advances have included the long-path absorption tube,[27] the nitrous oxide/acetylene flame,[28] and the graphite furnace atomization

[26] Walsh, A. (1955). *Spectrochim. Acta* **7,** 110.
[27] Fuwa, K., and Vallee, B. L. (1963). *Anal. Chem.* **35,** 942.
[28] Willis, J. B. (1965). *Nature (London)* **207,** 715.

cells.[29] Unfortunately, the great bulk of the applications has been concerned with determinations in inorganic sample matrices. Here, the method *does* fulfill most of its earlier promises, but when extended to biological samples and other complex matrices, rather serious limitations appear. Invariably, these are due to overextrapolations from simple and well-defined matrices to measurements made in far more complex systems. For example, it is one thing to demonstrate infallibility in simple inorganic systems, but it is quite another to boldly extrapolate these results linearly to samples ranging from sea water to whole blood.

Let us look briefly at some of the salient features of this widely used method, omitting for the moment detailed suggestions for working with biological materials. The latter is a problem common to the three techniques of atomic emission, absorption, and fluorescence, and will be discussed in a later section.

Basis of the Method[6,7]

Just as atoms can be excited thermally and emit radiation of discrete wavelengths, unexcited atoms can also absorb radiation at these same discrete energies. The absorption measurement is made with an external source which emits the narrow line spectrum of the desired element, with the flame or other source serving only as an atomization cell for the sample. The process can be illustrated as:

$$M + h\nu \text{ (from external source)} \rightarrow M^*$$

The decrease in the intensity of the appropriate emission line from the source due to absorption by analyte atoms is measured. The measurement must be made under conditions in which Beer's absorption law is valid. Beer's law may be expressed as:

$$\log I^0/I = A = abc$$

where I^0 is the source intensity measured in the absence of sample, i.e., when atomizing the blank solution, and I is the intensity measured when atomizing the sample or standard. The base-10 log of this ratio is the *absorbance* (A) and is equal to the product of the atomic absorption coefficient, a, a constant having a different value for each transition in each element, the absorption path length, b, and the concentration *of absorbing species*, c, within the atomization cell. Since a is a constant for a given line of a given element, analytical sensitivity can only be improved by altering the path length, b, over which absorption takes place, or the concentration,

[29] L'vov, B. V. (1961). *Spectrochim. Acta* **17,** 761.

c, of absorbing atoms within the atomization device. Path-length modification is employed routinely, as evidenced by the universal application of slot burners in commercial instrumentation. This approach is utilized to its fullest extent in the Fuwa-Vallee long-path absorption tube,[27] which by virtue of its long absorption path length and the use of a "total-consumption" burner provides improved analytical sensitivities by about a decade over those of conventional commercial systems for a number of biologically important elements.

Impressive sensitivity improvements have been achieved in recent years with the advent of graphite tube furnace atomizers. Due to their oxygen-free atmospheres, reducing environment (the carbon furnace walls), small volumes, and relatively long atom residence times in the optical path, sensitivity improvements of 2–3 orders over conventional flames can be realized with simple matrices. However, there are rather serious limitations to current designs of these furnace atomizers, especially when applied to samples with complex matrices, as will be discussed in a later section. An additional limitation of conventional atomic absorption spectrometry is its unsuitability for simultaneous analyses of multiple elements. This is due to two factors in current designs: (1) A separate narrow-line source is needed for each element to be measured, and these must have a common optical axis; and (2) the limited dynamic range of absorption measurements since absorption is a ratio measurement, and good precision requires that measurements be confined to the 10–90% transmitted light region, effectively limiting it to a concentration range of about one decade. Limitation (2) is inherent in the method, but we will later describe some ingenious methods of overcoming limitation (1).

Atomization Cells

Flames. The chemical flame is used most widely by far as an atomization device for atomic absorption spectrometry. All of the desirable features cited earlier for the flame as an emission atomization/excitation source apply to atomic absorption spectrometry, with the added advantage that the requirement for *excitation* is eliminated. Between the popular air/acetylene and nitrous oxide/acetylene flames, a large number of elements can be determined at suitably low levels. Temptingly, the graphite furnaces are offered by the manufacturers, with quoted detection limits of 2–3 *orders* lower than those of flame atomizers. Unless one *needs* that kind of sensitivity, it is best to restrain oneself; working at the concentration levels attainable by the furnace systems presents serious problems in contamination control, not to mention matrix effects, which are poorly understood at present.

Furnace Atomizers. Following the earlier work of L'vov[29] with resistively heated graphite tube furnace atomizers for atomic absorption spectrometry, numerous alternatives to the chemical flame were investigated. These included rods, tubes, filaments, and boats fabricated of graphite, tantalum, tungsten, and platinum in a variety of configurations. The devices which have survived to the present are all basically small-volume graphite tube furnaces designed to atomize small, individual samples.

Commercially available versions all operate in a similar manner. Sample solutions ranging in size from about 5–50 μl are placed in the furnace and dried by briefly raising the tube to an appropriate temperature. In applications where the sample matrix is partly organic in nature, the tube is briefly raised to a temperature of several hundred degrees to "ash" the sample, thus reducing undesirable smoke formation during the atomization and measurement step. The sample is then atomized by rapidly heating the tube to a controlled high temperature up to about 3000°.

The oxygen-free atmosphere, the reducing conditions afforded by the presence of hot carbon, the small furnace volume, and the relatively long residence time of atoms in the optical path, all serve to greatly increase the analytical sensitivity relative to flame atomization. For most elements, this increase is on the order of 100-fold, on an absolute basis. In addition, the utilization of microvolume amounts constitutes an important advantage in situations where very limited amounts of sample exist. This may be a *disadvantage* when the sample does not constitute the limit, since small samples *must* be used.

The primary limitations of these devices are due, in large part, to their mode of operation. The samples are vaporized and atomized under rapidly changing conditions while the analytical measurement is in progress. Evaporating and dissociating a sample in a complex matrix, while in contact with the furnace wall and in an atmosphere undergoing rapid temperature change, can lead to rather serious matrix effects. Any component of the matrix which affects the evaporation or dissociation processes, or their kinetics, will affect the accuracy of the method unless duplicated *exactly* in the standards. Thus, the difficulties associated with direct measurements in biological materials should be obvious, yet this is seldom recognized or appreciated. For this reason alone, it is imperative that analytical accuracy in applications involving direct measurements on biological materials be verified on the sample by one or more *independent* methods—or at least on samples which are closely similar when identical ones are precluded for some reason. This situation is improving with the increasing availability of biological standard reference materials certified for trace-metal content which are being issued by the U.S. National Bureau of Standards.

Volatilization losses of trace metals during the drying and ashing modes constitute another limitation encountered frequently. This is an especially important problem for elements like Hg, As, Se, and Cr in biological matrices, and for many others when the matrix includes, for example, appreciable quantities of halides, since under suitable conditions many elements can form relatively volatile compounds.

Ashing of biological samples in the furnace tube to remove organic material may be incomplete at the temperature used, which must be below that at which volatilization loss occurs, and when the sample is atomized considerable smoke formation occurs. Since this will obstruct the radiation emanated by the source to some extent, it will be measured erroneously as analyte atoms. To correct for this, as well as for any molecular absorption or scattering caused by the matrix components, as an option, most instrument manufacturers offer a "background corrector" which measures the attenuation of the signal from a continuum source over a relatively wide spectral interval bracketing the very narrow atomic absorption line profile and subtracts it from the atomic line source signal. The process actually constitutes a compensation rather than a true correction, and only works well within certain limits. This feature should be part of the instrument used by an investigator employing the commercial furnace atomic absorption systems!

There are means to circumvent most of these limitations of furnace atomic absorption for biological materials (see below). However, many of these also complicate the work considerably and introduce additional problems which must be controlled. At times, "the cure is worse than the disease"; hence, our earlier advice stands: These devices should not be used unless the sensitivity is *needed*.

New Developments. A novel multi-element, furnace atomic absorption instrument is under development, intended specifically for direct trace-metal determinations in biological materials;[30] it may overcome many of the limitations of other designs for this purpose. It employs a single continuum source and a high-resolution echelle spectrometer. Spectral bandwidths approach atomic absorption line widths, permitting the same analytical sensitivity with a continuum source as that obtained with separate line sources in conventional atomic absorption instrumentation. Wavelength modulation is employed to provide true background correction over a wide range of course attenuation by the background, and it does not affect the analyte signal. To overcome the dynamic range limitation of absorption measurements, two or more lines of each element with considerably different sensitivities can be monitored. With a single source, the

[30] Wolf, W. R. (1977) Nutrition Institute, USDA, Beltsville, Maryland; O'Haver, T. C. (1977) University of Maryland, College Park (personal communication).

optical limitation of multiple-line sources is unimportant. The instrument is capable of monitoring 20 lines simultaneously, and these can be interchanged rapidly to another set with a simple drop-in cassette. Certainly this system bears watching.

Atomic Fluorescence Spectrometry

Basis of the Method

Essentially, atomic fluorescence spectrometry is a combination of the absorption and emission processes, retaining most of the advantages of each and possessing some unique assets of its own. The process can be represented as:

$$M + h\nu \text{ (from external source)} \rightarrow M^* \rightarrow M + h\nu$$

Radiation from an external source is absorbed as atomic absorption, resulting in excitation of the analyte atoms, and a fraction of these excited atoms decay by radiative means as atomic emission. It is an absorption–reradiation process, hence the name "fluorescence." It differs from molecular fluorescence in that the absorption and re-emission usually occur at the same wavelength. It enjoys the advantages of atomic absorption in that the excitation process does not depend exponentially on temperature, and excellent specificity is realized. It enjoys the advantages of atomic emission in that a linear dynamic range of measurement of several decades is commonly available, and it is readily adaptable to simultaneous, multi-element determinations. The optical arrangement, with the light source off the optical axis, entails to atomic fluorescence two of its greatest advantages over atomic absorption: (1) The analytical sensitivity is *directly proportional* to the source intensity; and (2) line width considerations of the source are no longer important, since the instrument does not "see" the source, only the re-emitted radiation. Consequently, narrow-line sources for each element to be determined are not required and, in fact, a continuum source can be used without loss of inherent sensitivity.[31] Only the source *intensity* is important. Thus, several line sources can be arranged around the atomization cell, or a single continuum source can be used[32,33] when performing simultaneous, multi-element determinations, taking advantage of the very simple atomic fluorescence spectrum.

[31] Veillon, C., Mansfield, J. M., Parsons, M. L., and Winefordner, J. D. (1966). *Anal. Chem.* **38,** 204.

[32] Murphy, M. K., Clyburn, S. A., and Veillon, C. (1973). *Anal. Chem.* **45,** 1468.

[33] Clyburn, S. A., Bartschmid, B. R., and Veillon, C. (1974). *Anal. Chem.* **46,** 2201.

Scattered source radiation is the primary limitation of atomic fluorescence spectrometry. This is true of any fluorescence technique and is due to the fact that excitation and measurement are usually made at the same wavelength. Actually, this is only a limitation when *line* sources are employed, since, under these circumstances, scatter is not distinguishable from fluorescence. Using a continuum source, wavelength scanning techniques readily eliminate this problem. In practice, scattering seldom constitutes a serious problem, since efficient atomization systems are usually employed.

Atomization Cells

Both flames and furnace devices are used in atomic fluorescence spectrometry. The requirements of an atomization system are basically the same as those for atomic absorption, i.e., efficient conversion of the sample into ground-state, gaseous atoms. Only the geometric considerations differ, since long-path absorption is not practical.

Flames. Chemical flames are used widely in atomic fluorescence, and their advantages cited earlier for atomic emission and absorption apply equally for atomic fluorescence. Capillary burners have become popular, and the development of separated flames has been a major advance.[34] These exhibit drastically reduced background emission, important in fluorescence where relatively low light levels are measured.

Furnaces. The same advantages of furnace atomization systems realized in atomic absorption are also achieved in atomic fluorescence spectrometry. An additional advantage over atomic absorption is also attained if an *atomic* gas like argon is used in the atomization system. This is due to the improved quantum efficiency of the fluorescence process.[31,35]

Recently, some major advances in furnace atomizers for atomic fluorescence have been made, and the results have been applied to metalloenzymes. The optical arrangement of atomic fluorescence precludes utilization of conventional graphite tube furnace arrangements, which were designed for absorption measurements. In addition, the transient conditions described earlier, which lead to serious matrix effects, gave rise to a furnace system specifically designed for atomic fluorescence measurements.[36] The system which evolved employs continuous introduction of desolvated sample[37] and continuous operation, thus greatly reducing matrix effects[38] and the complexity of the power supply. A novel methane

[34] Larkins, P. L. (1971). *Spectrochim. Acta, Part B* **26,** 477.
[35] Veillon, C. (1976). *Chem. Anal. (N.Y.)* **46,** 123–181.
[36] Murphy, M. K., Clyburn, S. A., and Veillon, C. (1973). *Anal. Chem.* **45,** 1468.
[37] Veillon, C., and Margoshes, M. (1968). *Spectrochim. Acta, Part B* **23,** 503.
[38] Kantor, T., Clyburn, S. A., and Veillon, C. (1974). *Anal. Chem.* **46,** 2205.

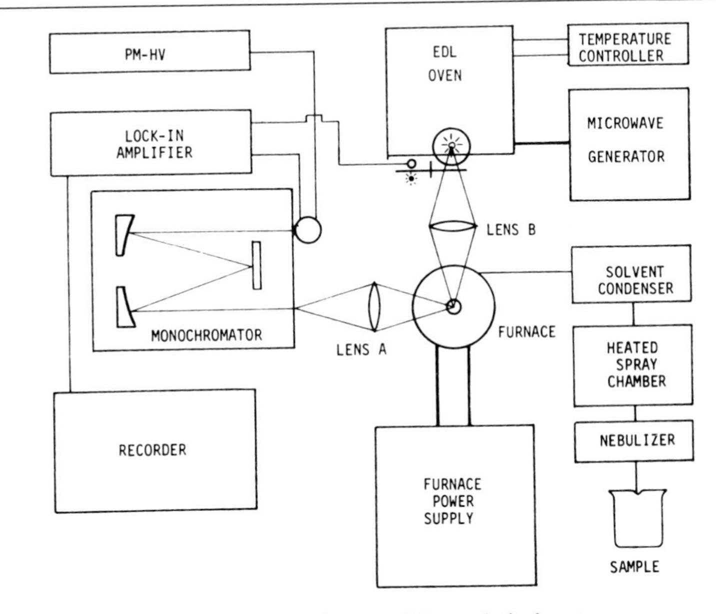

FIG. 4. Schematic diagram of the analytical system.

pyrolysis treatment was developed, leading to long-term stability and freedom from memory effects for the system.[39]

A schematic diagram of this system is shown in Fig. 4. The furnace is shown in Fig. 5.

The system was first applied to the direct determination of the zinc content and stoichiometry in DNA-dependent RNA polymerase isolated and purified from *E. coli*, using well-characterized carbonic anhydrase as the standard.[40] It was demonstrated that the inorganic and metalloenzyme standards gave comparable results, and that purified RNA polymerase from *E. coli* contains 1.99 ± 0.03 g-atom of Zn per 470,000 molecular weight.

Subsequently, the system has been applied to the direct determination of the metal stoichiometry of yeast alcohol dehydrogenase, bovine carboxypeptidase A, *E. coli* alkaline phosphatase, and human-liver alcohol dehydrogenase,[41] and in these instances the analyses were verified by two independent analytical methods employed on the same samples. On an absolute basis, detection limits for most elements are generally in the low

[39] Clyburn, S. A., Kantor, T., and Veillon, C. (1974). *Anal. Chem.* **46,** 2213.

[40] Clyburn, S. A., Serio, G. F., Bartschmid, B. R., Evans, J. E., and Veillon, C. (1975). *Anal. Biochem.* **63,** 231.

[41] Veillon, C., and Sytkowski, A. J. (1974). *Biochem. Biophys. Res. Commun.* **67,** 1494.

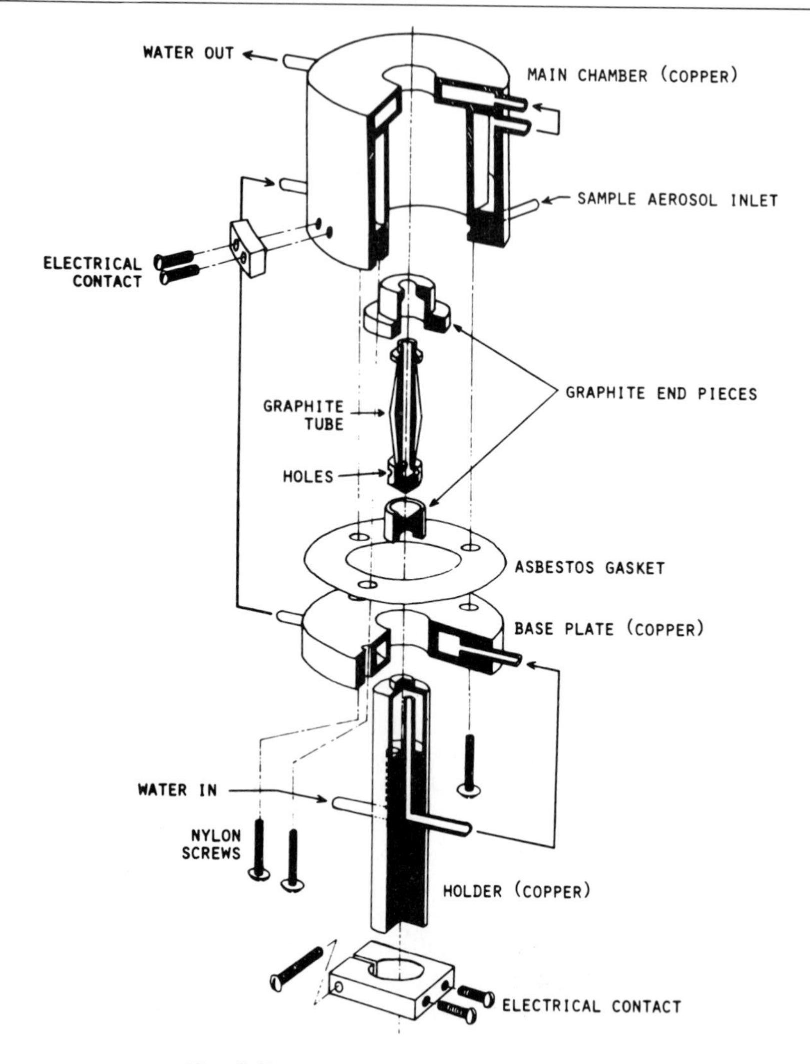

FIG. 5. Furnace, exploded view (not to scale).

picogram range.[34] Generally, work can be performed over several orders of magnitude above the detection limit, because the extraordinary sensitivity of the technique renders just the *measurement* of the detection limit an exercise in contamination control.

It should be emphasized that the method, though of high potential, has not as yet been applied widely, largely owing to the lack of readily available commercial instrumentation.

Trace-metal Determinations of Enzyme Preparations

As the saying goes, "That was the good news. . . ." If the reader is still with us at this point, he: (1) has a need for—or at least an interest in—trace-metal determinations in biological materials; (2) has selected, is contemplating purchasing, or is limited to one or more of the analytical systems described earlier; (3) finds his present system and/or procedures fraught with difficulties and is disappointed with it; (4) is simply curious; or (5) a masochistic streak may require attention.

This section will outline the procedures for the successful performance of this task, and the reasons why these procedures are important. We draw upon many man-years of experience in this field, and many lessons learned "the hard way." We draw upon the old—observations that have withstood the test of time—and the very new. We are convinced that the widespread adoption of these analytical methods and techniques may go a long way toward the elimination of an unwelcome percentage of unsuccessful experiments and inadequate data being generated by biochemical efforts based on less familiarity, and with and without access to information pertinent to trace-metal analysis, and *vice versa*.[42] We trust that this remark will not be misinterpreted. The nature of our work has forced on us what may now best be described as experience and conventional wisdom. We pass them along together with our opinions for work in both areas, much of it performed by trial and error, with the full knowledge that a great deal more of both lies ahead for all concerned, ourselves included.

Enzyme Purification

Whenever a new enzyme is isolated, purified, and characterized, it should be ascertained whether or not it is a metalloenzyme; not by adding metals or metal-chelating agents and noting their effects upon its properties, but by the analysis of its content of metal which is the sole, definitive criterion of their presence in the stoichiometrically significant quantities. The emission spectrograph with photographic detection[21] is especially valuable to narrow the choice of possible candidates and has often resulted in decisive data.[9]

It is important that the enzyme be reasonably pure by physicochemical criteria. Its homogeneity must be established by accepted and proven transport methods like ultracentrifugation, gel electrophoresis, and chromatographic procedures. Once enzyme purity is established, other intrinsic properties like specific activity, dry weight, the absorption coefficient, and molecular weight can be determined. Chromophoric properties of certain transition metal complexes can be valuable guides, as may be the

[42] Thiers, R. E. (1957). *Methods Biochem. Anal.* **5**, 273.

results of inhibition studies with chelating agents. These can often point to the probable presence of a metal and even greatly reduce the number of likely choices regarding its identity. They never substitute for direct metal analysis, however. Claims to the contrary are cause for distrust of the data under discussion.[43]

Reagents and Purification

We have often referred to the conditions required for the analysis of metalloenzymes as "chemical metal sterility" to make the obvious analogy to bacteriological work. We generally remind those colleagues dismayed by first exposures to the need that no microbiology could be performed without prior sterilization of equipment, and having decided to work in that field this "inconvenience" is universally accepted as self-understood by investigators of that discipline. Like considerations pertain to work on metalloenzymes.[42]

Water. By far, the single most important reagent used in most laboratories is *water*. This reagent is used everywhere and in large quantities for washing containers, cleaning apparatus, and preparing buffers, analytical standards, and other solutions. It is essential that it be available in large amounts and be of consistent, known high quality.

The least suitable arrangement, much as it is the most common, is a central still or demineralizer for a building, with the processed water piped to numerous outlets in the individual laboratories. This is an architectural holdover from the "chemistry lab" and needs of yesteryear, much like the design of most laboratory furniture. If one assumes that this water is "pure," more by definition than fact, erroneous data and artifacts are inevitable—from a trace analysis standpoint. It is best to assume that this is merely *pretreated* water, piped into the laboratory for subsequent purification, a meritorious insight which is rewarded by avoidance of specious results and acrimonious literature exchanges. The main limitations of these central systems are the vast surface area presented by the piping arrangement and material, uncertain maintenance and storage conditions, bacterial growth in the system, and low reliability of supply in some installations. The fact that the host institution pays for its maintenance is limited comfort.

The next best approach is to take the "house" pretreated water and redistill or demineralize it. Lacking these facilities or depending on an unreliable local house supply which could then be pretreated, distillation—or further demineralization—is indicated. Conscientiously

[43] Vallee, B. L., and Wacker, W. E. C. (1970). *Proteins, 2nd Ed.* **5**, 1–192.

applied, this approach can provide good-to-excellent water, depending on the final treatment system. For trace-metal work, water of reasonable quality can be obtained from "all-glass" distillation systems. However, the production of this water is expensive, leaching of contaminants from the glass surface can be a problem, and the low rate of water production requires a storage container with attendant opportunities for contamination and bacterial growth. Distillation systems employing a quartz condenser are only slightly better but eliminate the leaching problem.

Metal-free water can be produced easily from pre-treated house water in reasonable quantities and at modest rates using two disposable mixed-bed ion-exchange cartridges in series (e.g., "research"-grade cartridges available from I.W.T. or Cole-Parmer, to name just two suppliers with whose products the authors have had personal experience and on which they have obtained analytical data). Because of the modest production rate, a storage container must still be employed, serving peak demand periods. Recommended is an all-plastic handling and storage system, for which polypropylene is serviceable with the storage container sealed. Air venting into the storage container on withdrawal is passed through a 0.2 μm filter to prevent airborne contamination.

All of the above water purification–storage systems have one or more disadvantages. These include inadequate purity, dependence on a "house" supply system—when it is shut down, the work is shut down—or the necessity of storing the treated water with the attendant problems of bacterial growth and possibility of contamination.

All of these problems can be avoided easily. Water of exceptional purity, both in terms of ultra trace-metal and trace-organic contaminants, can be prepared on demand at high rates from tap water, eliminating the need for storage and dependence on a central system. The apparatus is illustrated schematically in Fig. 6. Brand names are indicated, reflecting our personal experiences, but not necessarily to indicate our preference to the exclusion of comparable systems.

Beginning with tap water, the Continental pretreatment system provides 1 MΩ water, equivalent to triple-distilled, which is fed to a 4-cartridge system which continuously supplies 18 MΩ water at the rate of 1.5 liters/min ($\sim$0.5 gal/min) free of particulate matter, organic contaminants, and trace metals, as judged by the most sensitive analytical criteria presently available. Our only modification of the system constituted the discard of the brass pressure gauges, P_1 and P_2, supplied with the Millipore unit and replacing them with similar gauges, 0–60 psi, constructed of monel (Matheson).

This system will produce 18 MΩ water within 1–2 min after turn-on. As of this writing the system has proved to be trouble-free over the initial

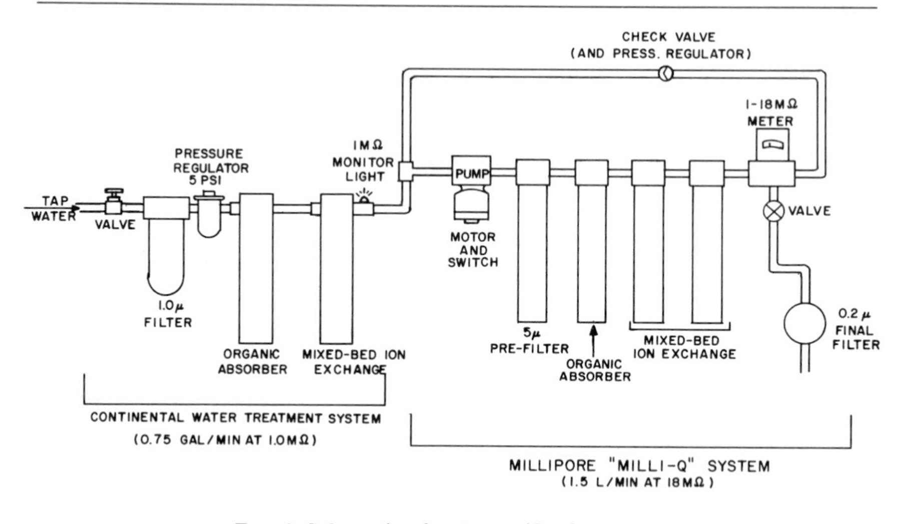

FIG. 6. Schematic of water purification system.

6-month period of use, requiring no cartridge or filter replacement. Analysis of the tap water feed indicates that the pretreatment system should provide about 400–500 gallons of 1 MΩ water before servicing is necessary. Supplied with 1 MΩ water, the Millipore system will require very infrequent cartridge replacement, >6 months.

Before you rush out to buy these systems, we must indicate the ever-present limitations. When using a lot of water, or a little *every day*, then this system seems ideal. If several days pass without use of water, as in small laboratories or during vacations, etc., problems may be encountered. The manufacturer specifies that periods of nonuse longer than 100 hr may induce bacterial growth within the system that cannot be overcome completely by subsequent operation, requiring a change of unexpended cartridges. If water must be sterile and pyrogen-free, this is perhaps not the system to employ unless needs for this water are intermittent and allowance is made for the extra expense of a sterile final-filter change just before withdrawal, or for subsequent sterilization of the water by other means. Given a regular need for very pure metal-free water, then this system is state-of-the-art.

Chemicals. These include chemicals used to prepare buffer solutions, and acids and bases used to adjust their pH. When working with metalloenzymes and/or performing trace-metal analyses, it is imperative that "reagent-grade" chemicals not be relied upon, without subsequent analysis and/or purification. Like "distilled water," they are not "pure" in the trace-element analysis sense. They, or their resultant solutions, can be made "metal-free" (i.e., free of undesirable metal contaminants) by one of several means, all of which should be verified experimentally for the particular application (see below).

Buffers. Buffers are made from chemicals and water, and their pH adjusted with acids or bases. Opportunities for trace-metal contamination at any one of these points abound. The water may be contaminated (*vide supra*), the reagents may be contaminated, or the acids and/or bases may be contaminated with undesirable or unknown trace elements.

There are two choices: (1) prepare the buffers from reagents *known* not to be contaminated, and adjust the pH with acids or bases *known* not to contain adventitious trace metals; or (2) prepare the buffers from ordinary reagent-grade chemicals, adjusting pH with ordinary acids and bases, then extracting contaminants with agents known to remove undesired adventitious metals.

Choice (1) is obvious, though often ignored in practice, while (2) may be achieved readily in several ways. Materials may be prepared from high-purity materials under controlled environmental conditions. Ordinary chemicals can be used, then the solution rendered free of trace-metal contamination by solvent extraction, ion-exchange chromatography, or electrochemical techniques.

When using inorganic buffers, like phosphate or acetate, it is often sufficient to use salts sold as "ultrapure" by several companies—not to be confused with "reagent grade." Alpha Inorganics/Ventron Corp., and Bakers "Ultrex" line, are but two examples of commercial sources. When prepared under carefully controlled conditions with metal-free water and the pH is adjusted with high-purity acids or bases (see below), they are usually comparable to buffers subsequently purified by other means. However, when organic buffer systems are called for, e.g., Tris, MES, etc., purification is invariably needed since these reagents are not usually available in sufficient purity.

Probably the most widely used and successful buffer purification methods employ solvent extraction techniques.[42] A large volume of a concentrated buffer is prepared and its pH adjusted near to the value desired. It is then extracted repeatedly with an organic solvent containing a very *non*specific broad-spectrum chelating agent, e.g., dithizone in CCl_4, followed by several extractions with the solvent alone to remove excess chelating agent from the aqueous phase, followed by bubbling *filtered* air through the solution to remove remnants of the immiscible organic solvent used. Stored in carefully cleaned (see below) polypropylene containers, this concentrated stock solution may then be diluted as needed with metal-free water and the exact final pH adjusted with small additions of metal-free acid or base (see below).

Another purification method, frequently employed, is to pass the solution over a column of an appropriate cation-exchange resin, Chelex-100 resin representing a popular example. While this represents a theoretically ideal, rapid means of processing large volumes of buffer solutions, less hard data is available than that accumulated to date for solvent extraction techniques. Thus, the reader is advised to evaluate this approach for his particular application.

Largely unexplored in biochemical applications, electrochemical techniques would appear to be excellent, rapid, and broad-spectrum means of removing trace elements from buffer systems. For example, controlled potential electrolysis of solutions against a stirred Hg pool electrode should quantitatively remove many of the trace-metal contaminants of interest. An application might be the preparation of metal-free strong base solutions used for pH adjustment, since purification of strong base solutions is not achieved so readily as that of acids.

Acids and Bases. As indicated above, rendering strong bases (NaOH, KOH, etc.) free of trace-metal contamination is not a simple task. Yet these reagents are used commonly to adjust the pH of buffer solutions. Whenever possible, the buffer should be prepared so that the pH adjustment is made by adding *acid*.

Volatile acids and bases, like HCl, acetic acid, and ammonia, can be prepared in modest amounts and with exceptional purity by isothermal distillation.[44] Carefully cleaned plastic vessels, containing ordinary "reagent-grade" concentrated acid and metal-free water are placed in a closed container and allowed to stand for several days. The volatile acid or base then distills isothermally from the higher concentration to the lower, with virtually no possibility of contaminant transfer. The process is not rapid since it is diffusion-controlled; on long standing for equal volumes the water only approaches one-half the molarity of the starting material. Changing the depleted acid or base several times could speed things along, but at increased risk of contamination on opening and closing the chamber. The system is simple, inexpensive, and can operate unattended.

If amounts needed are greater than can be prepared conveniently by this method, or if concentrations are required greater than several molar, other choices are available. For example, concentrated HCl of very high purity is prepared readily by dissolution of the anhydrous gas in metal-free water. Ammonium hydroxide can be prepared in a similar manner. Other nonvolatile strong acids, like HNO_3, H_2SO_4, and $HClO_4$, can be prepared in

[44] Alvarez, R. A., Paulsen, P. J., and Kelleher, D. E. (1969), *Anal. Chem.* **41,** 955.

high purity by sub-boiling distillation in quartz or Teflon apparatus, but it is much more convenient to purchase these acids having the required purity from commercial sources. The same is true for strong bases, like NAOH and KOH. The purification and subsequent analysis of these materials at ultratrace levels is no easy task, and the reputation and capabilities of the supplier must be relied upon. Wariness is the proper posture to assume when "ultrapure" reagents, especially bases, are shipped and stored for long periods in glass containers.

Glassware and Plastics

By far one of the greatest sources of misery and mystery in trace-metal determinations is due to "glassware," i.e., to containers. Like laboratory furniture, we are graduating from a "chemistry lab" environment to an ultramicro, trace analysis environment, with holdovers and instincts that are causing grief. Two major schools of thought have evolved, best illustrated by examples. Situation: a recent change in analytical requirements and/or the availability of suitable instrumentation has moved the trace-metal research level with which a given investigation is concerned down in concentration by a factor of 100.

Reaction 1: One should become 100-times more concerned, careful, and compulsive. One has always used glassware before, cleaned in some acid solution; now one should clean it 100 times longer, or in acid 100 times stronger, or switch to plastics, and clean them like one previously cleaned the glassware. One should switch from HNO_3/H_2SO_4 to boiling aqua regia; after all, more must be better.

Reaction 2: One won't change the way one has always done things. One will just dilute everything 100-fold further in the same apparatus cleaned in the same manner. After all, it's just a logical extension of what has transpired before.

What seems facetiousness in fact reflects the usual reactions which we have encountered, and these seem quite reasonable, natural, and are quite common. Faced with trace-metal determinations at very low concentration levels, one has probably switched from glass to plastics for much of the apparatus and containers used. Conventional wisdom has it that glass should be cleaned by violent means for the removal of trace-metal contamination, and often this is effective, particularly when the piece is of unknown prior history due to multiple users. Another reason is that glass will withstand this repeated assault. Plastics will also withstand almost any cleaning procedure, but not without serious consequences in some cases. If plastic ware is cleaned routinely in HNO_3/H_2SO_4 in the manner glassware was cleaned, most trace-metal contamination will surely have been removed. However, many workers are unaware that as a consequence of this very procedure the plastic container will now be unsuitable for the intended purpose. Solutions having very low trace-metal levels cannot be stored in them except for very short time periods. This is due to the fact that acid treatment has changed what was inert material into a surface which has become an ion exchanger.

From a trace analytical standpoint we have found the following procedures to be completely satisfactory for cleaning glassware. In a laboratory situation where there is complete control over the history of the containers used, a simple, 24-hr soaking in a detergent/chelating agent solution (e.g., "Count-off" or its equivalent) followed by thorough rinsing in metal-free water will suffice for *new* apparatus never exposed before to violent cleaning conditions. This is true of glassware as well as plastics. If plastics have been cleaned in baths like HNO_3/H_2SO_4, they should be discarded or used only for noncritical work.

After containers have been exposed to proteins, or to buffers like phosphate, which are adsorbed on the surface, cleaning by the detergent/chelating agent followed by rinsing with metal-free water will render them clean again. When pressed for time, they can be scrubbed with brushes in detergent/chelating agent mixtures omitting the soak cycle. From a trace-metal standpoint disposable plastic containers, including sample cups, beakers, pipettes, and micropipette tips, are usually "clean" as received. This is especially true of sterile apparatus. Here, a simple rinsing with metal-free water to remove electrostatistically attracted dust particles is usually sufficient to render it free from contamination. The cleanliness of container surfaces can be estimated visually: clean plastic surfaces are extremely hydrophobic, while glass surfaces are hydrophylic and show no "beading" in aqueous solutions.

Dialysis Tubing

Dialysis tubing is a convenient cellulose material often used in preparing enzymes for the purification of high-molecular-weight materials by dialysis against several changes of metal-free buffer. This material is usually supplied in a flexible form, primarily due to glycerine absorbed in the tubing. The literature describes rather drastic and sometimes arbitrary means of purifying this material, e.g., boiling it in multiple changes of EDTA solution. This is contraindicated for two reasons. First, cellulose dialysis tubing should not be exposed to temperatures above 70°C, since its molecular weight cut-off properties may change above this temperature. Second, the EDTA is not necessary, since the primary contaminants are organic, and EDTA may bind to the membrane, presenting problems later.

We have found that heating dialysis tubing to 70° in four changes of metal-free water renders the material metal-free. It can then be stored in metal-free water in plastic containers at 4° for use as needed. If there is concern about possible bacterial growth, although bacteria do not prosper under truly metal-free conditions, tubing can be stored in a 0.01% sodium azide solution. The azide must be removed before use, however, since it is a rather effective chelating agent.

Airborne Contamination

In the typical laboratory, one need only observe the appearance of, for example, a He/Ne laser beam to fully appreciate the amount of particulate matter constantly present in the atmosphere. This same atmosphere is in contact with samples and can seriously affect the analytical results. The contamination problem becomes increasingly severe as one tries to work at ever-lower trace element concentration levels.

Working at the microgram level, common in trace analysis, it is essential to be both conscious of and careful about preventing contamination. With modern techniques, work at the microgram, nanogram, and even picogram levels is feasible. This calls for "paranoia" about contamination. At picogram levels, obtainable, for example, with microwave emission or atomic fluorescence techniques, fanaticism is desirable.

Assuming the adherence to earlier recommendations for eliminating contamination from the reagents, water, and containers, the remaining source of contamination is airborne. The best solution to this problem is to conduct critical operations (e.g., sample preparation) in laminar flow hoods.

In these work stations a laminar flow of particle-free air at about 100 ft/min velocity shields the work area. Air recirculated by the unit passes through a HEPA absolute filter, which removes 99.97% of all particulate matter greater than 0.3 μm in diameter. This is termed a class-100 environment, with less than 100 particles per cubic meter at any given time. An *excellent* laboratory environment would be class-10,000, but most ordinary laboratories are several orders of magnitude above this.

Sample Preparation

Few analytical determinations at the ultratrace level can be made directly on biological materials without pretreatment of the sample usually to remove the bulk of the organic material. A few analytical systems, like the microwave emission and atomic fluorescence systems described earlier, have been designed specifically to obviate this requirement and have been shown thus far to give accurate results, particularly when applied to enzymes. Pretreatment allows conversion of samples to a relatively simple inorganic matrix, and the concentration of the samples for analysis by less sensitive means. The increased sensitivity afforded by these procedures permits subsequent sample dilution and, consequently, reduced matrix effects.

Two procedures are usually applied, separately or in sequence, to reduce the sample size and to reduce samples to an inorganic matrix: drying, and ashing.

Drying. In this procedure the primary goal is to remove water or other solvents, particularly if this also concentrates the sample and/or facilitates its further processing. Samples are commonly dried (solvent removal) in one of three ways: heating, i.e., evaporation; heating under vacuum, i.e., evaporation at lower temperatures; and lyophilization, i.e., "evaporation" at very low temperatures.

Ashing. Serious analytical errors are often encountered when trace metals are determined in samples containing large amounts of organic material, such as biological matter, owing to interactions between the organic material and the analyte atoms, relative to the inorganic standard solutions usually employed for calibration of the measurement system. Hence, it is unfortunately necessary to remove the organic matrix for most trace-element procedures. This is usually accomplished in one of two ways: dry ashing, or wet digestion.

Dry ashing can be accomplished either at high temperatures or at low temperatures, the latter in an oxygen plasma device. High-temperature dry ashing requires placing a known amount of sample into a clean platinum dish, drying the sample, then placing it in a muffle furnace at 450° for 24 hr or until ashing is complete, with the hope that nothing falls into the sample from the furnace walls or from the airflow through the furnace. The procedure must be *verified by independent means* to show that no elements of interest are lost by volatilization or by irreversible adsorption onto the container walls. Volatilization losses are especially serious for elements like mercury and selenium. Under certain conditions, and especially in biological materials, elements like chromium, arsenic, cadmium, iron, lead, zinc, and others, particularly at temperatures higher than 450°, may be lost. Anions, particularly the halides, may lead to the formation of volatile metal compounds and to their subsequent loss.

In some cases, the use of an ashing aid can be employed to good advantage. Pretreatment of the sample with H_2SO_4 can char the organic material and speed the ashing process or allow it to proceed at a lower temperature. Also, the volatile compounds may be converted to less-volatile sulfates.

Dry biological materials can usually be ashed with less risk of volatilization loss in low-temperature oxygen plasma devices. Dry samples in quartz, glazed porcelain, or open platinum containers are placed in a glass or quartz vacuum chamber which is maintained at a pressure of about 1 torr with oxygen. Surrounding the chamber(s) is a radio-frequency coil which maintains a glow-discharge plasma within the chamber. The oxygen atoms and free radicals gently oxidize the organic material while the sample remains at a relatively low temperature. Although quite efficient, and while greatly reducing volatilization losses, the process is rather slow when large samples are to be ashed.

Wet digestion constitutes another means of "ashing" a biological sample. In wet digestion, the sample is heated, or better yet, refluxed, with mineral acids and/or strong oxidizing agents. Oxidizing conditions are maintained and volatilization losses are minimized because of the relatively low temperatures (100–200°). Depending on the specific system used and the nature of the sample, it is usually faster than dry ashing. Since rather reactive reagents are added to the sample in appreciable quantities, it is imperative that they be of the highest available purity. Oxidizing acids like HNO_3 and $HClO_4$ are employed frequently. With these acids, certain elements—particularly at ultratrace levels—can be lost by volatilization, and the explosive nature of many perchlorate compounds and special-handling precautions contraindicate the use of this acid in digestions except by personnel well-versed in its dangers and precautions for handling. When perchlorate is used, this should always be done in a hood designed specifically for this acid. Sulfuric acid is relatively safe and nonvolatile, and serves well to char organic materials. Charring alone could lead to the development of reducing conditions, thus leading to volatilization losses of several elements, so the use of H_2SO_4 alone is not recommended, aside from the fact that it is a nonoxidizing acid. When combined with an oxidant, like H_2O_2, HNO_3 (see precautions above), or permanganate, if available in sufficient purity, it provides an excellent system, particularly under reflux conditions.[45] Hydrochloric acid presents special problems, in that it is nonoxidizing and, more importantly, increases the possibility of volatile halide formation. Thiers has presented data which indicate that losses of Sb, As, Cr, Ge, Se, and Sn can occur when acid mixtures containing HCl and H_2SO_4 or $HClO_4$ are boiled.[41]

Now before the advocates and enthusiasts of various other digestion or ashing procedures protest, we would like to issue a warning about "recovery" studies. These reflect additions of a metal (in a simple, inorganic matrix) or radioisotope to a biological sample, and the data are intended to

[45] Wolf, W. R. (1977). *J. Chromatogr.* **134,** 159.

show that all of the material added is "recovered" by the method, the implication being that this "proves" the procedure accurate. It does not. It merely proves that one can recover what one added in the procedure. Unless *by independent means* the additive can be demonstrated to exchange fully with analyte in the sample, the results are not conclusive if full recovery is obtained. Recovery of less than 100% proves the method invalid; 100% recovery only *suggests* that the method might be valid, until separately shown by an independent method(s). The literature shows that these considerations are often overlooked and grossly abused (see also the standard addition method, below).

Standards

Most quantitative determinations require calibration of the measurement system with standards. Trace-element determinations require that the reponse of the instrumental system be calibrated with samples of known analyte concentration, quantity, stoichiometry, or ratio. Highly specific trace-metal determinations, as afforded by atomic emission, absorption, or fluorescence spectrometry, greatly reduce the requirement that samples and standards be matched exactly in composition, *provided that matrix effects are not encountered*. For most applications, where the sample matrix is primarily inorganic, i.e., inorganic samples, or biological materials reduced to an inorganic matrix by ashing, digestion, etc., the preparation of suitable standards is relatively straightforward. Where matrix effects *are* encountered, they must be both recognized and compensated for by some means.

In ultratrace work, one routinely deals with solutions having a very low concentration of a particular element or elements and must prepare standard solutions accordingly. The preparation and storage of these standards can be a major problem, from the standpoint of contamination and stability, respectively. Naturally then, these problems are related to the purity of the materials used, the storage containers, and their cleaning procedures, as described earlier. Additionally, the availability of suitable standard reference materials, with highly similar matrices and/or independently verified matrix effects, is required.

Only general recommendations can be made. Chemicals used to prepare standard solutions and/or used in dialyses, buffers, etc. must be free of contamination by adventitious metals. Chemicals proven to be pure by spectrographic analysis, like those supplied by Johnson-Matthey, Ltd. and others, are a good choice. It must be borne in mind that, like most "reagent-grade" chemicals, these materials are certified to be free of trace-metal contaminants, but not as regards their stoichiometry. Therefore, after preparation of a solution of these chemicals, an accurate analysis is recommended to establish the exact stoichiometry.

Standard reference materials, like those supplied in the United States by the National Bureau of Standards, are recommended highly for a procedural check on, and quality control of, any method used. For example, if one were measuring the levels of a trace element in human liver biopsies and the results of the procedures used did not agree with those certified for that element in the N.B.S. Bovine Liver Standard Reference Material (SRM), the analytical procedure in use should be suspect. Similarly, an analytical procedure developed for the determination of zinc in enzyme preparations, which did not yield accurate stoichiometries for well-characterized, physicochemically homogenous zinc metalloenzymes of known stoichiometry and molecular weight, would likewise be suspect.

Unfortunately, the great majority of trace-metal determinations on biological materials have to be made on materials and/or media for which no obvious standard reference material exists. For this reason, it is of the utmost importance that the procedure and measurement system selected be verified by independent means whenever possible.

Regarding both samples and standards, it should be pointed out that very dilute solutions (<1 $\mu g/ml$) of many elements are unstable over relatively short time periods. It is primarily a container problem, with very dilute solutions gaining or losing analyte due to desorption and adsorption, respectively. The latter is more common in clean, new containers. Dilute standard solutions prepared in the same manner and repeatedly stored in the same containers may gradually approach equilibrium with the container, thus permitting relatively long storage times after a while. In any event, sample and standard stability must be continually checked.

Blanks. Ideally, a blank solution contains everything contained in the sample except for the analyte. Since this is seldom available, a reagent blank is usually employed, i.e., a sample containing all of the chemicals added to or used to treat the sample. In enzyme preparations involving procedures like dialysis or gel filtration chromatography, the final dialysis buffer or the elution buffer after having passed through the column serves as a reasonable blank. In ashing or digestion procedures, blanks should be carried through the same procedures and the same reagents added. The blank signal is subtracted from the sample readings and thus corrects for solvent, reagent, and container contamination.

Method Verification

As indicated earlier, perhaps the greatest failing in trace-metal research of biological systems has been the lack of suitable standards, and given this limitation, the lack of verification of widely used methodology by independent means. In this regard, a large amount of misleading and sometimes erroneous information has been repeated in the scientific literature, most obtained and communicated innocently, but nevertheless with counterproductive consequences.

The widely used "standard additions" method has been an important contributor to this unfortunate state of affairs. The theory behind this method is that if something in the sample (e.g., the matrix) interferes with the measurement, one can correct for this by successive additions of the analyte element to aliquots of the sample. Theory assumes that the interferent will affect the analyte added in the same way as the intrinsic analyte atoms. While probably true for relatively simple inorganic matrices, this approach rapidly fails with biological materials. Here the assumption of equilibration with the sample analyte "pool" is frequently invalid, and the method merely proves that one can get back what one added.

In the standard addition method, several conditions must be met to render the technique valid: (1) The analytical curve must be linear over the region of interest; (2) the added analyte *must be in the same chemical form* as the intrinsic analyte *or,* be *known* to exchange fully with the analyte and its matrix *or,* undergo the same effects as does the sample matrix. Obviously, a standard addition *known* to meet those criteria is valid, but if all of this information is *known,* then the method is probably not necessary for quantitation. So in practice, the standard addition method is most valid under conditions where it is least needed in the first place!

Generally, these criteria are met for simple inorganic matrices, and their validity in these systems has been amply demonstrated. Unfortunately, this success in one application has been extrapolated to an entirely different matrix, i.e., biological materials, where these criteria are *usually invalid.* This maleficent practice has resulted in, and is responsible for, a great deal of the current misinformation which has appeared in the literature regarding the role of trace metals in biological systems.

For any method to be accepted fully for a particular application, it must be verified by independent means. Once verified, it may be applied unchanged with relative impunity. But verification means getting the same answer by a completely independent method *on the same sample,* or, at least, successfully comparing the method to similar standard reference materials of known composition and separately verified.

[26] Spectrophotometric Assay of Oxygen Consumption

By OCTAVIAN BÂRZU

I. Introduction

The measurement of oxygen consumption, most currently using manometric and oxygraphic procedures,[1] is still an important tool for investigating the oxidative processes in a wide variety of biological preparations such as tissue fragments, isolated cells or organelles, and purified enzymes. Various improvements of these procedures, such as an increase in the sensitivity and a decrease in the lag of response to the variations of oxygen uptake, make them applicable to a large number of analytical purposes.

The spectrophotometric assay of oxygen consumption (uptake) is based on the utilization of oxyhemoglobin[2] to serve both as a source of oxygen and as an indicator of respiration.[3-7] Deoxygenation of HbO_2 is accompanied by changes in optical density at different wavelengths. The advantages of the spectrophotometric method consist in its wide range of sensitivity as well as in its relative simplicity. By decreasing the reaction volume down to 1 nl it has been possible to measure the oxygen uptake of one single cell.[8-11] Any spectrophotometer or photometer supplied with a mercury lamp and corresponding interference filter can prove suitable for oxygen-consumption measurements. The use of commercial oxyhemoglobin analyzers might be extended to the oxygen-consumption assays.[12]

[1] See W. W. Kielley, Vol. 6 [33], E. C. Slater, Vol. 10 [3], and R. W. Estabrook, Vol. 10 [7].

[2] *Abbreviations*: HbO_2, oxyhemoglobin; Hb, hemoglobin in the deoxygenated form; Hb_t, total hemoglobin (HbO_2 + Hb); MetHb, methemoglobin; Mb, myoglobin; MetMb, metmyoglobin; Hb-CPA and Hb-CPB, hemoglobin treated with carboxypeptidase A or carboxypeptidase B; β-PMB, isolated β chain of hemoglobin, with *p*-mercurybenzoate blocked SH groups; QO_2, oxygen uptake; K_m^0, Michaelis-Menten constant for oxygen of the considered oxidase.

[3] O. Bârzu and V. Borza, *Anal. Biochem.* **21,** 344–357 (1967).

[4] O. Bârzu, L. Mureşan, and C. Tărmure, *Anal. Biochem.* **24,** 249–258 (1968).

[5] O. Bârzu and M. Satre, *Anal. Biochem.* **36,** 428–433 (1970).

[6] O. Bârzu and P. Cioara, *Anal. Biochem.* **43,** 630–632 (1971).

[7] O. Bârzu, L. Mureşan, and G. Benga, *Anal. Biochem.* **46,** 374–387 (1972).

[8] R. Hultborn, *Anal. Biochem.* **47,** 442–450 (1972).

[9] H. Herlitz and R. Hultborn, *Acta Physiol. Scand.* **90,** 594–602 (1973).

[10] R. Hultborn and H. Hyden, *Exp. Cell Res.* **87,** 346–352 (1974).

[11] R. Hultborn, personal communication.

[12] D. Glick, *Clin. Chem.* **23,** 1465–1471 (1977).

ISBN 0-12-181954-X

TABLE I
EQUILIBRIUM AND KINETIC CONSTANTS FOR THE REACTION OF SOME HEMOPROTEINS WITH OXYGEN AT pH 7 AND 20–23°

Compound	Equilibrium constant K (M^{-1}) × 10^{-5}	Dissociation rate constant K (sec^{-1})	Hill number
Human hemoglobin[a]	0.56	34	2.9–3.0
Myoglobin[a]	5.0	10	1.0
Hb-CPB[b]	5.6	23	2.7
Isolated α chain[a]	12	28	1.0
Isolated β chain[a]	14	16	1.0
Hb-CPA[b]	17	16	1.0
Yeast hemoglobin[c]	500	17	1.0

[a] From Antonini and Brunori.[17]
[b] From Antonini *et al.*[20]
[c] From R. Oshino, N. Oshino, and B. Chance, *FEBS Lett.* **19,** 96–100 (1971).

II. Theoretical Considerations

The most important feature of the spectrophotometric method is that it operates at low concentrations of molecular oxygen. For human hemoglobin at pH 7 and 20° this range is situated approximately between 8 and 40 μM O_2. Use of other natural donors of oxygen or their derivatives may offer various ranges of oxygen concentration as shown in Table I. Since the affinity for oxygen of different O_2-consuming systems varies within large limits (10^{-8}–$10^{-4}M$),[13,14] it is clear that the choice of a suitable oxygen donor is essential for obtaining a correct relationship between oxygen consumption and variations in optical density. The ideal conditions are established when the affinity for oxygen of the respiratory system is at least two orders of magnitude higher than the affinity for gas of the oxygen donor.[7] Such conditions are met for instance in the case of mitochondrial respiration in the presence of human hemoglobin. If the ideal conditions cannot be created, we can overcome some difficulties by using a definite part of the dissociation curve of the oxygen donor and a corresponding calibration.

A particular situation is created by enzymic systems generating H_2O_2 during oxygen uptake. It is well known that Hb is sensitive, while HbO_2 is not, to the oxidative action of H_2O_2.[15] As a consequence, a subsequent

[13] O. Hayaishi, *Annu. Rev. Biochem.* **31,** 25–46 (1962).
[14] B. Chance, *J. Gen. Physiol.* **49,** 163–188 (1965).
[15] A. White, P. Handler, and E. L. Smith, *in* "Principles of Biochemistry," 3rd ed., p. 199. McGraw-Hill, New York, 1959.

TABLE II

EFFECT OF pH, CATALASE, AND METHANOL (MeOH) ON THE RATE OF HbO_2 DEOXYGENATION AND MetHb FORMATION DURING THE OXIDATION OF GLUCOSE BY GLUCOSE OXIDASE[a]

pH	HbO_2 deoxygenated		MetHb formed		(MetHb formed / HbO_2 deoxygenated) × 100	
	−MeOH	+MeOH	−MeOH	+MeOH	−MeOH	+MeOH
5.9	1.52	3.09	0.87	0.51	56	17
6.7	1.49	2.98	0.31	0.28	21	9
7.6	1.16	2.36	0.14	0.07	12	3
8.1	0.62	1.01	0.09	0.03	14	3

[a] The reaction medium (final volume 3 ml) contained in a 1-cm path cuvette and at 23°: 0.1 *M* K phosphate buffer, 0.02 m*M* HbO_2, 1.67 m*M* glucose, 300 units catalase, and 0.07 *M* methanol. After deoxygenation of HbO_2 by about 30% the reaction was started by addition of 8 μg glucose oxidase (a Sigma preparation). The rate of HbO_2 deoxygenation (430 nm) and MetHb formation (420.6 nm) was followed with a Gilford 2400 Spectrophotometer and was expressed as nmoles/min.

conversion of Hb into MetHb is expected to take place during the deoxygenation of HbO_2 by these oxygen-consuming systems. Addition of catalase alone will not improve the situation too much due to its low affinity for H_2O_2.[16] To avoid almost completely the risk of Hb conversion into MetHb, catalase is added together with methanol or ethanol to the reaction mixture (Table II).

$$\text{Glucose} + O_2 \xrightarrow{\text{glucose oxidase}} \text{Gluconolactone} + H_2O_2$$
$$H_2O_2 + RCH_2OH \xrightarrow{\text{catalase}} 2\ H_2O + RCHO$$

In this way the stoichiometry of the oxygen-consuming system is improved, i.e., 1 mole of oxidized substrate corresponds to 1 mole of oxygen taken up, whereas when catalase is added alone, 1 mole of oxidized substrate corresponds to 1/2 mole of oxygen.

III. Preparation and Conservation of Oxygen Donors

Human hemoglobin has proven to be the best oxygen donor, being easily obtainable, well characterized from a physicochemical point of view,[17,18] and ideally suited to the function of "oxygen buffer" because of

[16] P. Nicholls and G. R. Schonbaum, *in* "The Enzymes" (P. D. Boyer, H. Lardy, and K. Myrbäck, eds.), 2nd rev. ed., Vol. 8, pp. 147–225. Academic Press, New York, 1963.

[17] E. Antonini and M. Brunori, *Annu. Rev. Biochem.* **39**, 977–1042 (1970).

[18] O. W. Assendelft, "Spectrophotometry of Haemoglobin Derivatives," Van Gorcum, Assen, Netherlands, 1970.

the sigmoidal shape of its dissociation curve. However, it is of interest to note that some other oxygen donors might be successfully used, providing they are easy to obtain and store. Thus, myoglobin, whose affinity for oxygen is higher than that of Hb, might substitute for Hb in measuring the mitochondrial oxygen uptake.[7] However, its rather laborious preparation as well as its rapid oxidation to MetMb are the major inconveniences for its use as an oxygen donor in the spectrophotometric method. The same applies to the isolated hemoglobin chains. Instead, carboxypeptidase-treated Hb, which is easy to prepare, combines the properties of both myoglobin (high oxygen affinity) and hemoglobin; for these reasons we have recently chosen to use it, especially in the assay of mitochondrial respiration.[19] Among the numerous procedures available in the literature for preparation of Hb and its derivatives we considered those given below because of their acceptable yield, simplicity, and rapidity.

Erythrocytes were separated by centrifugation of whole human blood (10 ml) to which heparin had been added. The cells were washed 3 times with NaCl solution (0.9%). Double-distilled water (2 volumes) was added to 1 volume of packed red cells, lysis being enhanced by shaking. After 2–4 hr the hemolysate was centrifuged at 50,000 *g* for 30 min to remove the stroma. To the clear supernatant K phosphate buffer (pH 8) and NaCl were added in order to attain the concentration of 0.18 and 0.1 *M*, respectively. The hemolysate was then placed on a Sephadex G-25 column equilibrated with the same mixture, and fractions of 2–3 ml were collected. The HbO_2 concentration in the eluent (on the heme basis) was estimated by measuring the optical density at 577 nm ($\epsilon_{mM} = 15.4$).

Digestion of HbO_2 with carboxypeptidase A (diisopropylphosphate-treated enzyme, a commercial preparation supplied by Sigma Chem. Co.) was done essentially as described by Antonini *et al.*[20] and Zito *et al.*[21] To 10 ml of HbO_2 solution eluted from the Sephadex column (about 1.0–1.3 m*M*), 75 μl of carboxypeptidase A were added and the mixture incubated at 28° between 3 and 6 hr with continuous stirring. At different time intervals samples of 0.1 ml were withdrawn and precipitated with 0.5 ml of 0.5 *M* cold trichloroacetic acid. Then 0.5 ml of the clear supernatant was used for tyrosine determination as described by Udenfriend.[22] When 1 mole of tyrosine was released per ($\alpha\beta$) unit (usually after 4–5 hr of incubation) the mixture was chilled at 0° and centrifuged at 20,000 *g* for 10 min to remove

[19] O. Bârzu, M. Dânşoreanu, R. Munteanu, and A. Ana, in preparation.

[20] E. Antonini, J. Wyman, R. Zito, A. Rossi-Fanelli, and A. Caputo, *J. Biol. Chem.* **236,** PC60-PC63 (1961).

[21] R. Zito, E. Antonini, and J. Wyman, *J. Biol. Chem.* **239,** 1804–1808 (1964).

[22] S. Udenfriend, Vol. 3 [87].

the denaturated proteins. The supernatant was dialyzed overnight against double-distilled water and adjusted to pH 6.5. In these conditions the released tyrosine was removed and practically all the carboxypeptidase was inactivated, the enzyme being unstable at low ionic strength and at pH below 7. The solution was centrifuged again at 20,000 *g* for 10 min to remove any insoluble material. The clear solution of Hb-CPA (0.8–1.0 m*M*) was divided into 0.5-ml aliquots which were kept frozen at $-20°$ until used. Untreated HbO_2, after dialysis against double-distilled water and adjustment at the same concentrations, also was kept at $-20°$. Under these conditions the pigment solutions are stable for at least 6 weeks.

Usually, no difficulties are encountered with the purity of HbO_2 since good results can be obtained often with nonpurified hemolysates. However, the integrity and the functional capacity of Hb or its derivatives are critical, and they should be checked by measuring the following ratios: $E_{577}/E_{542} \geqslant 1.05$ and $E_{577}/E_{560} \geqslant 1.72$. Knowing the HbO_2 concentration one can calculate the $\Delta\epsilon_{mM}$ value for any given wavelength if the values reported in the literature are not available. For HbO_2 deoxygenation one can use either N_2 (5 min of bubbling proved to be enough) or Na dithionite. Usually, either procedure is equally effective if care is taken to avoid a too large excess of Na dithionite which may denaturate Hb.

IV. Selection of Wavelengths, Cuvettes, and HbO_2 Concentration

Although these three factors are highly interdependent, a number of possible combinations are still available depending upon the measuring instrument or the properties of the biological material. Commercially available cuvettes having a path of 0.1–1.0 cm and a volume of 0.1–3.0 ml are perfectly suitable. An essential point is that the final optical density of the system should not be higher than 3.5, including the absorption due to biological material. At higher values, deviations from the Lambert-Beer's law are to be expected even when instruments with high-resolution monochromators are used. Thus if a λ of 430 nm is selected, with 0.1 m*M* HbO_2, a cuvette with a path equal to or smaller than 0.25 cm is recommended; if a λ of 577 nm is chosen (the sensitivity in this case is evidently lower), then a cuvette having a path of 0.5 cm will allow a 3-fold increase of the HbO_2 concentration. Generally, to make use of the advantages offered by the procedure, i.e., a high degree of sensitivity and versatility, we prefer to carry out the measurements at 430 nm (or at 436 nm for the photometers supplied with a mercury lamp). In this case cuvettes with a path between 0.2 and 0.4 cm and a concentration of HbO_2 between 0.15 and 0.07 m*M* are optimal. Of course, for ultramicroanalytical purposes the light path can be decreased from 0.3 mm to a few micrometers.[8,23] When changes in the optical density due to other secondary processes, like swelling or shrinkage of mitochondria, are expected to be significant, dual-wavelength spectrophotometers are recommended (λ_1 where $\Delta\epsilon_{mM}$ is maximum and λ_2 at isosbestic point).[7]

A factor considerably limiting the use of commercially available cuvettes is the optical inhomogeneity of many biological preparations, mainly due to their tendency to sediment, as in the case of whole tissue homogenates or isolated cells. In such cases it is possible to increase the viscosity of the reaction mixture (which, however, might affect negatively the oxygen uptake) or to use chambers with continuous stirring (see below) making sure that the diffusion of atmospheric oxygen is avoided and the accuracy of the spectrophotometric recordings is not affected by a high signal-to-noise ratio.

V. Special Cuvette and Device for Oxygen-consumption Measurements

As shown in Fig. 1, the thermostated cell holder *a* is a support which holds the glass cell *b* covered with a stopper *c* of glass or plastic. Under the cell there is a magnet *d*, driven by a microelectromotor *f* via a belt *g*. The energy of the microelectromotor is provided by a small low-voltage transformer. The speed of the electromotor and consequently of the magnet is adjusted by a potentiometer. In the cross-section the cell *b* is of square shape (in this example 4 × 4 mm), but at the base it has a cylindrical hole, 2.5-mm high and 6-mm diameter, which houses the magnetic stirrer *e*. The stopper has a depth of up to 12.5 mm from the bottom of the cell, so that the remaining volume of the cell is approximately 230–240 μl. The stopper is fitted with a longitudinal cylindrical inset which is directed downward and allows the injection of various additions via a Hamilton microsyringe. The whole device is designed to fit into the Eppendorf photometer by a simple replacement of its cell holder.[24] The size of this device was chosen so that the light beam of the photometer crosses the cell at a distance between 6 and 10 mm from the bottom. This reduces the noise to a minimum in the continuous optical-density recording due to stirring, and the atmospheric oxygen diffusion into the cell. The optimal wavelength of 436 nm is obtained from a low-pressure mercury lamp using an interference filter (Ealing Scientific Ltd., Watford).

[23] E. Raddatz and P. Kucera, *Experientia* **33,** 9 (1977).

[24] An excellent thermostated cell holder with a magnetic stirring system is now available from Eppendorf Gerätebau (Hamburg, GFR).

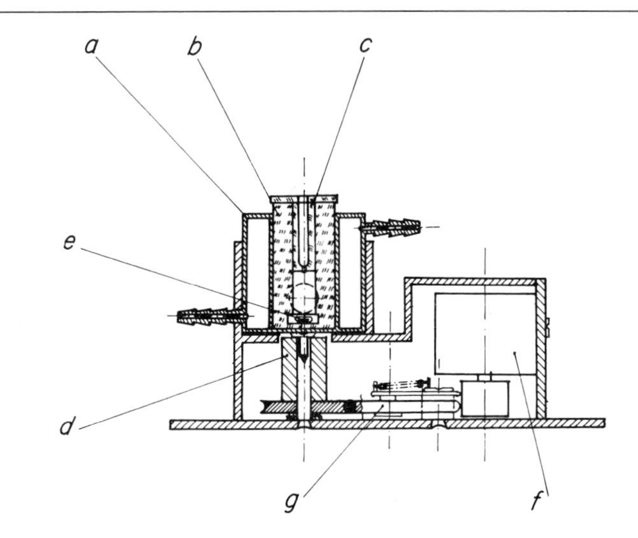

FIG. 1. Cuvette and device for oxygen-consumption measurements adapted to commercial photometers. (*a*–*g*) See text for description.

VI. Assay of Oxygen Uptake

Before starting the assay procedure, the following checks are to be made: (1) choice of optical density range in which a perfect linearity between hemoglobin concentration and optical density is achieved. A series of solutions of different HbO_2 concentrations (0.01–0.05 m*M*) are prepared and the optical density against water is read before and after Na dithionite addition. Use of instruments with high-resolution monochromators will allow a good linearity up to an optical density of 3.5 (high values are read by increasing the sensitivity of the photomultiplier, each step corresponding in the Eppendorf photometer to 0.25 units of optical density). Deviations from linearity are avoided by either lowering the HbO_2 concentration or decreasing the light path. (2) Diffusion of atmospheric oxygen into the cuvette is checked as follows: The optical density of the HbO_2 solution is read under chosen experimental conditions, after which N_2 is bubbled to produce a 40–60% deoxygenation. The cuvette is then sealed with either paraffin oil (previously treated with N_2) or covered with the stopper described in Fig. 1, and the optical density is recorded for a few minutes. Normally, the recording trace should not exhibit any changes; if a decrease in optical density at 430, 436, and 560 nm or an increase in optical density at 415, 542, or 577 nm is encountered, then these changes must be corrected in the calculation. When using cuvettes equipped with a firmly attached stopper, oxygen diffusion is negligible even for the cases of low respiratory activity and longer-term experiments. (3) Mixing of reactants in the cuvettes: When small reaction volumes (0.3 ml) and narrow cuvettes (0.2 cm) are employed, injection through the paraffin oil layer of small volumes of solutions (1–5 μl) together with air (5 μl) allows a good mixing of reactants providing their viscosity and density are not too different from those of the reaction medium. When inhomogenous suspensions or highly viscous reagents are being investigated, a good mixing of the reaction medium is achieved by using continuous magnetic stirring. The efficiency of stirring is checked by injecting colored solutions into the reaction medium and recording the jump of optical density at the maximum speed of the recorder. With the aid of the mixing device already described, a complete mixing is attained within 2 sec or less (Fig. 2A).

Three typical examples for assay of the activity of some oxygen-consuming enzymes, not published previously, are given below.

1. Assay of the Monoamine Oxidase Activity

In a 0.5-cm cuvette with 1.5-ml final volume the following reagents are pipetted: 0.05 *M* K phosphate buffer (pH 7.6), 0.1 *M* methanol, 300 units catalase, 0.05 m*M* HbO_2, and 25 μg of purified ox kidney monoamine oxidase.[25] The optical density at 430 nm is about 1.3; then its value is adjusted to zero by opening the slit of the instrument (Gilford 2400 spectrophotometer, equipped with a recorder having the full scale of 0.1 units of optical density). N_2 is now bubbled into the cuvette until the optical density becomes 0.5–0.6 (corresponding to 25–30% deoxygenation of HbO_2). The optical density is then brought again to between 0 and 0.1, and the cuvette is sealed with a layer of paraffin oil. After temperature and optical density equilibration, 10 μl of 30 m*M* benzylamine solution are injected. A linear increase of 0.3–0.5 units of optical density is recorded at 430 nm. In order to check whether a part of Hb was oxidized to MetHb due to H_2O_2 produced during the reaction, recordings are alternatively conducted at 430 and 420.6 nm (at this λ, $\epsilon_{mM}^{Hb} = \epsilon_{mM}^{HbO_2}$). From the rate of optical density decrease at 420.6 nm it is possible to calculate the rate of Hb oxidation to MetHb.

2. Assay of Mitochondrial Cytochrome Oxidase

In a cuvette as described in Fig. 1 the following reagents are pippeted: 100 μl of 25 μM cytochrome *c* in 0.2 *M* K phosphate buffer (pH 7.2), 135 μl water, and 15 μl Hb-CPA (final concentration 0.06 m*M*). The cuvette is

[25] The purified enzyme was kindly provided by Drs. L. Hellerman and D. R. Patek (Baltimore, Maryland).

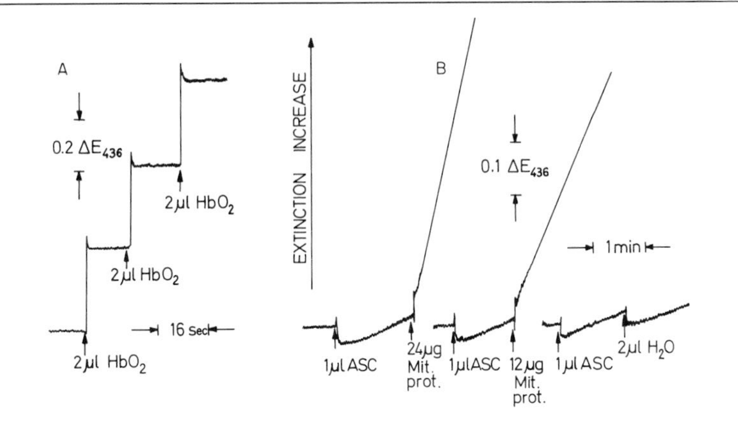

FIG. 2. Verification of the efficiency of the stirring system (A) and the assay of mitochondrial cytochrome oxidase from rat liver (B).

then placed in the thermostated cell holder, and the optical density at 436 nm is adjusted to zero. N_2 is bubbled until the optical density reaches a value in the range of 0.5–0.7 (corresponding to 30–40% deoxygenation); the cuvette is then recovered with stopper *c* (Fig. 1) so that an excess of 15 μl of reaction medium fills the longitudinal channel of the stopper thus preventing the diffusion of atmospheric oxygen into the cuvette. The optical density is again adjusted to between 0 and 0.2, and now both the magnetic stirrer and the recorder are switched on. After temperature and optical density equilibration of the system, 1 μl of 0.25 *M* Na ascorbate (pH 6) is injected with a Hamilton microsyringe. After about 2 min 2 μl of mitochondrial suspension (previously lysed with lubrol WX) are added. The increase in optical density is recorded on the full scale of 0–1. The reaction is linear up to 0.7–0.9 units of optical density (Fig. 2B).

3. *Determination of Oxygen Consumption of Polymorphonuclear Leucocytes during Phagocytosis*

In a cuvette as above, 210 μl of Dulbecco buffer without glucose (pH 7.4), 15 μl of HbO_2 (0.06 m*M* final concentration), 1 μl of methanol, and 50 units of catalase are added. The optical density is adjusted to zero as above, and N_2 is bubbled until the optical density at 436 nm is in the range of 1.0–1.4. Next 5 μl of leucocyte suspension (containing 8.85×10^5 cells) and 20 μl of human autologous serum are added. N_2 is then bubbled for a few seconds in order to homogenize the mixture, and the cuvette is covered as above. In this case a higher degree of HbO_2 deoxygenation is necessary since the reaction mixture is supplied with an additional amount of oxygen brought by the leucocyte suspension and the human serum. Leucocyte oxygen consumption is then recorded under the resting state conditions; next 5 μl of zymosan suspension are added. Since this suspension creates a considerable degree of absorption, before continuing the recording it is necessary to increase the photomultiplier sensitivity and bring the optical density to the level displayed before zymosan addition. After a lag period of a few minutes an acceleration of oxygen consumption is recorded. The determination may be repeated after the complete deoxygenation of HbO_2 by addition of 3–5 μl of oxygen-saturated water (Fig. 3).

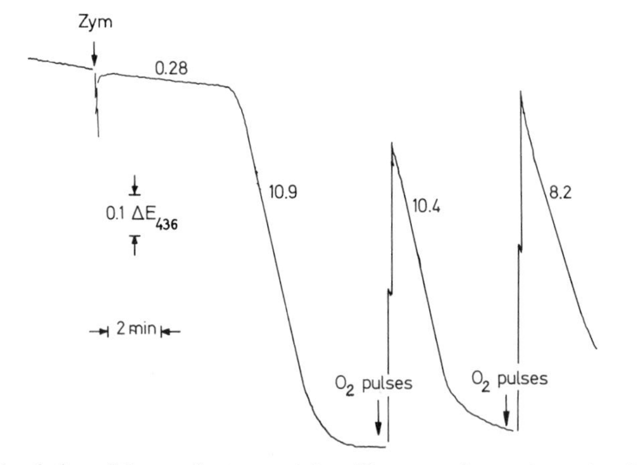

FIG. 3. Stimulation of the respiratory activity of human polymorphonuclear leucocytes at 37° by a suspension of zymosan (experiment done in cooperation with Dr. M. Markert, Lausanne, Switzerland). Numbers on the tracing express oxygen consumption as ng-atoms/min/10^6 cells.

VII. Calculation of the Oxygen Consumption

The correspondence between the experimentally measured variation of optical density and oxygen consumption of biological samples is established according to the following general formula, keeping in mind that 2 atoms of oxygen correspond to 1 mole of deoxygenated HbO_2:

$$QO_2 \ (\mu\text{g-atoms/min/sample}) = \frac{2V \cdot f \cdot \Delta E/\text{min}}{\Delta\epsilon_{mM} \cdot d}$$

where V represents the reaction volume (ml); d the light-path length (cm); $\Delta\epsilon_{mM}$ the difference of the millimolar coefficients of extinction of HbO_2 and Hb at the chosen wavelength; and f represents a correction factor depending on hemoglobin concentration and the affinity for oxygen of the respiratory system. For the experimental evaluation of f there are at least three possibilities, applicable depending on the K_m^0 of the respiratory system, the stability of the enzymes during the assay, and the hemoglobin concentration in the spectrophotometric system.

1. Oxygraphic Estimation of the Oxygen Consumption in the Presence of HbO_2 at the Concentration Used in the Spectrophotometric Experiment

The method is applicable to samples having high oxygen affinities (K_m^0 $\leq 0.4\ \mu M$), stable for at least 5–10 min during the oxygraphic assay and with HbO_2 concentration not smaller than 0.1 m*M*. In this case, the oxygraphic trace has two distinct linear regions (Fig. 4A) and their slopes (*m*

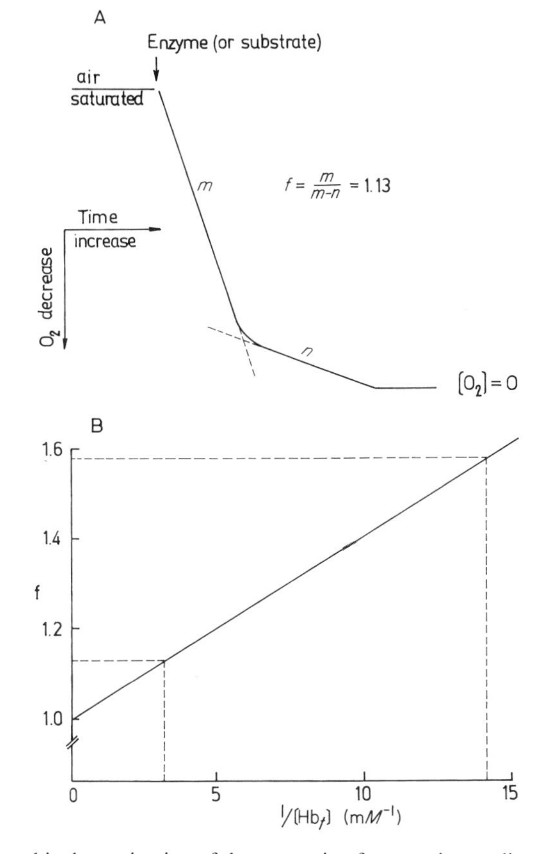

FIG. 4. Oxygraphic determination of the correction factor using rat liver mitochondria at 25° and HbO_2 as oxygen donor. (A) Typical oxygraphic trace indicating mitochondrial respiration in the presence of 0.3 m*M* HbO_2. (B) The plot of the correction factor as a function of $1/Hb_t$.

and n) directly yield the correction factor, $f = m/m - n$. Through extrapolation, the method can be extended to HbO_2 concentrations below 0.1 mM by determining the value of f at 3–4 HbO_2 concentrations (e.g., 0.1, 0.15, 0.2, and 0.3 mM). The plot of f as a function of $1/Hb_t$ is a straight line (Fig. 4B). For example, for 0.07 mM HbO_2 one obtains f equal to 1.58.

2. *Spectrophotometric Estimation of Oxygen Consumption at Different HbO_2 Concentrations*

Only the first condition required by the oxygraphic method needs to be obeyed in this case. One plots $1/\Delta E$ as a function of $1/Hb_t$; in other words, one determines the value of K_m^{Hb}. From the plot, or by direct calculation, one then determines the value of f for any HbO_2 concentration. As shown in Fig. 5, for a HbO_2 concentration of 0.07 mM, the correction factor is equal to 1.56, remarkably close to the value 1.58 derived from the oxygraphic experiment. Using Hb-CPA instead of HbO_2, the same pigment concentration yields a lower correction factor, equal to 1.20.

For both procedures the assumption is that $f = 1$ for infinite HbO_2 concentration, and this is indeed so in the case of the mitochondrial respiratory systems, from mammalian cells (K_m^0 of cytochrome oxidase ranges between 10^{-8} and 10^{-7} M).

3. *Comparative Measurements of HbO_2 Deoxygenation and Disappearance (or Formation) of One of the Reactants*

In the case of the oxidases with low oxygen affinity, $f > 1$ even at infinite HbO_2 concentrations. Under these circumstances, it is necessary to

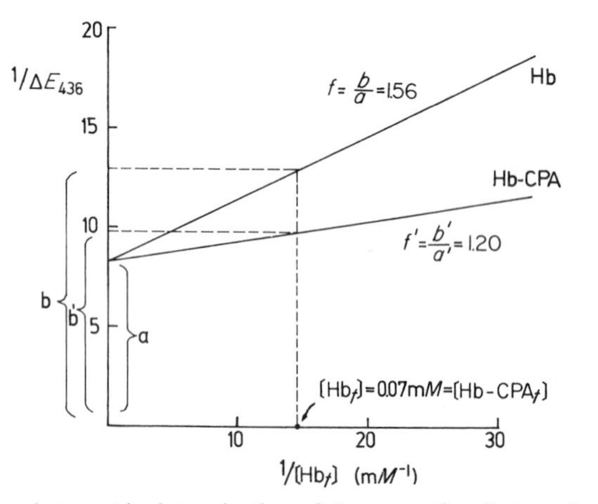

FIG. 5. Spectrophotometric determination of the correction factor using rat liver mitochondria at 25° and HbO_2 or Hb-CPA as oxygen donors.

make comparative measurements in *identical experimental conditions* of the rate of HbO_2 deoxygenation as well as the rate of disappearance (or formation) of one of the reactants. For example, measuring at 250 and 430 nm the activity of the monoamine oxidase from ox kidney, in 0.05 *M* K phosphate buffer (pH 7.6), 0.2 m*M* benzylamine, and 0.05 m*M* HbO_2 one obtains a value of f equal to 2.18. ϵ_{mM} of benzaldehyde at 250 nm is 12; at this wavelength the difference of the millimolar coefficients of extinction of HbO_2 and Hb is very low and does not interfere with benzaldehyde formation. For infinite HbO_2 concentration this correction factor is equal to 1.24.[26]

After the calculation of f, the other parameters are inserted in the equation. The following examples illustrate the experiments described in Section VI.

Example 1: Estimation of Monoamine Oxidase Activity.

$$QO_2\ (\mu\text{g-atoms/min/sample}) = \frac{2 \cdot 1.5 \cdot 2.18 \cdot \Delta E/\text{min}}{0.5 \cdot 81} = 0.161\ \Delta E/\text{min}$$

ΔE is measured at 430 nm. If there is a decrease in optical density at 420.6 nm due to the formation of MetHb, the following correction applies:[27]

$$QO_2\ (\mu\text{g-atoms/min/sample}) = 0.161 \cdot (\Delta E_{430}/\text{min} + 1.8\ \Delta E_{420.6}/\text{min})$$

Example 2: Determination of Cytochrome Oxidase with Hb-CPA as Oxygen Donor.

$$QO_2\ (\mu\text{g-atoms/min/sample}) = \frac{2 \cdot 0.24 \cdot 1.21 \cdot \Delta E/\text{min}}{0.4 \cdot 65} = 0.0223\ \Delta E/\text{min}$$

By substracting the oxygen consumption due to the auto-oxidation of the ascorbic acid, one obtains 0.386 μg-atoms oxygen/min/mg protein when 24 μg of mitochondrial protein are used and, respectively, 0.368 μg-atoms oxygen/min/mg protein with 12 μg of mitochondrial protein.

[26] It is important to note that for enzymes such as glucose oxidase or monoamine oxidase, the kinetic parameters determined by the spectrophotometric method are considerably different from those obtained by other methods in which the reaction medium is generally saturated with air or oxygen. Thus, for purified ox kidney monoamine oxidase the values of K_m (μM) for benzylamine and V_m (μmoles/min/mg protein) are 222 and 2.08 in an oxygen-saturated medium; 147 and 1.37 in an air-saturated medium; and 25 and 0.23 under the conditions of the spectrophotometric method, where the oxygen concentration in the medium is approximately equal to 40 μM.

[27] The last relation is derived from the following considerations: $-\Delta HbO_2/\Delta t = \Delta Hb/\Delta t + \Delta MetHb/\Delta t$. Since $\Delta MetHb/\Delta t = \Delta E_{420.6}/\Delta\epsilon_{420.6}^{Hb\text{-}MetHb}$, the rate of HbO_2 deoxygenation is as follows:

$$\frac{-\Delta HbO_2}{\Delta t} = \left(\frac{\Delta E_{430}}{\Delta t} + \frac{\Delta\epsilon_{430}^{Hb\text{-}MetHb}}{\Delta\epsilon_{420.6}^{Hb\text{-}MetHb}}\Delta E_{420.6}\right)\frac{1}{\Delta\epsilon_{430}^{Hb\text{-}HbO_2}}$$

Since $\Delta\epsilon_{430}^{Hb\text{-}MetHb} = 115$ and $\Delta\epsilon_{420.6}^{Hb\text{-}MetHb} = 64$, then the factor is 115/64 or 1.8.

Example 3: Oxygen Consumption of a Leucocyte Suspension in the Presence of 0.06 mM HbO_2 ($f = 1.67$).

$$QO_2\ (\mu\text{g-atoms/min/sample}) = \frac{2 \cdot 0.24 \cdot 1.67 \cdot \Delta E/\text{min}}{0.4 \cdot 65} = 0.031\ \Delta E/\text{min}$$

VIII. Applications and Future Prospects of the Spectrophotometric Method for Assay of Oxygen Consumption

The expected applications of the spectrophotometric method may be divided into two groups, depending on the degree of technical complexity involved:

1. Assays in the usual range of volumes (0.2–1.0 ml), using general-purpose instruments, such as described above. One may estimate the oxygen consumption of whole-tissue homogenates, isolated cells, and mitochondria,[28-31] the results being comparable, though of higher sensitivity, to those obtained with the oxygraphic method. In particular it is possible to assay oxydative enzymes from crude enzymic samples relevant to diagnosis (e.g., liver phenylalanine hydroxylase for the diagnosis of phenylketonuria).

2. Special assays, requiring equipment of higher technical standard: (a) single-cell respiration, by adapting this technique to the microscale measurement of oxygen consumption, ingeniously accomplished by Hultborn;[8-11] (b) continuous measurement of oxygen consumption of *in vitro* cultured embryo; microspectrophotometric assays combined with a scanning system revealed regional variations of tissue activity;[23] (c) rapid oxygen consumption using stop-flow methods.[32] In this latter case, the limiting factor is the rate of dissociation of oxygen from HbO_2. For human hemoglobin, the value of k equal to 34 sec^{-1} (Table I) allows the assay of reactions with $t_{1/2}$ considerably larger than 20 msec. A promising candidate as oxygen donor in such cases seems to be β-PMB,[17] which has a dissociation rate of oxygen considerably higher ($k = 156\ sec^{-1}$). This donor may allow in principle the estimation of oxygen consumption, at low gas concentration, with $t_{1/2}$ around 10 msec.

Acknowledgments

The author is grateful to Professors P. V. Vignais (Grenoble, France), J. Frei (Lausanne, Switzerland), R. W. VonKorff (Baltimore, Maryland), Th. Bücher (München, GFR), and S. Papa (Bari, Italy) for support and technical facilities. Special thanks are due to Eppendorf Gerätebau (Hamburg, GFR) for making available excellent possibilities for some new developments described in this work.

[28] G. Benga, L. Mureşan, A. Hodârnău and S. Dancea, *Biochem. Med.* **6,** 508–521 (1972).
[29] O. Bârzu, G. Benga, L. Mureşan, S. Dancea, and R. Tilinca, *Enzyme* **12,** 433–448 (1971).
[30] A. Hodârnău, S. Dancea, and O. Bârzu, *J. Cell Biol.* **59,** 222–227 (1973).
[31] P. Nessi, S. Billesbølle, M. Fornerod, M. Maillard, and J. Frei, *Enzyme* **22,** 183–195 (1977).
[32] S. Papa and O. Bârzu, unpublished results.

[27] Sensitive Oxygen Assay Method by Luminous Bacteria

By BRITTON CHANCE, REIKO OSHINO, and NOZOMU OSHINO

This article describes a sensitive and convenient method to detect changes of oxygen in a range below 10^{-6} using luminous bacteria.[1]

It is well known that the intensity of luminescence of marine bacteria is oxygen-dependent. Hastings first quantitated the oxygen dependence of luminescence and showed the half-maximal luminescence intensity was obtained at $8 - 11 \times 10^{-8} M\ O_2$.[2] Bacterial luminescence was first applied by Schindler to study the kinetics of cytochrome oxidase–O_2 reaction.[3]

The reaction sequence for light emission is:

$$\mathrm{FMN + NADH \rightarrow FMNH_2 + NAD}$$

$$\mathrm{FMNH_2 + RCHO + O_2 \xrightarrow[\text{luciferin}]{\text{luciferase}} FMN + RCOOH + H_2O + light}$$

where RCHO and RCOOH are a long-chain fatty aldehyde and acid, respectively. Since the emission spectrum of bacterial luminescence has a broad peak around 475 nm, oxygen concentration can be calibrated by measurement of the light intensity near this wavelength using a suitable optical filter. An advantage of the use of intact luminous bacteria, and not using the isolated luciferin–luciferase system, is that otherwise the latter system requires oxygen concentrations above $10^{-6} M$. Application of the method using the bacteria is most useful for enzyme systems for which spectrophotometric or fluorometric methods for their activity measurement can be used. Also the use of intact bacteria helps to avoid direct interaction of the oxygen indicator (the bacteria) and the oxidase system (see below). The 1–3% NaCl and 0.5 *M* mannitol required for the bacteria may interfere with the enzyme system under study, and the mannitol is not harmful to mitochondria.

Culture of Luminous Bacteria

Photobacterium phosphoreum, *P. fisheri*, and *P. fisheri* strain MAV are used mostly for biochemical researches. They are grown in McElroy's

[1] R. Oshino, N. Oshino, M. Tamura, L. Kobilinsky, and B. Chance, *Biochim. Biophys. Acta* **273,** 5 (1972).

[2] J. W. Hastings, *J. Cell. Comp. Physiol.* **39,** 1 (1952).

[3] F. J. Schindler, Ph.D. Dissertation, University of Pennsylvania, Philadelphia (1964); University Microfilms, Inc., Ann Arbor, Michigan.

ISBN 0-12-181954-X

medium:[4] NaCl, 3 g; $Na_2HPO_4 \cdot 7\,H_2O$, 7 g; KH_2PO_4, 1 g; $(NH_4)_2HPO_4$, 0.5 g; $MgCl_2$, 0.1 g; glycerol, 3 ml; bactopeptone, 5 g; and yeast extract, 3 g in 1 liter of distilled water. The pH is adjusted to 7.0–7.2 with NaCl. They are cultured in 2-liter flasks containing 100 ml of the culture medium at 15–25° with vigorous shaking for 15–25 hr. Because biosynthesis of luciferase depends on cell growth, luminescence increased steeply during late log phase in a growth curve of the bacteria. In the case of *P. phosphoreum* luminescence reaches maximal intensity at 10^8–10^9 cells/ml of culture medium. They were harvested by centrifugation at 6000 *g* for 10 min, washed with 3% NaCl–0.1 *M* phosphate buffer (pH 7.0), and then suspended in the same medium at a concentration of about 500 mg wet weight cells per milliliter. This bacterial preparation could be kept at 0–5° for a few days without significant loss of activity. Occasionally, during the repeated culture, brightness of the cells decreases. In this case it becomes difficult to get sufficient luminescent intensity with a minimum amount of bacteria. It is recommended that the whole culture process be started again by selecting brilliant colonies on agar slants.

Optical System

The optical system consists of a dual wavelength spectrophotometer and a luminescence detector. The latter consists of a photomultiplier and a DC amplifier as shown schematically in Fig. 1. In most experiments for cytochromes, hemoglobins, or myoglobins, the luminescence change can be measured at 470 nm by using a glass filter with a half bandwidth of 10 nm.

Absorbance measurements ordinarily are not disturbed by luminescence, if an appropriate wavelength pair is selected for a dual wavelength spectrophotometer. The wavelength pairs, 555–540 nm and 436–410 nm, were used for the measurements of the redox changes of cytochrome *c* and oxygenation–deoxygenation of hemoglobins, respectively, with satisfactory results.

To minimize interference of absorbance measurements with luminescence, the exit slits for absorbance measurements were guarded with a suitable Kodak Wratten filter, No. 36, which cut out light having wavelengths longer than 460 nm for measurement in the Soret region; filter number 24 which cuts out light having wavelengths shorter than 600 nm was used for measurements of absorbance of cytochrome oxidase at 605–630 nm. In the case of measurements of absorbance between 540 and 580 nm a glass filter that was transparent between these wavelengths was used.

[4] A. A. Green and W. D. McElroy, this series, Vol. 2, p. 857.

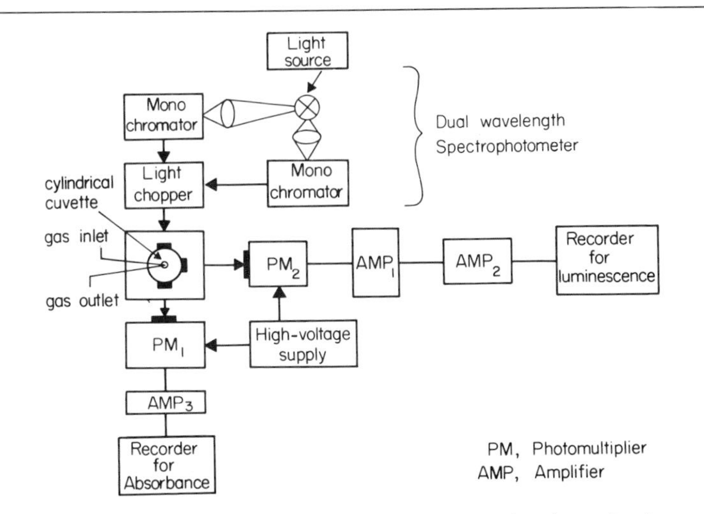

FIG. 1. A schematic diagram of the measuring system used to detect luminescence and absorbance changes.

Reaction Cuvette

Although any size of glass cylinder can be used as the reaction cuvette, a cylinder with a total volume of 100–150 ml is convenient. A 120-ml cylinder covered with black tape except for two windows (0.7 × 1.5 cm) and with a 5-cm diameter was used. The cuvette was closed with a tight-fitting rubber stopper. The two metal needles (diameter, about 2 mm) were thrust into the rubber stopper for inlet and outlet for the gas streams and/or addition of substances.

Assay Method

The Reaction Mixture. The reaction mixture consists of 3% NaCl, 0.1 *M* phosphate buffer (pH 6–8), 10 m*M* glucose, and a small amount of catalase to destroy H_2O_2. The 3% NaCl or 0.5 *M* mannitol must be present in the reaction mixture to provide an appropriate osmolarity for the bacteria.

Oxygen Calibration Curve. Prior to the experiment the reaction mixture is bubbled with prepurified argon gas under stirring (with a tank of N_2 gas it is very hard to get the oxygen concentration below 10^{-6} *M*). Yeast cells in a dialyzing tube hanging into the cuvette are used in order to attain hypoxia in the reaction mixture more rapidly. The yeast cells are removed and the bacterial suspension is added under constant flow of argon. The intensity of luminescence attains its maximum and then disappears quickly as oxygen is consumed by bacterial respiration, indicating 10^{-8} *M*. This procedure is recommended to obtain good reproducibility in luminescence assay since it is not necessary to incubate the bacteria for long periods in the cuvette.

The calibration curve between oxygen concentration and luminescence intensity is obtained by additions of various volumes of an oxygenated solution to the anaerobic bacterial suspension using microsyringes. An example for calibration is shown in Fig. 2. Upon addition of oxygen (10^{-5}) the maximum intensity of luminescence develops within a second and then declines slowly as the bacteria consume oxygen. Subsequent additions of oxygen ranging from 2×10^{-8} to 5×10^{-6} *M* in initial concentration to the anaerobic reaction mixture produce proportional increases in luminescence intensity (100%); for a given oxygen concentration added to the cuvette, the ratio of the luminescence intensity (I_i) to the intensity of saturation (I_{max}) is constant. While the intensity of I_{max} varies from one bacterial preparation to another, the ratio of I_i/I_{max} remains the same. The oxygen concentration is calculated by applying the appropriate dilution factor to oxygen concentrations of 0.25 m*M* and 1.2 m*M* for air- and O_2-saturated water at 20°, respectively.

Figure 3 shows the calibration curve obtained by plotting I_i/I_{max} as a function of oxygen concentration.

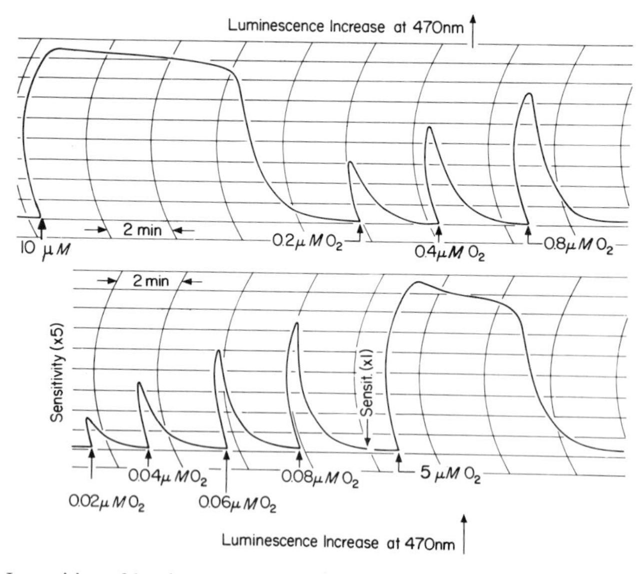

FIG. 2. Intensities of luminescence responding to various oxygen concentrations. The reaction mixture, in a final volume of 120 ml, contained 100 mg wet weight of the bacteria, 3% NaCl, 0.1 *M* phosphate buffer (pH 7.0), 10 m*M* glucose, and 10^{-8} *M* beef liver catalase. In the lower trace of the figure the full scale was expanded 5 times as indicated by the mark. From Oshino *et al.*[1]

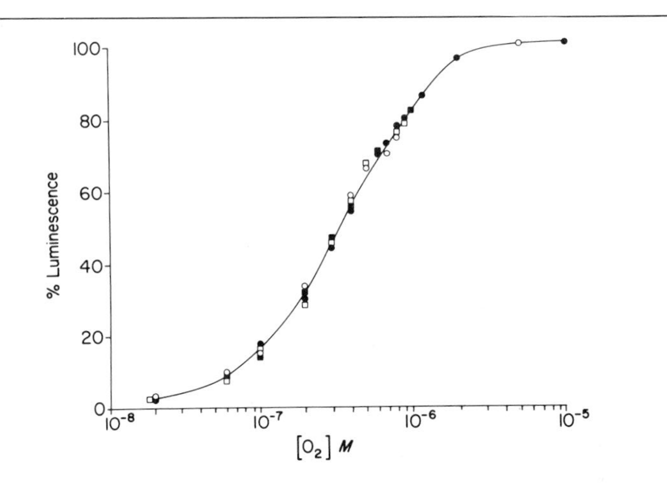

FIG. 3. Oxygen calibration curve. The reaction conditions were the same as described for Fig. 2, except that the pH of phosphate buffer was changed as follows: pH 6.0 (○), pH 7.0 (●), pH 8.0 (□); or 0.5 *M* mannitol was used instead of 3% NaCl (■). From Oshino *et al.*[1]

Properties of the Assay System

Effect of the Bacterial Concentration. The intensity of the luminescence was not influenced by the bacterial concentrations of 50–100 mg wet weight bacteria per 120 ml. It is necessary to choose suitable amounts of the bacteria depending on the assay range of oxygen concentration.

Effect of pH and Salt Concentration. Within the range of pH 6.0–8.0 the curve is unaltered in the salt concentration; if the osmolarity is maintained constant the calibration curve is not influenced significantly. In experiments with mitochondria such a high salt concentration has an inhibitory effect on the mitochondrial respiration. Instead of 3% NaCl, 0.5 *M* mannitol, in which mitochondria function normally, can be used. The stability of the reaction system was affected by mitochondrial inhibitors or uncouplers, such as cyanide and pentachlorophenol. If the luminescence intensity is expressed in terms of the ratio of I_i/I_{max}, the calibrated relation between oxygen concentration and the ratios of I_i/I_{max} is also applicable to this particular case.

Examples of Applications of the Method

There are two procedures for measuring oxygen concentrations:

1. Non-steady-state System. In principle the decrease of oxygen concentration by respiration is accompanied by a corresponding decrease in luminescence intensity. Therefore, it is possible to follow the change of oxygen concentration by luminescence as a function of time. Typical traces

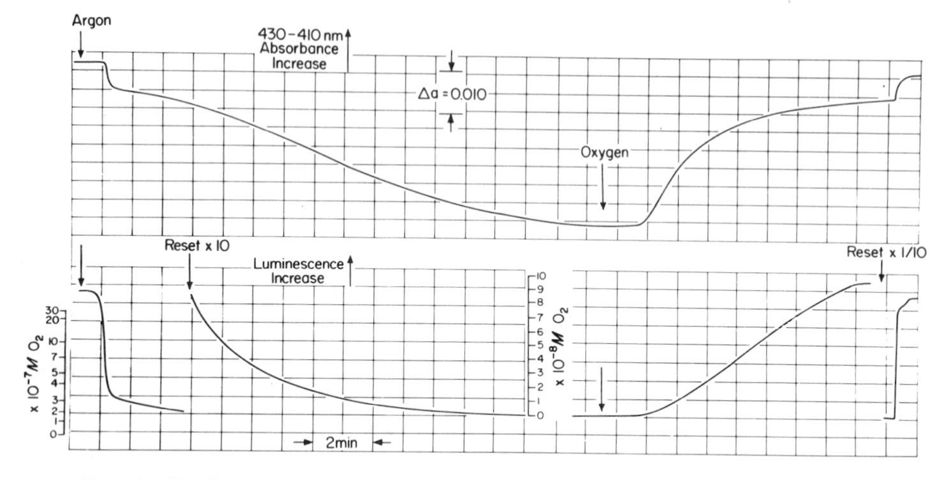

FIG. 4. Simultaneous recordings of luminescence and absorbance changes during oxygenation–deoxygenation of yeast hemoglobin in yeast cells. The reaction mixture was the same as described for Fig. 2, except that 50 mg wet weight of the bacteria and 200 mg wet weight of yeast cells (*Candida mycoderma*) were used. The luminescence of the bacteria was followed at 475 nm. Yeast hemoglobin was measured at 436–410 nm. From Oshino *et al.*[5]

of simultaneous measurements of luminescence and absorbance are shown in Fig. 4.[5] In this particular case the oxygen affinity of yeast hemoglobin in intact cells was measured in the Soret region using a wavelength pair of 436–410 nm, at which the absorbance change is mainly due to oxygenation–deoxygenation of yeast hemoglobin. A decrease in oxygen concentration, which is indicated by a decrease in the intensity of luminescence (bottom trace), causes a decrease in the absorbance at 436–410 nm, which is due to deoxygenation of yeast hemoglobin (upper trace). All of the absorbance and luminescence changes observed are reversible, and, hence, a slow introduction of oxygen produces increases in both absorbance and luminescence.

2. *Steady-state Titration.* In reaction systems which consume oxygen rapidly, changes of oxygen concentration cannot be followed with accuracy. Since the oxygen concentrations in the solution are affected by both rates of oxygen supply and rates of oxygen consumption, various steady states of oxygen concentrations in the systems can be obtained by changing gas flow rates or changing oxygen concentrations in gas flow. This method is recommended for systems with high respiration rates. Application of this method to the isolated pigeon heart mitochondria is shown in Fig. 5. As shown in this figure, both the redox state of cytochrome *c* and the oxygen

[5] R. Oshino, N. Oshino, B. Chance, and B. Hagihara, *Eur. J. Biochem.* **35,** 23 (1973).

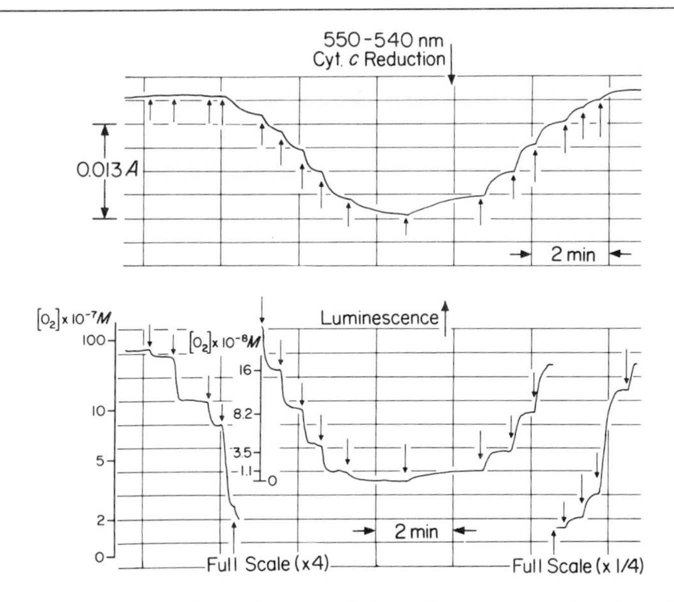

FIG. 5. The redox states of cytochrome *c* of pigeon heart mitochondria depending on the oxygen concentration. The reaction mixture contained, in a final volume of 120 ml, 300 mg wet weight of the bacteria, 60 mg protein of pigeon heart mitochondria, 0.5 *M* mannitol, 0.1 *M* phosphate buffer (pH 7.5), 10 m*M* glucose, 2.5 μM rotenone, 2.5 μM pentachlorophenol, 5 m*M* succinate, 5 m*M* glutamate, and 50 μg antimycin A. At the points indicated by arrows, oxygen concentrations in the gas flow were changed. From Oshino *et al.*[1]

concentration in the solution attain various states as the oxygen concentration in the gas flow is changed.

Limitations of the Method

1. Concentration range. The method is not suitable for oxygen concentrations greater than 10^{-6} *M* or less than 5×10^{-9} *M* O_2.

2. Interfering chemicals. Inhibitors of the luminescence reaction are of course not usable.

3. Effects of temperature. A calibration curve must be run for each temperature.

4. Relative amounts of bacteria to mitochondria. The mitochondria must be present in sufficient amounts to measure one of their cytochromes, and the bacteria must be sufficient to measure their luminescence, which usually means a vast preponderance of mitochondria over bacteria.

[28] Principles of Ligand Binding to Heme Proteins

By HANS FRAUENFELDER

In this article we will consider studies of ligand binding to heme proteins in the temperature range from 2–350 K. We stress the principles that underlie the evaluation of the data and treat specific cases only as illustrations. A question is asked by nearly every biochemist and biologist who looks at these extended-temperature investigations: "Are experiments at unphysiological temperatures meaningful and necessary?" Indeed, life processes are generally considered to stop near the freezing point of water, and some remarkable phenomena such as the production of "antifreeze" in the blood of certain types of hybernators remind us of the fact that life usually does not proceed below the freezing point. Nevertheless, two observations force us to the inescapable conclusion that low-temperature studies are not a luxury, but are essential to understand biomolecular processes. The first observation is that processes in biomolecules do proceed well below the freezing point of water. The pioneering experiments of Chance and Spencer,[1a] of Strother and Ackerman,[1b] of Tyler and Estabrook,[1c] and of Douzou and his colleagues,[1d] show that many enzyme reactions can be observed at temperatures as low as 4 K. The second observation can be illustrated with an analogy. Sophisticated electronic systems, such as computers and color TVs, usually work at physiological temperatures. However, to understand the mechanisms responsible for the function of solid-state devices, experiments over wide temperature ranges are required. Investigations with biomolecules also demonstrate that temperatures down to at least 4 K are necessary to obtain a complete picture of their reaction mechanisms. At higher temperatures, processes are *intermolecular* and the phenomena involving interactions among biomolecules, or biomolecules and solvent, can be explored. At low temperatures, processes are *intramolecular*, and the phenomena occurring near and at the active center become visible. The borderline between high and low depends on solvent, but usually lies between 200 and 270 K. Many experiments demonstrate that both high and low temperatures are compulsory for detailed investigations of the complex processes that take place in even the simplest biomolecules. In the present article, we first describe a hypothetical example, derived from an investigation of carbon monoxide

[1a] B. Chance and E. L. Spencer, *Faraday Soc. Disc.* **27,** 200 (1959).
[1b] G. K. Strother and E. Ackerman, *Biochim. Biophys. Acta* **47,** 317 (1961).
[1c] D. D. Tyler and R. W. Estabrook, *J. Biol. Chem.* **241,** 1672 (1966).
[1d] P. Douzou, R. Sireix, and F. Travers, *Proc. Nat. Acad. Sci. USA*, **66,** 787 (1970).

METHODS IN ENZYMOLOGY, VOL. LIV

ISBN 0-12-181954-X

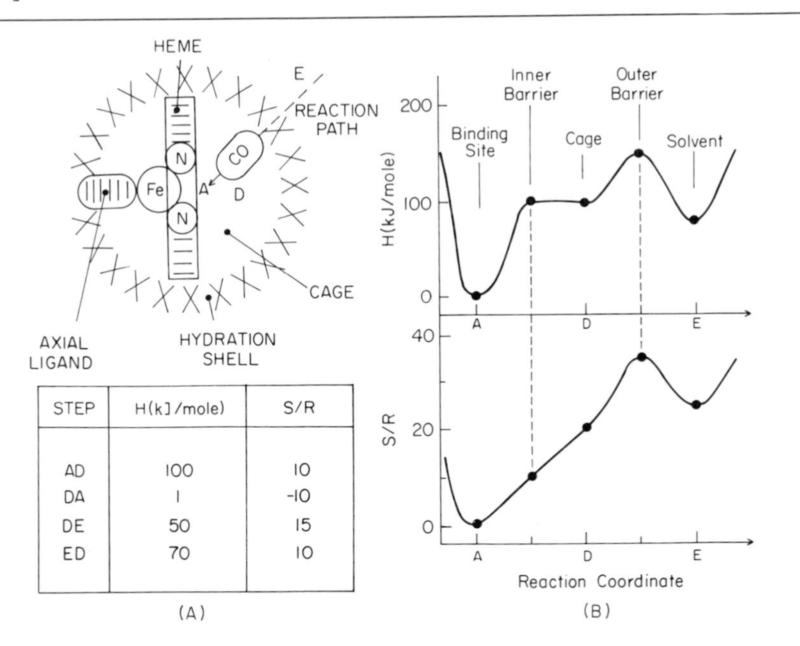

STEP	H(kJ/mole)	S/R
AD	100	10
DA	1	-10
DE	50	15
ED	70	10

FIG. 1. (A) Heme embedded in a solvent, with a solvent cage near the binding site A at the heme iron. A ligand molecule coming from the solvent E hops over the outer barrier to reach cage D and from there over the inner barrier to bind at A. The solvent cage is exaggerated; in reality, solvent molecules will fill the space near the heme. (B) Enthalpy H and entropy S (in units of the gas constant R) as a function of the reaction coordinate. Activation values for all steps are given in the insert in (A).

(CO) binding to protoheme,[2] and then we treat the systematic evaluation of experimental data.

Consider the situation shown in Fig. 1A in which a heme molecule is embedded in a solvent and surrounded by a hydration shell. A denotes the binding site at the heme iron, D the cage adjacent to A, E the solvent, and we assume that a CO molecule encounters enthalpy and entropy barriers as shown in Fig. 1B. In the bound state, the CO molecule sits in well A. Upon photodissociation, the bond between CO and the heme iron is broken and the CO moves to cage D. From there it can either rebind or move over the outer barrier into the solvent. All CO molecules in the solvent can then compete for the vacant binding site. The behavior of a CO molecule after photodissociation can be calculated by solving the relevant differential equations.[2] Denoting with $N_A(t)$ the probability of finding a CO molecule in well A at time t, and with k_{DA} the rate coefficient for the transition $D \rightarrow A$,

[2] N. Alberding, R. H. Austin, S. S. Chan, L. Eisenstein, H. Frauenfelder, I. C. Gunsalus, and T. M. Nordlund, *J. Chem. Phys.* **65,** 4701 (1976).

the fraction of heme molecules that have not rebound CO at time t after the photodissociation becomes

$$N(t) = 1 - N_A(t) = N_{\mathrm{I}}(t) + N_{\mathrm{IV}}(t) \tag{1}$$

where

$$N_i(t) = N_i(0)\, e^{-\lambda_i t} \qquad i = \mathrm{I,IV} \tag{2}$$

$$N_{\mathrm{I}}(0) = k_{DA}/(k_{DA} + k_{DE}) \qquad N_{\mathrm{IV}}(0) = k_{DE}/(k_{DA} + k_{DE}) \tag{3}$$

$$\lambda_{\mathrm{I}} = k_{DA} + k_{DE} \qquad \lambda_{\mathrm{IV}} = k_{ED}k_{DA}/(k_{DA} + k_{DE}) \tag{4}$$

These equations are valid if $k_{AD} \ll k_{DA}, k_{DE}, k_{ED}$ and $k_{ED} \ll k_{DA}, k_{DE}$, and if the CO concentration [CO] in the solvent is sufficiently larger than the heme concentration so that the second-order reaction $E \rightarrow D$ can be treated as pseudo-first order, with rate coefficient

$$k_{ED} = k'_{ED}[\mathrm{CO}] \tag{5}$$

The prime here denotes a second-order rate. We assume the rate coefficients to be given by Arrhenius relations of the form[3]

$$k_{DA} = \nu\, e^{-G_{DA}/RT} = \nu\, e^{S_{DA}/R} e^{-H_{DA}/RT} \tag{6}$$

where G_{DA}, S_{DA}, and H_{DA} are activation Gibbs energy, activation entropy, and activation enthalpy for the step $D \rightarrow A$. $R = 8.314\ \mathrm{J\ K^{-1} mol^{-1}}$ is the gas constant, and ν is a constant taken to be $10^{13}\ \mathrm{sec^{-1}}$. The notation H_{DA}, for instance, implies that H_{DA} is the enthalpy *difference* between the bottom of well D and the top of the barrier between D and A.

The Arrhenius relation describes classical over-the-barrier hopping. At very low temperatures, typically below about 20 K, quantum mechanical tunneling can set in.[4] Tunneling is normally characterized by a temperature dependence of the form

$$k^t_{DA} = \mathrm{const}\ T^m\, e^{-\gamma d[2M(H_{DA} - E_D)]^{\frac{1}{2}}/\hbar} \tag{7}$$

Here, M is the mass of the tunneling system, d the thickness of the barrier, E_D the thermal energy of the system in state D, and γ a numerical factor which depends on the shape of the barrier and lies normally between 1 and 2. The temperature dependence is no longer exponential, but given by a power law. The exponent m becomes 0 in the limit $T \rightarrow 0$, and tunneling becomes temperature independent.

[3] S. Glasstone, K. J. Laidler, and H. Eyring, "The Theory of Rate Processes." McGraw-Hill, New York, 1941; R. A. Marcus, *Techn. Chem. (N.Y.)* **6**, Part 1, 13, Wiley (1974); V. M. Fain, *Phys. Status Solidi B* **63**, 411 (1974).

[4] V. I. Goldanskii, *Annu. Rev. Phys. Chem.* **27**, 85 (1976).

With Eq. (6) and the values of the activation entropies and enthalpies shown in Fig. 1B, the rate coefficients k_{AD}, k_{DA}, k_{DE}, and k_{ED} can be computed and are plotted in Fig. 2A and B. The tunneling rate is difficult to calculate since the values of the constant, the exponent m, and the thickness d are hard to estimate; experimentally observed values[2] are therefore used in Fig. 2B.

In Fig. 2A, $\log k$ is plotted versus inverse temperature; in Fig. 2B it is plotted versus the logarithm of T. If measurements are performed over a wide temperature range, both plots are useful. In a log-lin plot, an exponential appears as a straight line, but power law behavior is hard to recognize and only a limited temperature range can be accommodated. In a log-log plot, a wide temperature range can be displayed. Power laws become straight lines, can be easily identified, and their slope directly yields the power. Exponentials are curved, but the shapes of all exponentials are identical. In order to see if a curve represents an exponential, a cardboard is cut into the shape of an exponential and moved in the $\log k$–$\log T$ plane to fit the observed data.

With the rates as given in Fig. 2 and with Eqs. (1)–(4), the fraction $N(t)$ of hemes that have not rebound CO at time t after photodissociation can be computed at any desired temperature. Typical rebinding curves are given in Fig. 3, where $\log N(t)$ is plotted versus $\log t$. As in Fig. 2B, exponentials

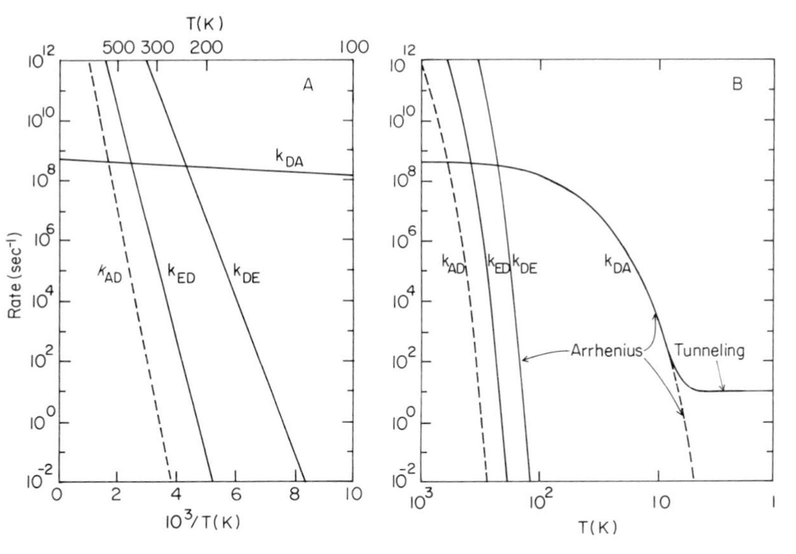

FIG. 2. Rates for the various transitions in the model of Fig. 1. The rate coefficients are calculated from Eq. (6) with the values of H and S/R given in Fig. 1A. Log k is plotted versus $10^3/T$ in (A) and versus log T in (B). The tunneling rate is taken from experimental data.

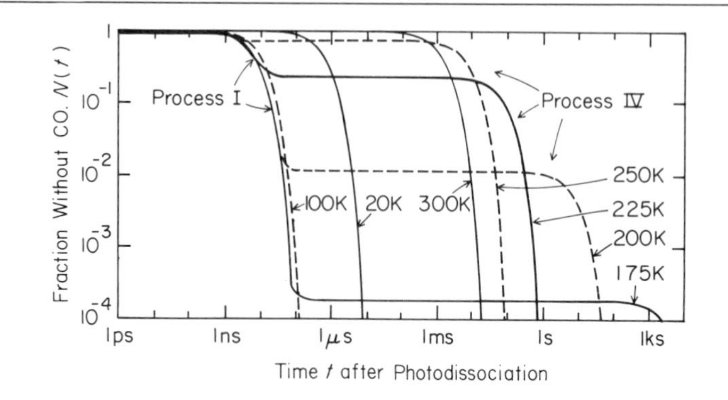

FIG. 3. Rebinding after photodissociation for the model of Fig. 1, calculated with the rate coefficients from Fig. 2 and Eqs. (1)–(4). $N(t)$ denotes the fraction of heme molecules that have not rebound a ligand at time t after the flash. Note that log $N(t)$ is plotted versus log t. The rapidly dropping curves are exponentials. A superposition of two exponentials yields two "shoulders," as for instance at 225 K, and can be recognized easily.

appear in a "universal" shape—constant up to nearly the mean rebinding time ($1/k$) and then rapidly dropping. Figs. 2 and 3 show kinetic features that are important in studying ligand reactions with heme:

(1) Processes extend over a wide range in temperature and time. In order to disentangle the various steps and determine all activation enthalpies and entropies, an experimental system should ideally observe the binding kinetics over 10–15 orders of magnitude in time and 2–4 orders in $N(t)$. Observations in a small temperature interval, say from 270–320 K, cannot elucidate the reaction mechanism.

(2) Figure 3 shows two processes, I and IV, which differ in their intensities and temperature dependences. Figure 1B implies that the inner barrier is entropy, the outer enthalpy controlled. Consider first low temperatures, where Gibbs energy differences, $\Delta G = \Delta H - T\,\Delta S$, are approximately equal to enthalpy differences, ΔH. A ligand molecule, occupying cage D in Fig. 1 after photodissociation, then sees an inner barrier that is much smaller than the outer one so that $k_{DA} \gg k_{DE}$. The ligand molecule will therefore rebind without first moving into the solvent; Eqs. (2) and (3) indicate that only process I is observable and IV is negligibly small. Indeed, Fig. 3 shows that only process I can be seen below about 175 K in the example characterized by Figs. 1 and 2. At low temperatures, photodissociation can thus be used to investigate intramolecular phenomena. As the temperature is increased, the entropy contribution to ΔG becomes more important. In the example of Fig. 1B, the outer Gibbs energy barrier decreases, the inner increases. At some temperature, a few ligand molecules will overcome the outer barrier and move into the solvent

E. Return from the solvent is a second-order process and gives rise to process IV. In Fig. 3, the observable onset of IV occurs at about 175 K. As the temperature is further increased, more ligands move into the solvent, process I decreases, and IV increases. As Fig. 2 shows, $k_{DA} = k_{DE}$ at 230 K, and I and IV are equally intense at this temperature. At still higher temperatures, the inner barrier becomes higher than the outer one. Ligand molecules shuttle many times between cage *D* and solvent *E*, setting up a preequilibrium between the two states. Binding occurs through a small leakage over the inner barrier. The total rate is determined by λ_{IV} in Eq. (4). In this high-temperature regime, rebinding seems to consist of only one process, IV, and the existence of the transition over the inner barrier is hidden. Experiments in this temperature range therefore do not reveal the complexity of the system.

(3) Figure 3 demonstrates that the observed quantum yield depends on the shortest observation time of the experimental set-up. At 175 K, for instance, a quantum yield of 2×10^{-4} is observed at 1 μsec, while it rises to 1 at 1 nsec.

Experimental Approach and Typical Results

The wide range of temperatures required to investigate complex processes implies that studies must be performed in the glasseous or solid state. Rapid-mixing techniques, while adaptable to low-temperature work,[5] cease to function when the sample becomes solid. Reactions must be initiated with an event that works nearly independently of the state of the sample. The best tool at present is flash photolysis—initiation of a reaction by a light pulse.[6] The heme protein with bound ligand is placed in a cryostat with optical windows at a temperature *T* and photodissociated with a laser flash. The subsequent rebinding of the ligand to the heme protein is followed optically. The absorption spectra of the ligand-bound and free heme protein differ, and the change in absorbance $\Delta A(t) = \log [I(t < 0)/I(t)]$ is measured at a suitably chosen wavelength λ_0 as a function of time *t* after photodissociation. Here $I(t)$ is the transmitted intensity of the monitoring light; $t < 0$ denotes the time before the photodissociation. $N(t)$, the fraction of heme protein molecules that have not rebound a ligand after time *t*, is connected to $\Delta A(t)$ by

$$N(t) = \frac{\Delta A(t)}{(\epsilon' - \epsilon)\, cl} \tag{8}$$

[5] S. I. Chan and R. C. Gamble, this volume [19].

[6] P. M. Renzepis, this volume [2]; Q. Gibson, this volume [8].

where c is the protein concentration, ϵ the extinction coefficient for bound and ϵ' for free heme proteins, and l the path length. Equation (8) is valid if only two optical states are involved and all ligand molecules are dissociated by a single flash.

The ideal system, in which the laser flash has a width of a few picoseconds and in which the absorbance is monitored from picoseconds to kiloseconds after each flash, does not yet exist, and the total range is subdivided. The part from about 1 μs to 1 ks can be handled efficiently by using a transient analyzer with a logarithmic time base[7] which records the entire interval in a single sweep and has a time constant that is continuously optimized as time progresses. No information is lost and $\Delta A(t)$ is measured over nearly four orders of magnitude in intensity. Instead of a hard-wired device,[7] a minicomputer may be used.[8]

A few examples of rebinding after photodissociation are displayed in Figs. 4–8.[2,9,10] Photodissociation was initiated by a laser pulse of 1 μs duration and rebinding observed with the transient analyzer with logarithmic time base.[7] $N(t)$ is defined by Eq. (1) and $\Delta A(t)$ by Eq. (8). All data are represented in log-log plots, as in Fig. 3. The techniques to evaluate such curves are discussed in the following section.

Data Evaluation

Figures 4–8 contain features that show up in many other heme proteins, namely multiple barriers, nonexponential kinetics, conformational relaxation, and molecular tunneling.[9,10]

Multiple Barriers

In the model discussed in the first subsection we have assumed two barriers in sequence. With barrier properties as given in Fig. 1A, the rebinding curves after photodissociation are expected to look as in Fig. 3. The real data, for protoheme-CO and shown in Figs. 4 and 5, do not extend below 2 μs, but where they can be observed, they behave as the ones in Fig. 3: At high temperatures ($\sim$300 K), one exponential process, called IV, can be seen. With decreasing temperature, IV becomes slower and less intense. The speed is characterized by the rate λ_{IV}, and the intensity by the initial

[7] R. H. Austin, K. W. Beeson, S. S. Chan, P. G. Debrunner, R. Downing, L. Eisenstein, H. Frauenfelder, and T. M. Nordlund, *Rev. Sci. Instrum.* **47**, 445 (1976).

[8] M. Sharrock, *Rev. Sci. Instrum.* **48**, 1202 (1977).

[9] R. H. Austin, K. W. Beeson, L. Eisenstein, H. Frauenfelder, and I. C. Gunsalus, *Biochemistry* **14**, 5355 (1975).

[10] N. Alberding, R. H. Austin, K. W. Beeson, S. S. Chan, L. Eisenstein, H. Frauenfelder, and T. M. Nordlund, *Science* **192**, 1002 (1976).

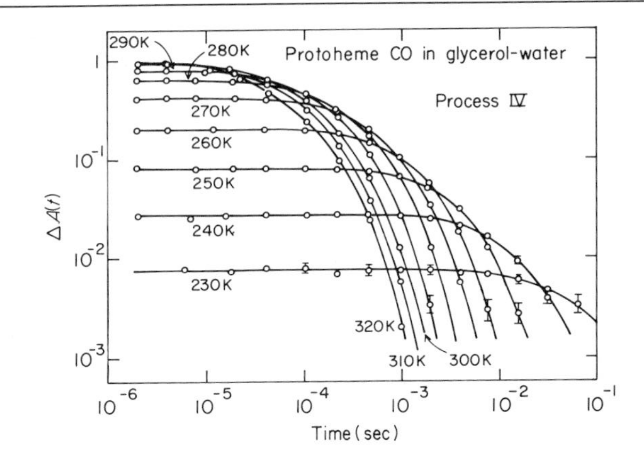

FIG. 4. Binding of CO to protoheme after photodissociation, temperatures between 230 and 320 K. The solvent is glycerol–water (3:1, v/v), 10 mM NaOH, $[CO] = 4 \times 10^{-4} M$. $\Delta A(t)$, the change in absorbance after photodissociation, is given in a log-log plot. Process IV shown here is exponential, $\Delta A^{IV}(t) = \Delta A^{IV}(0) \exp(-kt)$, and proportional to CO concentration. A few typical errors are shown. For $\Delta A(t) > 0.02$, the errors are about the size of the data circles. [N. Alberding, R. H. Austin, S. S. Chan, L. Eisenstein, H. Frauenfelder, I. C. Gunsalus, and T. M. Nordlund, *J. Chem. Phys.* **65,** 4701 (1976).]

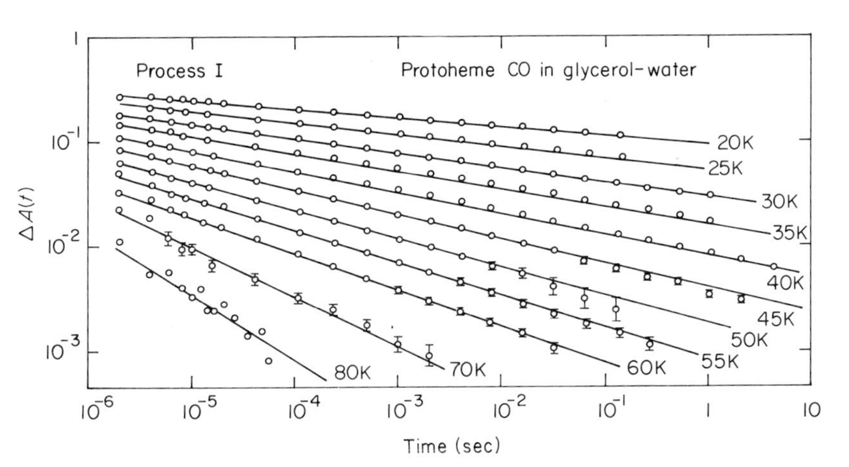

FIG. 5. Binding of CO to protoheme (in glycerol–water, 3:1, v/v) after photodissociation at temperatures below 80 K. The process shown here, called I, is independent of CO concentration and *not* exponential. The solid lines are computed with Eqs. (27)–(30) and $H_{BA}^{peak} = 1.05$ kJ mol^{-1}, $\alpha = 0.48$ mol kJ^{-1}, $T_c = 180$ K, and $n(T_f) = 0.04$. The remarkable agreement between the theoretical curves and the experimental points implies that rebinding occurs by classical over-the-barrier motion, with the barrier enthalpy and entropy distributed as shown in Fig. 11 and characterized by Eq. (24). [N. Alberding, R. H. Austin, S. S. Chan, L. Eisenstein, H. Frauenfelder, I. C. Gunsalus, and T. M. Nordlund, *J. Chem. Phys.* **65,** 4701 (1976).]

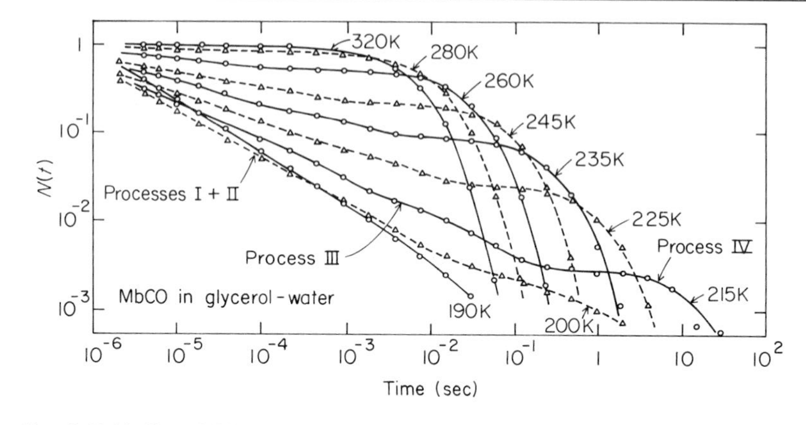

FIG. 6. Rebinding of CO to Mb after photodissociation. Solvent: glycerol–water, 3:1, v/v. [CO] = $3 \times 10^{-5} M$. Three different processes, II–IV, can be recognized directly. The fourth, called I, is identified by extrapolating the data taken at lower temperatures and shown in Fig. 7. [R. H. Austin, K. W. Beeson, L. Eisenstein, H. Frauenfelder, and I. C. Gunsalus, *Biochemistry* **14,** 5355 (1975).]

absorbance change $\Delta A(0)$ or the corresponding value $N_{IV}(0)$. Below about 230 K, IV can no longer be observed. Comparison of Figs. 3 and 4 makes it plausible that the experiments of Fig. 4, extended to shorter times, would display also the fast process I. As the temperature is further lowered, I indeed appears (Fig. 5). Comparison of Figs. 3 and 5 reveals an unexpected feature: Process I is not exponential in time, but closer to a power law. The information contained in the deviation of I from an exponential will be

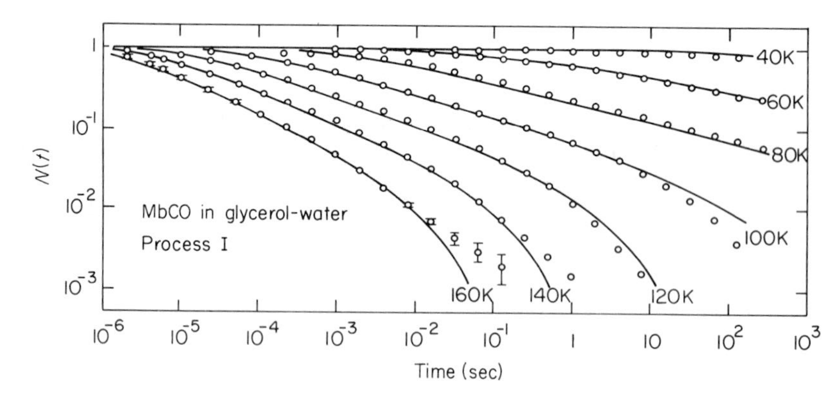

FIG. 7. Rebinding of CO to Mb after photodissociation. The solvent is glycerol–water, 3:1, v/v. The curves are not exponential in time, but can be approximated by the power law Eq. (17). The solid lines are theoretical fits, based on the enthalpy spectrum shown in Fig. 12. [R. H. Austin, K. W. Beeson, L. Eisenstein, H. Frauenfelder, and I. C. Gunsalus, *Biochemistry* **14,** 5355 (1975).]

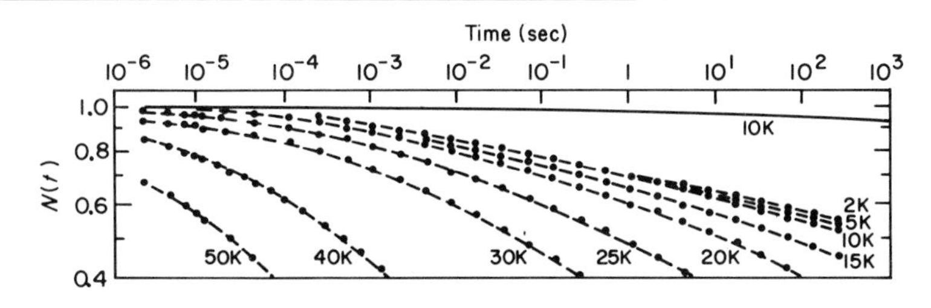

FIG. 8. Rebinding of CO to the separated beta chain of hemoglobin. These curves provide evidence of molecular tunneling: As the temperature is lowered below about 15 K, binding does not become much slower, but tends towards a limit. The solid line labeled 10 K is the theoretical prediction extrapolated from the curves above 20 K by assuming the binding occurs only by motion over the barrier. The theoretical expectation falls considerably below the experimental values; the difference is ascribed to tunneling.

discussed below; here we disregard this feature and assume I to be exponential in time. Equations (1)–(5) can then be used to extract all rate parameters from the measured curves; with an Arrhenius plot, activation enthalpies and entropies for both barriers are obtained.[2] The rate k_{AD}, however, cannot be found, and the depth of well A in Fig. 1 must be taken from equilibrium data.

In a protein, more than two successive barriers may occur, and the analysis must be generalized. Figures 6 and 7 represent rebinding of CO to myoglobin after photodissociation.[9] Four different processes occur. To evaluate the data, the curves for $\Delta A(t)$ are decomposed into components, as shown in Fig. 9 for 215 K. At this temperature, three processes can be

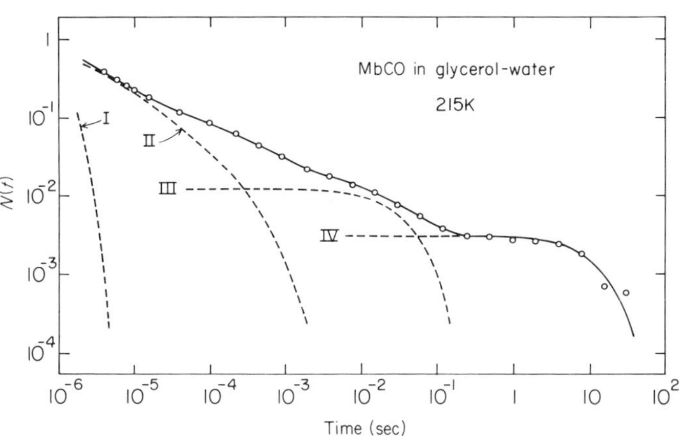

FIG. 9. Decomposition of $N(t)$ for binding of CO to Mb after photodissociation into the components II–IV. Component I is obtained by extrapolation from lower temperatures. [R. H. Austin, K. W. Beeson, L. Eisenstein, H. Frauenfelder, and I. C. Gunsalus, *Biochemistry* **14**, 5355 (1975).]

seen directly; the fourth (I) is obtained by extrapolation from lower temperatures. Decomposition is not always easy. High-quality data at closely spaced temperatures and taken with different ligand concentrations are required. The ligand-concentration dependence separates first- and second-order processes. The direct result of the experiments are the rates λ_i and the intensities $N_i(0)$ of the various processes at all temperatures. The number of processes depends on protein and ligand; for the following discussion we assume that four processes are seen. The simplest way to account for the presence of four different processes is to postulate that the ligand, coming from the solvent, encounters four barriers in succession. Symbolically, the situation is represented by the chain

$$\underset{\substack{\text{CO bound}\\ \text{to heme}}}{A} \underset{k_{BA}}{\overset{k_{AB}}{\rightleftharpoons}} B \underset{k_{CB}}{\overset{k_{BC}}{\rightleftharpoons}} C \underset{k_{DC}}{\overset{k_{CD}}{\rightleftharpoons}} D \underset{k'_{ED}}{\overset{k_{DE}}{\rightleftharpoons}} \underset{\substack{\text{CO in}\\ \text{solvent}}}{E} \tag{9}$$

The next step in the data evaluation consists in educing the rate coefficients in Eq. (9) from the observed quantities λ_i and $N_i\ (0)$, i = I–IV. The differential equations for the motion of a ligand in the system described by Eq. (9) are[9,11]

$$\begin{aligned}
\frac{dN_A}{dt} &= -k_{AB}N_A + k_{BA}N_B \\
\frac{dN_B}{dt} &= k_{AB}N_A - k_{BA}N_B - k_{BC}N_B + k_{CB}N_C \\
\frac{dN_C}{dt} &= k_{BC}N_B - k_{CB}N_C - k_{CD}N_C + k_{DC}N_D \\
\frac{dN_D}{dt} &= k_{CD}N_C - k_{DC}N_D - k_{DE}N_D + k_{ED}N_E \\
\frac{dN_E}{dt} &= k_{DE}N_D - k_{ED}N_E
\end{aligned} \tag{10}$$

The ligand concentration in the solvent is assumed to be large enough so that the step $E \rightarrow D$ can be treated as pseudo-first order, but not so large that the probability of finding two or more ligand molecules inside the biomolecule must be considered.[12] Before photodissociation, the ligand is

[11] K. W. Beeson, unpublished thesis (1975) (available as a Technical Report from the Department of Physics, University of Illinois at Urbana-Champaign, Urbana).

[12] It is likely that in many enzyme reactions linear deterministic equations of the form of Eq. (10) are inappropriate and that nonlinear equations are required. For a stochastic treatment of the more general case see N. Alberding, H. Frauenfelder, and P. Hänggi, *Proc. Nat. Acad. Sci. USA*, **75,** 26 (1978).

bound at the heme iron in well A; photodissociation breaks the bond so that the initial conditions after photodissociation are

$$N_B(0) = 1, \quad N_A(0) = N_C(0) = N_D(0) = N_E(0) = 0 \tag{11}$$

While the case of three wells can be solved explicitly and leads to Eqs. (1)–(4), the five-well situation must be handled by computer. Eq. (10) is written in matrix form,

$$\frac{d\vec{N}}{dt} = M\vec{N} \tag{12}$$

where

$$M = \begin{bmatrix} -k_{AB} & k_{BA} & 0 & 0 & 0 \\ k_{AB} & -k_{BA} - k_{BC} & k_{CB} & 0 & 0 \\ 0 & k_{BC} & -k_{CB} - k_{CD} & k_{DC} & 0 \\ 0 & 0 & k_{CD} & -k_{DC} - k_{DE} & k_{ED} \\ 0 & & 0 & k_{DE} & -k_{ED} \end{bmatrix}, \quad \vec{N} = \begin{bmatrix} N_A \\ N_B \\ N_C \\ N_D \\ N_E \end{bmatrix} \tag{13}$$

The general solution of Eq. (12) is found by determining the eigenvalues λ_i and eigenvectors $\vec{v}_i$ of the eigenvalue equation

$$(M - \lambda_i I)\vec{v}_i = 0 \tag{14}$$

and expanding,

$$\vec{N}(t) = \sum_{i=1}^{5} c_i \vec{v}_i \, e^{\lambda_i t} \tag{15}$$

I in Eq. (14) is the unit matrix. The coefficients c_i are determined by the initial conditions [Eq. (11)]. The solution of Eq. (12) is simplified by the special form of the matrix M: All diagonal elements of M are negative, and the sums of elements in each column vanish. The eigenvalues of such matrices, with rate coefficients k_{ij} different from zero, have a number of interesting properties[13] of which two are particularly relevant here: (1) One eigenvalue, say λ_1, is zero. (2) All other eigenvalues λ_i ($i > 1$) are simple (nondegenerate), real, and negative definite. Equation (15) can therefore be written as

$$\vec{N}(t) = \vec{v}_1 + \sum_{i=2}^{5} c_i \vec{v}_i \, e^{\lambda_i t} \tag{15'}$$

A further simplification comes from the fact that the ligand is usually bound tightly in well A. Thermal dissociation can be neglected in photodissociation experiments and $k_{AB} \approx 0$. The state A then is called "absorbing" and $\vec{v}_1$ takes on the form

[13] W. Ledermann and G. E. H. Reuter, *Philos. Trans. R. Soc. London, Ser. A* **246,** 321 (1954); P. Hänggi and H. Thomas, *Helv. Phys. Acta* **49,** 803 (1976).

$$\vec{v}_1 = \begin{pmatrix} 1 \\ 0 \\ 0 \\ 0 \\ 0 \end{pmatrix}$$

The rebinding function $\bar{N}(t) = 1 - N_A(t)$ is determined by $\vec{N}(t)$, and Eqs. (12)–(15) provide the connection between the experimentally observed eigenvalues λ_i and intensities $N_i(0)$, and the rate coefficients k_{ij}. To obtain the actual values of the rate parameters, the computer is instructed to fit $\bar{N}(t_i)$ as calculated from Eq. (15′) to the experimental points $N_{\text{exp}}(t_i)$ by minimizing

$$\chi^2 = \sum_{\text{all } i} \left[\frac{N_{\text{exp}}(t_i) - \bar{N}(t_i)}{\Delta N(t_i)} \right]^2 \tag{16}$$

Here $N_{\text{exp}}(t_i)$ is the value of $N(t)$ measured at time t_i, with error $\Delta N(t_i)$. Since $N_{\text{exp}}(t_i)$ is a time average over a logarithmically increasing time interval, the computer is told to perform the corresponding average also on $N(t_i)$; the bar symbolizes the average. In theory, the evaluation procedure described here is straightforward and leads to the rate parameters without additional problems. In practice, fitting can be very difficult, and the computer has to be helped by judicious choice of approximate values for the parameters. Details of the fitting procedure are given in footnote 11.

From the rate coefficients determined as a function of temperature, activation enthalpies and entropies are found with Eq. (6). Examples of Arrhenius plots over extended temperature ranges and of the corresponding barriers are given in Alberding *et al.*[2] and Austin *et al.*[9]

Nonexponential Kinetics

The model of Fig. 1 predicts exponential rebinding after photodissociation also at very low temperatures, where only process I is observed. In reality, however, I is not exponential, as is evident from Figs. 5, 7, and 8, but can often by approximated by a power law,

$$N(t) = [1 + t/t_0(T)]^{-n(T)} \tag{17}$$

Here, $t_0(T)$ and $n(T)$ are two temperature-dependent fitting parameters. (In passing we note that it is difficult to recognize deviations from exponential behavior in a log-lin plot; log-log plots are essential for diagnosis.)

The simplest way to explain nonexponential first-order kinetics is to assume that the innermost barrier does not possess a unique height, but must be described by a distribution. We begin by assuming that, at any temperature, the Gibbs energy for the step $B \rightarrow A$ is not sharp, but given by a spectral function $g(G_{BA})$, where $g(G_{BA})dH_{BA}$ denotes the probability of having a barrier with Gibbs activation energy between G_{BA} and $G_{BA} + dG_{BA}$, with $\int_0^\infty dH_{BA}\, g(H_{BA}) = 1$. The rebinding function then becomes

$$N(t) = \int_0^\infty dG_{BA}\, g(G_{BA})\, e^{-k_{BA}t} \tag{18}$$

where k_{BA} depends on G_{BA} in the Arrhenius regime through Eq. (6). Equation (18) alone is not very useful; it only replaces a distribution in time by one in Gibbs energy. Since $G_{BA} = H_{BA} - TS_{BA}$ depends explicitly on T, distributions at different temperatures are not simply related. To unify the descriptions at various temperatures, we express $g(G_{BA})$ in terms of distributions in activation enthalpy and entropy, H_{BA} and S_{BA}, and assume these to be temperature-independent. (Of course, H_{BA} and S_{BA} are also temperature-dependent through the heat capacity of activation, but we neglect this effect here.) Consequently, evaluation of nonexponential kinetics involves two steps, inversion of Eq. (18) to get $g(G_{BA})$ and determination of $g(H_{BA})$ and $g(S_{BA})$ from the temperature dependence of $g(G_{BA})$. Both steps are in general very difficult to perform rigorously[14] and we therefore introduce approximation methods.

Inversion of Eq. (18), with $N(t)$ given by Eq. (17), yields[9]

$$g(G_{BA}) = \frac{(\nu t_0)^{n(T)}}{RT\Gamma[n(T)]} e^{-n(T)G_{BA}/RT}\, e^{-(\nu t_0)\exp(-G_{BA}/RT)} \tag{19}$$

where $\Gamma[n(T)]$ denotes the gamma function. This approximate form of $g(G_{BA})$ rises steeply to a peak value, turns over, and then decreases exponentially with G_{BA}. The condition $dg\ (G_{BA})/dG_{BA} = 0$ gives the value of G_{BA} at which the distribution peaks as

$$G_{BA}^{\text{peak}} = RT\, ln\, [\nu t_0/n(T)] \tag{20}$$

The experimental parameters $n(T)$ and t_0 thus determine G_{BA}^{peak} at each temperature.

The transition from $g(G_{BA})$ to $g(H_{BA})$ or $g(S_{BA})$ becomes easier if we first determine which thermodynamic variable is distributed from the temperature dependence of $n(T)$. Equation (17) implies that $N(t)$ for $t \gg t_0$ is proportional to $t^{-n(T)}$. Figures 5, 7, and 8 show indeed that most curves can be fitted by a straight line over a major part of the time range. The corresponding $g(G_{BA})$ is found easily from Eq. (18) or Eq. (19):

$$N(t) \propto t^{-n(T)} \rightleftharpoons g(G_{BA}) \propto \exp\left\{-n(T)G_{BA}/RT\right\} \tag{21}$$

[14] D. E. Koppel, *J. Chem. Phys.* **57**, 4814 (1972); S. W. Provencher, *Biophys. J.* **16**, 27 (1976); *J. Chem. Phys.* **64**, 2772 (1976); U. Landman and E. W. Montroll, *ibid.* p. 1762.

A power-law dependence in $N(t)$ rigorously implies an exponential dependence of $g(G_{BA})$ on G_{BA} and vice versa. This fact, together with the assumption of temperature independence of the enthalpy and entropy distribution functions, leads to a criterion that distinguishes three cases:[15]

(1) H_{BA} distributed, S_{BA} single valued. We insert $G_{BA} = H_{BA} - TS_{BA}$ into Eq. (21), retain only the part of $g(G_{BA})$ that depends on H_{BA}, and write Eq. (21) as $g(H_{BA}) \propto \exp\{-n(T)H_{BA}/RT\}$. This enthalpy distribution is temperature independent if

$$g(H_{BA}) \propto \exp\{-\alpha_H H_{BA}\}, \quad n(T) = \alpha_H RT \tag{22}$$

where α_H is a constant. A pure enthalpy distribution is present if $n(T)$ is proportional to T.

(2) H_{BA} single valued, S_{BA} distributed. In this case we write $g(S_{BA}) \propto \exp\{n(T)S_{BA}/R\}$; temperature independence gives

$$g(S_{BA}) \propto \exp\{\alpha_S S_{BA}\}, \quad n = \alpha_S R \tag{23}$$

with α_S constant. A pure entropy distribution leads to constant n.

(3) H_{BA} and S_{BA} distributed. An independent determination of the distributions for H_{BA} and S_{BA} is extremely difficult. We therefore assume that each value of H_{BA} is associated with a unique value of S_{BA}, expand S_{BA} in terms of H_{BA}, and retain the linear term:

$$S_{BA} = S^0_{BA} + (1/T_c)\, H_{BA} \tag{24}$$

This relation is known as "compensation law," and the constant T_c is called compensation temperature.[16] For the present purpose, interpretation of Eq. (24) is not important. Inserting Eq. (24) into (21), retaining only terms that depend on H_{BA}, and demanding temperature independence, yields with constant α_{HS}

$$g(H_{BA};\, T_c) \propto \exp\{-\alpha_{HS} H_{BA}\}, \quad \frac{1}{n(T)} = \frac{1}{\alpha_{HS} R}\left\{\frac{1}{T} - \frac{1}{T_c}\right\} \tag{25}$$

Equations (22), (23), and (25) provide the method to distinguish the three different cases: $n(T)$ is determined from the experimental curves and $1/n(T)$ plotted versus $1/T$, as in Fig. 10. This plot yields three pieces of information. First, if the experimental points fall on a straight line over a wide temperature range, the approximations introduced above are corrobo-

[15] The actual distribution functions $g(H_{BA})$ and $g(S_{BA})$ very likely do depend on temperature. Assuming temperature independence is, however, a simple and satisfying first approximation. The agreement between the calculations based on temperature-independent distribution functions and the experimental data, shown for instance in Figs. 5 and 7, is very good and thus lends support to this approach. We therefore demand temperature independence of the distribution functions as a step to arrive at the criteria Eqs. (22), (23), and (25).

[16] R. R. Krug, W. G. Hunter, and R. A. Grieger, *J. Phys. Chem.* **80,** 2335 and 2341 (1976).

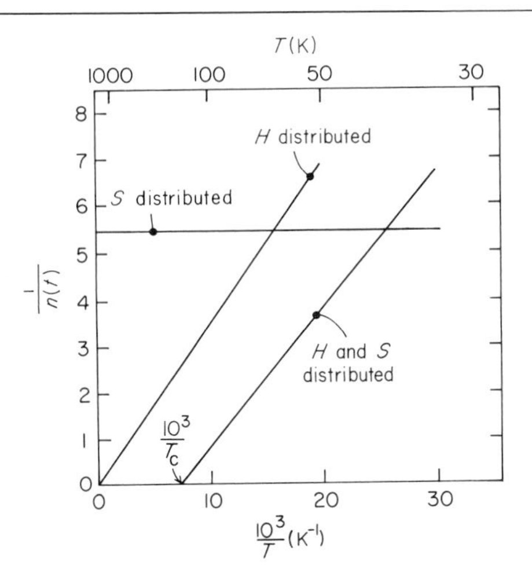

FIG. 10. Plot of $1/n(T)$ versus $10^3/T$; $n(T)$ is the exponent in Eq. (17). This plot shows if the nonexponential rebinding is due to distributed entropy, distributed enthalpy, or a combination of both.

rated. If deviations occur, the treatment must be generalized. Second, the intercept with the $(1/T)$ axis yields T_c and selects among cases (1)–(3); $T_c = \infty$ implies distributed enthalpy (1), $T_c = 0$ distributed entropy (2). If T_c is between these two limits, case (3) applies and the value of T_c characterizes the entropy spread. Third, the slope of the straight line yields the parameter α.

One remark is in order with respect to the distributions given in Eqs. (22), (23), and (25). These expressions do not describe the full spectrum, but only the tail, corresponding to $G_{BA} > G_{BA}^{\text{peak}}$.

For the remainder we assume that case (3) applies; H_{BA} and S_{BA} are both distributed and connected by Eq. (24). Equation (20) then becomes

$$H_{BA}^{\text{peak}} - T\, S_{BA}^{\text{peak}} = RT \ln \left\{\nu t_0/n(T)\right\} \tag{26}$$

A plot of $RT \ln \left\{\nu t_0/n(T)\right\}$ versus T, as given in Fig. 11, determines H_{BA}^{peak} and S_{BA}^{peak}. Since T_c is already known from Fig. 10, S_{BA}^0 can now be calculated from Eq. (24).

With the parameters gleaned from Figs. 10 and 11, we can determine $g(H_{BA})$ in a first approximation. To do so we note that the approximation Eq. (17) does not fit the experimental data equally well at all temperatures. Call the temperature at which the fit is best T_f, the "fitting temperature." At this temperature, we can write Eq. (19) in terms of H_{BA}; with Eqs. (20), (24), (25), and $\alpha_{HS} \equiv \alpha$, this $g(H_{BA})$ becomes

$$g(H_{BA}) = \text{const exp}\left\{\alpha(H_{BA}^{\text{peak}} - H_{BA}) - n(T_f)\right.$$
$$\left.\exp\left[\frac{\alpha}{n(T_f)}(H_{BA}^{\text{peak}} - H_{BA})\right]\right\} \quad (27)$$

The constant is fixed by the normalization

$$\int_0^\infty dH_{BA}\, g(H_{BA}) = 1 \quad (28)$$

In principle, all constants in Eq. (27) are now known, and the enthalpy and entropy distributions have been found. In practice, one additional refinement is useful. With $g(H_{BA})$ as given, $N(t)$ is calculated through

$$N(t) = \int_0^\infty dH_{BA}\, g(H_{BA})\, e^{-k_{BA}t} \quad (29)$$

with

$$k_{BA} = \nu \exp\left\{\frac{S_{BA}^0}{R} - \frac{H_{BA}}{R}\left[\frac{1}{T} - \frac{1}{T_c}\right]\right\} \quad (30)$$

Equations (29) and (30) together are valid in the temperature range where only process I occurs and tunneling can be neglected. The parameter T_f in Eq. (27) is varied till a best fit is obtained. A few distributions obtained in this manner are shown in Fig. 12. The solid lines in Fig. 5 are calculated from the protoheme enthalpy distribution in Fig. 12, with $T_c = 180$ K. The fit to the experimental points is extremely good. A similar fit to the low-

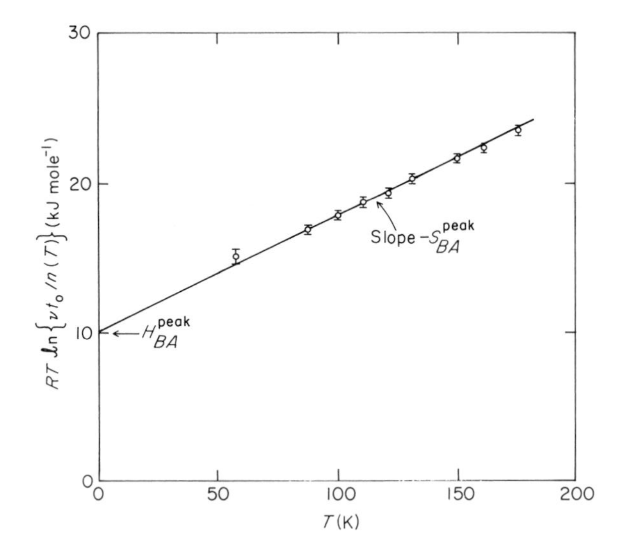

FIG. 11. The plot of $RT \ln\{\nu t_0/n(T)\}$ versus T determines H_{BA}^{peak} and S_{BA}^{peak}.

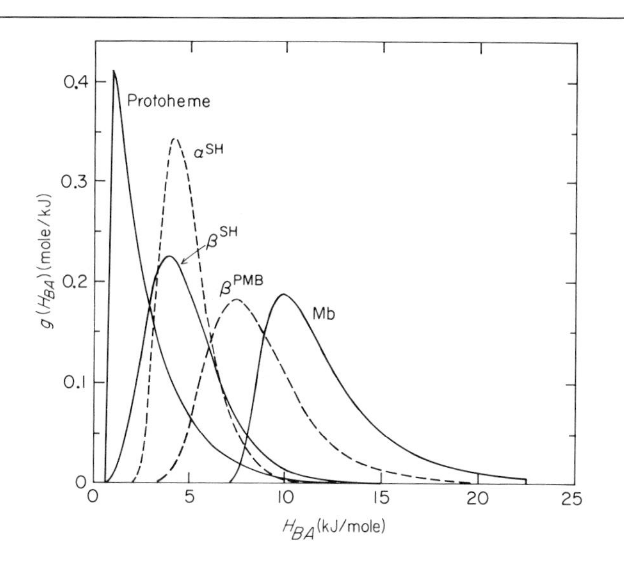

FIG. 12. Activation enthalpy distributions at the innermost barrier in heme proteins. Mb: myoglobin; α and β: separated hemoglobin monomers; β^{PMB}: cysteinyl-*p*-mercury benzoate derivative. For protoheme, the entropy is also distributed; the width of the distribution is characterized by $T_c = 180$ K.

temperature data for myoglobin is much less satisfactory; the experimental points at long times no longer follow a power law, but drop nearly exponentially. The departure from a power law can be explained if $g(H_{BA})$ breaks off at some maximum enthalpy, H_{BA}^{max}. A cut-off is easily incorporated in the data evaluation by setting the upper limit of the integral in Eq. (29) equal to H_{BA}^{max}. The solid lines in Fig. 7 are computed with the spectrum for myoglobin in Fig. 12 and a cut-off; again the agreement between theoretical fit and experimental data is very good.[9]

The discussion of nonexponential kinetics has so far been restricted to the innermost barrier, between wells B and A. The other barriers may, however, also be distributed,[9] and even a small enthalpy spread on a high barrier may result in large effects. Assume the activation enthalpy to be given by a distribution as in Fig. 12, with H_0 and $H_0 + \Delta H$ denoting the minimum and maximum, and ΔH the width, of the spectrum. The ratio of rate coefficients corresponding to H_0 and $H_0 + \Delta H$,

$$\frac{k^{max}}{k^{min}} = \frac{\exp(-H_0/RT)}{\exp[-(H_0 + \Delta H)/RT]} = \exp(\Delta H/RT)$$

is only determined by ΔH. Even a small spread on a high barrier can lead to markedly nonexponential rebinding.

In the evaluation of the multiple barriers and the nonexponential kinetics we have assumed validity of the Eyring relation Eq. (6). Application of this relation over the temperature range from 20–300 K may, however, exceed the range of validity of the activated-complex theory, and the interpretation of S as activation entropy becomes problematic. Nevertheless, Eq. (6) provides a convenient parameterization, and the technique of data evaluation does not depend on the specific interpretation of S. Moreover, the experimental data are not yet accurate enough to investigate the temperature dependence of the parameter ν. The assumption of a temperature-independent ν is a sufficient first approximation which can be modified when the data evaluation produces rate coefficients that no longer satisfy Eq. (6).

Conformational Relaxation

The experimental data in Figs. 4–8 show that binding is exponential in time above about 230 K, but follows a power law below about 200 K. The behavior can be understood by assuming that heme and heme proteins possess many different conformational states, with different barriers for the step $B \rightarrow A$.[9] Below about 200 K, each biomolecule remains frozen in a particular conformational state whereas above about 230 K in a glycerol–water solvent, a given molecule changes rapidly from one state to another. In the frozen state, the binding process reflects the activation enthalpy spectrum at the innermost barrier and is nonexponential; if rapid conformational relaxation is present, each molecule averages over conformational states and binding becomes exponential.

Conformational relaxation can be characterized by a rate coefficient k_r which gives the mean rate at which the protein conformation changes. Data evaluation as discussed earlier must now be modified depending on k_r. While the general solution of this problem has not yet been given, two extreme cases can be described easily.[9] *Relaxation is complete* if the relaxation rate coefficient is much larger than the maximum return rate over the innermost barrier,

$$k_r \gg k_{BA}^{\max} \tag{31}$$

Rebinding then will be exponential in time, and the rate coefficient k_{BA} in Eqs. (10)–(16) is taken to be the rate coefficient averaged over the distribution $g(H_{BA})$,

$$\langle k_{BA} \rangle = \int_0^\infty dH_{BA}\, g(H_{BA}) k_{BA} \tag{32}$$

In a good approximation, $\langle k_{BA} \rangle$ can be replaced by k_{BA}^{peak}, which is easier to calculate. *No relaxation* occurs if

$$k_r \ll k_{BA}^{\min} \tag{33}$$

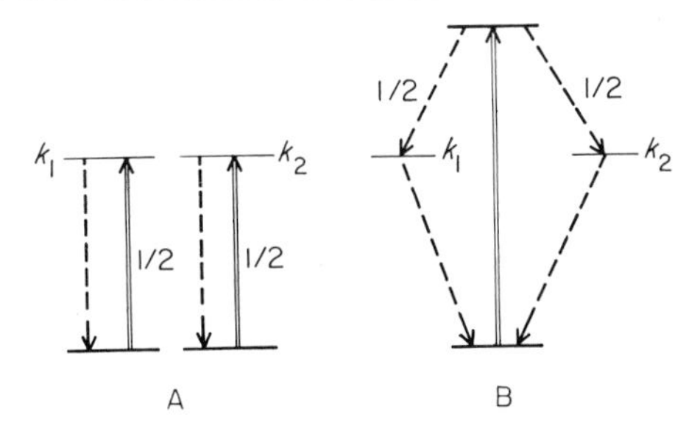

FIG. 13. Unrelated and related binding after photoexcitation. In (A), the two states, with rate coefficients k_1 and k_2, are excited and decay independently. In (B), the two states are fed from the same higher state.

and the distributed nature of the innermost barrier then affects all processes. The rebinding function is obtained from $\vec{N}(t)$, Eq. (15), by integrating $[1 - N_A(t)]$ with the proper weight $g(H_{BA})$,

$$N(t) = \int_0^\infty dH_{BA}\, g(H_{BA})[1 - N_A(t)] \tag{34}$$

The conditions Eqs. (31) and (33) are actually too restrictive, and a more careful discussion leads to weaker criteria.[9]

Equation (34) assumes that only the innermost barrier is distributed. For a complete fitting of experimental data, it is necessary to generalize the treatment and assume that other barriers are also distributed.[17]

Multiple Excitation

Whenever multiple processes appear in a kinetic investigation, as for instance in Figs. 4–8, questions as to their relationships arise. Consider for instance the nonexponential kinetics in Fig. 7 which, according to Eq. (29), is interpreted as a superposition of exponentials. Is each individual biomolecule in a given conformational state characterized by only one value of k_{BA}, or does each biomolecule contain a distribution? Some problems of this type can be resolved by multiple excitation experiments. Two specific possibilities are given in Fig. 13. In Fig. 13A, two independent systems are shown, and we assume that both are equally excited by a photoflash, but decay (rebind) with very different rate coefficients, $k_1 \gg k_2$. In Fig. 13B, two levels with decay coefficients k_1 and k_2 are fed with equal probabilities from the same excited state. After a single photoexcitation,

[17] N. Alberding, S. S. Chan, L. Eisenstein, H. Frauenfelder, D. Good, I. C. Gunsalus, T. M. Nordlund, M. F. Perutz, A. H. Reynolds, and L. B. Sorensen, *Biochemistry* **17,** 43 (1978).

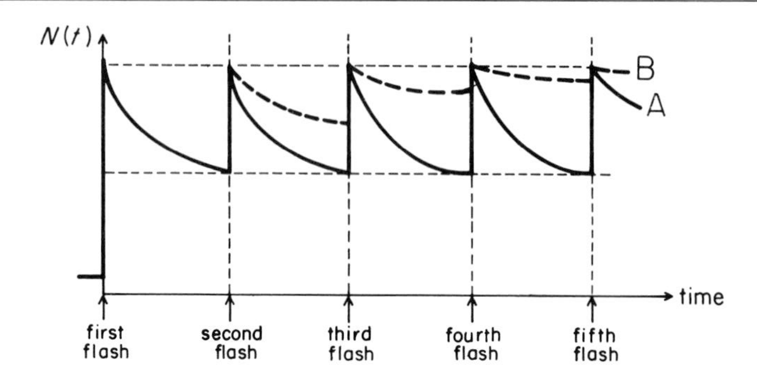

FIG. 14. The two cases shown in Fig. 13 can be distinguished by multiple excitation. The situation shown holds if k_{flash} is intermediate between k_1 and k_2, and $k_1, k_{flash} \gg k_2$. The case in Fig. 13A leads to a repetitive behavior, the case in Fig. 13B to "pumping."

the two cases are indistinguishable; both show two rebinding processes, with rate coefficients k_1 and k_2. Multiple excitation, with flashes repeated with a rate coefficient k_{flash} intermediate between k_1 and k_2, separates the two cases. Assume for simplicity also that $k_{flash} \gg k_2$. With independent systems (Fig. 13A), half of all molecules, namely the ones with the small rate coefficient k_2, will essentially all remain in the excited state after the first flash and are removed from further participation. The half with the large k_1 and hence short rebinding time will account for the signal seen after each successive flash. After the first flash, the rebinding curve no longer changes, as is indicated by the solid line in Fig. 14. With genetically related systems, as in Fig. 13B, the first decay is as in Fig. 13A. The second flash populates the top state again; half of all decays will lead to state 1, half to state 2. State 1 now contains only 1/4 of all molecules, and the observed rebinding thus is as shown by the dashed line in Fig. 14. Each successive flash will pump more ligands into the long-lived state 2, and rebinding from state 1 becomes progressively weaker. In reality, k_2 is usually not negligibly small and return from state 2 must also be considered, but the essential aspect of "pumping" remains. Multiple excitation thus is a valuable tool in distinguishing independent from related processes.

Molecular Tunnel Effect

At temperatures below about 20 K, the rebinding rate tends toward a temperature-independent value (Fig. 8). The simplest explanation for the departure from an Arrhenius behavior is the onset of quantum-mechanical tunneling, with rate as in Eq. (7).[2,4,10] Since other processes without activation energy are known to exist,[18] criteria for tunneling must be established. We take as primary criterion the existence, in the limit $T \rightarrow 0$, of a finite and temperature-independent transition rate between two well-defined states that are separated by a barrier. The existence of the barrier, and the barrier height, are established through observation of the Arrhenius transition at higher temperatures; initial and final states are distinguished by their optical spectra. A second criterion is the isotope effect; according to Eq. (7), the tunneling rate depends on the mass M of the tunneling system.

Tunneling effects are usually difficult to evaluate. The main difficulty is apparent from Eq. (7): The exponent ruling k^t_{BA} is temperature independent; tunneling of a molecule of given mass through a barrier of known height H_{BA} is characterized by *one* value of the rate coefficient and it is impossible to separate the two unknowns, the preexponential A^t_{BA} and the barrier width $d(H_{BA})$ unambiguously. The separation can be performed through the isotope effect by determining the rate for two isotopes. This approach works for proton reactions, where the rate for hydrogen and deuterium can be compared, but is difficult for heavy molecules. In the binding of small molecules to heme proteins, a different approach is possible and we describe the basic idea here.

Arrhenius and tunneling transitions are but two aspects of the same quantum-mechanical process[19] and consequently are governed by the same activation enthalpy spectrum $g(H_{BA})$. Indeed, Fig. 8 demonstrates that the low-temperature limit of CO binding to β hemoglobin is nonexponential in time and that the shape is similar to the one observed for Arrhenius transitions. In order to evaluate the data in the tunneling regime, we write the total transition rate coefficient for the step $B \rightarrow A$ as

$$k^{tot}_{BA} = k_{BA} + k^t_{BA} \tag{35}$$

where k_{BA} is given by Eq. (6). The first term in Eq. (35) dominates at high, the second at low temperatures. Above about 30 K, the distribution function $g(H_{BA})$ can be found as described in the subsection "Nonexponential Kinetics." This distribution is then used in Eq. (29) to find the total rate coefficient k^{tot}_{BA} from $N(t)$ measured as temperatures below about 30 K. The numerical extraction of k^{tot}_{BA} as a function of H_{BA} is done by computer. Equation (29) thus is used in two different ways. Above about 30 K, k_{BA} is assumed to be known and to obey the Arrhenius law and $g(H_{BA})$ is determined. Below about 30 K, $g(H_{BA})$ is assumed to be known and the unknown rate coefficient k^{tot}_{BA} is extracted as a function of H_{BA}. The tunneling rate coefficient k^t_{BA} is then found by subtracting k_{BA}, calculated with Eq. (6),

[18] H. S. Johnston and P. Goldfinger, *J. Chem. Phys.* **37,** 700 (1962).

[19] J. A. Sussmann, *J. Phys. Chem. Solids* **28,** 1634 (1967); *Ann. Phys.* (*Paris*) [14] **6,** 135 (1971); F. K. Fong, "Theory of Molecular Relaxation." Wiley, New York, 1975.

from k_{BA}^{tot}. The result is the tunneling rate coefficient k_{BA}^{t} as a function of barrier height H_{BA}. As final step, Eq. (7) is used to determine A_{BA}^{t} and $d(H_{BA})$. The evaluation is described in Frauenfelder.[20] The unique information obtained from tunneling data is d, the barrier width.

Pitfalls

Kinetic experiments extending over a very wide temperature range yield rich results and promise to provide insight into processes in the interior and at the active site of biomolecules. At the same time, pitfalls exist that do not occur in room-temperature measurements.

(1) The nonexponential kinetics that occur in all heme proteins at low temperatures provides a particularly nasty trap. Consider for instance photodissociation of CO from Mb, as shown in Fig. 7. Assume that an experiment has been performed at 40 K; a small signal has been observed. After 100 sec, the temperature has been changed to 60 K and rebinding after a second laser pulse is observed. Rebinding then will *not* follow the 60 K curve shown in Fig. 7, but will be much faster, and the signal will be small. The reason is the nonexponential rebinding: At 40 K, all CO molecules have been flashed off, but only very few have rebound; after heating to 60 K, only Mb molecules with small values of H_{BA} rebind CO. The second flash therefore probes only the small subset of all Mb molecules with small H_{BA}. In order to observe the complete $N(t)$ curve, the sample must be warmed up after each photoflash to a temperature where all ligand molecules rebind so that a uniform initial state is established.

(2) The nonexponential kinetics can lead to another artifact. With exponential binding, the monitoring light intensity can always be adjusted so that it produces negligible photodissociation during the observation time. In nonexponential kinetics, a single observation may extend from 1 μsec to over 1 ksec. Monitoring light that is too weak to perturb the measurement below, say, 1 sec may be too intense after that time. To show that the monitoring light does not falsify $N(t)$, $N(t)$ is measured for widely different monitoring light intensities.

(3) The choice of solvent for the heme protein is important.[21] On cooling, the solvent should not crystallize, crack, or precipitate the solute but form a glass with good light transmission. The solute should remain a monomer and not aggregate or denature. Glycerol–water and ethylene glycol–water mixtures have the desired properties. Even in these solvents, light transmission is lower in the glasseous than in the liquid state and a thin

[20] H. Frauenfelder, *in* "Tunneling in Biological Systems" (B. Chance, ed.). Academic Press, New York, 1978.

[21] P. Douzou, *Methods Biochem. Anal.* **22,** 401 (1974).

cell, with light-path length of 0.5–1 mm, is advisable. The nature of the solvent affects the outer barriers of the biomolecules markedly, but disturbs the intramolecular processes only slightly.

(4) The transition of the ligands from the solvent into the biomolecule can be controlled either by diffusion or by a barrier at the outside of the biomolecule. A decision between the two alternatives is not as straightforward as appears on first sight.[2] Experiments with widely different solvents may sometimes be needed to establish the control mechanism at the outermost barrier.

(5) In the interpretation of low-temperature data, conformational relaxation must be considered. Different states, for instance oxidized and reduced Mb, may involve different protein conformations. At temperatures above about 200 K, the conformation can adjust and each state will reside in the native environment. Below about 200 K, however, the overall conformation may be frozen, and a change in state of the active center may not be accompanied by the corresponding relaxation of the protein conformation.[22]

Examples

The general principles and techniques treated up to this point may appear remote from the usual problems of biochemistry. We therefore sketch here examples that indicate the type of results that experiments over an extended temperature range can yield. The discussion is far from complete and is only meant to show how the approaches outlined above can be used and interpreted.

Myoglobin[9]

Myoglobin, because of its apparent functional simplicity and well-established structure, serves as prototype and test case. It provides a prime example of how access to the active center is governed by multiple barriers. The data in Figs. 6 and 7 show the existence of four different kinetic processes. The simplest interpretation assumes that a CO molecule encounters four barriers between solvent and binding site at the heme iron. A possible model for the reaction path, based on these kinetic data and on X-ray and neutron structures, is shown in Fig. 15. Evaluation of the kinetic data over the temperature range from 40 to about 320 K with the approach

[22] L. A. Blumenfeld, R. M. Davydov, N. S. Fel', S. N. Magonov, and R. O. Vilu, *FEBS Lett.* **45,** 256 (1974); L. A. Blumenfeld, R. M. Davydov, S. N. Magonov, and R. O. Vilu, *ibid.* **49,** 246 (1974); L. A. Blumenfeld, D. S. Burbaev, R. M. Davydov, L. N. Kubrina, A. F. Vanin, and R. O. Vilu, *Biochim. Biophys. Acta* **379,** 512 (1975).

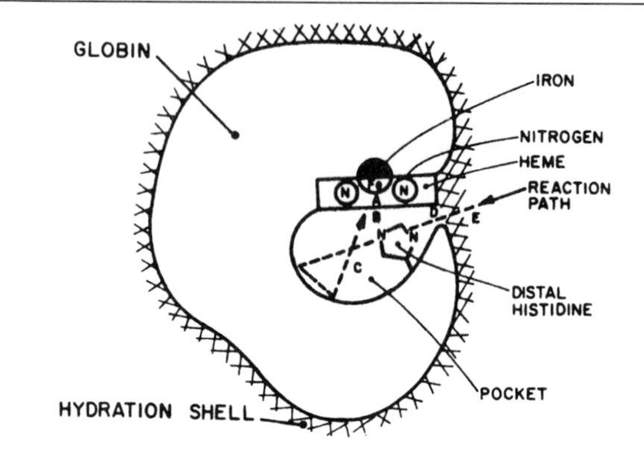

FIG. 15. Binding of a ligand molecule to myoglobin. A possible reaction path is indicated by the dashed line.

described in the subsection "Multiple Barriers" yields the enthalpy and entropy profiles in Fig. 16. Also shown is the Gibbs energy profile at 310 K. The detailed discussion of the barriers[9] demonstrates that specificity, namely kinetic rejection of CO and easy acceptance of O_2, is achieved by the joint action of all barriers. One particular point should be stressed here: Figure 6 shows only one process at 320 K, and it would be tempting to say that only one barrier governs CO binding at this temperature. The agreement of the calculations based on the model of Fig. 15 and the barriers of Fig. 16 with the experimental data implies, however, that even at 320 K all four barriers are still in action. The one observable rebinding process comes from the interplay of all barriers. Only by following photodissociation over a wide temperature range can the separate barriers be recognized and investigated.

Separated Hemoglobin Chains[17]

The separated hemoglobin chains also possess multiple barriers that determine binding. We will not treat this aspect here, but instead will describe some properties of the innermost barrier, between wells B and A. Comparison of flash photolysis experiments with myoglobin and protoheme indicates that the innermost barrier resides at the heme group.[2] In state A, CO is bound to the iron which has spin 0 and lies in the heme plane. In state B, CO is not bound, the iron has spin 2 and lies out of the heme plane, and the heme group is very likely domed. The transition from state B to A involves simultaneous motion of iron and CO, spin change of the iron, and change in shape of the heme group.[23] The activation enthalpy H_{BA} for

[23] B. R. Gelin and M. Karplus, *Proc. Natl. Acad. Sci. U.S.A.* **74,** 801 (1977); A Warshel, *ibid.* p. 1789.

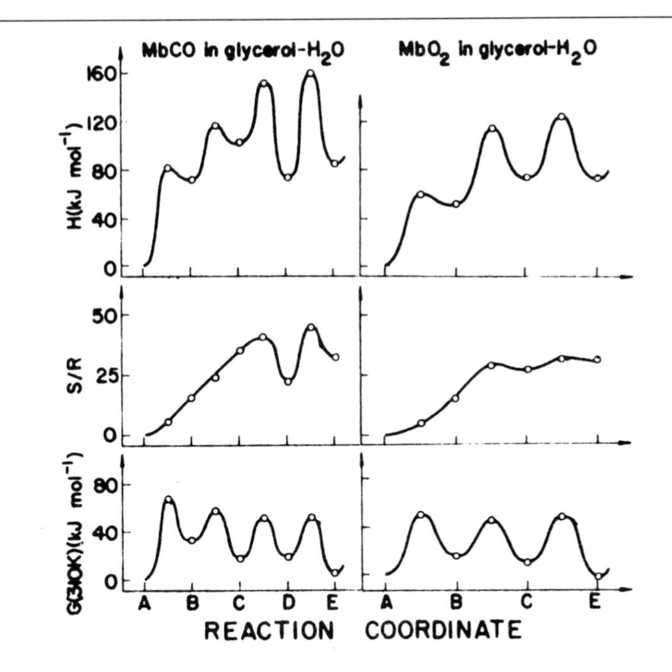

FIG. 16. Enthalpy, entropy (in units of R), and Gibbs energy (calculated at $T = 310$K) as a function of reaction coordinate for CO and O_2 binding to myoglobin. The values in well E correspond to concentrations of $[CO] = 30\ \mu M$, $[O_2] = 60\ \mu M$. [R. H. Austin, K. W. Beeson, L. Eisenstein, H. Frauenfelder, and I. C. Gunsalus, *Biochemistry* **14,** 5355 (1975).]

this complex process depends on the conformational state of the protein and is affected by the protein structure. Figure 12 demonstrates *how* strongly the protein can influence the last step in ligand binding. A full understanding of the distribution $g(H_{BA})$ in terms of protein structure is not yet available, but some preliminary remark can be made. The simultaneous motion of iron, CO, and heme will be linked to changes in the tertiary structure of the polypeptide segments surrounding the heme, and $g(H_{BA})$ will be influenced by the rigidity of that structure. Heme is more rigidly held in Mb than in Hb,[24] and $g(H_{BA})$ is indeed shifted to higher values in Mb than in the Hb chains. Another interesting comparison involves the Hb chains with and without paramercuribenzoate (PMB), denoted for instance by β^{PMB} and β^{SH}. Reaction with PMB leaves $g(H_{BA})$ essentially unchanged in the α chain, but shifts it strongly in the β chain. In the α chain, PMB reacts only with cysteine G11(104) which is far removed from the heme. In the β chain, reaction occurs with cysteines G14(112) and F9(93). Cys F9, the

[24] T. Takano, *J. Mol. Biol.* **110,** 533 and 569 (1977); R. C. Ladner, E. J. Heidner, and M. F. Perutz, *ibid.*

residue that follows the heme-linked histidine, can influence the heme because its side chain can take up two alternative configurations, with looser and tighter tertiary structure.[25] PMB fixes the protein in the tighter structure and the enthalpy distribution shifts to higher values, as displayed in Fig. 12.

Cytochrome Oxidase[26,27]

Cytochrome oxidase plays a vital role in biological oxidation. Low-temperature flash photolysis studies of the binding of O_2 and CO to the cytochrome a_3 in cytochrome oxidase reveal some similarities and some pronounced differences compared to myoglobin and hemoglobin.[26,27] Multiple barriers and intermediate states exist, but their properties are profoundly different: The innermost barrier is entropy-controlled in Mb, but enthalpy-controlled and much higher in cytochrome a_3. State *C* in Mb (Fig. 16) communicates only with one heme group and is normally empty at CO pressures below a few atmospheres; the corresponding state (*R*) in cytochrome a_3 forms a reservoir that connects to a number of heme groups and is occupied by about 30 CO molecules at 1 atm.

The examples presented here demonstrate that low-temperature experiments form an essential and indispensable part of the technique required to explore the kinetics and dynamics of heme proteins.

Acknowledgments

Part of this work was supported by the National Science Foundation under Grant BMS 74-01366 and the U.S. Department of Health, Education, and Welfare under Grant GM 18051. N. Alberding, R. H. Austin, K. W. Beeson, S. S. Chan, B. Chance, L. Eisenstein, D. Good, I. C. Gunsalus, P. Hänggi, T. M. Nordlund, A. H. Reynolds, and L. B. Sorensen have contributed much, and I would like to thank them for collaboration, discussions, and many valuable comments.

[25] E. J. Heidner, R. C. Ladner, and M. F. Perutz, *J. Mol. Biol.* **104,** 707 (1976).

[26] B. Chance, C. Saronio, and J. S. Leigh, Jr., *Proc. Natl. Acad. Sci. U.S.A.* **72,** 1635 (1975); *J. Biol. Chem.* **250,** 9226 (1975).

[27] M. Sharrock and T. Yonetani, *Biochim. Biophys. Acta* **434,** 333 (1976); *ibid.* **462,** 718 (1977).

Author Index

Numbers in parentheses are footnote reference numbers and indicate that an author's work is referred to although his name is not cited in the text.

C

D

E

L

N

O

S

T

U

V

X

Y

Z

Subject Index

A

B

C

F

G

J

K

L

M

N

O

Q

R

U

V

W

X

A
B
C 8
D 9
E 0
F 1
G 2
H 3
I 4
J 5